Genetically Engineered Organisms

Assessing Environmental and Human Health Effects

Genetically Engineered Organisms

Assessing Environmental and Human Health Effects

Edited by
Deborah K. Letourneau
Beth Elpern Burrows

CRC PRESS

Boca Raton London New York Washington, D.C.

Senior Editor: John Sulzycki
Project Editor: Christine Andreasen
Production Coordinator: Patricia Roberson
Marketing Manager: Carolyn Spence
Cover Design: Dawn Boyd

Library of Congress Cataloging-in-Publication Data

Genetically engineered organisms : assessing environmental and human health effects /
edited by Deborah K. Letourneau and Beth Elpern Burrows.
    p. cm.
  Includes bibliographical references and index.
  ISBN 0-8493-0439-3 (alk. paper)
    1. Transgenic organisms—Risk assessment. I. Letourneau, Deborah Kay. II. Burrows,
Beth Elpern.

QH442.6 .G466 2001
306.4'6—dc21

2001037635

Visit the CRC Press Web site at www.crcpress.com

Dedication

To mother Ethel Marie and daughter Nicole Marie
for their respective wisdom and promise

D. K. L.

To those who offer a more poetic perspective:
Ed, Charlie, and Freida

B. E. B.

Preface

This book is a compendium of research and review articles of some of the latest developments in genetic engineering. Here you will find reports of research, reviews of research, and calls for research, as well as discussion about risk assessment and risk management. Here you will discover the work of scientists from a broad range of disciplines including botany, entomology, plant pathology, microbial ecology, virology, fish and wildlife, evolutionary biology, genetics, public health, and environmental studies. Here too you will find a volume that has benefited from the insights and comments of the many people who have helped ensure that each chapter was peer reviewed.

Here you can read about pollen movement, spread of transgenes in natural communities, fitness effects, resistance development, and unpredicted impacts on target and non-target organisms — topics explored in contexts ranging from *Bt*-corn events and viral resistant oats to transgenic salmon and altered malarial vectors.

This book was inspired by a symposium co-organized by the book's co-editors at the 1999 national meeting of the Ecological Society of America (ESA), held in Spokane, Washington. The symposium was entitled "Theoretical and Practical Considerations for Assessing Ecological and Human Health Effects of Genetically Engineered Organisms" and featured presentations by some of those who wrote chapters in this book (D. Andow, A. Kapuscinski, T. Klinger, A. Spielman).

The ESA symposium in turn had its roots in the creation of a *Manual for Assessing Ecological and Human Health Effects of Genetically Engineered Organisms* (Scientists' Working Group on Biosafety 1998). The Manual was the work of a group of scientists who, convened by the Edmonds Institute, a public interest organization following the biosafety deliberations of the Convention on Biological Diversity, undertook to help consumers and policy-makers evaluate likely impacts of genetically engineered organisms in a variety of settings and applications.

What the Manual, the symposium, and this book have in common were these understandings on the part of the instigators of all three events:

- Genetic engineering, like other technologies, not only suggests new avenues for constructing useful products, but also poses hazards to the health of the environment and the public (see, e.g., Ellstrand 2001).
- Science can provide insight into the benefits and the risks of the products of genetic engineering. We need not argue about the ecological and human health effects of any event if we are willing to undertake the painstaking research necessary to identify and understand those effects.
- The pursuit of biosafety requires a wide "spectrum of expert knowledge consonant with the specific molecular biological, organismal, ecological, and applied dimensions of the proposed genetically engineered organism" (Scientists' Working Group 1998).

- Research, data, and analysis can and should be made understandable and accessible to the general reader; such accessibility benefits expert practitioners of other disciplines, as well as the general public.
- Science and other scholarly endeavors can and must inform public policy but the public is the ultimate decision maker about which problems are most pressing, which risks are most worth taking, and which technologies are most desirable, given all the available alternatives (see Liebhardt 1993, Raffensperger and Tickner 1999, Thornton 1999, O'Brien 2000).
- The pursuit of biosafety is necessarily a shared endeavor of science and the general public (see van Dommelen 1998, Wright 1998, Levidow and Carr 2000).
- Members of the scientific community and the public interest community can cooperate to the benefit of both science and the public interest, but such cooperation requires continuous exercise of patience, care, and respect, occasional adjustments in language and perspective, and initial positive judgments about the other party's good will and intelligence. A sense of humor is also essential.

Those understandings acknowledged, it is hoped that it will come as no jarring surprise to the reader that one of the editors of this volume is a scientist (Deborah Letourneau) and the other (Beth Burrows) is not.

I first met Deborah when the organization I headed invited a group of scientists that included Deborah to write a manual about biosafety. Since then, we have worked together on several occasions to bring broad-based thinking and cooperation to bear on questions raised about genetically engineered organisms. This book is the culmination of our efforts, as well as the work of many generous authors and peer reviewers. It is hoped that the collection is a useful contribution not only to science but also to the public interest.

Beth Elpern Burrows

Literature cited

Ellstrand, N. C., When transgenes wander, should we worry? *Plant Physiol.*, 125, 1543, 2001.

Levidow, L. and Carr, S., Trans-Atlantic regulatory disputes over GM crops, *Int. J. Biotechnol.*, 2(1/2/3), 257, 2000.

Liebhardt, W. C., Ed., The Dairy Debate: Consequences of Bovine Growth Hormone and Rotational Grazing Technologies, University of California Sustainable Agriculture Research and Education Program, Davis, 1993.

O'Brien, M., *Making Better Environmental Decisions: An Alternative to Risk Assessment*, MIT Press, Cambridge, MA, 2000.

Raffensperger, C. and Tickner, J., Eds., *Protecting Public Health and the Environment*, Island Press, Covelo, CA, 1999.

Scientists' Working Group on Biosafety, *Manual for Assessing Ecological and Human Health Effects of Genetically Engineered Organisms, Part One: Introductory Materials and Supporting Text for Flowcharts; Part Two: Flowcharts and Worksheets*, The Edmonds Institute, Edmonds, WA, 1998.

Thornton, T., *Pandora's Poison: Chlorine, Health, and a New Environmental Strategy*, MIT Press, Cambridge, MA, 2000.

van Dommelen, A., *Hazard Identification of Agricultural Biotechnology: Finding Relevant Questions*, International Books, Utrecht, the Netherlands, 1999.

Wright, S., Molecular politics in a global economy, in *Private Science: Biotechnology and the Rise of the Molecular Sciences*, Thackray, A., Ed., University of Pennsylvania Press, Philadelphia, 1998.

Biographies of the Editors

Deborah K. Letourneau is full professor in the Department of Environmental Studies at the University of California, Santa Cruz. Environmental Studies is an interdisciplinary department with graduate and undergraduate courses and faculty research focus in Agroecology, Conservation Biology, Political Economy, and Public Policy. Dr. Letourneau received her B.S. in Zoology and her M.S. in Biology from the University of Michigan in 1976, and her Ph.D. in Entomology from the University of California, Berkeley in 1983. She has served as an ecologist on scientific advisory committees regarding the biosafety of genetically engineered organisms, including the Agricultural Biotechnology Research Advisory Committee for the U.S. Department of Agriculture, the Scientists' Working Group for the Edmonds Institute, and Sub-Committee on Environmental Impacts Associated with Commercialization of Transgenic Plants for the National Research Council, National Academy of Sciences.

Dr. Letourneau's field research programs in ecology involve the interactions among plants, plant-feeding insects, and their natural enemies (predators and parasitoids) in agricultural fields and natural forests. She has investigated these multitrophic-level interactions in a variety of contexts and communities, including tropical rain forests in Costa Rica and Papua New Guinea, subsistence farmers' fields in Africa, and tomato agroecosystems in California. Dr. Letourneau's ongoing research in Costa Rica involves the web of interactions among rain forest plants, the insects that feed on them, and predators of these herbivorous insects, including specialized ants and the insects that prey on these predators. This rich natural system of mutualistic and agonistic interactions on multiple trophic levels is the basis for her current field and laboratory research on trophic cascades, the principles of which can contribute to our understanding of community structure in natural systems, stability in managed ecosystems, and conservation biology.

Dr. Letourneau's honors and accomplishments include several Fulbright Fellowship awards for international research, editorial board nomination for Environmental Biosafety Research, organization of numerous national and international research symposia, publication of two co-edited books and an Ecological Applications Invited Feature, as well as more than 50 articles in scientific journals including *Ecology, Proceedings of the National Academy of Sciences*, and *Science*.

Beth Elpern Burrows is the founder, president, and director of The Edmonds Institute, an award-winning public interest, non-profit organization concerned with issues related to environment and technology. The Institute has been called "an environmental think tank in the grassroots mode — lean on spending and long on accomplishments." Current program emphasis is on biosafety and the legally binding international regulation of modern biotechnologies; intellectual property rights and just policies for the maintenance and protection of biodiversity, including policies that foster recognition and sustenance of agricultural biodiversity; and exploration of the ethical implications of new technologies.

Ms. Burrows has been on the board and served as officer of many public interest and environmental organizations including the Abundant Life Seed Foundation, National Boycott News, the Peace Campaign, Northwest Coalition for Alternatives to Pesticides, Pilchuck Audubon Society, and the Biotechnology Working Group — to name a few. She was a co-founder and member of the founding steering committee of Diverse Women for Diversity, an international organization that promotes biological and cultural diversity. She served for 4 years on the state-appointed Northwest Regional Citizen's Advisory Committee for the Washington State Department of Ecology Model Toxic Control Act. A former Master Gardener, Master Composter/Recycler, and member of the year-long (Washington State University) "Initiatives for Century 2" Task Force to "Maintain a Safe and Abundant Food Supply," she was also one of the creators of the Master Birders program.

Ms. Burrows received her B.A. in Political Science from the University of Chicago. She received her M.A. from the University of Michigan, where she also did doctoral coursework. She holds certificates in a variety of subjects ranging from education to horticulture.

Prior to her work in the public interest community, Ms. Burrows was a public broadcaster and a writer/producer of many nationally broadcast, award-winning documentaries and instructional series. In recent years, she has lectured on trade and technology issues to professional, academic, and community audiences all over the world. She is the author of the self-study guide *On Becoming Sexist*, several teachers' manuals, a pamphlet entitled *An Activist's Introduction to Intellectual Property in the GATT*, a chapter in *In Our Back Yard: State Action to Govern the Release of Genetically Engineered Organisms into the Environment*, a chapter in the *U.S. Citizens Analysis of the North American Free Trade Agreement*, a chapter in *Conserving the Sacred for Biodiversity Management*, three chapters in *Redesigning Life?: The Worldwide Challenge to Genetic Engineering*, and articles in many other publications, including *The Ecologist, Prairie Fire Journal, Earth Island Journal, Third World Resurgence, Synthesis/Regeneration, Food and Water Journal, British Bulletin of Medical Ethics, Lola, Earthcare, The Seattle Times, PCC Sound Consumer, Boycott Quarterly, Alternatives, Works in Progress, Washington Wildfire, The Gene Exchange, Gene Watch, Organic Farmer, Global Pesticide Campaigner, Terrain, Rachel's Environment and Health Weekly*, and *Journal of Pesticide Reform*.

Ms. Burrows was the subject of a chapter in Mary Kenady's book, *Pacific Northwest Gardener's Almanac*, a featured technology critic in Stephanie Mills' book, *Turning Away from Technology*, and one of ten environmentalists profiled in a Patagonia catalog several years ago. Concerned with the ecological, social, and ethical impacts of new technologies, she has followed the biosafety deliberations of the Convention on Biological Diversity and other forums for years.

Contributors

David A. Andow
Department of Entomology and
  Center for Community Genetics
University of Minnesota
St. Paul, Minnesota, U.S.A.

John C. Beier
School of Tropical Medicine
Tulane University
New Orleans, Louisiana, U.S.A.

Joy Bergelson
Department of Ecology and Evolution
University of Chicago
Chicago, Illinois, U.S.A.

Henk R. Braig
School of Biological Sciences
University of Wales
Bangor, U.K.

Beth Elpern Burrows
The Edmonds Institute
Edmonds, Washington, U.S.A.

Claudia Eckelkamp
Institute for Applied Ecology
Freiburg, Germany

Joy A. Hagen
Department of Environmental Studies
University of California
Santa Cruz, California, U.S.A.

Angelika Hilbeck
Geobotanical Institute
Swiss Federal Institute of Technology
Zurich, Switzerland

Richard D. Howard
Department of Animal Sciences
Purdue University
West Lafayette, Indiana, U.S.A.

Ruth A. Hufbauer
Department of Bioagricultural Sciences
  and Pest Management
Colorado State University
Fort Collins, Colorado, U.S.A.

Manuela Jäger
Institute for Applied Ecology
Freiburg, Germany

Anne R. Kapuscinski
Department of Fisheries and Wildlife
  and Institute for Social, Economic,
  and Ecological Sustainability
University of Minnesota
St. Paul, Minnesota, U.S.A.

Anthony E. Kiszewski
School of Public Health and
  Center for International Development
Harvard University
Boston, Massachusetts, U.S.A.

Terrie Klinger
University of Washington
Friday Harbor Laboratories
Friday Harbor, Washington, U.S.A.

Deborah K. Letourneau
Department of Environmental Studies
University of California
Santa Cruz, California, U.S.A.

John E. Losey
Department of Entomology
Cornell University
Ithaca, New York, U.S.A.

William M. Muir
Department of Animal Sciences
Purdue University
West Lafayette, Indiana, U.S.A.

John J. Obrycki
Department of Entomology
Iowa State University
Ames, Iowa, U.S.A.

Alison G. Power
Department of Ecology and
  Evolutionary Biology
Cornell University
Ithaca, New York, U.S.A.

Colin B. Purrington
Department of Biology
Swarthmore College
Swarthmore, Pennsylvania, U.S.A.

Gaden S. Robinson
Department of Entomology
The Natural History Museum
London, U.K.

Andrew Spielman
School of Public Health and
  Center for International Development
Harvard University
Boston, Massachusetts, U.S.A.

Guenther Stotzky
Department of Biology
New York University
New York, New York, U.S.A.

Beatrix Tappeser
Institute for Applied Ecology
Freiburg, Germany

Terje Traavik
Norwegian Institute of Gene Ecology
Department of Microbiology and Virology
University of Tromsö
Tromsö, Norway

Guiyun Yan
Department of Biological Sciences
State University of New York
Buffalo, New York, U.S.A.

Contents

chapter one

Variability and uncertainty in crop-to-wild hybridization

Terrie Klinger

Contents

1.1 Crop–wild hybridization

Spontaneous hybridization is known to occur between conventional crops and their wild relatives, and the introgression of crop genes into populations of related wild plants has been demonstrated for several crop–wild pairs. While introgression is the variable of evolutionary and ecological importance, hybridization is one measure of the potential for introgression to occur.

1.1.1   Incidence

A review of 13 of the most important food crops worldwide showed that the majority are known to hybridize spontaneously with wild relatives (Ellstrand et al. 1999). Of these 13 crops, genetic data indicate that crop–wild hybridization occurs in wheat, rice, soybean, cottonseed, sorghum, beans, canola, sunflower, and sugarcane. Morphological data imply that crop–wild hybridization also occurs in barley, pearl millet, and the maize–teosinte complex. Spontaneous hybridization occurs in numerous other crop–wild pairs. For example, cultivated beet and sea beet are known to hybridize in Europe (Boudry et al. 1993, Bartsch and Schmidt 1997) and are suspected of hybridizing in the western United States (Bartsch and Ellstrand 1999). Strawberry (Rosskopf 1999), blueberry (Rosskopf 1999), quinoa (Wilson and Manhart 1993), radish (Klinger et al. 1991, 1992), and poplar (Strauss 1999) are all known to hybridize with their wild relatives in areas where they co-occur. Thus, the tendency for spontaneous crop-to-wild hybridization may be widespread and occurring more often than is recognized.

Crop-to-wild hybridization can occur across relatively large spatial scales (at least 1 km; Klinger et al. 1991, Wilkinson et al. 1995) and across a range of genetic incompatibilities (Snow and Moran-Palma 1977). Rates of hybridization have been estimated for relatively few crop–wild pairs, but experimental evidence in small-scale trials indicates that crop-to-wild gene flow typically exceeds 1% (Ellstrand et al. 1999) and can be as great as 100% when the crop and wild plants occur in close proximity (Klinger et al. 1991). Rates of introgression will necessarily be smaller than those of hybridization. Even so, gene flow resulting from hybridization can influence the genetic structure of populations by counterbalancing the effects of natural selection and genetic drift (Slatkin 1987). In the most extreme cases, introgressive gene flow can confer an increased risk of extinction among small recipient populations via outbreeding depression or genetic assimilation (Fryxell 1979, Kiang et al. 1979, Rogers et al. 1982, Ellstrand 1992).

1.1.2   Transgenic crops

Like conventional crops, transgenic crops are capable of hybridization with wild relatives (e.g., Jørgensen et al. 1996). Consequently, pollen-mediated transgene escape is frequently cited as one of the hazards associated with the environmental release of genetically engineered organisms. Transgenes that confer resistance to herbicides, pests, or pathogens have received considerable attention (e.g., Rissler and Mellon 1996, Trainor and Westwood 1999) because they represent a large fraction of the genetic constructs in use and under development (Hails 2000) and because their release is thought to pose some degree of hazard to wild populations. For example, the unintentional transfer of resistance to wild plants could make them better competitors or release them from such population-regulating factors as predation and disease. The expansion of populations of transgenic hybrids resulting from ecological release could, in theory, threaten the persistence of co-occurring wild populations.

1.1.3   Importance of hybidization rates

Crop-to-wild hybridization, resulting in fertile hybrid offspring, represents the first step in the spread of transgenes through wild populations. Depending on the frequency with which it occurs, hybridization may or may not limit rates of introgression and spread. In either case, rates of spread cannot be calculated without some estimate of hybridization rate. Therefore, an understanding of hybridization, its associated variance, and the uncertainty conferred by such variance is critical to considerations of the risks of trangene spread through wild populations.

1.1.4 Factors affecting hybridization rates

Ecologically based field studies have demonstrated that rates of crop-to-wild hybridization generally decline with distance from the crop, but most studies indicate a high degree of variability associated with distance-dependent hybridization. For example, Hokanson et al. (1997) estimated rates of hybridization within and between treatments of cultivated cucumber. In one treatment, one of eight replicate plots exhibited a hybridization rate of 38%, while hybridization among all other replicates within the treatment was nil. Large amounts of variability associated with distance-dependent rates of hybridization additionally have been reported for radish (Klinger et al. 1991, 1992, Goodell et al. 1997), sunflower (Arias and Rieseberg 1994), canola (Morris et al. 1994), and sorghum (Arriola and Ellstrand 1996).

A number of factors can contribute to variability in rates of hybridization. These include genetic factors such as genotype, mating system, and compatibility of the cultivar and wild plants (Tonsor 1985, Langevin et al. 1990, Darmency 1994); size, number, density, and geometry of donor and recipient populations (Levin and Kerster 1969, Handel 1989, Ellstrand 1992, Klinger et al. 1992, Ellstrand and Elam 1993, Goodell et al. 1997, Tufto et al. 1997); pollinator behavior (Schmidt 1980, Handel 1989, Real 1989, Umbeck et al. 1991); and spatial and temporal variability in the environment (Palmer et al. 1988, Campbell and Wasser 1989, Hokanson et al. 1997). Thus, while average rates of hybridization typically are described as monotonically declining functions of the distance between donors and recipients, actual rates can vary widely. Variability in distance-dependent hybridization is important because it increases uncertainty in the estimation of hybridization rate.

1.2 Variability in distance-dependent hybridization in Raphanus sativus

Members of the Brassicaceae, especially those belonging to the genus *Brassica*, are widely cultivated and are the subject of intensive genetic engineering for crop improvement (Neeser 1999). In addition, many species within the family function as weeds of agricultural systems. Members of the Brassicaeae have been studied extensively with regard to mating system, population biology, and population genetic structure (e.g., Ellstrand et al. 1999, Neeser 1999 and references therein). Radish (*Raphanus sativus* L.) is of minor agricultural importance compared with *Brassica*. Even so, *R. sativus* provides a good model system with excellent applicability to the cultivation of *Brassica* and other outcrossed, insect-pollinated species and cultivars.

1.2.1 Gene flow and hybridization

A number of studies have addressed issues of gene flow and hybridization in the Brassicaceae (e.g., Klinger et al. 1991, 1992, Manasse 1992, Eber et al. 1994, Jørgensen et al. 1996, Goodell et al. 1997, Ellstrand et al. 1999). These studies have shown that spontaneous hybridization between crops and wild plants occurs in natural settings. Rates of outcrossing in this group are affected by a number of factors including the size, shape, and position of donor and recipient populations, pollinator behavior, and compatibility.

Hybridization rates in the Brassicaceae and other outcrossed taxa often have been reported as mean values plus some measure of the associated variance (e.g., Klinger et al.

1991, 1992, Arias and Rieseberg 1994, Morris et al. 1994, Arriola and Ellstrand 1996, Goodell et al. 1997, Hokanson et al. 1997). However, average values are inappropriate for use in risk assessment because they tend to underestimate long-distance hybridization events, as demonstrated below.

1.2.2   Raphanus sativus: *A case study*

Here I use data from small-scale field trials of *R. sativus* to illustrate the relationship between average and maximum rates of hybridization in wild and cultivated radish and to compare variability in hybridization rates across spatial and temporal scales. The results indicate that analyses based on average rates of hybridization will generally underestimate the frequency of long-distance hybridization events. A precautionary approach to safe release and effective post-release monitoring of new agricultural products therefore will depend on choosing measures of gene flow that are least likely to underestimate actual rates of hybridization at all spatial scales.

1.2.2.1   *Experimental system*
I performed a new analysis of data reported previously (Klinger et al. 1991, 1992). The experiments were initially designed to detect wild–crop hybridization at distances of up to 1 km (Klinger et al. 1991) and to test the effects of recipient population size on rates of wild–crop hybridization (Klinger et al. 1992). In this new analysis, I explored the *variability* associated with distance-dependent hybridization rather than the incidence of hybridization.

The design of both experiments is fully described in Klinger et al. (1991, 1992). A donor population of commercial radish cultivar was sown and maintained to simulate breeder's blocks for seed multiplication plantings (George 1985). Seeds were drilled at a spacing of 2.5 cm in 16 double rows 92 m in length. Small recipient populations (n = 2–9) of wild radish (*R. sativus* L.) grown from wild-collected seed were planted at the crop margin (1 m) and at distances of 200, 400, 600, and 1000 m. This design was replicated at two agricultural field stations in California (South Coast and Moreno Valley) in 1989, and was repeated at the South Coast station in 1990, with the exception that no 600 or 1000 m treatments were planted at South Coast in 1990. The South Coast and Moreno sites are separated by about 100 km, and differ in a number of physical and biological factors, including elevation, annual rainfall, daily and seasonal temperatures, and in the composition of the local agricultural and natural flora.

Bolting in the crop occurred simultaneously with flowering in the wild plants. The crop and weeds were allowed to hybridize. Once harvested, seeds from wild plants were assayed for the presence of an isozyme (LAP-6) that was fixed in the crop but absent from wild populations. Rates of hybridization were inferred from the frequency of the crop marker among the progeny of the wild plants. More detailed field and laboratory methods are given in Klinger et al. (1991, 1992).

Average rates of weed–crop hybridization were calculated for individuals and populations and compared within and between populations, distances, and sites. Resampling techniques were used to generate distributions of mean rates of hybridization. Ratios of maximum to average hybridization rates were plotted as a function of distance and compared within and between sites and years. In addition, the observed rates of hybridization were compared with expected rates based on extrapolation from an exponential model.

1.2.2.2   *Variability in hybridization rates*
Rates of hybridization were greatest along the crop margin (Table 1.1). However, hybridization varied greatly within treatments at all distances from the crop. For example, the frequency of hybrid seeds along the crop margin spanned an order of magnitude, ranging

Table 1.1   Frequency of Hybrid Seed Progeny among
Experimentally Propagated Wild Plants

Distance from crop	South Coast 1989	Moreno Valley 1989	South Coast 1990
Crop margin	0.69	1.00	0.89
	0.64	1.00	0.84
	0.64	1.00	0.83
	0.63	1.00	0.83
	0.61	1.00	0.65
	0.58	1.00	0.62
	0.57	1.00	0.56
	0.56	0.86	0.44
	0.50	0.75	0.42
	0.50	0.67	0.39
	0.48	0.92	0.37
	0.42	0.70	0.37
	0.30	0.62	0.37
	0.27	0.58	0.31
	0.20	0.69	0.28
	0.19	0.57	0.26
	0.19	0.54	0.24
	0.19	0.42	0.23
	0.18	0.33	0.20
	0.17	0.27	0.17
	0.14	N/A	0.17
	N/A	N/A	0.14
	N/A	N/A	0.11
	N/A	N/A	0.09
	N/A	N/A	0.08
200 m	0.03	0.28	0.30
	0.03	0.14	0.26
	0.00	0.10	0.17
	0.00	0.09	0.06
	0.00	0.09	0.03
	0.00	0.06	0.03
	0.00	0.05	0.03
	0.00	0.00	0.03
	0.00	0.00	0.00
	0.00	0.00	0.00
	0.00	0.00	0.00
	0.00	0.00	0.00
	0.00	0.00	0.00
	0.00	0.00	0.00
	0.00	0.00	0.00
	0.00	0.00	0.00
	N/A	0.00	0.00
	N/A	0.00	0.00
	N/A	0.00	0.00
	N/A	0.00	0.00
	N/A	N/A	0.00
	N/A	N/A	0.00
	N/A	N/A	0.00
	N/A	N/A	0.00

Table 1.1 (continued)   Frequency of Hybrid Seed Progeny
among Experimentally Propagated Wild Plants

Distance from crop	South Coast 1989	Moreno Valley 1989	South Coast 1990
	N/A	N/A	0.00
	N/A	N/A	0.00
	N/A	N/A	0.00
400 m	0.39	0.17	0.24
	0.19	0.00	0.09
	0.17	0.00	0.07
	0.17	0.00	0.06
	0.11	0.00	0.06
	0.03	0.00	0.03
	0.00	0.00	0.00
	0.00	0.00	0.00
	0.00	0.00	0.00
	0.00	0.00	0.00
	0.00	0.00	0.00
	0.00	0.00	0.00
	0.00	0.00	0.00
	0.00	N/A	0.00
	N/A	N/A	0.00
	N/A	N/A	0.00
	N/A	N/A	0.00
	N/A	N/A	0.00
	N/A	N/A	0.00
	N/A	N/A	0.00
	N/A	N/A	0.00
	N/A	N/A	0.00
	N/A	N/A	0.00
	N/A	N/A	0.00
	N/A	N/A	0.00
	N/A	N/A	0.00
	N/A	N/A	0.00
	N/A	N/A	0.00
	N/A	N/A	0.00
	N/A	N/A	0.00
600 m	0.00	0.00	N/A
	0.00	0.00	N/A
	0.00	N/A	N/A
	0.00	N/A	N/A
	0.00	N/A	N/A
	0.00	N/A	N/A
	0.00	N/A	N/A
	0.00	N/A	N/A
	0.00	N/A	N/A
	0.00	N/A	N/A
	0.00	N/A	N/A
	0.00	N/A	N/A
	0.00	N/A	N/A
	0.00	N/A	N/A
	0.00	N/A	N/A
	0.00	N/A	N/A
	0.00	N/A	N/A

Table 1.1 (continued) Frequency of Hybrid Seed Progeny among Experimentally Propagated Wild Plants

Distance from crop	South Coast 1989	Moreno Valley 1989	South Coast 1990
1000 m	0.03	0.11	N/A
	0.00	0.00	N/A
	0.00	0.00	N/A
	0.00	0.00	N/A
	0.00	0.00	N/A
	0.00	0.00	N/A
	0.00	0.00	N/A
	0.00	0.00	N/A
	0.00	0.00	N/A
	0.00	0.00	N/A
	0.00	0.00	N/A
	0.00	0.00	N/A
	0.00	0.00	N/A
	0.00	0.00	N/A
	0.00	0.00	N/A
	0.00	0.00	N/A
	0.00	0.00	N/A
	0.00	0.00	N/A
	0.00	0.00	N/A
	0.00	0.00	N/A
	0.00	0.00	N/A
	0.00	N/A	N/A
	0.00	N/A	N/A
	0.00	N/A	N/A
	0.00	N/A	N/A
	0.00	N/A	N/A
	0.00	N/A	N/A

Note: For details regarding experimental design, see Klinger et al. (1991, 1992).

from 0.08 to 0.89 (South Coast, 1990), and the frequency of hybrid seed at 400 m ranged from 0 to 0.39 (South Coast, 1989). In addition, hybridization did not show a consistent decrease with distance. For example, hybridization at 400 m greatly exceeded that at 200 m (South Coast, 1989), and hybridization at 1000 m slightly exceeded that at 600 m (South Coast, 1990, and Moreno Valley, 1989). Thus, considerable variability exists independent of distance from the crop, and hybridization rates cannot be predicted adequately by crop–weed distance alone.

Resampling techniques were used to generate distributions of mean hybridization frequencies (Figures 1.1 through 1.3). The results show that neither the shapes nor the modes of the distributions are constant between distances, sites, or years, indicating that variability is high across all three factors. Discrepancies between average and maximum rates of hybridization were small at the crop margin but increased dramatically with distance (Figure 1.4). At 1000 m, the maximum rate of hybridization exceeded the average rate by a factor of about 30.

1.2.2.3 Expected vs. observed hybridization rates

To explore further the relationship between expected and observed rates of hybridization, I used an exponential model to estimate the expected number of hybrids at four distances,

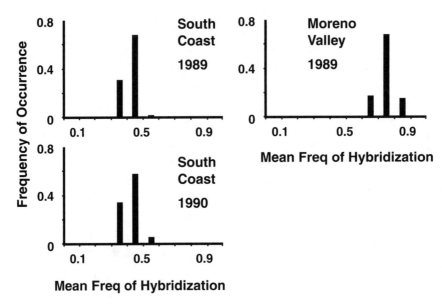

Figure 1.1 Distribution of mean hybridization frequencies generated by resampling ($n = 100$ for each distribution): Crop margin.

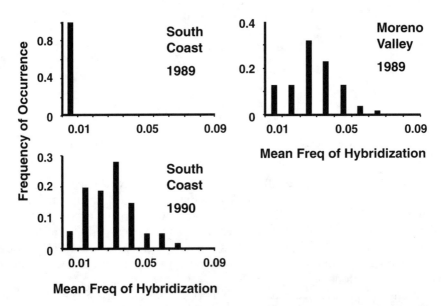

Figure 1.2 Distribution of mean hybridization frequencies generated by resampling ($n = 100$ for each distribution): 200 m.

and compared the results with the observed number of hybrid seeds at each distance. I used data from Klinger et al. (1991, 1992) to calculate the mean distance traveled by pollen. For pollen recipients positioned at distances of 1, 200, 400, and 1000 m from a single pollen source, and based on a sample of 50 seeds per recipient plant, the mean distance traveled by pollen across all treatments was 35.2 m (S.E. = 2.6 m). I then estimated the expected number of hybrids at each experimental distance using an exponential model,

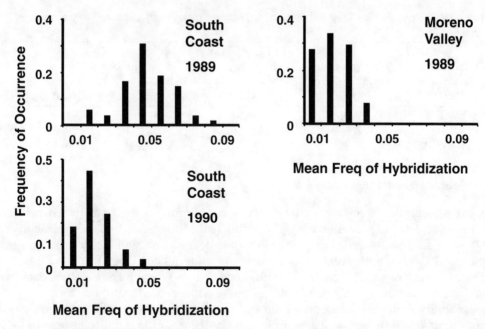

Figure 1.3 Distribution of mean hybridization frequencies generated by resampling ($n = 100$ for each distribution): 600 m.

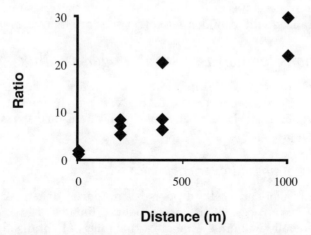

Figure 1.4 Maximum frequency of hybridization/average frequency of hybridization vs. distance. Each point represents the ratio of maximum frequency/average frequency for a single treatment.

based on a mean hybridization distance of 35.2 m. Results indicated that the observed number of hybrids greatly exceeded the expected number of hybrids at 1, 400, and 1000 m (Table 1.2). These discrepancies were reduced but still remained large when the upper 95% confidence interval was used instead of the mean expected value.

1.2.2.4 Consequences of variability in hybridization rates

These findings indicate that the use of average values to estimate the probability of hybridization events in risk assessment can cause risk to be underestimated. Although it

Table 1.2 Expected vs. Observed Values for Number of Hybrids Expected in a Sample of 1900 Seeds

Distance from source (m)	Mean number hybrids expected (A)	Upper 95% C.I. for number hybrids expected (B)	Mean number hybrids observed (C)	C/A	C/B
1	53.26	46.54	1701.5	31.95	36.56
200	1842.97	1842.76	88.38	0.05	0.05
400	6.48	13.35	105.86	16.32	7.93
1000	0.02	0.095	7	314.54	73.67

Note: See text for explanation. C.I. = confidence interval.

may be possible to identify covariates that would improve the fit between predicted hybridization rates and rates observed in nature (e.g., Tufto et al. 1997, Nurminiemi et al. 1997), such analyses often require large amounts of data and will likely produce meaningful results only after specific experiments are undertaken. Thus, results will be very difficult to extrapolate to uncontrolled situations or those for which few prior data exist. Therefore, the use of covariates may not improve general predictability.

Variability in distance-dependent rates of hybridization have been reported by other investigators (cited in Section 1.1.4). In fact, high degrees of variability appear to be the rule rather than the exception in distance-dependent hybridization rates. This is especially true for insect-pollinated species in which complex behavioral patterns among pollinators contribute to variability in hybridization. Consequently, hybridization rates are not adequately described by measures of central tendency, nor can rates realized at one site in one year be generalized with confidence to other sites or other years.

1.3   Implications for risk assessment of genetically engineered crops

The above example of crop-to-wild hybridization in radish demonstrates that distance-dependent rates of hybridization can be highly variable. To the extent that variability contributes to uncertainty, variability will have implications for risk assessment of genetically engineered crops.

1.3.1   Employing a precautionary approach

Measures of central tendency will be poor descriptors of distance-dependent rates of hybridization for the purposes of risk assessment. Estimates of hybridization based on average values generally will underestimate actual rates of hybridization at intermediate and long distances from the crop. In addition, discrepancies between actual and average values will tend to increase with distance. Therefore, using average rates of hybridization rates in risk assessment is not precautionary.

Use of the upper 95% confidence interval will provide a greater margin of safety compared with the use of average values in risk assessment. However, even this approximation typically will underestimate actual rates of hybridization at distance. The use of maximum or "worst-case" rates of hybridization in risk assessment will reduce but not eliminate the number of unexpected long-distance hybridization events.

An alternative approach is to assume that ecologically relevant rates of hybridization and introgression will occur and are inevitable wherever crops and sexually compatible wild relatives co-occur within mating distance. This constitutes a more precautionary approach to the problem of transgene spread. Once the assumptions of inevitable hybridization and introgression are made, strategies for safe release then can be based on the nature of the engineered trait(s) themselves. For example, the release of crops bearing hazardous or undesirable traits might be disallowed wherever sexually compatible wild plants occur within mating distance. The implementation of such a strategy would have the effect of shifting the burden of proof to those promoting specific releases. In such cases, those promoting release would be required to demonstrate that the worst-case scenarios for distance-dependent hybridization were too conservative. A management strategy of this sort has the additional benefit of requiring detailed hybridization studies only when the default assumption of panmixis is challenged.

1.3.2   Numerical models

Numerical models can be used to describe the expected relationship between distance and the probability of hybridization. For example, maximum likelihood techniques have been used to model the dispersal of pollen from source populations (Kareiva et al. 1994, Tufto et al. 1997, Nurminiemi et al. 1997). However, as noted above, the capabilities of models or projections based on average rates of hybridization are inadequate to predict low-frequency events, even if they are of relatively large magnitude. Low-frequency hybridization events are likely to lead to unexpected "hot spots" of hybridization that contribute disproportionately to spread. Kareiva et al. (1996) have suggested that the utility of such models is in guiding the design of monitoring studies rather than in predicting invasions.

1.3.3   Case-specific studies

Many have called for evaluating the effects of trangenic cultivars on a case-by-case basis, and this is clearly an ideal strategy. However, the large number of trangenic products and the wide range of applications makes this impractical. In the case of crop-to-wild hybridization, multiple local biological and environmental factors will affect distance-dependent hybridization rates in each new application, and it will be impossible to assess adequately each new release in each new environment. Therefore, in the absence of case-specific information regarding hybridization, the approach to risk assessment should be sufficiently precautionary to protect the environment from irreversible harm. As noted above, an important first step in assessment is to determine if compatible wild plants occur within mating distance.

1.3.4   Monitoring and management strategies

For systems in which compatible wild plants occur within mating distance of trangenic crops, monitoring and management efforts should focus on detecting and managing rates of transgene spread.

1.3.4.1   Short- vs. long-distance gene flow

Crop genes will be introduced into wild populations by both short-distance and long-distance gene flow. Short-distance gene flow will occur consistently and at relatively high frequencies, and is likely to contribute to slow but consistent rates of transgene spread. Long-distance gene flow will occur at lower frequencies and will be highly unpredictable. The dynamics of short- and long-distance gene flow will be difficult to account for in a single model.

Trade-offs in the estimation of short- vs. long-distance hybridization events will occur depending on the exact model chosen (e.g., Kareiva et al. 1994). Consequently, selection of inappropriate models could bias the design of a monitoring program, allowing hybridization events to remain undetected because they were unexpected. The environmental hazard imposed by lack of detection will depend on multiple factors, including the nature of the crop trait as expressed in a wild-type background. Risk could range from very low for an innocuous trait with low probabilities of hybridization and introgression to very high for a noxious trait with high probabilities of hybridization and introgression. In addition, risk will be a function of the probability of hybridization, as well as the number of trials imposed. Therefore, even when hybridization rates are low, a very large number of trials (i.e., very large donor and recipient populations, or repeated plantings of the same donor crop over time) could result in elevated risk.

1.3.4.2 *Multiple source populations*
As transgenic crops become fully commercialized, wild populations will be exposed to pollen from multiple sources. The presence of multiple source populations likely will increase rates of transmission and also increase the variance associated with these rates, making attempts to quantify and manage risk more difficult. In addition, multiple source populations likely will carry and transmit multiple transgenic constructs, again increasing the difficulty of detection and risk management.

1.3.4.3 *Variability and uncertainty*
The variability associated with distance-dependent hybridization will confer a great deal of uncertainty in estimates of hybridization. Some of this variability might be reduced by determining and managing the sources of variation. For example, areas of high environmental heterogeneity might be avoided in selecting sites for the release of transgenic cultivars; this conceivably would reduce variability in pollinator behavior and hence in hybridization rates. However, regardless of the precautions that are taken, an element of persistent, irreducible scientific uncertainty will remain in the estimation of hybridization rates. Where a high level of uncertainty is associated with a large probability of hybridization, the environmental risk of release could be substantial, especially for releases bearing hazardous or undesirable transgenes.

Typically, hybridization rates will be greatest close to the crop. Hybridization occurring close to the crop will be the easiest to detect because the total area affected will be smaller and the allocation of monitoring efforts larger than those at distance. The opposite situation will obtain at longer distances, where hybridization will be more difficult to detect because the area affected will be larger and because the allocation of monitoring efforts often will be smaller. In most cases, the ability to detect and remedy the occurrence of hybridization events will be inversely proportional to distance from the crop. Thus, when hybridization distances are large, the potential for detection and remedy is likely to be small.

1.3.4.4 *Monitoring*
Because of potential hazards associated with the transmission of transgenic constructs to wild populations, the release of genetically engineered organisms into the environment calls for vigilant monitoring and management. At a minimum, monitoring programs should be designed to detect hybridization events at relatively high probabilities (e.g., 90%). Further, controlled experiments on the *effects* of crop-to-wild hybridization should be completed prior to wide-scale release of new trangenic products. The considerable cost of monitoring programs and associated research should be borne by the individuals or groups that will benefit directly from their release.

1.3.4.5 *Adaptive management*
Even the most aggressive monitoring programs may not detect environmental problems quickly enough to avoid long-term or irreversible harm. In particular, species additions

have proved difficult to manage, in part because the cessation of causative human activities (i.e., new introductions) will not necessarily reduce the magnitude of the existing problem. This is because once introduced, organisms can continue to increase exponentially, independent of efforts to control new introductions. The literature on invasive species is rich with examples of introductions that have proved resistant to management, some of which have led to substantial losses of biodiversity or had significant negative economic consequences. Transgenes introduced into wild populations may or may not behave in the same manner as invasive species, and relevant data are still lacking. Even so, it is reasonable to expect that some transgenes could spread rapidly within wild populations, making them impossible to recall, and rendering conventional management strategies ineffective.

Adaptive management techniques, which consist of testing the effects of various management strategies in an experimental context and which have been applied to problems of species removals and extirpations, may prove difficult to apply to problems of trangene release. This is because the less-restrictive management strategies established as part of an adaptive management regime could increase the likelihood of transgene escape and long-term negative effects. Therefore, the application of adaptive management techniques to problems of transgene release should be made only after careful consideration of the outcome of worst-case management scenarios.

1.4   A conservation perspective

Because the introduction of transgenes to wild populations could negatively impact recipient populations and their ecosystems, the problems of unintentional gene transfer and introgression must be addressed in the context of conservation biology. According to Mangel et al. (1996), "The goal of conservation should be to secure present and future options by maintaining biological diversity at genetic, species, population, and ecosystem levels; as a general rule, neither the resource nor other components of the ecosystem should be perturbed beyond natural boundaries of variation." In some cases, the introgression of transgenes into wild populations will almost certainly cause significant changes in the genetic structure of recipient populations. In the short term, the introgression of transgenes could increase genetic diversity within a population by the addition of novel genes. For example, increased genetic diversity has been demonstrated in wild sea beet growing in proximity to crops of non-transgenic sugar beet, relative to wild populations growing in areas remote from cultivated beet (Bartsch and Ellstrand 1999). The effects of increased genetic diversity are not known for either this specific case or more general cases. Typically, conservation biologists have been concerned with *decreases* in genetic diversity within populations; increased diversity has yet to be studied in the context of conservation.

Even though genetic diversity might increase in the short term, diversity could decline in the long term as a result of intraspecific competition and the loss of native alleles. In either case, the natural boundaries of genetic variation within wild populations could be exceeded, and the abundance or distribution of the wild species consequently could be altered. Additional losses in genetic or species diversity could be brought about by secondary or higher-order effects, especially for cases in which wild plants acquire pest resistance. The loss of insect populations or the development of resistance by pest populations is a likely outcome in at least some instances. In all cases, the loss of alleles, populations, or species reduces present and future options for resource conservation and utilization.

Thus, to guard against unacceptable losses in genetic or species diversity, hazards must be identified and their attendant risks quantified and managed. Further, uncertainties inherent in estimates of crop-to-wild hybridization and the potential or realized effects of transgenes on wild populations call for caution on the part of those developing and using the technology and its products. In the context of conservation, users should minimize the risk of irreversible or long-term damages to species, populations, or ecosystems. As uncertainty increases, management strategies should become increasingly risk averse and include appropriate safety margins.

Literature cited

Arias, D. M. and Rieseberg, L. H., Gene flow between cultivated and wild sunflowers, *Theor. Appl. Genet.*, 89, 655, 1994.

Arriola, P. E. and Ellstrand, N. C., Crop-to-weed gene flow in the genus *Sorghum* (Poaceae): spontaneous interspecific hybridization between johnsongrass, *Sorghum halepense*, and crop sorghum, *S. bicolor*, *Am. J. Bot.*, 83, 1153, 1996.

Bartsch, D. and Ellstrand, N. C., Genetic evidence for the origin of California wild beets (genus *Beta*), *Theor. Appl. Genet.*, 99, 1120, 1999.

Bartsch, D. and Schmidt, M., Influence of sugar beet breeding on populations of *Beta vulgaris* spp. *maritima* in Italy, *J. Veg. Sci.*, 8, 81, 1997.

Boudry, P. et al., The origin and evolution of weed beets: consequences for the breeding and release of herbicide-resistant transgenic sugar-beets, *Theor. Appl. Genet.*, 87, 471, 1993.

Campbell, D. R. and Wasser, N. M., Variation in pollen flow within and among populations of *Ipomopsis aggregata*, *Evolution*, 43, 1444, 1989.

Darmency, H., The impact of hybrids between genetically modified crop plants and their related species: introgression and weediness, *Mol. Ecol.*, 3, 37, 1994.

Eber, F. et al., Spontaneous hybridization between a male-sterile oilseed rape and two weeds, *Theor. Appl. Genet.*, 88, 362, 1994.

Ellstrand, N. C., Gene flow by pollen: implications for plant conservation genetics, *Oikos*, 63, 77, 1992.

Ellstrand, N. C. and Elam, D. R., Population genetic consequences of small population size: implications for plant conservation, *Annu. Rev. Ecol. Syst.*, 24, 217, 1993.

Ellstrand, N. C., Prentice, H. C., and Hancock, J. F., Gene flow and introgression from domesticated plants into their wild relatives, *Annu. Rev. Ecol. Syst.*, 30, 539, 1999.

Fryxell, P. A., *The Natural History of the Cotton Tribe (Malvaceae, Tribe Gossypieae)*, Texas A&M University Press, College Station, 1979.

George, R. A. T., *Vegetable Seed Production*, Longman, London, 1985.

Goodell, K. et al., Gene flow among small populations of a self-incompatible plant: an interaction between demography and genetics, *Am. J. Bot.*, 84, 1362, 1997.

Hails, R. S., Genetically modified plants — the debate continues, *Trends Ecol. Evol.*, 15, 14, 2000.

Handel, S. N., Pollination ecology, plant population structure, and gene flow, in *Pollination Biology*, Real, L., Ed., Academic Press, Orlando, FL, 1989, 163.

Hokanson, S. C., Grumet, R., and Hancock, J. F., Effect of border rows and trap/donor ratios on pollen-mediated gene movement, *Ecol. Appl.*, 7, 1075, 1997.

Jørgensen, R. B. et al., Spontaneous hybridization between oilseed rape (*Brassica napus*) and weedy relatives, *Acta Hortic.*, 407, 193, 1996.

Kareiva, P., Morris, W., and Jacobi, C. M., Studying and managing the risk of cross-fertilization between transgenic crops and wild relatives, *Mol. Ecol.*, 3, 15, 1994.

Kareiva, P., Parker, I. M., and Pascual, M., Can we use experiments and models in predicting the invasiveness of genetically engineered organisms? *Ecology*, 77, 1670, 1996.

Kiang, Y., Antonovics, J., and Wu, L., The extinction of wild rice (*Oryza perennis formosana*) in Taiwan, *J. Asian Ecol.*, 1, 1, 1979.

Klinger, T., Elam, D. R., and Ellstrand, N. C., Radish as a model system for the study of engineered gene escape rates via crop-weed mating, *Conserv. Biol.*, 5, 531, 1991.

Klinger, T., Arriola, P. E., and Ellstrand, N.C., Crop-weed hybridization in radish (*Raphanus sativus*): effects of distance and population size, *Am. J. Bot.*, 79, 1431, 1992.

Langevin, S., Clay, K., and Grace, J. B., The incidence and effects of hybridization between cultivated rice and its related weed red rice (*Oryza sativa* L.), *Evolution*, 44, 1000, 1990.

Levin, D. A. and Kerster, H. W., Density-dependent gene dispersal in *Liatris*, *Am. Nat.*, 103, 61, 1969.

Manasse, R. S., Ecological risks of transgenic plants: effects of spatial dispersion on gene flow, *Ecol. Appl.*, 2, 431, 1992.

Mangel, M. et al., Principles for the conservation of wild living resources, *Ecol. Appl.*, 6, 228, 1996.

Morris, W. F., Kareiva, P. M., and Raymer, P. L., Do barren zones and pollen traps reduce gene escape from transgenic crops? *Ecol. Appl.*, 4, 157, 1994.

Neeser, C., Report of the *Brassica* Crops Working Group, in *Ecological Effects of Pest Resistance Genes in Managed Ecosystems*, Trainor, P. L. and Westwood, J. H., Eds., Information Systems for Biotechnology, Blacksburg, VA, 1999.

Nurminiemi, M. et al., Spatial models of pollen dispersal in the forage grass meadow fescue, *Evol. Ecol.*, 12, 487, 1997.

Palmer, M., Travis, J., and Antonovics, J., Seasonal pollen flow and progeny diversity in *Amianthium muscaetoxicum*: ecological potential for multiple mating in a self-incompatible, hermaphroditic perennial, *Oecologia* (Berlin), 77, 19, 1988.

Real, L., Microbehavior and macrostructure in pollinator-plant interactions, in *Pollination Biology*, Real, L., Ed., Academic Press, Orlando, FL, 1989, 287.

Rissler, J. and Mellon, M., *The Ecological Risks of Engineered Crops*, MIT Press, Cambridge, MA, 1996.

Rogers, C. E., Thompson, T. E., and Seiler, G. J., *Sunflower Species of the United States*, National Sunflower Association, Bismarck, ND, 1982.

Rosskopf, E., Report of the Berry Working Group, in *Ecological Effects of Pest Resistance Genes in Managed Ecosystems*, Trainor, P. L. and Westwood, J. H., Eds., Information Systems for Biotechnology, Blacksburg, VA, 1999.

Schmitt, J., Pollinator foraging behavior and gene dispersal in *Senecio* (Compositae), *Evolution*, 34, 934, 1980.

Slatkin, M., Gene flow and the geographic structure of natural populations, *Science*, 236, 787, 1987.

Snow, A. A. and Moran-Palma, P., Commercialization of transgenic plants: potential ecological risks, *BioScience*, 47, 86, 1997.

Strauss, S., Report of the Poplar Working Group, in *Ecological Effects of Pest Resistance Genes in Managed Ecosystems*, Trainor, P. L. and Westwood, J. H., Eds., Information Systems for Biotechnology, Blacksburg, VA, 1999.

Tonsor, S. J., Intrapopulation variation in pollen-mediated gene flow in *Plantago lanceolata* L., *Evolution*, 39, 775, 1985.

Trainor, P. L. and Westwood, J. H., *Ecological Effects of Pest Resistance Genes in Managed Ecosystems*, Information Systems for Biotechnology, Blacksburg, VA, 1999.

Tufto, J., Engen, S., and Hindar, K., Stochastic dispersal processes in plant populations, *Theor. Popul. Biol.*, 52, 16, 1997.

Umbeck, P. F. et al., Degree of pollen dispersal by insects from a field test of genetically engineered cotton, *J. Econ. Entomol.*, 84, 1943, 1991.

Wilkinson, M. J. et al., Problems of risk assessment with genetically modified oilseed rape, *Brighton Crop Protection Conference — Weeds*, 1995, 1035.

Wilson, H. and Manhart, J., Crop/weed gene flow: *Chenopodium quinoa* Willd. and *C. berlandieri* Moq., *Theor. Appl. Genet.*, 86, 642, 1993.

chapter two

Factors affecting the spread of resistant Arabidopsis thaliana populations

Joy Bergelson and Colin B. Purrington

Contents

2.1 Transgenic crops and selection for "superweeds"

The testing and market release of genetically engineered plants has ballooned in recent years, fulfilling expectations that transgenic agriculture would revolutionize farming. Public skepticism over large-scale commercialization of transgenic crops has increased to unexpectedly high levels, prompting many biotechnology corporations to make expensive revisions to their research and marketing strategies. Some opponents of genetically engineered crops have specific concerns that include health issues, environmental risks, and economic consequences, while others, such as Prince Charles (Anonymous 2000), oppose any manipulation on religious grounds. Together, these concerns represent a public-relations nightmare for producers of genetically engineered organisms (GEOs). It will be interesting to see how and if these fears are addressed for transgenic crops that are developed in the future. In this chapter, we focus on a single environmental concern, the "superweed threat." In particular, we examine the chance that genetic enhancement will create crops that indirectly spawn invasive weeds by transferring the transgene, via errant pollen, to nearby weedy relatives. The thought of kudzu-like superweeds ruining our already-shrunken natural areas has captured the imagination of the public and is one of the many stumbling blocks to the widespread acceptance of transgenic technologies.

Like many of the above-mentioned issues, concern over creating a superweed is reasonable in certain situations. Natural selection has been marvelously successful at producing truly noxious weeds (from the point of view of the farmer tending cropland or the steward overseeing natural areas). The evolution of the weedy "habit" has occurred independently in many taxonomic lineages, thereby demonstrating that weediness occurs even when humans *do nothing* to manipulate the genetic constitution of crops. With the techniques of artificial selection and introgression, it became possible for new weed cultivars to be created inadvertently. However, this possibility was thought to be relatively low because the selected traits were already present in conspecifics, or related species, with no observed weed problems (e.g., NRC 1989). Transgenic manipulations now allow the possibility that genes will be introduced into a crop without us first knowing that they are safe, at least in related species. Thus, many believe that eventually, and innocently, an environmental menace will be created.

2.1.1   Enhanced weediness of wild relatives

It is seldom appreciated that the creation of a superweed depends upon a *series* of events, all of which must occur before the invasiveness of a weed is enhanced. First, there must be a conduit for the introduced gene to enter the weed population. This requires that the weed species can hybridize with a transgenic crop and produce fertile offspring (e.g., wild and crop squash species), or more trivially requires that the weed and the crop are the same species (e.g., *Brassica rapa*). For a comprehensive review of gene flow between crops and weeds, see Ellstrand et al. (2000). Second, expression of the introduced gene must confer a fitness advantage that allows the gene, over multiple generations, to spread through the weed population. While such gene spread would certainly alter the genetic composition of the weed population, this change would not be sufficient to cause a weed problem. For the weed to become even weedier, a third important criterion must be met: the introduced gene must enable the weed to expand its geographical range or elevate its densities. Only when all three of these criteria have been met will a weed problem be created.

In the early days of biotechnology risk assessment, the majority of research focused on the first step, gene escape, in this three-step process (reviewed in Ellstrand 1988, Snow and Palma 1997). The risk of transgene spread through pollen movement was found to be a function of pollinator visitation rates (e.g., Cresswell 1994, McPartlan and Dale 1994), proximity of pollen recipients (e.g., Ellstrand 1992, Klinger et al. 1992, Morris et al. 1994, Lavigne et al. 1996, Timmons et al. 1996), and the probability that transgenic pollen will fertilize nontransgenic ovules (e.g., Wilson and Payne 1994). Although this research was important and interesting, it has been argued (Kareiva et al. 1994) to be irrelevant to predicting ultimate risk, since a transgene will escape whenever the probability of hybridization is non-zero, which will be the case when hybrids are viable and species co-occur. Indeed, commercialization involves plantings of transgenic crops as large as a small state, potentially for many decades. In situations like these, even an excruciatingly low probability event will have sufficient opportunity to occur.

2.1.2   Relative fitness vs. invasiveness

Thus, today, there is a greater interest in fitness effects of introduced genes, both as they might affect the rate of initial gene spread through a population and the invasiveness of the final, transgene-dominated populations. It is important to emphasize that these two "fitness effects" are distinct, and that conflation of the two within the risk assessment literature leads to confusing (and wrong) predictions. For some — evolutionary biologists in particular — fitness refers to the *relative* survival and reproductive output of particular genotypes. Within the context of risk assessment, differences in relative fitness are critically important because they indicate whether certain genotypes will spread through a population at the expense of other genotypes. For example, a favorable gene might place individuals at such an advantage (i.e., with increased survivorship, increased fecundity, or both) that no individuals without the transgene would be expected to remain after a few years of natural selection. Selective "sweeps" such as these are associated with the loss of genetic variation at loci linked to the favored allele (Kaplan et al. 1998), and thus this outcome might diminish the ability of the recipient species to survive periods of environmental change. However, selective sweeps do not automatically lead to the production of superweeds. Instead, there is the additional requirement that the transgene must enhance the population growth of the recipient species. This increase in the population growth of populations might result from an increase in competitive ability and/or an increase in the environmental tolerances of the species. Changes in the reproductive capacity of individuals in a population, such as those associated with the growth of the population (sometimes called *absolute fitness*), corresponds to the ecological use of the word *fitness*.

2.1.3 Conflicting outcomes

One illustration of the ecological and evolutionary concepts of fitness is provided by genetically engineered medaka (a fish species). Muir et al. (1996) transformed medaka with the human growth hormone gene driven by a growth hormone promoter cloned from Atlantic salmon. The transformed medaka experienced a dramatically increased juvenile growth rate but also suffered a 49% reduction in the viability of fry. Muir and colleagues used mathematical models to predict that, if rapid early growth translates to early maturity, the transgene can spread to fixation in populations containing transgenic and nontransgenic individuals, despite severe reduction in viability among transgenic fish. Furthermore, the genetic load imposed on the population should doom the fish to extinction within 28 generations despite strong selection for the spread of the transgene. This case study provides one example in which a transgene with a strong positive effect on relative fitness does not similarly enhance the absolute fitness of individuals. Indeed, in this dramatic example, the favored allele is predicted to cause the *extinction* of the population.

The preceding example demonstrates the various levels of fitness that should be explored to estimate the potential for "weediness." We have used transgenic *Arabidopsis thaliana* (Brassicaceae) expressing a gene that confers resistance to the herbicide chlorsulfuron as a model system to examine some of the critical ecological issues surrounding transgenic crops. In this work, we have focused on whether there exist "unintended effects" of this modification on outcrossing rates (i.e., gene transfer via pollen), relative fitness, and population growth. We found an increase in the outcrossing rate of transgenic *A. thaliana* relative to mutants carrying the same resistance allele (Bergelson et al. 1999). We additionally found positive effects of the introduced gene on the relative fitness of plants in the presence of herbicide (as would be predicted) and negative effects in the absence of herbicide (Bergelson et al. 1996, Purrington and Bergelson 1997). Surprisingly, just as in the medaka example above, these fitness effects did not translate into changes in population growth rates. In this chapter, we summarize our characterization of this system, the primary results, and discuss in particular the finding that fitness effects of the transgene are dramatically different among the several replicate transgenic lines. These replicate lines differ only in the position in which the introduced genes reside. This latter finding has important implications for risk assessment because companies often produce hundreds of transgenic lines, but, in practice, risk assessment analyses would rarely catch the unusual line.

2.2   *Chlorsulfuron resistance in* Arabidopsis thaliana

Our work has focused on *A. thaliana*, a small crucifer with a cosmopolitan distribution. In the Midwest of the United States, where our work was completed, *A. thaliana* is a winter annual that overwinters as a rosette and flowers in the spring. It occurs in marginal habitats, such as those on the edges of agricultural fields and along abandoned train tracks (Bergelson et al. 1997), and therefore is not considered an aggressive weed. Indeed, it is outcompeted by many early successional species and exists only in gaps (Bergelson 1994) and at disturbed sites. We were interested in studying resistance to the sulfonylurea herbicide chlorsulfuron, which is produced by DuPont and marked under the trade name, Glean®. Resistance to chlorsulfuron is known to occur naturally in many weed species (reviewed in Saari et al. 1994), although naturally occurring resistance has not yet been identified in *A. thaliana*.

A mutant *A. thaliana*, generated by ethyl methane sulfonate mutagenesis, was isolated that is resistant to the herbicide chlorsulfuron (Haughn and Somerville 1986). This resistant line, designated *GH50*, was demonstrated to contain a point mutation in the gene that codes for acetolactate synthase (ALS), an enzyme important for the synthesis of branched-chain amino acids (Haughn et al. 1988). Resistance stems from the inability of chlorsulfuron to bind to mutant ALS molecules (i.e., resistance is conferred by a binding site modification; Bright 1992). In our work, we attempted to understand how this point mutation affects the ecology and evolution of *A. thaliana* and to understand whether the ecological effects of the mutation differ between mutant and transgenic lines. All these questions have implications for the design and interpretation of risk assessment protocols.

We began by creating a series of transgenic *A. thaliana* lines in which the mutant ALS allele in *GH50* was subcloned into a pBin transformation vector (Bevan 1984, Frisch et al. 1995), and introduced through standard root co-cultivation with *Agrobacterium tumefaciens* (Valvekens et al. 1988). Four such transgenic *Arabidopsis thaliana* lines, each representing an independent transformation event, were created, each with a single copy insertion of the mutant ALS allele (Bergelson et al. 1996). Both the transgenic lines and *GH50* were in the Columbia genetic background and both lines were backcrossed to Columbia in order to purge mutations (six generations for *GH50* and two generations for the transgenic lines). Columbia is a particular genotype of *A. thaliana*, originally collected in Columbia, Missouri, that has become the predominant laboratory strain for molecular genetic studies. Thus, the transgenic lines differed from the mutant lines in the following ways:

1. The transgenic lines underwent tissue culture whereas the mutant line did not.
2. The transgenic lines are resistant to both chlorsulfuron and the antibiotic kanamycin because of the incorporation of a selectable marker in the transformation vector.
3. The transgenic lines contain the mutant allele in a random chromosomal position rather than the native position at which ALS is typically found.
4. The transgenic lines contained both wild-type and mutant copies of ALS, whereas *GH50* contained only the mutant copy of ALS.

These are standard differences between plants produced through mutagenesis and recombinant genetic manipulation.

Table 2.1   Lines of *A. thaliana* Used in the Field Experiments

Line designations	Modification	Resistance phenotype	No. of independent comparisons
GH50	EMS mutagenesis	Chlorsulfuron resistant	1
Columbia (Wt)	None	Susceptible	
pGH8	Transformed and tissue cultured	Chlorsulfuron/kanamycin resistant	4
Null-segregant	Tissue cultured	Susceptible	
Vector control	Transformed and tissue cultured	Kanamycin resistant	3
Null-segregant	Tissue cultured	Susceptible	
Multiple ALS alleles	Transformed and tissue cultured	Kanamycin resistant	4
Null-segregant	Tissue cultured	Susceptible	

Note: Each line was created in the *A. thaliana* genetic background, Columbia, and thus all lines are isogenic except for mutations and inserts resulting from our manipulations. Transformation was performed through root co-cultivation with *Agrobacterium*, and all lines were backcrossed at least twice prior to the creation of null-segregants through selfing. Resultant line names and characteristics are given. The number of independent comparisons corresponds to the number of independent transformed lines of each type.

In addition to these transgenic lines, we also created a variety of control lines that are useful in addressing particular questions. First, we created independent "vector control" lines into which an empty transformation vector was randomly inserted (again, only lines with single insertion sites were used). These vector control lines are resistant to kanamycin, and therefore enable us to address potential deleterious effects of expressing nptII (Bevan 1984, Frisch et al. 1995), which confers resistance to this antibiotic. Second, we created transgenic lines in which an additional copy of a wild-type ALS allele was introduced. These chlorsulfuron-susceptible lines have two copies of ALS and therefore enable us to factor out potential effects of expressing multiple ALS alleles. Third, to control for the mutations or hormonal abnormalities that might have been induced during tissue culture *and* that remained after backcrossing, we took our heterozygous, backcrossed transgenic lines and "selfed" them, producing homozygous wild-type, hemizygous (= heterozygous) transgenic, and homozygous transgenic progeny all within a single fruit. By identifying and selfing the two types of homozygotes, we created thousands of seeds for pairs of lines that are homozygous for the presence and absence of the insert. We called the lines lacking the insert "null-segregants" and compared them with their transgenic counterparts containing the transgene. Finally, to randomize across insertion sites, we created independent transformants using each construct. In total, we considered four independently transformed lines of pGH8, three independently transformed vector control lines, and four independently transformed lines of pALS. A summary of our experimental lines is given in Table 2.1. Although this might seem an inordinate number of controls, we discuss below why such an approach can be useful.

2.2.1   *Differences in outcrossing between mutant and transgenic lines*

There has been considerable work on the rates of outcrossing for various crops and on agricultural strategies, such as buffer zones, that might better contain pollen within the borders of agricultural fields (e.g., Ellstrand 1988, 1992, Klinger et al. 1992, Cresswell 1994, Kareiva et al. 1994, McPartlan and Dale 1994, Morris et al. 1994, Lavigne et al. 1996, Timmons et al. 1996; reviewed in Snow and Palma 1997). In most of this work, the authors assume that mutants and/or distinct genotypes can be used as models for the reproductive behavior of transgenic crops. This assumption that transformation does not alter mating behavior extends to the regulatory process, where evidence of low rates of outcrossing in *conventional* crops is taken as evidence that their *transgenic* counterparts will exhibit similar (if not identical) low rates of outcrossing. While this assumption is intuitively appealing, it has been subjected to surprisingly few tests.

Arabidopsis thaliana is a highly selfing species, with outcrossing rates estimated to be less than 1% (Snape and Lawrence 1971, Redei 1975, Abbott and Gomes 1989, Bergelson et al. 1998). The conduit for outcrossing has never been proved definitively, although visitation by syrphid flies was apparent at our field site and during the experiments of Snape and Lawrence (1971).

In 1996, we performed a field experiment to measure the outcrossing rate of mutant and transgenic *A. thaliana* (see Bergelson et al. 1999). Unlike the experiments on fitness described below, we were not interested in the effect of our ALS allele per se. Our motivation was to ask specifically whether transformation influences rates of outcrossing. To answer this question, it was therefore important to control for the presence of the ALS mutation in our comparison. We designed an experiment to compare rates of outcrossing in transgenic, pGH8 lines with those of the GH50 mutant. These two types of lines both contain the mutant allele, but differ in how that allele was introduced. The question that our experiment addressed, therefore, was whether the method of genetic modification has an effect on gene flow.

For simplicity, our experimental design (Bergelson et al. 1999) focused on outcrossing rates calculated through male function. In the design, 144 plants were placed in a grid at natural densities (based on nearby populations). A quarter of these plants were homozygous mutants, and contained the mutant allele *Csr1-1* that confers resistance to the herbicide chlorsulfuron in a Columbia genetic background. Half of the plants in the field were homozygous, transgenic plants expressing *Csr1-1* in a pBin transformation vector (designated the pGH8 vector). These transgenic plants were equally divided between representatives from each of two independent insertion events, and were also in the Columbia genetic background. The remaining quarter of these plants were untransformed Columbia, which is susceptible to chlorsulfuron. Plants were randomly allocated to positions in the grid, grown at our field site in central Illinois, in the absence of herbicides, and allowed to grow and reproduce naturally.

At the end of the growing season, we collected seeds produced by the 36 Columbia untransformed plants and germinated them on agar plates containing lethal concentrations of chlorsulfuron. This allowed us to identify all progeny that were fathered by either mutant or transgenic *A. thaliana* in the field. All resistant progeny were carefully transplanted from the chlorsulfuron plates to soil, and their selfed seeds were collected in the greenhouse. Cross pollination was prevented through the use of pollination bags that protected all buds from airborne pollen. By germinating these selfed seeds on plates containing kanamycin, we were able to distinguish progeny that had been fathered by transgenic *A. thaliana* (resistant to chlorsulfuron and kanamycin) from those sired by mutant *A. thaliana* (resistant to chlorsulfuron only). A survey of almost 100,000 seeds allowed us to estimate the outcrossing as 0.34% for mutant fathers and 5.82% for transgenic fathers. That is, the transgenic *A. thaliana* were, *on average*, 17 times more likely to outcross than the mutant fathers. It is important to note, however, that the difference between the two transgenic lines was large.

To determine which transgenic line had fathered the transgenic, resistant progeny, we developed PCR (polymerase chain reaction) markers specific to each insertion site. Two pairs of primers that specifically amplify DNA from each insertion site were used to determine paternity for a subset of 281 transgenic, resistant progeny from the outcrossing experiment. We calculated the outcrossing rate of pGH8(2) and pGH8(3) as 0.8 and 10.7%, respectively. Both transgenic genotypes demonstrated a significant increase in their out-crossing rates relative to the mutant plant, although the two transgenic plants also significantly differed from one another in their outcrossing rates, with only one showing a dramatic increase in its propensity to outcross.

We repeated an outcrossing experiment in 1998 using all four of the pGH8 transgenic lines. In this experiment, we introduced an additional modification that greatly facilitated our identification of outcrossed progeny. Specifically, the susceptible plants used in the 1998 experiment were Columbia plants homozygous for a recessive mutation that causes plants to lack trichomes. Thus, every progeny with trichomes is the result of an outcrossing event and large numbers of rosettes can be easily genotyped with the naked eye. There were two important results obtained from this experiment:

1. The absolute levels of outcrossing for pGH8(2) and pGH8(3) were very similar to rates determined in the earlier study [1998 rates: 1.3% for pGH8(2) and 8.6% for pGH8(3)].
2. The additional pGH8 lines were variable in their outcrossing rates. One of these new lines, pGH8(1), displayed the highest outcrossing rate of the four lines (12.4%), and the second new line, pGH8(4), displayed an intermediate outcrossing rate (1.9%). Thus, two of the four lines show dramatic increases in their propensity to outcross.

2.2.2   Effects of chlorsulfuron resistance on relative fitness

The probability that an introduced gene will persist in a population depends on fitness effects of the gene. If individuals expressing a particular gene are strongly favored, then this gene is likely to sweep relatively quickly to a high frequency in the population, leading to two effects. First, genetic variability in the population can be reduced as a consequence of the selective sweep (Kaplan et al. 1989). Such selective sweeps raise concerns about the future adaptability of populations because there may be little extant variation to provide genetic material for change (e.g., Regal 1986, Tiedje et al. 1989, Keeler and Turner 1991). Second, a population composed of individuals expressing (or primarily expressing) an introduced gene might have dynamics different from that of populations of plants without the gene. That is, the population might be able to achieve higher population density, might be able to expand into new regions, and/or might be able to outcompete other species that it could not outcompete before. These latter effects are at the heart of concern about the introduction of new genes enhancing the invasiveness of weeds and creating so-called superweeds.

In this context, it is important to consider the fitness effect of genes in a variety of environments. In the case of resistance genes, for example, resistance to chlorsulfuron, two obviously important environments would be those with and without the herbicide. Although it is reasonable to imagine that resistance will be favored in the presence of herbicides, and that resistant weeds might be more likely than susceptible weeds to invade habitats that are regularly sprayed, it is also possible that resistant weeds will be disadvantaged in habitats lacking herbicides. A position paper on biotechnology risk assessment, written by Tiedje et al. (1989) on behalf of the Ecological Society of America, states:

> It has often been assumed that the addition of extra genes reduces the
> competitiveness of an organism due to the added cost of synthesizing
> additional nucleic acids and proteins ... many engineered organisms
> will [therefore] probably be less fit than the parent organism.

In other words, there is an assumption that costs of resistance will be present and of sufficient magnitude to prevent the population growth of resistant weeds in environments

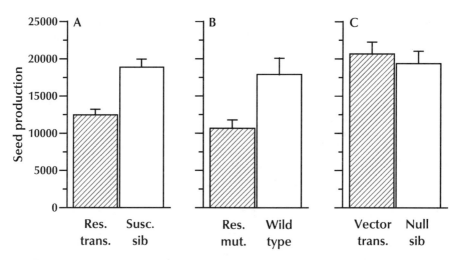

Figure 2.1 Lifetime seed production given for (A) transgenic lines and their null-segregants; (B) the mutant line, GH50, and its susceptible counterpart, Columbia; and (C) the vector control lines and their null-segregant. The results in (A) and (C) represent pooled results for replicate, independently transformed lines. All data were obtained under field conditions in central Illinois. (Data from Bergelson et al. 1996.) (From Purrington, C. B. and Bergelson, J., *Am. Nat.*, 154, 582, 1999. With permission of University of Chicago Press.)

lacking the agent of selection (in our case, the herbicide). Despite this assumption, many empirical studies have failed to detect measurable fitness costs of resistance (e.g., Simms and Rausher 1987; reviewed in Bergelson and Purrington 1996, Snow et al. 1999). In fact, one of the most conspicuous patterns regarding costs of resistance is their almost complete absence in trials using transgenic crops. This is probably due, not to the absence of costs in transgenic systems, but, rather, either to the fact that fitness costs may be small or to the lack of statistical power and genetic control in many of these experiments (as discussed in Bergelson and Purrington 1996).

In our work on chlorsulfuron resistance, we have provided a clear demonstration that herbicide resistance is costly to the lifetime seed production of plants in the absence of herbicides. Indeed, costs of resistance to chlorsulfuron in *A. thaliana* are often substantial and have been detected in a number of experiments and environmental conditions (Bergelson 1994, Bergelson et al. 1996, Purrington and Bergelson 1997, 1999, Bergelson and Purrington, unpublished data). Because we have never detected effects of chlorsulfuron on the survival or germination rates of resistant *A. thaliana* (Bergelson et al. 1996, Purrington and Bergelson 1997), measures of seed production provide the key fitness parameter (Figure 2.1). The cost of resistance in the absence of herbicide counteracts the clear benefits of resistance in the presence of herbicides. The transgenic lines we studied varied in their level of resistance, suffering between a 0 and 41% reduction in lifetime seed production when sprayed with chlorsulfuron at recommended rates. This contrasts with susceptible plants that suffered extreme fitness reductions of approximately 90% in terms of seed production.

In summary, there are several important results from these field experiments, all of which were conducted in the absence of herbicide. Transgenic, resistant *A. thaliana* carrying *Csr1-1* displayed an approximate 30% reduction in lifetime fitness (averaged across four replicate lines) when grown in the absence of herbicides in a field characteristic of natural *A. thaliana* habitats. This result has now been repeated in several experiments performed over 3 years (Bergelson et al. 1996, Purrington and Bergelson 1997, unpublished data). The fitness cost of resistance is due solely to the expression of the ALS mutation. Our experiments enable us to eliminate such factors as the potential costs associated with expression of the selectable marker (kanamycin resistance), the possible disruption due to the insertion of the plasmids, the possible cost of expressing an extra copy of ALS, and the action of mutations induced during transformation. Interestingly, the cost of resistance in transgenic (Figure 2.1A) and mutant (Figure 2.1B) lines was similar, and there was no cost associated with expression of the pBIN vector (Figure 2.1C).

It has been suggested that fitness costs are less likely to be detectible in nutrient-rich, agricultural conditions than in natural habitats (e.g., Gulmon and Mooney 1986, Bazzaz et al. 1987). Further, sensitivity of costs of resistance to levels of stress have been observed in a number of systems (reviewed in Bergelson and Purrington 1996). Although one goal of ecological risk assessment is to predict invasions into *natural* habitats, most studies are performed only in an agricultural setting. Thus, it is important to determine whether fitness costs are sensitive to environmental conditions for particular transgenic systems of concern. We have completed two field experiments using our transgenic and mutant lines to examine how stress imposed by either nutrient limitation or enhanced competition may influence the magnitude of fitness costs of resistance. We find that costs of resistance increase with the level of stress in the environment. The addition of fertilizer or the removal of competitors is sufficient to remove detectable fitness costs of resistance (Bergelson 1994, Purrington and Bergelson 1997, Bergelson and Purrington, in preparation).

Further, we have found significant variation among lines in the *magnitude* of costs of resistance. For example, when grown in the presence of competition and absence of fertilizer, fitness costs ranged from 23 to 51% (Bergelson and Purrington, unpublished data). This variation among lines is statistically significant (as evidenced by a significant effect of "line" in an ANOVA), and is presumably due to position effects associated with the random integration of the introduced ALS allele. These position effects can be the consequence of variation in the expression of the introduced gene and/or can be due to disruption of other genes. Because we have detected a positive association between the cost of resistance demonstrated by a particular line and the activity of ALS in that line (Purrington and Bergelson 1999), we believe that the position effects that we observed are due, at least in part, to variation in the level of ALS activity.

2.2.3 *Population-level effects of chlorsulfuron resistance*

Chlorsulfuron resistance provides a unique opportunity to test the assumption that differences in the seed production among lines have population-level consequences. In our system, there are very large fitness differences between resistant and susceptible lines in both the presence and absence of herbicide. Thus, if fitness effects extend across scales, it is likely that we will be able to observe them. In our study, we addressed two questions:

1. In plots that contain mixtures of resistant and susceptible lines, do we see a change in gene frequency between generations?
2. Do we see differences in the between-generation population growth rate of plots containing only resistant or only susceptible plants?

The first question is relevant to discussions about *relative* fitness and the second question is relevant to discussions about *absolute* fitness. As discussed earlier in this chapter, these two types of "fitness" have different implications for risk assessment, with only absolute fitness being directly related to changes in invasiveness.

To address these questions, we performed a field experiment in which replicate 0.25-m^2 plots were established at our field site in central Illinois. Each plot was inoculated with four *A. thaliana* plants, and plots were equally divided among those containing four resistant plants, four susceptible plants, or a 50:50 mixture of resistant and susceptible lines. In this experiment, we used two of our four pGH8 lines [pGH8(2) and pGH8(3)] and used the relevant null-segregants as susceptible plants in these plots. Plots had a naturally occurring assemblage of weeds. Half of the plots received regular applications of chlorsulfuron, and the other half of the plots were left unmanipulated.

We found that differences in relative fitness had a dramatic effect on the representation of each genotype in subsequent generations. When resistant and susceptible plants were grown in the absence of herbicide, the resistant genotypes quickly dropped in frequency (from 50 to approximately 25%) after one generation. Although we do not have similar data for gene frequencies in the presence of herbicides after one generation, we did find a substantial increase in the frequency of resistant individuals in the presence of herbicides after three generations (from 50 to approximately 95%). These changes are not surprising, given the large fitness costs and benefits associated with expression of the *Csr1-1* allele.

Despite these differences in relative fitness, there were no significant differences in the plant numbers after one generation. At first, this seems surprising, especially given the widespread assumption that changes in the performance of individuals will be reflected in changes in the invasiveness of species (e.g., OTA 1988, Dunwell and Paul 1990). However, the population sizes of some weed species appear to be regulated by the availability of space rather than by the production of seeds (e.g., Bergelson 1990, 1994). In these instances, it is reasonable to imagine that even dramatic changes in seed number (from, for example, 20,000 seeds to 10,000 seeds produced per plant) will have few repercussions at the population level. It remains to be determined how often weed species, in general, are seed-limited, and what regulates weeds that can hybridize with transgenic crops. In any event, our results provide one clear demonstration in which the relationship between individual and population-level performance is not straightforward.

2.3   Relevance to risk assessment

Astute readers may have guessed that chlorsulfuron-resistant *A. thaliana* is not a looming threat to environmental integrity. Our findings showed that the addition of this herbicide-resistance gene results in plants that are reproductively handicapped relative to normal individuals and that decrease in frequency when grown with the nontransgenic variety. Certainly, the only risk that could be imagined is fostering the presence of chlorsulfuron-resistant *A. thaliana* in agricultural settings that are sprayed frequently with chlorsulfuron. Natural areas, on the other hand, should not be affected. Indeed, biologists familiar with *A. thaliana* are sometimes skeptical that we could assess *any* real risk with such a small, nondescript plant — a species that is considered preadapted for growth only in small growth chambers. To a certain degree this criticism is fair, but if we had hoped to place a transgene into crabgrass, for example, we probably would not have obtained the necessary permission from APHIS (the U.S. Animal and Plant Health Inspection Service) to perform the experiments under natural field conditions.

Thus, we do not mean to imply that our example of "probably low risk" should be interpreted to mean that all types of genetic modification are going to be risk free. In fact, we would be hesitant to say that the chlorsulfuron-resistance transgene gene will cause a fitness reduction if it were inserted into a crop or transferred to a wild species. Our experience in analyzing the constructed lines in numerous experiments has impressed upon us that insertion position of the transgene can produce dramatic differences in phenotype and reproductive capacity among transgenic lines. Our major conclusion from the above-described case study is that risk assessment should be applied to individual lines, not to the "average risk" of lines created through the independent insertion of a novel gene. Specifically, if a company is interested in commercializing a transgenic crop, we would recommend that risk be determined independently (through experimentation) for each transformed line that goes to market (discussed in Purrington and Bergelson 1995).

The calculation of weediness risk by averaging or pooling multiple lines will, of course, decrease the probability that real risk will be detected. For example, if 1 line out of 30 (not an unreasonable number initially to produce and field test) had twice the seed production of the parental variety, but the other 29 lines produced fewer seeds than the parental variety, then the average transgenic seed production might reveal no overall risk. In this regard, it is instructive to note how risk is evaluated in an actual "Petition for Determination of Nonregulated Status," the document produced for APHIS by biotechnology companies wishing to commercialize a transgenic crop. In examining the petitions received during the early 1990s, we found that biotechnology corporations often requested blanket approval of multiple, independently transformed lines, yet in many instances reported data that were *averages* of those lines (Purrington and Bergelson 1995). Although this method of reporting data may be standard practice, it makes science-based risk assessment difficult. The worst-case scenario, to continue with the example of the hypothetical lines described above, would result in the release of the "weedy" line into the environment. Moreover, it is likely that such a line would be marketed because seeds for higher-yielding crops are *exactly* what farmers want to buy.

A cautious regulator would be aware that even if the initial lines did not present any weed risk, the product that eventually was commercialized might. Because companies producing transgenic crops must obtain nonregulated status years before the crops are commercialized, the breeders at those companies have several years in which to cross the "low-vigor" transgenic lines with the parental variety (backcrossing) or with cultivars that are distantly related to the parental strain. Both of these procedures have the effect of minimizing any yield-reducing effects that are caused by the introduction of mutations during transformation and tissue culture. Indeed, many APHIS petitions document that such breeding efforts are already under way, and express optimism that the undesirable "low-yield" characteristics will eventually be completely eliminated. Therefore, if the initial lack of weed risk is caused by mutations *other* than the transgene per se, then traditional methods of crop breeding eventually will be able to remove these mutations. To perform the risk assessment correctly, it is the *final* product of these breeding regimes that should be tested for weed risk (e.g., high seed production, high survivorship), not the initial transformants (which often exhibit low fitness).

Unfortunately, petitions for nonregulated status suffer from far more basic problems than the confusion over line variation and backcrossing. Foremost among the problems is the largely descriptive and speculative nature of the documents. They cannot logically answer the question of whether a new crop poses a weed risk to the environment. For example, 14% of petitions submitted prior to 1994 failed to present any data whatsoever on seed dormancy, a trait that might have real importance to weed risk (and require data that would be simple to collect, as well). When data are presented in these petitions, they are usually incorrectly analyzed (e.g, lacking comparisons to a control) and almost always suffer from poor experimental design (e.g., insufficient sample size to show differences even when differences are truly present) (discussed in Purrington and Bergelson 1995). In the place of data are heuristic arguments. For example, there are statements that the addition of a transgene cannot possibly create a superweed because the recipient cultivar was not a weed initially, or that the transgene was not cloned from a weed, and thus it cannot act to increase weediness in another species. This type of reasoning is insufficient to address the single most important question that is posed to the petitioner: Is there a weed risk? To date, it does not appear that APHIS is alarmed by this lack of data, as no petition yet has been listed as "denied."

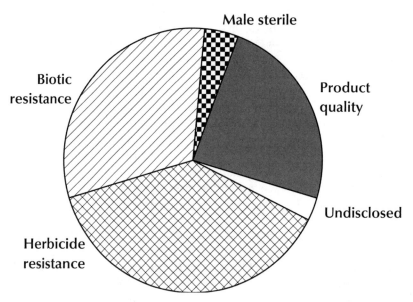

Figure 2.2   Percentage of 71 petitions for nonregulated status, submitted to the USDA/APHIS through 19 February 2000, that falls within various categories of phenotype alteration. Of the 22 herbicide-resistant transgenics, 6 also had a second transgene that conferred either male sterility or resistance to a pest or pathogen. Phenotypes that are "undisclosed" on the APHIS database are typically from petitions that are still being analyzed for completeness. (Data tabulated from the APHIS database, available at www.aphis.usda.gov/biotech/petday.html.)

In addition, even when petitions do contain adequate data comparing the characteristics of the transgenic line(s) to the parental line, the results, at best, only allow comparison of *relative* fitness values. Usually lacking is any quantitative measurement of the growth potential of a population of the transgenic variety, the "ecological" fitness (described in case study above) that can only be estimated by measuring whether population size of a particular genotype *increases*. Therefore, until the guidelines for such petitions are rewritten to emphasize quantitative data, especially and specifically data relating to evolutionary and ecological types of fitness, it is probably safe to conclude that current risk assessment has a limited, albeit nonzero, ability to identify a weed risk to the environment.

We are not saying that the currently deregulated crops are likely to develop into true ecological problems, but rather that some crop — 5, 10, or 100 years in the future — *will* have such a potential, and that current procedures could easily miss this risk. Part of our pessimism over this future event is based on an examination of the type of phenotypes that are currently being commercialized (Figure 2.2), and the belief that future crops will have much more ambitious alterations of their phenotypes. Because the vast majority of crops already deregulated or awaiting status determination are in some way resistant, fitness costs of these resistance phenotypes might tend to minimize risk, assuming that the resistance phenotype does not influence fitness in natural areas (an admittedly optimistic assumption, especially for herbivore and disease resistance). Alterations in product quality such as ripening characteristics and nutritional content are also probably not particularly worrisome, because they seem less likely to confer a fitness advantage than do genes for resistance traits. Relatively underrepresented among these petitions are crops that have had an "agronomic property" such as male fertility manipulated. In this category would also be seed number, growth rate, geographical range, and average life span, all of which are intimately associated with weediness and could be altered by genetic modification. That this category is being "ignored" by biotechnology corporations is likely more an indication of the immediate economic attractiveness of resistant crops than an indication of any inherent problems in engineering such agronomic properties as higher yield or elevated growth rate. Another interesting trend, not shown by Figure 2.2, is the increase in the number of petitions for lines that contain multiple transgenes. If dozens or hundreds of genes encoding particular traits are integrated simultaneously, the modified crop might have considerably increased fitness and invasive potential when released into nonagricultural areas. Thus, we expect to see an increase in petitions for transgenic lines that will, or should, elicit greater concern within regulatory agencies.

2.4   Concluding remarks

Although our experiments involving *A. thaliana* were not exclusively aimed at evaluating risks of weediness, we believe that our work can act as a valuable guide to such an endeavor. In particular, we have demonstrated that regulatory agencies should be aware of line variation in transgenic crops and the potential for such variation to obscure underlying risk. In our research we have also highlighted the difference between relative fitness and ecological fitness, and, we hope, have illustrated that "superweeds" are plants whose ecological fitness has been increased. Recognizing that evaluation of only relative fitness is easier, we have emphasized that it can never replace experiments that test whether *population* growth rates of transgenic organisms have been inadvertently increased.

Finally, we have found that current risk assessment strategies in the United States are surprisingly light on data. Perhaps this is an inevitable consequence of the U.S. Department of Agriculture (USDA) being charged with both the regulation and promotion of transgenic plant technology. Interestingly, the Food and Drug Administration (FDA) and the Environmental Protection Agency (EPA), other U.S. regulatory agencies with mandates to oversee transgenic products, seem to be far more quantitative in the types of assurances they require from biotechnology companies (NRC 2000). This difference may be due to the fact that APHIS has a much more difficult question to rule on (Is there a weed risk?) than do the other agencies (e.g., Will the product make people sick?). To remedy this situation, scientists familiar with both the molecular aspects of transgenic technology and the ecological/evolutionary factors will have to take a much more active role in shaping regulatory rules. And APHIS, in turn, will have to become more rigorous in evaluating whether a transgenic crop is likely to create a weed risk.

Acknowledgments

We thank Gale Wichmann for sharing her outcrossing results and the Packard Foundation and USDA for funding this research.

Literature cited

Abbott, R. J. and Gomes, M. F., Population genetic structure and outcrossing rate of *Arabidopsis thaliana* (L.) Heynh., *Heredity*, 62, 411, 1989.

Anonymous, The green man, editorial, *N. Sci.*, 166, 3, 2000.

Bazzaz, F. A., Chiariello, N. R., Coley, P. D, and Pitelka, L. F., Allocating resources to reproduction and defense, *BioScience*, 37, 58, 1987.

Bergelson, J., Life after death: site pre-emption by the remains of *Poa annua, Ecology*, 71, 2157, 1990.

Bergelson, J., Changes in fecundity do not predict invasiveness: a model study of transgenic plants, *Ecology*, 75, 249, 1994.

Bergelson, J. and Purrington, C. B., Surveying patterns in the cost of resistance in plants, *Am. Nat.*, 148, 536, 1996.

Bergelson, J., Purrington, C. B., Palm, C. J., and López-Gutiérrez, J. C., Costs of resistance: a test using transgenic *Arabidopsis thaliana*, *Proc. R. Soc. London B*, 263, 1659, 1996.

Bergelson, J., Stahl, E., Dudek, S., and Kreitman, M., Genetic variation within and among populations of *Arabidopsis thaliana*, *Genetics*, 148, 1311, 1998.

Bergelson, J., Purrington, C. B., and Wichmann, G., Promiscuity in transgenic plants, *Nature*, 395, 25, 1999.

Bevan, M., Binary *Agrobacterium* vectors for plant transformation, *Nucl. Acids Res.*, 12, 8711, 1984.

Bright, S. W. J., Herbicide-resistant crops, in *Biosynthesis and Molecular Regulation of Amino Acids in Plants*, Singh, B. K., Flores, H. E., and Shannon, J. C., Eds., American Society of Plant Physiologists, Rockville, MD, 1992, 184.

Cresswell, J. E., A method for quantifying the gene flow that results from a single bumblebee visit using transgenic oilseed rape, *Brassica napus* L. cv. Westar, *Transgenic Res.*, 3, 134, 1994.

Dunwell, J. M. and Paul, E. M., Impact of genetically modified crops in agriculture, *Outlook Agric.*, 19, 103, 1990.

Ellstrand, N. C., Pollen as a vehicle for the escape of engineered genes? *Trends Ecol. Evol.*, 3, S30, 1988.

Ellstrand, N. C., Gene flow among seed plant populations, *N. For.*, 6, 241, 1992.

Ellstrand, N. C., Prentice, H. C., and Hancock, J. F., Gene flow and the introgression from domesticated plants into their wild relatives, *Annu. Rev. Ecol. Syst.*, 30, 539, 1999.

Frisch, D. A., Harris-Haller, L. W., Yokubaitis, N. T., Thomas, T. L., Hardin, S. H., and Hall, T. C., Complete sequence of the binary vector Bin 19, *Plant Mol. Biol.*, 27, 405, 1995.

Gulmon, S. L. and Mooney, H. A., Costs of defense and their effects on plant productivity, in *On the Economy of Plant Form and Function*, Givnish, T. J., Ed., Cambridge University Press, Cambridge, U.K., 1986, 681.

Haughn, G. W. and Somerville, C., Sulfonylurea-resistant mutants of *Arabidopsis thaliana*, *Mol. Gen. Genet.*, 204, 430, 1986.

Haughn, G. W., Smith, J., Mazur, B., and Somerville, C., Transformation with a mutant *Arabidopsis* acetolactate synthase gene renders tobacco resistant to sulfonylurea herbicides, *Mol. Gen. Genet.*, 211, 266, 1988.

Kaplan, N. L., Hudson, R. R., and Langley, C. H., The "hitchhiking effect" revisited, *Genetics*, 123, 887, 1989.

Kareiva, P., Morris, W., and Jacobi, C. M., Studying and managing the risk of cross-fertilization between transgenic crops and wild relatives, *Mol. Ecol.*, 3, 15, 1994.

Keeler, K. H. and Turner, C. E., Management of transgenic plants in the environment, in *Risk Assessment in Genetic Engineering: Environmental Release of Organisms*, Levin, M. and Strauss, H., Eds., McGraw-Hill, New York, 1991, 189.

Klinger, T., Arriola, P. E., and Ellstrand, N. C., Crop-weed hybridization in radish (*Raphanus sativus*): effects of distance and population size, *Am. J. Bot.*, 79, 1431, 1992.

Lavigne, C., Godelle, B., Reboud, X., and Gouyon, P. H., A method to determine the mean pollen dispersal of individual plants growing within a large pollen source, *Theor. Appl. Genet.*, 93, 1319, 1996.

McPartlan, H. C. and Dale, P. J., An assessment of gene transfer by pollen from field-grown transgenic potatoes to non-transgenic potatoes and related species, *Transgenic Res.*, 3, 216, 1994.

Morris, W. F., Kareiva, P. M., and Raymer, P. L., Do barren zones and pollen traps reduce gene escape from transgenic crops? *Ecol. Appl.*, 2, 431, 1994.

Muir, W. M., Howard, R. D., Martens, R., and Bidwell, C., Use of multigenerational studies to assess genetic stability, fitness, and competitive ability of transgenic Japanese medaka: II. Production of transgenic fish and preliminary studies, in *Proceedings of the 7th Symposium on Environmental Releases of Biotechnology Products: Risk Assessment Methods and Research Progress*, Levin, M., Grim, C., and Scott, J. S., Eds., University of Maryland Biotechnology Institute, College Park, 1996.

NRC (National Research Council), Field Testing Genetically Modified Organisms: Framework for Decision, National Academy Press, Washington, D.C., 1989.

NRC (National Research Council), Genetically Modified Pest-Protected Plants: Science and Regulation, National Academy Press, Washington, D.C., 2000.

OTA (Office of Technology Assessment, U.S. Congress), New Developments in Biotechnology. (3) Field Testing Engineered Organisms: Genetic and Ecological Issues, U.S. Government Printing Office, Washington, D.C., 1988.

Purrington, C. B. and Bergelson, J., Assessing weediness of transgenic crops: industry plays plant ecologist, *Trends Ecol. Evol.,* 10, 340, 1995.

Purrington, C. B. and Bergelson, J., Fitness consequences of genetically engineered herbicide and antibiotic resistance in *Arabidopsis thaliana, Genetics,* 145, 807, 1997.

Purrington, C. B. and Bergelson, J., Exploring the physiological basis of costs of herbicide resistance in *Arabidopsis thaliana, Am. Nat.,* 154, S82, 1999.

Redei, G. P., Arabidopsis as a genetic tool, *Annul. Rev. Genet.,* 9, 111, 1975.

Regal, P. J., Models of genetically engineered organisms and their ecological impact, in *Ecology of Biological Invasions of North American and Hawaii,* Mooney, H. A. and Drake, J. A., Eds., Springer-Verlag, New York, 1986, 111.

Saari, L. L., Cotterman, J. C., and Thill, D. C., Resistance to acetolactate synthase inhibiting herbicides, in *Herbicide Resistance in Plants,* Powles, S. B. and Holtum, J. A. M., Eds., Lewis Publishers, Boca Raton, FL, 1994, 83.

Simms, E. L. and Rausher, M. D., Costs and benefits of plant defense to herbivory, *Am. Nat.,* 130, 570, 1987.

Snape, J. W. and Lawrence, M. J., The breeding system of *Arabidopsis thaliana, Heredity,* 27, 299, 1971.

Snow, A. A. and Palma, P. M., Commercialization of transgenic plants: potential ecological risks, *BioScience,* 47, 86, 1997.

Snow, A. A., Andersen, B., and Jorgensen, R. B., Cost of transgenic herbicide resistance introgressed from *Brassica napus* into weedy *B. rapa, Mol. Ecol.,* 8, 605, 1999.

Tiedje, J. M., Colwell, R. K., Grossman, Y. L., Hodson, R. E., Lenski, R. E., Mack, R. N., and Regal, P. J., The planned introduction of genetically engineered organisms: ecological considerations and recommendations, *Ecology,* 70, 298, 1989.

Timmons, A. M., Charters, Y. M., Crawford, J. W., Burn, D., Scott, S. E., Dubbels, S. J., Wilson, N. J., Robertson, A., O'Brien, E. T., and Squire, G. R., Risks from transgenic crops, *Nature,* 380, 487, 1996.

Valvekens, D., Van Montagu, M., and Van Lijsebettens, M., *Agrobacterium tumefaciens*-mediated transformation of *Arabidopsis thaliana* root explants by using kanamycin selection, *Proc. Natl. Acad. Sci. U.S.A.,* 85, 5536, 1988.

Wilson, H. D. and Payne, J. S., Crop/weed microgametophyte competition in *Cucurbita pepo* (Cucurbitaceae), *Am. J. Bot.,* 81, 1531, 1994.

chapter three

Bt-crops: Evaluating benefits under cultivation and risks from escaped transgenes in the wild

Deborah K. Letourneau, Joy A. Hagen, and Gaden S. Robinson

Contents

3.1   Bt-*crops: Benefits and risks*

Crops transformed with genes derived from the bacterium *Bacillus thuringiensis* (*Bt*) to express insecticidal proteins (insect-resistant *Bt*-crops) constitute one of the most common genetically engineered organisms in the environment. Widely known across a range of sectors including the agricultural, environmental, and regulatory communities, *Bt*-crops are the subject of contradictory claims, promises, and warnings. As such, they present a compelling case for the analysis of environmental benefits and risks. *Bt*-insect resistance traits potentially increase yield while substituting for the chemical insecticide applications used to control certain pests. However, *Bt*-toxin-based pest resistance also represents an excellent example of a trait that, should it escape (be transferred to wild plants), potentially increases weediness for the very same reasons it is expected to cause increased yields in crops. In the first section of this chapter, we present an analysis of claims regarding yield increases and insecticide usage and discuss potential environmental advantages and disadvantages of *Bt*-crops. The second section presents a biosafety analysis for potential transgene escape and introgression of these genes in wild plants. In both sections we pose critical questions, present available data, and call attention to important knowledge gaps that impede rigorous biosafety assessment.

3.1.1   Bt-*crops: An overview*

Small-scale field trials of genetically engineered crops containing *Bt*-genes to protect the plants against insect pests began in 1986, and large-scale trials have been carried out since the early 1990s (Krattiger 1997). By the mid-1990s, the U.S. Department of Agriculture Animal and Plant Health Inspection Service (USDA–APHIS) had deregulated three transgenic crops with *Bt*-endotoxins (potato, cotton, and corn) constituting approval for them to be grown commercially. These *Bt*-crops were also determined safe for commercialization by the Environmental Protection Agency (EPA) and the Food and Drug Agency (FDA). Currently, corn, cotton, and potato are the only commercialized *Bt*-crops, and they are undergoing a review by the EPA for renewal of their deregulated status. Many other *Bt*-plants have been field-tested and are at various stages in the regulatory process (Table 3.1).

In 2000, genetically engineered crops were grown on 109.2 million acres worldwide (James 2000), with nearly three quarters of this acreage occurring in the United States. Of worldwide transgenic crop acreage, 26% (28.4 million acres) is planted with *Bt*-protein-based, insect-resistant crops. The total acreage of *Bt*-crops has increased tenfold since 1996. Much of this increase is due to rapid adoption of *Bt*-corn and *Bt*-cotton by U.S. farmers, whose nearly 20 million acres accounts for the vast majority of global acreage in *Bt*-crops. Peak acreage for *Bt*-crops occurred in 1999, with a slight decrease in 2000. This decrease was likely due to several factors, including a low infestation of the target pest of *Bt*-corn in 1998 and 1999 (resulting in lower benefit for farmers planting *Bt*-corn) and consumer concern (resulting in perceived market insecurity for *Bt*-crops).

Records of *Bt*-crop plantings since 1996 (Table 3.2) show that *Bt*-corn dominates the acreage planted in genetically engineered, pest-resistant crops. *Bt*-corn is grown on over 20 million acres worldwide and accounts for just over 70% of the global area devoted to *Bt*-crops. *Bt*-cotton is the second most widely grown *Bt*-crop, with plantings quickly increasing to 67% of U.S. cotton acreage by 2000, more than doubling the U.S. *Bt*-cotton

Table 3.1 Status of Various Genetically Engineered Crops with *Bt*-Derived Genes for Toxic Protein Expression as a Source of Insect Pest Resistance

Cry toxin	Crop	Status	Year[a]	Company
CryVIA	Alfalfa	Field tested	1994	Mycogen
CryIA(b)	Apple	Field tested	1995	Dry Creek
CryIA(c)	Apple	Field tested	1991	UC Davis
	Broccoli	Field tested	1995	Cornell University
CBI[b]	Broccoli/cabbage	Field tested	1999	ELM (Emprezas La Moderna)/ Asgrow/Seminis
	Corn	Field tested	1996	Agracetus
Cry9C	Corn	Deregulated[c]	1997	Aventis[d]
CryIA(b)	Corn	Field tested	1997	Aventis[d]
CBI[b]	Corn	Field tested	2000	Aventis[d]
	Corn	Field tested	1995	Cargill
CryIA(b)	Corn	Field tested	1996	Cargill
CBI[b]	Corn	Field tested	1994	Ciba-Geigy, now Syngenta[e]
CryIA(b)	Corn	Deregulated[c]	1994	Ciba-Geigy, now Syngenta[e]
CBI[b]	Corn	Field tested	1997	Northrup King/Syngenta[e]
Cry IA(b)	Corn	Deregulated[c]	1995	Northrup King/Syngenta[e]
CBI[b]	Corn	Field tested	1994	DeKalb
CryIA(b)	Corn	Field tested	1996	DeKalb
CryIA(c)	Corn	Field tested	1992	DeKalb
	Corn	Deregulated[c]	1996	DeKalb
	Corn	Commercialized	1996	Delta and Pine Land
	Corn	Field tested	1995	DowElanco
CBI[b]	Corn	Field tested	1995	ELM/Asgrow/Seminis
CryIA(b)	Corn	Field tested	1996	ELM/Asgrow/Seminis
	Corn	Field tested	1994	Hunt-Wesson
CryIA(a)	Corn	Field tested	1996	Limagrain
CBI[b]	Corn	Field tested	1994	Monsanto
Cry IA(b)	Corn	Deregulated[c]	1995	Monsanto
Cry IA(c)	Corn	Field tested	1992	Monsanto
CBI[b]	Corn	Field tested	1997	Mycogen
Cry IA(b)	Corn	Field tested	1993	Mycogen
CBI[b]	Corn	Field tested	1997	Plant Genetic Systems
	Cotton	Field tested	2000	Aventis[d]
CryIA(c)	Cotton	Field tested	1993	Delta and Pine Land
CBI[b]	Cotton	Field tested	2000	Dow
	Cotton	Field tested	1998	Monsanto
Cry IA(c)	Cotton	Commercialized	1996	Monsanto
CryIA(a)	Cotton	Deregulated	1994	Monsanto
	Cotton	Commercialized	1995	Monsanto
CryIA(b)	Cotton	Commercialized	1996	Monsanto
CryIA(b) and Cry IA(c)	Cotton	Field tested	1992	Monsanto
CryIIIA(a)	Cotton	Commercialized	1995	Monsanto
CBI[b]	Cotton	Field tested	1997	Mycogen
CryIA(b)	Cotton	Commercialized	1996	Mycogen
CryIA(b) and Cry IA(c)	Cotton	Field tested	1990	Northrup King/Syngenta[e]
CBI[b]	Cotton	Field tested	2000	Syngenta[e]
CryIIA	Cotton	Field tested	1996	Delta and Pine Land
CBI[b]	Creeping Bentgrass	Field tested	1998	Scotts
CryIIIA	Eggplant	Field tested	1994	Rutgers University

Table 3.1 (continued)   Status of Various Genetically Engineered Crops with *Bt*-Derived Genes
for Toxic Protein Expression as a Source of Insect Pest Resistance

Cry toxin	Crop	Status	Year[a]	Company
CryIA(c)	Grape	Field tested	1997	University of California
CBI[b]	Lettuce	Field tested	2000	ELM/Asgrow/Seminis
CryIA(c)	Peanut	Field tested	1997	University of Georgia
CBI[b]	Poplar	Field tested	2000	Oregon State University
CryIII(a)	Poplar	Field tested	1998	Oregon State University
CryIA(c)	Poplar	Field tested	1993	University of Wisconsin
CBI[b]	Potato	Field tested	2000	Monsanto
CryIIIA	Potato	Field tested	1990	Monsanto
	Potato	Deregulated[c]	1994	Monsanto
CryIA(b)	Potato	Field tested	1996	Plant Genetics
	Potato	Commercialized	1995	Syngenta[e]
CBI[b]	Rapeseed	Field tested	1998	Aventis[d]
CryIA(b)	Rapeseed	Field tested	1991	Agrigenetics
CBI-Btk	Rapeseed	Field tested	1998	Calgene/Monsanto
CryIA(c)	Rapeseed	Field tested	1995	University of Chicago
	Rapeseed	Field tested	1995	University of Georgia
CryIA(a)	Rice	Field tested	1991	Louisiana State University
CryIIIA	Rice	Field tested	1999	Louisiana State University
CBI[b]	Rice	Field tested	2000	Syngenta[e]
	Soybean	Field tested	1997	Monsanto
	Soybean	Field tested	2000	Pioneer Hi-Bred International
CryIA(c)	Soybean	Field tested	1997	University of Georgia
CBI[b]	Sunflower	Field tested	1999	Ohio State University
	Sunflower	Field tested	1999	Pioneer Hi-Bred International
Cry1F	Sunflower	Field tested	2000	Pioneer Hi-Bred International
CBI[b]	Sunflower	Field tested	1999	University of Nebraska
CryIA(b)	Tobacco	Field tested	1988	Sandoz, now Syngenta[e]
	Tomato	Field tested	1999	Calgene/Monsanto
CryIA(b) and Cry IA(c)	Tomato	Field tested	1992	Campbell
CBI[b]	Tomato	Field tested	1999	ELM/Asgrow/Seminis
CryIA(b)	Tomato	Field tested	1988	Monsanto
	Tomato	Deregulated[c]	1997	Monsanto
CryIA(b) and Cry IA(c)	Tomato	Field tested	1990	Monsanto
CryIA(b)	Tomato	Field tested	1991	Northrup King/Syngenta[e]
Cry1A(c)	Walnut	Field tested	1993	University of California, Davis

[a] Examples are listed by the gene for the toxic Cry protein, with its status in the regulatory process as of June 2001, and the earliest year the specific crop, toxin, and company combination was approved for that status.

[b] CBI (Confidential Business Information) is shown when the company has not released the source of the gene that is responsible for the insect pest resistance trait in the crop.

[c] Deregulated means that the crop has been removed from the regulatory process (as a step toward commercialization) and can be grown without any permit or notification with either the EPA or the USDA–APHIS.

[d] Previously AgrEvo.

[e] Previously Novartis Seeds, Inc.

Source: USDA–APHIS Biotechnology Permits Database, available at http://www.nbiap.vt.edu/ (March 2001).

acreage since 1998 (NASS–USDA 2000). The acreage of *Bt*-potatoes remains relatively small (less than 5% of U.S. acreage), with only a slight increase since commercialization (Table 3.2). Some commercialized *Bt*-crop varieties are also herbicide tolerant or virus resistance (containing stacked transgenic traits). Farmers have widely adopted stacked herbicide and

Table 3.2 Area of Commercialized *Bt*-Crops Worldwide

Year	Potato	Corn	Cotton[a]	Total *Bt*-crop acreage
1996	14,500 acres	0.74 million acres	1.7 million acres	2.41 million acres
1997	34,600 acres	6.9 million acres	2 million acres	8.93 million acres
1998	50,000 acres	16.6 million acres	6 million acres	22.65 million acres
1999	68,800 acres	23.7 million acres	6.14 million acres[b]	29.17 million acres
2000	64,800 acres	20.3 million acres	8 million acres[b]	28.4 million acres

[a] Does not include *Bt*-cotton acreage in China because only tentative estimates are available before 1999. James (1998) reports a tentative figure of less than 0.25 million acres of *Bt*-cotton acreage in China for 1998.

[b] Data from China are included for 1999 (0.74 million acres of *Bt*-cotton in 1999 and 1.24 million acres in 2000) (James 2000).

Sources: James (1998, 1999, 2000), Monsanto (1999), and the U.S. Department of Agriculture Acreage report (NASS–USDA June 30, 2000).

insect resistance in corn and cotton, increasing global usage by 21% in 1999 compared with just a 10% increase in plantings with only insect resistance. Crops with stacked traits are being planted on increasing acreage, but still accounted for less than 7.5 million acres, or 7% of the global area of transgenic crops in 1999 (James 1999).

In addition to corn, cotton, and potato, a wide range of other plants have been engineered to express *Bt*-toxins and are in various stages of development and regulatory assessment (see Table 3.1). *Bt*-tomato was field-tested in 1988, and Monsanto has scheduled insect-resistant tomato lines for commercial release. Although not yet commercialized, *Bt*-broccoli and *Bt*-canola were among the first insect-resistant crops field-tested (Metz et al. 1995), and field trials for lepidopteran-resistant broccoli and cabbage by Seminis Vegetable Seeds, Inc. are currently under way (USDA–APHIS 2001). A variety of other crops are in development, including rice, poplar, and walnut (see Table 3.1). Future innovations of genetically engineered *Bt*-crops include:

1. The addition of transgenes for performance or quality traits, e.g., aluminum tolerance (de la Fuente et al. 1997) or iron-enriched rice (Goto et al. 1999);
2. The incorporation of *Bt*-based resistance traits in new plants, e.g., coffee (Leroy et al. 2000);
3. The inclusion of novel fusion protein genes (Tu et al. 2000) or more than one type of *Bt*-transgene to provide multiple levels of plant protection, e.g., *Bacillus thuringiensis* cry2a endotoxin stacked with cry1a in cotton); and
4. The synthetic modification of hybrid *Bt*-proteins to have broad insecticidal activity across several orders of arthropod pests (Malvar and Gilmer 2000, De Cosa et al. 2001).

3.1.2 *Comparison of* Bt-*crops and conventionally bred varieties*

Genetically engineered crops are designed to provide a benefit beyond what could be attained through traditional breeding practices. In the case of *Bt*-based insect-resistant crops, the use of recombinant DNA techniques to incorporate *Bt*-genes into the crop genome is meant to provide specific control of target pests while substantially reducing reliance on conventional applications of chemical pesticides (Llewellyn et al. 1994). Farmers pay a premium price for genetically engineered *Bt*-crop seed. Farmer adoption and satisfaction with these crops depend mainly on increased returns (from higher quality and yields) and decreased input costs (lowered insecticide, labor, and equipment usage), increasing the farmer's net returns despite higher seed costs. Acceptance also involves a number of other factors, including cultivation ease, management restrictions, pest levels, and existing treatment options. For example, a new insecticide introduced just prior to

Bt-potato provides an effective alternative for growers, and may have reduced adoption rates of *Bt*-potato. *Bt*-crops are a logical new pest control option if (1) existing options are ineffective for controlling pest problems, (2) *Bt*-crops represent a more effective technology offering a better (more efficient, convenient, or environmentally sound) alternative to current practice, or (3) *Bt*-crops are less expensive than current technology with equivalent control. In this section, we consider how different *Bt*-crops compare as alternative insect control options and discuss the problems with measuring benefits, including *Bt*-crop performance with respect to increased yields, decreased chemical inputs, lowered grower costs, and increased farmer income.

3.1.2.1 Similarities of transgenic and conventional crop varieties

Commercial, transgenic insect-resistant crops generally perform as well or better than their traditionally bred (non-genetically engineered) counterparts because, in most cases, they represent high-yielding, commercial varieties with an added, genetically engineered, trait. In the development of *Bt*-crop parents, the site of insertion of the *Bt*-transgene in the original plant transformation is unpredictable, as is the successful transfer of the complete genetic construct needed for proper transgene expression. Because the transformation is unpredictable, many individual transformed plants are produced and those with lower genetic stability or inferior performance (for example, those in which transgene expression and the capacity to reduce insect damage are coupled with undesirable agronomic characteristics such as lower yield) are culled. Only a small proportion of the remaining plants are field-tested and very few transformation events (individual, successful insertions of the transgene into a crop parent) are moved forward in product development and then on to regulatory assessment and production of foundation seed (original transformants are used to produce hybrids through standard plant-breeding techniques for commercialization). The genetically modified trait is inherited in a predictable Mendelian fashion, as required by U.S. regulatory guidelines, enabling seed companies to add a transformation event to many of their commercially important crop varieties, so that the same variety can be available to growers with or without the new trait.

3.1.2.2 Farmer adoption

Adoption rates for this new crop technology are high, increasing more than 300% from less than 9 million acres in 1997 to more than 29 million acres globally in 1999. This increase in acreage planted with *Bt*-crops reflects the attraction (for farmers) of technology that provides more convenient pest management, potentially higher productivity, and possible increased net returns. Intense marketing also influences rapid adoption of *Bt*-crops initially. After initial marketing success, however, farmers purchasing genetically engineered crop seed from biotechnology companies decide if genetically engineered varieties provide a cost-effective solution for pest problems that would lower yields if left uncontrolled.

According to Agricultural Resource Management Studies (ARMS) conducted by the Economic Research Service (ERS) and the National Agricultural Statistics Service (NASS) of the U.S. Department of Agriculture (ERS–USDA 1999, Fernandez-Cornejo and McBride 2000), the main reasons that farmers are adopting *Bt*-crops are (1) to increase yields through improved pest control and (2) to decrease pesticide costs. These economic benefits must outweigh the substantially higher cost of *Bt*-crop seed. The premiums for *Bt*-seed (often called "technology fees") are approximately $8 to $10/acre for *Bt*-corn, $32/acre for *Bt*-cotton, and $30 to $46/acre for *Bt*-potatoes. The recent decline in *Bt*-crop plantings is generally attributed to market pressure, but much of this decrease could also result from the failure of farmers to recoup the higher up-front costs of planting *Bt*-seed (through increased yields and decreased pesticide inputs). Different studies report contradictory results for insecticide use, yield advantages, and overall economic benefit, reflecting the

Table 3.3 Examples of Yield Studies Comparing *Bt*-Field Corn to Non-*Bt*-Varieties

Number of Locations[a]	Number of years[b]	Year	Change in yield bushels/acre (year)	Ref.
Not reported	3	1997	+11.7[c]	Gianessi and Carpenter 2001
		1998	+4.2	
		1999	+3.3	
14	1	1997	−3.9 to +17.3[d]	Rice 1998
3	1	1997	+23.8[e]	Brunoehler 1998
1	3	1996–1999	+6.8 to 10.4[f]	Novartis 1998
84	1	1998	+2.9[g]	Farnham 1998
1048 (1997)	2	1997	+9.4[h]	Novartis (Dornbos 1998)
580 (1998)		1998	+4.6[h]	
200	1	1997	+14.9[h]	Monsanto (Davis et al. 1997)
310	2	1997	+10.8[h]	Monsanto 1999
		1998	+2.4[h]	
		1997–1998	+1.5 to 17.4[i]	
15	1	2000	+2.36 to 44.08[j]	Wiebold et al. 2000

Note: Subscripts denote different comparative approaches among the studies to illustrate some of the reasons for variability in results.

[a] "Sites" in this column refers to different locations used for yield trials.

[b] Most of these studies were carried out in years with relatively high insect pressure (high population levels of ECB, 1997).

[c] Aggregated average from a variety of studies.

[d] Range represents comparisons of different hybrids in two or more of the 14 locations. The low (negative) value is the result of no significant ECB damage (low ECB tunneling observed).

[e] Average yield at three locations in northern Minnesota.

[f] Value is average yield increase for same location (site) for multiple growing seasons. Range of yield values for different years not reported.

[g] Value is an average of 84 comparisons between *Bt* and non-*Bt*-lines. *Bt*-corn showed a yield advantage in only 61% (51 out of 84) of comparisons.

[h] Average yield increase for a wide sample of locations in one growing season.

[i] Range represents highest (Illinois, 1997) and lowest (Illinois, 1998) in-state averages for the corn-belt.

[j] Range represents highest and lowest differences between the mean yield of all *Bt*-varieties tested and all non-*Bt*-varieties tested for four different regions, plus two irrigated regional trials (regions represent two to seven different locations).

difficulty in establishing a baseline benefit across diverse growing regions and farming regimes. In the next three sections we discuss the effects of *Bt*-crop plantings on yields and insecticide use, as well as considering overall economic benefits as justification for *Bt*-crop adoption by farmers.

3.1.2.3 *Yield gain and insecticide savings*

Accurate regional predictions about yield advantages and insecticide savings due to the substitution of nontransgenic crop varieties with *Bt*-crops are difficult to make because pest levels vary spatially and temporally. Further, rigorous, large-scale comparisons cannot always be made based on data from experimental field trials. Experimental plot studies may not accurately reflect results for commercial operations, whereas wide-ranging farm surveys confound a number of important variables (such as irrigated and non-irrigated acreage, farm management practices, etc.) (ERS–USDA 1999). Ideally, yields of *Bt*-crops should be compared to non-*Bt*-isolines at the same geographical location and under the same management conditions to eliminate all variables beyond the *Bt*-toxin. For *Bt*-corn, many locations are often lumped when reporting yield increases, especially in subsequent literature (Table 3.3). Yield trials (where different crop varieties are grown in different locations) and yield comparisons (where *Bt*-varieties and their non-*Bt*-counterparts are grown in the same locations and yield differences are measured) are given a degree of generalizability in most published benefit estimates (e.g., Carpenter and Gianessi 2001) that is inherently absent in yield studies. Yield studies are very dependent on location and growing condition. Their aim is to provide a performance measure to local farmers, who grow their crops in the same region, with similar soil and climatic conditions. Results are reported as the mean of the plot replications for each variety at each location and make a useful comparison for farmers in this format (e.g., Wiebold et al. 2000; Table 3.3). There is some value to bulking many locations together, because a high-yielding variety that performs well across many growing regions will withstand year-to-year variations in climate and give the farmer more confidence of a good crop (e.g., Dornbos 1998; Table 3.3). Bulking together yield studies from only a few locations, however, will not give an accurate indication of performance. Yield advantage predictions for a particular growing region are not necessarily accurate when based on results from other regions.

Most local yield trials do not compare *Bt*-isolines vs. non-*Bt*-isolines. Rather, they compare a range of commercial crop varieties, and because *Bt*-toxins are engineered into a company's top varieties, they will inevitably rank among the highest-yielding varieties in a yield trial. This yield advantage is not necessarily due to the *Bt*-trait, but can also be attributed to the higher agronomic quality of the variety being transformed (e.g., Wiebold et al. 2000). *Bt*-crops are selected for yields that compare with their non-*Bt*-isolines. Thus, it is reasonable to expect, on average, a higher yield for *Bt*-crops whenever some farmers avoid pest damage that would be sufficient to decrease yields in non-*Bt*-varieties. Insecticide usage trends are somewhat more complex than are yield comparisons for several reasons (discussed below), and estimates of net economic gain depend on a comparison of yield gain vs. seed cost premiums (minus any other savings, such as reduced pesticide costs). Thus, an analysis of net financial gain can suffer from a combination of problems.

Broad generalizations about yields, insecticide reductions, and economic benefits, although compelling, are subject to certain assumptions, and must be viewed as static descriptions of something that is dynamic over time and varies by crop and region. For example, the total cost for insecticides that could be replaced by *Bt*-technology is projected to be $2.69 billion annually (Krattiger 1997), but there is substantial disparity between projections and actual measurements of on-farm pesticide use or reductions in pesticide sales. Insecticide usage, and therefore estimated reduction due to adoption of commercialized *Bt*-crops, also differs among crops (Table 3.4). It is a problematic issue to isolate insecticide reduction due specifically to the adoption of *Bt*-crops for the following reasons:

1. The lack of data reporting in formats that discern pesticide use by individual target pest;
2. The use of a single insecticide for several different target pests (making it difficult to determine how much, if any, of that insecticide would no longer be used on the *Bt*-crop);
3. Yearly fluctuation in populations of insect pests (making it difficult to observe trends); and
4. Other dynamics of insecticide use such as new product introduction and development of resistance in pest populations.

It is virtually impossible to remove confounding variables and isolate the adoption of *Bt*-crops as an independent variable; thus, cause and effect can be supported, but not demonstrated, using survey data on rates of adoption of *Bt*-crops and rates of pesticide reduction.

Given these caveats, the major environmental benefit of *Bt*-crops involves the substitution of insect-resistant crops for insecticide applications. Beginning in 1996, several corn seed

Table 3.4 Amounts of Chemical Insecticides Used
to Control Lepidopteran Pests on Corn and All Pests
of Cotton and Potato

Crop	Active ingredient/acre/yr (lb)
Fresh market sweet corn	2.90
Processing sweet corn	0.32
Field corn	0.02
Cotton	1.77
Potato	2.43

Table 3.5 Commercially Available *Bt*-Corn Seed Listed by the Licensed
Transformation Event[a] and the *Bt*-Toxin Expressed in the Crop Line

Company	Transformation event/ Cry toxin	Brand name and crop
Syngenta[b]	Bt 11 Cry1A(b)	YieldGard® field corn and Attribute® sweet corn
Syngenta and Mycogen/Dow	176 Cry1A(b)	Knockout® (Novartis) Nature Gard® (Mycogen)
Monsanto	Mon 810 Cry1A(b)	YieldGard®
Dekalb Genetics	DBT 418 Cry1A(c)	BT-Xtra®
Aventis[c]	CBH 351 Cry 9C	Starlink®

[a] A more detailed treatment of events can be found in Andow, Chapter 4 of this volume.

[b] Syngenta was previously Novartis Seeds, Inc.

[c] Aventis was previously AgrEvo.

companies began to offer *Bt*-field corn expressing the delta endotoxin cry 1A(b) effective against European corn borer (*Ostrinia nubilalis*) (ECB) and corn ear worm (*Helicoverpa zea*) (Table 3.5). *Bt*-corn represents an opportunity to control ECB, a pest that is difficult to manage with other methods (including ineffective insecticide application and limited success of biological control agents) and one that has caused yield losses ranging from 33 million bushels to over 300 million bushels/year in the United States (savings equivalent to 500,000 acres in 1999, when yield gain in the United States was 66 million bushels). Reports of average yield advantage for *Bt*-corn hybrids range from –3.9 to 44.1 bushels/acre (see Table 3.3). The highest yield advantages were seen in 1996 and 1997. Since 1997, populations of ECB have been low, accounting for a drop in measured yield benefits for *Bt*-corn. Still, an additional 55.8 million bushels to 66.4 million bushels total in the United States were attributed to the use of *Bt*-corn in 1998 (Gianessi and Carpenter 1999).

Similarly, yield advantages are documented for *Bt*-cotton in the United States. The ERS–USDA reported significantly higher yields in most years and most regions for *Bt*-cotton as compared with non-transgenic varieties (Fernandez-Cornejo and McBride 2000). Carpenter and Gianessi (2001) also report higher yields across seven states from five different sources, ranging from 2 to 19% yield increases for *Bt*-cotton. Several previous survey and experimental studies have found that *Bt*-cotton yields are equivalent or greater than those of conventional varieties (Gibson et al. 1997, ReJesus et al. 1997, Stark et al. 1997, Bryant et al. 1998, and Marra and Hubbell 1998). One analysis of *Bt*-cotton showed an aggregated yield increase of 85 million lb for southeastern states in 1998 (Gianessi and Carpenter 1999). Average yields would be expected to be higher whenever some *Bt*-crops in the survey experienced higher yields as a result of reduced pest damage.

Much less information is available for the third commercialized *Bt*-crop, *Bt*-potato, which is transformed with the cry IIIA toxin from *B. thuringiensis* var. *tenebrionis* to control Colorado potato beetle (Coleoptera: Scarabaeidae: *Leptinotarsa decemlineata*). As with vegetable (and tuber) crops in general, genetically engineered potato is the subject of fewer studies than field crops such as cotton and corn, which far outstrip its acreage (corn accounts for $1/4$ of all crop acreage in the United States) and value (in terms of sales by seed companies). Annual surveys of yields by Monsanto for *Bt*-potatoes consistently report equivalent yields for NewLeaf® and conventional varieties. There are few independent (non-company) comparisons regarding *Bt*-potatoes, so trends are difficult to evaluate further.

Economic, health, and environmental benefits can stem from the reduction of insecticide use in agriculture, and *Bt*-crops were developed as an alternative pest control measure that targets certain pests and can potentially reduce some use of broad-spectrum insecticides. The amount of insecticide that is actually substituted by *Bt*-crops is some portion of the amount of insecticide applied on the non-*Bt*-crop. The quantity of insecticide substituted depends minimally upon pest infestation rates, the level of grower responsiveness (in terms of altering pest control practices), and the specificity of insecticide application targets. The relative weight of each of these factors differs with the particular crop. For *Bt*-corn, the amount of insecticide applied to the crop is relatively low. Less than 5% of U.S. corn acreage was sprayed for ECB before adoption of *Bt*-corn (Gianessi and Carpenter 1999). This upper restraint on the amount of insecticide savings possible is a result of the relatively high costs of insecticides and their application, their low effectiveness for ECB control, and minimal concern for cosmetic damage in field corn. Greater reductions of insecticide usage may have accrued from the development of *Bt*-sweet corn, which is sprayed more intensively than field corn for lepidopteran pests (see Table 3.4). However, available data for *Bt*-corn are not clearly divided to measure these potentially higher pesticide reductions for sweet corn. The acreage in *Bt*-sweet corn is comparatively small at 31,800 acres in 1999, and is often included in overall estimates for *Bt*-corn acreage (as in Table 3.2).

The performance of insecticides in controlling ECB is variable because of the following factors:

1. Egg laying by ECB moths extends over a 3-week period;
2. Larvae are protected from insecticide applications once they bore into cornstalks;
3. Most insecticides are active for only a 7- to 10-day period; and
4. Insecticides are difficult to apply when corn is tall, and aerial applications are imprecise.

Still, the small percentage of corn acreage sprayed with insecticides amounted to an estimated 1.47 million lb of active ingredient for clorpyrifos, lambdacyhalothrin, methyl parathion, and permethrin for field corn in the United States (Gianessi and Carpenter 1999). After widespread adoption of *Bt*-corn in 1998 (accounting for 18% of corn acreage), the total corn acreage sprayed for ECB was reduced to about half of the previous 5% level (~2.5% of the total acreage in corn was sprayed with insecticides in 1998). This reduction meant that 2 million fewer acres were sprayed for ECB, and some estimates attribute reduced chemical usage on 15 million acres in 1998 to *Bt*-corn (ERS–USDA 1999). However, the role of *Bt*-corn in accounting for this reduction is unclear. There was a very low rate of ECB infestation in 1998, reducing the impact of insecticides overall and thereby decreasing the predicted benefit of *Bt*-corn over non-*Bt*-varieties. A second variable that is important for determining the amount of insecticide savings due to *Bt*-corn substitution would be multiple target pest spraying in field and sweet corn. That is, some or all of the insecticides that are applied for control of ECB are also used for the control of corn rootworm or other pests. Thus, an unknown proportion of the total quantity of insecticide (plus any other types used specifically for ECB control) was applied to control only the pests that would otherwise be controlled by *Bt*-based toxins in the plant.

Yield advantages of *Bt*-corn may be substantial in high-ECB years, even if insecticide savings are rather low. *Bt*-corn overcomes some of the limitations of insecticide control through its constitutive presence and persistence in corn plant tissues, resulting in 95 to 99% (with *Bt*-11 event) control of ECB larvae in the first generation, compared with well-timed applications of insecticides typically providing between 60 and 95% control of first-generation ECB larvae and 40 to 80% control of second-generation larvae (Fawcett 1998). When the *Bt*-toxin is expressed only in the plant's green tissues and pollen (all varieties using *Bt* event 176), control of second-generation ECB larvae is lowered to 50 to 75% because some second-generation larvae feed on silks and developing kernels (Gianessi and Carpenter 1999). Better control is possible with varieties using the *Bt*-11 event, expressing toxin in all tissues (see Table 3.5).

What other technologies exist for improving control of ECB and how does their effectiveness compare to *Bt*-corn? Traditional breeding efforts did result in the development of varieties in the mid-1970s with intermediate resistance to ECB. These varieties were replaced by higher-yielding varieties that gave increased yields despite heavier losses from ECB. Although pest resistance may have been made more durable had the *Bt*-transgenes been incorporated into traditionally bred resistant varieties, the relative low yields of the traditionally resistant varieties may have made them less attractive. Thus far, 24 insects that parasitize and kill ECB have been introduced in the United States, 6 of which are established and provide limited control of ECB. The direct effects of *Bt*-corn on populations of these beneficial insects are not yet known; yet any reduction in pesticides toxic to these species would be compatible with successful biological control. Scouting for ECB can increase the effectiveness of conventional insecticide application, at a cost of $3 to 10/acre, compared with the cost of the two recommended insectide sprays at $14/acre each. However, relatively few field corn growers send scouts into the field to monitor the development of the pest population compared with an economic threshold level to time spray applications. A survey of Iowa and Minnesota corn growers showed that 66% of growers perceived ECB to be a serious threat, but only 35% had scouted their fields even once (Rice and Ostlie 1997). The ten-state corn-belt average for scouting was only 6.7% for 1994 (Gianessi and Carpenter 1999). Thus, because *Bt*-field corn represents a new technology addressing an insect control problem otherwise lacking sufficiently simple or effective control options, it tends to increase yields for more growers rather than benefiting them through any dramatic reduction in insecticide applications.

When contrasting lepidopteran pest control in cotton to that of corn, *Bt*-cotton offers an alternative to high-input insecticide spray regimes for the control of tobacco budworm, cotton bollworm, and pink bollworm. As such, it represents an opportunity for substantial insecticide savings, benefiting both the environment (cotton is among the most intensely sprayed crops worldwide) and the farmer's overall cost (through reduction in labor and equipment costs related to insecticide application). In the case of cotton, where insecticide sprays can effectively control the target pests, the major benefit of *Bt*-cotton is not necessarily increased yield, as with *Bt*-corn, but lower insecticide use and correspondingly lower input costs for the farmer.

By several accounts, the adoption of *Bt*-cotton has resulted in an insecticide reduction and corresponding cost savings for farmers (Table 3.6). However, the advantages of *Bt*-cotton are highly dependent on geographical region and bollworm/budworm infestation for the year surveyed, and on the amount of insecticides applied to control pests not controlled by *Bt*-toxins. According to survey results reported by Stark (1997) and Mullins and Mills (1999), growers applied less insecticide to *Bt*-cotton fields than to conventional cotton.

Table 3.6 Effects of *Bt*-Corn and Cotton on Farm Income, Including Yield Benefits and Estimated Savings from Reducing Pesticides

Crop	Net change in revenue $/acre (year)	Seed cost $/acre	Yield increase	Change in input costs (pesticide use)	Ref.
Corn	+18 (1997)	+10	3–33 bushels/acre	−$28/acre[a]	Carpenter and Gianessi 2001
	−1.81 (1998)	+10	3–33 bushels/acre	−$28/acre[a]	
	−1.73 (1999)	+8	3–33 bushels/acre	−$28/acre[a]	
	+3.31 (1999)	+10	Not reported	NR	EPA 1999
Cotton[b]	+39.86	+32	37 lbs. lint/acre	−$38.24/acre[c]	Mullin et al. 1999
	+2.21 (1996–1999)	+19	9.4 lbs. lint/acre	−$13.35/acre[d]	Bachelor 2000
	+38.19 (1995–1998)	+32	NR*	NR	Gianessi and Carpenter 1999[e]

Note: Pesticide groups included in each study are listed (superscripts) to illustrate different approaches to analyzing insecticide reductions.

[a] Insecticides include chlorpyrifos, lambdacyhalothrin, methyl parathion, and permethrin. Reduction was calculated by attributing 1.5% of the decrease in use of these pesticides to adoption of *Bt*-corn.

[b] There are a few studies showing negative net return for farmers growing *Bt*-cotton, although they are the minority (e.g., ReJesus 1997, Bryant et al. 1998, Seward et al. 2000).

[c] Total insecticide costs for all spray applications including those not targeted toward bollworm and budworm.

[d] Reduction from 2.53 insecticide applications to 0.75 applications of pyrethroid insecticides in North Carolina.

[e] This net benefit is an average of studies by nine authors, showing economic benefits in 14 out of 17 cases.

Carpenter and Gianessi (2001) report that insecticide applications decreased by 2.7 million lb (of insecticide, not active ingredient) and 15 million insecticide applications between 1995 and 1999. *Bt*-cotton controls 70 to 90% of cotton bollworm and tobacco budworm (Gianessi and Carpenter 1999); however, because bollworm and budworm are often controlled with the same insecticides used against aphids, thrips, and other pests not controlled by *Bt*-toxins, overall pesticide application is still high (Fernandez-Cornejo and McBride 2000). In fact, despite claims of overall pesticide reduction with *Bt*-cotton, the ERS found that an increase in adoption of *Bt*-cotton did not significantly decrease the acreage treated with organophosphate or pyrethroid insecticides throughout the Southeast (Fernandez-Cornejo and McBride 2000).

NewLeaf potato, transformed with the cry IIIA toxin from *B. thuringiensis* var. *tenebrionis*, is reported to be so effective that no Colorado potato beetle (CPB) larvae have ever been found alive on the plants and growers are informed that they should never see the larvae (Monsanto 2001). Each year, 2.5 million lb of insecticide active ingredient are used on potatoes (Gianessi and Carpenter 1999), and to the extent that these control measures are restricted to CPB, they can be eliminated with use of *Bt*-potatoes. Monsanto Company data from 1997 to 1999 show that growers of NewLeaf potatoes made an average of one fewer insecticide application for CPB compared with conventional growers (two vs. three insecticide applications; Gianessi and Carpenter 1999). The proportion of insecticide applications that will actually be substituted with *Bt*-potato is lower because most growers apply the same insecticides to control other foliage-feeding insect pests along with CPB. NewLeaf Plus® potato growers can reduce insecticide applications further because this *Bt*-potato also protects against potato leaf roll virus (PLRV), which is transmitted by aphids. The addition of stacked virus resistance and *Bt*-toxin genes can increase the tolerance for aphids on crops, thus reducing insecticide use for aphids.

3.1.2.4 *Measuring economic benefit*

Although predictions and projections for great farmer savings were made as they had been for insecticide savings and yield advantage, net revenues realized by farmers have been more modest (see Table 3.6). Price fluctuations, variation of insect populations over time and space, and continued dependency on chemical inputs vs. alternative management practices (scouting, trap crops, refugia for beneficial predators, etc.) can effectively reduce any on-farm benefit of *Bt*-varieties. Whole-farm studies can be difficult to perform because of the complexity of each farm, farmer, and management scheme. Nonetheless, data have been analyzed in several cases to isolate and compare net economic benefits of *Bt*-crops over non-*Bt*-crops. The most difficult aspect of this analysis is the estimation of pesticide reduction and the corresponding cost savings to farmers, which are highly variable among geographical locations.

Bt-corn, the most widely grown *Bt*-crop, has not been an overriding financial success for farmers. Recent declines in acreage planted to *Bt*-corn varieties may be a result of negative farmer returns for the 1998 and 1999 growing seasons. *Bt*-corn provided a net benefit for farmers in 1997, but since then has resulted in losses due to low grain prices and low ECB infestation. Although insecticide applications (estimated at $14/acre, with two applications per year) are recommended for ECB control, few farmers spray for ECB for reasons discussed in the previous sections. Effectively, the difference in cost between ECB control using *Bt*-corn ($10/acre premium for seed) and the common alternative management practice (no control) is $10/acre. Although this is less than the estimated cost for spraying insecticide to control ECB, low infestation years, since 1997, may be particularly disillusioning for growers paying higher seed prices. With less than a 2-bushel/acre benefit over non-*Bt*-varieties in years with low populations of ECB, and with low corn prices, farmers do not recoup higher seed prices associated with *Bt*-corn varieties.

The economic benefit for cotton farmers has been much more substantial. Mullins and Mills (1999) report in a survey of 109 sites in the southeastern United States that insecticide control costs were still $15.43 lower on *Bt*-crop fields (including the technology fee) despite an increase in insecticide application for pests not controlled by *Bt*-toxins. This increase is probably due to the impact of secondary pests that grow in numbers in the absence of the pests controlled by *Bt*-cotton. The overall economic advantage of *Bt*-cotton vs. non-*Bt*-cotton was approximately $40/acre (Mullin and Mills 1999). These data, extrapolated to all growers across the cotton belt, translate into an increase in farmer's profits by more than $92 million in 1998. However, farmer practices, problems with secondary pests, and yield benefits due to *Bt* are geographically highly variable throughout the Cotton Belt. Other survey results, including a USDA survey from 1997 (McBride 1999) and a combination of data from several studies (Carpenter and Gianessi 2001) show higher total seed and insecticide costs on *Bt*-cotton acreage compared with conventional cotton crops. In most cases, this higher input cost is associated with higher yields that compensate for the increased costs (see Table 3.6). A comparison of several studies reporting net returns to farmers growing *Bt*-cotton varieties gives an average +$20.81/acre (Carpenter and Gianessi 2001).

Reports of increased production and economic benefit for growers of *Bt*-crops are mixed, with mostly slight or modest advantage over conventional practices. Farmer's reasons for growing *Bt*-crops, increased yields and decreased pesticide costs, are largely justified, although it does not provide as large of an economic margin as most would hope. Does economics alone explain the continued large acreage planted to *Bt*-crops, even with markets under pressure by consumer groups opposed to genetically engineered foods? *Bt*-crops are being promoted and given preference for research funding over alternative, effective management options that may also provide modest economic benefit for farmers.

3.1.2.5   *Consumer acceptance of* Bt-*crops*

The costs and benefits of growing *Bt*-crops affect many different groups — farmers, consumers, and agribusiness — at different levels (Table 3.7). Consumers have added very little support to the demand for development of *Bt*-crops, so it is not surprising that consumers perceive fewer benefits from the adoption of *Bt*-crops than do growers and industry. Although reduction in pesticide use certainly benefits consumers, *Bt*-crops are one of several options for achieving this goal and have not been shown to be more effective than other management options, to which they are generally not even compared (Section 3.1.3.3). Biotechnology and seed companies have not endeavored to sell *Bt*-technology to consumers. Until recently, their efforts have been focused on their own customers — farmers. Since the introduction of genetically engineered crops into the food system, there has been steady erosion in trust of genetically engineered foods and the regulatory bodies responsible for overseeing (and not labeling) them (May 1999). In 1997, Novartis found that only 25% of Americans "would be likely to avoid labeled GE [genetically engineered] foods." Only 2 years later, a *Time* magazine poll indicated that 58% of American consumers "would avoid purchasing [labeled GE foods]" (Center for Food Safety 1999).

3.1.3   *Environmental effects of* Bt-*crops*

Farmer adoption of *Bt*-crops has increased in the United States as a result of a variety of factors, including the experience of increased yields and/or decreased net costs of pest controls and marketing efforts. Consumer confidence and overseas markets have become more volatile, accounting, to some degree, for the recent truncation of the rapid expansion in the acreage of *Bt*-varieties. To the extent that *Bt*-crops reduce the application of biocides that contaminate soil ecosystems, leach into the groundwater, and eliminate or cause sublethal effects to non-target animals, they will provide substantial environmental benefits

Table 3.7 Distribution of the Potential Risks and Benefits of *Bt*-Crops for Farmers, the Public, and Seed Companies

	Concerns	Opportunities
Farmers	Market acceptance: • Market segregation and reduced prices for engineered food • Decreased consumer demand and uncertainty of export markets Decreased freedom in farm management practices: • Seed sold and grown under contracts with management restrictions • Planting of refugia with prescribed management for insect resistance prevention • No seed saving Environmental concerns: • Increased weediness and higher costs of weed control for new cultivars or weeds with escaped transgenes • Development of pest resistance to *Bt*-toxin • Impacts on non-target insects • Increasing populations of secondary pests	Lowered input cost Increased yields Decreased insecticide exposure Increased flexibility of cultivation practices: • Decreased dependence on weather conditions • Not limited by spray timing Greater compatibility with biological control and integrated pest management strategies
Consumers	Environmental concerns: • Increased weediness of new cultivars or weeds with escaped transgenes • Development of pest resistance to *Bt*-toxin • Impacts on non-targets Health concerns: • New foods may cause unexpected reactions, allergenicity • Antibiotic resistance, perceived problems controlling infectious bacterial illness Ethical concerns No decrease in food costs	Decreased pesticide residues in food and in the environment Cosmetic improvement of food: • Lower necrosis in potato • Fewer corn ear worms Health concerns: • Reduced aflatoxins in grain
Biotechnology and seed companies	Market acceptance: • Seed sales to farmers • Adequate return on research investments Variable product performance Lawsuits Environmental concerns: • Development of pest resistance to *Bt*-toxin Regulatory hurdles and costs	Proprietary protection of modified varieties Decrease in seed-saving by farmers (potential increase in seed sales) Collection of fees: • License fees from other companies • Technology fees/seed premiums from farmers Increased brand recognition

in conventional agricultural fields. However, these benefits, which in part spurred the creation of *Bt*-crops, have been found to be more sporadic than promised, and not as simple as early analyses conceived (Sharma and Ortiz 2000).

Studies on the environmental fate of transgenic crops and their genetic constructs have begun to accumulate in the last decade. There have been advances in ecological research on the interface of agricultural and natural systems, partly in response to public concerns about large-scale use of *Bt*-crops. Basic ecological knowledge has been needed to assess and, if necessary, mitigate four types of potential negative environmental consequences of *Bt*-crops:

1. Increased weediness of the crops (Colwell et al. 1985);
2. The development of resistant pests (Gould 1988);
3. The disruption of natural enemies (Hoy et al. 1998); and
4. Transfer of genetic material that may increase the weediness of wild relatives of crops (Goodman and Newell 1985, Regal 1988, Hoffman 1990, Abbott 1994).

The latter concern, a major issue discussed in Wolfenbarger and Phifer (2000) and considered the major environmental risk perceived as resulting from the use of genetically engineered plants (NRC 1989), is the focus of the second half of this chapter.

3.1.3.1   *Advantages of* Bt-*crops*

Advantages of *Bt*-crop technology for the general populace, aside from any economic and management advantages for growers (see Table 3.7), lie mostly in the potential reduction of broad-spectrum insecticides and the relative specificity of plant-produced *Bt*-toxins, compared with many of the organophosphates and pyrethroids currently used for pest control. To the extent that a field of *Bt*-crops is actually sprayed with a lower rate, fewer kinds, or fewer total applications of broad-spectrum insecticides, it accrues several advantages, including increased diversity of natural enemies, higher predation rates of pests, and fewer secondary pest problems (e.g., Van Hamburg and Guest 1997; but see Hilbeck, Chapter 7 of this volume). The actual range of pesticide reduction for a given field when a *Bt*-variety replaces a non-*Bt*-variety is from 0 to 100%. That is, no reduction in insecticide usage is expected if either (1) no pesticides were applied previously to the non-*Bt*-crop or (2) the same pesticides are still applied on the non-*Bt*-crop because those pesticides are needed to control other pests that are not susceptible to *Bt*-toxins. The latter situation would occur if yields increased sufficiently as a result of improved lepidopteran control with the *Bt*-crop. We see this in the case of *Bt*-corn. Greater pesticide reductions are possible when most or all of the pesticides used on the non-*Bt*-crop were targeted at susceptible lepidopterans. Reduced usage of pesticides provides corresponding reductions in worker exposure, pesticide residues, spray drift, mortality of natural enemies and other beneficial organisms, groundwater contamination, fossil fuel consumption, spills, and waste from pesticide application. Betz et al. (2000) also pointed out that lower invasion levels of some fungal pathogens accompany the high efficacy of *Bt*-crops in reducing feeding wounds on crops. Thus, fungicide use could be reduced as well.

Comparative studies of arthropod biodiversity in conventionally managed crop fields with broad-spectrum insecticides vs. their conventional counterparts substituting *Bt*-crops support the notion that biodiversity is promoted by the latter condition (e.g., Lacey et al. 1999). Greater arthropod abundance and diversity in *Bt*-crops seem to emerge from the absence of added insecticide sprays, not from differences in the cultivars themselves (e.g., Lozzia 1999). Possibly as a result of conserving natural enemies and preventing secondary pest outbreaks, both cotton and potato growers have reported that beneficial insects have reduced the need to spray for non-target pests, such as aphids, in their

Bt-fields. Of course, the environmental benefits associated with reduced pesticide can be achieved in a variety of ways. Although some spray reduction has certainly occurred in commercialized Bt-crops, a number of alternative approaches to crop and pest management reduce the need for pesticides. Some of these approaches are discussed below
(Section 3.1.3.3).

[handwritten note: ADVATAGES, DISADVANTAGES]

... *of* Bt-*crops*

volume discuss potential negative environmental conse-
e purpose of describing what we do and do not know about
elopment of resistant pests (Andow, Chapter 4 of this volume),
soil environments (Stotzky, Chapter 8 of this volume), the
rbivores (Losey, Obrycki, and Hufbauer, Chapter 6 of this
es (Hilbeck, Chapter 7 of this volume), and the escape of
opulations in agricultural and "natural" habitats (this chapter).
ssment, and mitigation of these and other potential disadvan-
ered crops pose a variety of challenges to ecologists, and dif-
r levels of predictability and potential severity. For example,
resistance to Bt-toxins was extremely predictable (lauded as
past; Rosset 1989), monarch butterfly larval mortality from
s was something that did not even occur to most of us, despite
omplex array of questions on indirect effects and interactions
ted in Scientists' Working Group on Biosafety 1998). Estimating
d of hazards, once identified, poses other scientific challenges.
xins will eventually become useless in controlling lepidopteran
impact of toxic pollen on monarch populations or the likelihood
uperweed will make the deployment of a Bt-crop regrettable?
emed current and potential hazards associated with worldwide
nlikely or insignificant in terms of biosafety. Other scientists are
rent and future Bt-crops to identify new risks, assess hazard
ective mitigation schemes. An Expert Panel Report on the Future
repared by the Royal Society of Canada (2001), made a series of
the latter point of view, regarding the need for industry and
) take more seriously and (2) allocate more resources to investi-
ironmental risks of genetically engineered crops. Certainly, added
would help meet current challenges, and strengthen the scientific
k assessment of genetically engineered crops.

s *of comparison for advantages and disadvantages of* Bt-*crops*

es and disadvantages of genetically engineered crops measured?
f Bt-crops, environmental and otherwise, have been routinely
ng the performance of the conventionally grown Bt-crop or the
ed in conventionally grown Bt-crops with the performance or pes-
es of the same crop, grown under similar conditions, with conven-
chemical inputs. This type of comparison is reasonable in practice,
of conventional agriculture practices. It is also scientifically satisfy-
iables as possible are controlled. That is, the comparison consists of
p variety under similar management practices, with vs. without the
sociated marker and promoter genes. Yields are often comparable
varieties, except for any reductions due to lepidopteran pest damage,
nd the non-Bt-varieties have been selected for excellent agronomic

performance under conditions of high input, conventional agriculture. However, conventionally managed isolines (some of which are highly susceptible to *Bt*-target pests) are clearly not the only relevant comparison for scientifically rigorous testing in either commercial or subsistence agriculture. Thriving commercial agricultural operations and local production throughout the world fall under various alternative management categories, such as traditional, organic, biological, sustainable, and ecological agriculture, which rely neither on *Bt*-crop varieties nor on the application of synthetic insecticides. Environmental benefits due to reduced insecticide use will accrue for some *Bt*-crops in some regions when compared with their conventional counterparts. Environmental benefits, and even yield or net income advantages, however, may disappear when these same *Bt*-crops in conventional systems are compared with other crop management systems where a host of other factors (from pest-resistant traditional varieties to well-developed arthropod communities that serve in pest suppression) may operate to regulate pests. Although we know of no such comparisons that have been made specifically for the target pests of *Bt*-crops, alternative farming practices have resulted in comparable pest (pathogen/arthropod) management and yields for a wide variety of crops (e.g., Drinkwater et al. 1995, Settle et al. 1996, Zhu et al. 2000).

Possible risks of *Bt*-crops have been assessed often by comparing the *Bt*-crop with its parental organisms: the non-*Bt*-crop isoline and the bacterial toxin as it occurs in *Bt*-spray formulations. However, when viewed as a single organism, a crop plant expressing *Bt*-toxin, the *Bt*-crop has emergent qualities that may highlight concerns involving novel conditions and characteristics. For example, the exposure of arthropods to the toxic effects of *Bt*-sprays differs from exposure with *Bt*-plants. First, spatial and temporal scaling-up is required from the levels of exposure *Bt*-sprays could reasonably provide. Second, only those species that are able to cause activation of the bacterial protoxin in their guts are really exposed to the bacteria-produced toxin. In *Bt*-crops that express truncated *Bt*-toxins, however, exposure is greatly expanded across a range of taxa (see Chapter 8 by Stotzky and Chapter 7 by Hilbeck, this volume). In many cases, then, standards of comparison can determine not only the outcome of our tests but the questions that we pose from the initiation of the assessment process.

In other cases, especially in the context of assessing the possibility of increased weediness in the crop or in its wild relatives (via hybridization), transgenic *Bt*-crops are compared with pest-resistant crops developed with traditional breeding techniques (not involving recombinant DNA technology) (e.g., NRC 2000). In corn, for example, heritable pest resistance has been developed by selecting for mechanical traits such as tight husks and chemical properties such as high concentrations of digestive toxins (DIMBOA and related hydroxamic acids) (Houseman et al. 1992). Indeed, pest resistance in crops has been encouraged routinely using the gene pool of the crop plant and its wild relatives. The use of traditionally bred, pest-resistant crop varieties as a standard for comparison merits serious consideration. However, we lack comparative experiments and have few, if any, studies from which to draw conclusions about the presence or lack of environmental effects of traditionally bred, pest-resistant crops. Therefore, various contradictory arguments have been made, including

1. *Bt*-crops have undergone undue scrutiny compared with their pest-resistant predecessors, which can serve as appropriate models, and for which there is little or no evidence of environmental disruption;
2. The genes introduced that confer traits for fundamentally different and highly effective antibiosis, such as *Bt*-endotoxin production, are evolutionary and ecological novelties, sufficiently different from their traditionally bred predecessors to warrant extra scrutiny; and

3. Assessing environmental and economic risks is appropriate for any pest resistance trait(s), whether introduced through traditional breeding techniques (including mutagenesis) or introduced from outside the taxonomic boundaries of the crop plant's interfertility.

While any of these arguments could be correct, the scientific basis, in terms of experimental evidence needed to bolster the theoretical foundation of these arguments, is lacking. Thus, we suggest that they make for excellent questions with poor answers concerning risk assessment of *Bt*-crops. In the next section we pose a number of questions related to a specific potential hazard of *Bt*-crops — weediness in their wild relatives. Our goal is not to assess this risk for any particular *Bt*-crop; rather, we will examine the usefulness and limitations of currently available data in making such assessments.

3.2   Bt-*crops: Impacts on wild relatives*

Although some crop plants are very much dependent on human assistance for their survival and reproduction, other crops, and many of the wild relatives that can hybridize successfully with crop plants, are already somewhat weedy in habit. Primary concerns of ecologists have been that genes conferring novel forms of pest resistance might increase fitness, competitive ability, and invasiveness of the crop itself, or cause increased invasiveness of wild crop relatives that may obtain the trait through hybridization and subsequent gene flow (Colwell et al. 1985, Tiedje et al. 1989, Hoffman 1990, Kareiva et al. 1994, Darmency 1994, Hails 2000). These early concerns were based on the contrast between a trait like insect pest resistance (which might produce phenotypes with a selective advantage under field conditions) to many of the traits bred into crop plants over millennia (which increase production of food or feed, but which would be detrimental to plants in natural conditions), such as fruit indehiscence, low dormancy of seeds, or dwarf stature (Ellstrand 1988). The extremely high levels of pest resistance possible with the toxins expressed in *Bt*-crops may also be more likely to increase plant fitness even when compared with pest-resistant lines attained through traditional breeding. The specific issue here is whether or not the novel resistance trait of *Bt*-endotoxin expression against lepidopterans (and, by extension, target pests in other orders) could have the potential to cause a change in invasiveness. If lowered herbivory leads to increased fitness through various avenues, including increased photosynthetic capacity, vigor, seed output, and seed weight, these changes might result in greater levels of invasiveness. Whether or not this potential is realized depends on the probability of the trait being transferred to a particular wild relative of the crop, the fate of the gene(s) in the population, and any resulting hazard associated with that trait in its accessible environment.

Within the U.S. regulatory framework, the responsibility of assessing the risk of environmental or public health hazards of *Bt*-crops is shared among three agencies: the Environmental Protection Agency (EPA), the U.S. Department of Agriculture Agricultural Plant Health Inspection Service (USDA–APHIS), and the Food and Drug Administration (FDA). The determinations of these agencies have been based on assessments of *Bt*-endotoxins in plant tissues as plant pesticides, possible plant pests, and possible food additives, respectively. Both the EPA and USDA–APHIS are concerned with the introduction of new traits to wild relatives, thereby causing the formation of new or more noxious weeds. The first tier of oversight, for field-testing *Bt*-crops, has been less concerned with assessing the possible consequences of transgene transfer than with assuring that barriers to gene flow are incorporated in the design of the small-scale study. These prevention methods include cessation of the trial and destruction of the plants before flowering occurs, as well as other physical and biological barriers.

The granting of nonregulated status (USDA) or the registration of the event as a plant pesticide (EPA) to allow commercial production of *Bt*-crops, however, depends on a favorable environmental assessment (EA). This assessment comprises several areas of risk, including the possibility that (1) the introduction of the *Bt*-resistance trait could cause *Bt*-crop plants to become more weedy and (2) the *Bt*-resistance trait could transfer to wild relatives of the crop and cause higher fitness and/or invasiveness in those plants. There are several approaches to predicting *a priori* if and when a crop or a wild plant population could experience some form of ecological release via the introduction of a trait that provides those plants protection from a certain suite of herbivores. The most direct one, relying on the fewest assumptions, is to do the experiment. This rather complex experiment would include, for a range of conditions and wild relatives, measures of the following variables:

1. The potential for gene flow and introgression of crop transgenes to wild relatives;
2. The range of expression levels of the trait after random mating, under a range of conditions;
3. The fitness of individuals under different herbivory regimes;
4. Recruitment success of hybrid and backcrossed individuals; and
5. Demographic modeling of the resultant populations, with validation.

Hybridization trials may need to have very large numbers of replications and, ideally, include diverse representative genotypes if hybridization is a rare event.

   If hybridization trials resulted in fertile offspring expressing the *Bt*-endotoxin, the next steps would require field testing. However, without presupposing that no hazard exists, such an experiment to assess risk would be impossible to perform safely. That is, the use of flowering *Bt*-crops for field testing in a range of habitats to determine rates of spontaneous hybridization, relative levels of plant fitness, or tracking plant demography would present an opportunity for the kind of gene flow into natural populations that is itself being evaluated for potential hazards. Therefore, predictions must rely on other approaches, such as inference from available data on non-transgenic plants, simulations with topical *Bt*-applications (Delannay et al. 1989), reasonably contained, but necessarily constrained, experiments with *Bt*-plants, monitoring after commercialization, and theoretical considerations, including simulation models.

   Although regulatory bodies do infer the level of risk from these available data, the decision process often lacks sufficient transparency to trace the scientific arguments underlying their conclusions. For example, a typical determination statement for the deregulation of a *Bt*-crop, as prepared by the USDA–APHIS (1995), reads as follows:

> Cotton lines bred by traditional means, which should be no more or less likely to interbreed with [the wild relative] *G. [Gossypium] tomentosum* than lepidopteran-resistant cotton, are not considered to pose a threat to the wild cotton and are not subject to particular State or Federal regulation on this basis. Neither the weediness nor the survival of *G. tomentosum*, therefore, will be affected by the cultivation of lepidopteran-resistant cotton because the transgenic variety poses no increased weediness itself and the two species are unlikely to successfully cross in nature.

We use the remaining portion of the chapter to examine the assumptions underlying such statements, and to evaluate their scientific basis. Our aims are to (1) illustrate critical questions for assessing the biosafety of a *Bt*-crop, (2) characterize the data available to answer these questions, and (3) comment on knowledge gaps that may preclude rigorous risk assessment at this time. We present a simplified flowchart to illustrate the main intermediary steps needed to assess risks associated with gene flow from *Bt*-crops (Figure 3.1).

We explore here the same overall questions posed in environmental assessments used by the USDA–APHIS or EPA about risks of increased weediness; but instead of framing them as an overview, we have partitioned the process of biosafety assessment into several individual decisions (Figure 3.1). In this way, strengths and weaknesses can be discussed for each area where data may be needed to identify what hazards exist and how likely they would be to occur. The pathway we present reflects the Precautionary Principle in its treatment of knowledge gaps. That is, if data to make a decision are lacking, the conservative pathway is taken rather than an assumption of low environmental risk. Finally, if scientifically defensible arguments for anticipating low risks are not possible, an assessment of the consequences (severity) of various hazards is required (Figure 3.1).

Some of the critical questions for assessing potential hazards of lepidopteran-resistant *Bt*-crops and their probabilities of occurrence are as follows. Which wild plant populations could receive transgenes from these crops through pollen transfer? Which lepidopterans feed on these host plants? What subset of these lepidopterans is susceptible to the endotoxins produced by the related *Bt*-crop? Do lepidopterans play a role in regulating plant populations? Which plants, under what circumstances, when released from herbivory, will exhibit higher fitness? Over time, will the plant population become more invasive? What effects will such plants have on the community? We have synthesized information from databases, published studies, and our preliminary experiments to examine these questions for several of the crops that have been transformed to express *Bt*-endotoxins, and for *Bt*-broccoli in particular. In most cases, we focus on lepidopteran-resistant *Bt*-crops, and suggest that many of the questions and analyses we present will have a parallel development when assessing *Bt*-crops resistant to Coleoptera, Diptera, or a broader range of pest taxa.

3.2.1 Bt-*crops and their wild relatives*

All the *Bt*-crops, and most of the major and minor crops in the world, either exist in the wild themselves or hybridize with wild relatives somewhere in their range (Snow and Moran-Palma 1997, Ellstrand et al. 1999, Klinger, Chapter 1 of this volume). Whether or not a fertile hybrid between a crop plant and its wild relative will occur depends on a number of factors, each of which is associated with spatiotemporal variability. The plants must be compatible so that cross-pollination can occur, and close enough to each other, in both space and phenological time, for pollen vectors to move between flowering plants (Ellstrand et al. 1999). Sexual compatibility between *Bt*-crops and their wild relatives poses a potential risk that *Bt*-transgenes will move via pollen to their wild relatives, escape through hybridization, and become established (via introgression) in wild populations. The first step, then, in assessing risks of gene flow is to compile a comprehensive list of plant species known, or potentially able, to produce fertile offspring that express the *Bt*-transgene (e.g., Table 3.8). Not included in Table 3.8 are many plants of interest for which little or nothing is known about their potential for producing fertile hybrids with related *Bt*-crops. Neither barriers to successful hybridization nor results of experimental attempts have been documented for those plants. Instead, Table 3.8 shows *Bt*-crops and those relatives for which we have either good evidence for hybridization or some reasons preventing us from dismissing the risk of hybridization.

The latter category suffers from data gaps, and lack of experimentation, so could be expanded or contracted based on different baseline assumptions. For example, we include *Brassica oleracea* and *B. rapa* as possible interbreeding species. Many modern cruciferous

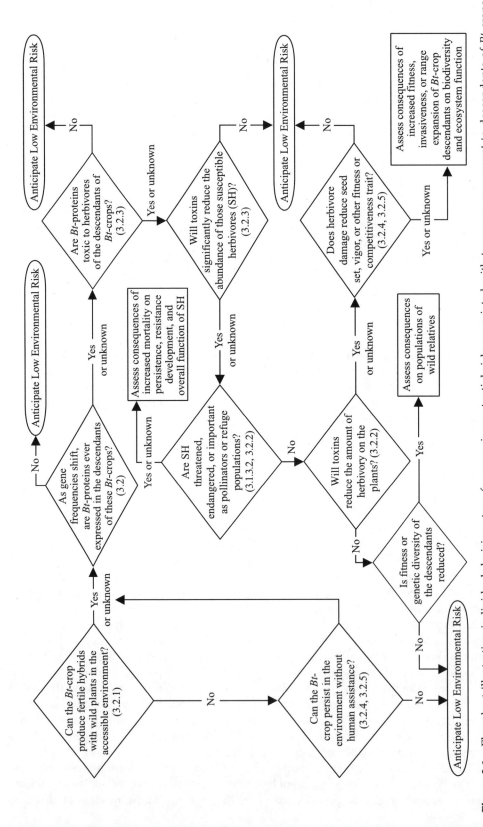

Figure 3.1 Flowchart illustrating individual decision steps for assessing potential risks associated with transgene movement to descendants of *Bt*-crops. Numbers in parentheses refer to sections in the chapter text that discuss related questions and describe the kinds of data needed for that decision step.

Table 3.8 Sexually Compatible Wild Relatives of Selected *Bt*-Crops (not exhaustive)

Beta vulgaris (sugar beet)
  B. vulgaris var. *maritima* (wild sea beet)[a]
  B. macrocarpa[c]

Brassica napus (rapeseed, canola)
  B. rapa (previously *B. campestris*)[d,g]
  B. nigra[e]
  B. juncea[b,g]
  Raphanus raphanistrum[b]
  B. oleracea
  Sinapis arvensis (previously *B. kaber*)[i]
  Hirschfeldia incana[b]
  Erucastrum gallicum[b]

B. oleracea (broccoli, cabbage, cauliflower,
  Brussels sprouts, collards, etc.)
  B. napus[b]
  B. rapa

Gossypium hirsutum (cotton)
  G. barbadense[b]
  G. tomentosum[b]
  G. arboreum
  G. haerbaceum
  G. thurberi
  G. darwinii[h]
  G. mustelinum[h]

Lycopersicon esculentum (tomato)
  L. pimpinellifolium
  L. hirsutum
  L. lycopersicoides[c]

Oryza sativa (rice)
  O. sativa (wild red rice)[d]
  O. rufipogon[d,f]
  O. punctata[f]
  O. longistaminata[e]
  O. glaberrima[e]

Solanum tuberosum (potato)
  S. tuberosum spp. *andigena*[a]
  S. fendleri
  S. jamesii
  S. pinnatiscetum
  S. longiconicum
  S. woodsonii
  S. demissum[b]

Zea mays (corn)
  Z. mays spp. *mexicana*[a]
  Z. mays spp. *huehuetenangensis*
  Z. diploperennis[a]
  Z. perennis[a]
  Z. luxurians[a]
  Tripsacum sp.[e]

Note: Compatibility was determined by forced pollinations unless otherwise noted.

[a] Fully cross-compatible.

[b] Fertile hybrids obtained.

[c] Sterile hybrids obtained.

[d] High rates of hybridization reported.

[e] Low rates of hybridization reported, high sterility of hybrids.

[f] U.S. Federal Noxious Weed List.

[g] Hybrids occur naturally in the field.

[h] Never tested for hybridization with crop.

[i] No hybrids obtained (Bing 1991).

crops have evolved from ancestral, natural crosses between two species that are considered sexually incompatible. *Brassica napus* (rapeseed or canola) is one of these, arising from ancient crosses of the diploid species *B. oleracea* (many cole crops such as broccoli, cabbage, kale, etc. CC $2n = 18$) and *B. rapa* (field mustard, AA $2n = 20$). Therefore, although crosses between *B. oleracea* and *B. rapa* (an invasive weed) have not been obtained in modern times, we do not dismiss such a cross as a possible avenue for gene escape from genetically engineered broccoli or cabbage. Whether or not particular plant species are considered possible sources of spontaneous crop–wild hybrids is yet another provocative question for risk assessment, and should be a transparent part of the process (Figure 3.1). Further, if an argument is made that rare hybridization events represent acceptable levels of gene escape risk, we suggest that it should be accompanied with a separate evaluation of the consequences of the rare event of gene flow into wild populations (Figure 3.1).

Many factors beyond strict sexual incompatibility may reduce the potential for gene escape to wild relatives of *Bt*-crops. Some of these are (1) physical or morphological barriers such as co-occurrence and overlapping flowering seasons; (2) transgene insertion site and chromosome homology; (3) crop breeding to reduce fertility of offspring, such as male sterile crops; and (4) engineering to reduce gene escape risk, such as chloroplast transformation (most often maternally inherited). Although these factors can be extremely useful, sometimes seemingly straightforward gene flow reduction factors work in complex ways under field conditions. For example, physical and morphological barriers to gene escape would be expected to reduce greatly any risks associated with escaped transgenes from *Bt*-cotton to the Hawaiian native *Gossypium tomentosum*. That is, although hand pollination has produced fertile crosses, *G. tomentosum* is naturally pollinated by moths, not bees like cultivated *G. hirsutum*. Flowers of *G. tomentosum* are receptive only at night, where other *Gossypium* species flower during the day. Furthermore, *G. tomentosum* does not produce nectar and is therefore less desirable to many potential pollinators. This would seem to preclude natural crossing between *G. tomentosum* and cultivated cottons, but Fryxell (1979) described *G. tomentosum* as threatened by genetic introgression by cultivated cotton varieties by unknown means. The strength of natural barriers to cross-pollination in *Gossypium* is not yet certain (DeJoode and Wendel 1992), but this example suggests that direct data from the field can be necessary for scientifically defensible conclusions.

General statements about the potential for gene flow have been based on the transgene insertion site and chromosome homology. While, again, these assumptions are valid for many cases, they too can be subject to important exceptions. Using oilseed rape or canola (*B. napus*) as an example: Cultivated *B. napus* is an allotetraploid with the chromosomes AACC. Wild *B. rapa* (field mustard, a dominant *Brassica* weed) produces fertile offspring with *B. napus*, resulting in relatively rapid gene escape potential (Mikkelson et al. 1996, Jorgensen et al. 1996, Hauser et al. 1998, Snow et al. 1999). The insertion of transgenes on the C chromosome of *B. napus* has been suggested as a means of reducing the risk of transgene escape (Metz et al. 1997), since the transgenes would then be less likely to be retained in backcross generations in *B. rapa* (due to selection in favor of the homeologous A chromosomes). If the transgene were inserted on the C chromosome in canola, then successful hybridization between canola and wild mustard and subsequent backcrossing would be less likely to result in progeny expressing the transgene than if the transgene were expressed on the A chromosome in canola. However, this assumption has been contradicted by findings that A and C chromosomes are equally transferred to subsequent generations (with 50% probability) under conditions of no selection (Tomiuk et al. 2000).

Tools such as male sterile crop hybrids and engineered reduction in fertility (e.g., chloroplast transformation) are exploited in genetically engineered crop development to reduce risk of hybridization and to exert marketing control over high-value agronomic lines. Inheritance of chloroplast chromosomes is maternal in most angiosperms, so plastid transformation reduces pollen transmission of transgenes in most cases. However, chloroplasts are not always maternally inherited in many plant families (N. C. Ellstrand, personal communication). Paternal inheritance, although rare, must be taken into account in all cases. Indeed, none of these factors is absolutely reliable as a barrier to gene flow. For example, transgenic chloroplasts will continue to segregate so that they still provide the possibility of irregular inheritance (Stewart and Prakash 1998). Even male sterility (crops that produce no functional pollen or are devoid of staminate flowers), which is clearly a useful barrier to gene flow, is not always dependable. Male-sterile hybrids are maintained by restorer or maintainer breeding parents and are a traditional tool for conserving proprietary germ plasm of crop hybrids. Clearly, it is effective for many crops, such as potato. Potato also has incompatible or unequal endosperm balance numbers, leading to endosperm failure and embryo abortion, multiple ploidy levels, and incompatibility mechanisms that do not express reciprocal genes and allow fertilization to proceed. Although transgene transfer barriers are not guaranteed even with potato, these mechanisms reduce considerably any risk of successful hybridization and gene introgression.

The risk of transgene escape can exist even between two species determined to be sexually incompatible. Crosses between incompatible relatives can occur via intermediate bridging. Bridging is an important consideration when evaluating gene flow risk from *Bt*-crops to their wild relatives because it increases the potential pool of sexual compatibility. Two non-compatible plants that have a third relative in common with which they both can cross opens the possibility for gene flow between those non-compatible relatives. Using *Brassica* as an example; *B. oleracea* is not known to cross with its related weed species. However, it could cross with *B. napus* and subsequently the resulting hybrid can cross with *B. rapa*, an important weed. Therefore, spatial relationships between crops and their related weeds become important considerations not just for crop-to-weed gene flow, but gene flow between all relatives (Klinger, Chapter 1 of this volume). When asking questions about the ecological relationships and population dynamics of weeds, we are not always met with a ready knowledge base. The case of gene flow from traditionally bred crops into related crop–weed populations is an example. If solid information about gene flow between non-transgenic crops and their wild relatives were already available, risk evaluation for *Bt*-crops could be better informed by those studies. Not knowing the dynamics of gene flow between traditionally bred crops and their wild relatives does not *a priori* negate the need for evaluating the risk of gene flow between *Bt*-crops and their wild relatives.

Finally, then, are there any general life history characteristics of *Bt*-crop relatives that could help in assessing risk of exacerbating weed or invasion problems? Other than clear cases of genetic incompatibility, the answer is likely to be no. Case-by-case examination is necessary because these plants represent a complex array of life history characteristics and habits. Some species of crop plant relatives occur in waste places, others are already weeds in crop fields, some are threatened, and some are extremely abundant. The plants include wind- and insect-pollinated species, obligate outcrossers, and facultative self-pollinators. To predict the likelihood of transgene transfer resulting in the expression of *Bt*-toxin in the tissues of wild relatives, reliable data on frequency of crosses that result in subsequent gene flow are needed for each crop–wild relative pair. To identify hazards that might be associated with the expression of *Bt*-toxin in the tissues of wild relatives, we need to know if one of the regulating factors for plant fitness in any of these species is herbivory by susceptible Lepidoptera. However, little or nothing is known about what forces regulate the population size of these species in natural habitats. Thus, risk assessments are currently based on a combination of strong, data-rich foundations and extrapolations— correct or not — that serve to bridge the existing knowledge gaps (Figure 3.1).

3.2.2 *Lepidopteran herbivores of* Bt-*crops and crop relatives*

To assess the impact of *Bt*-endotoxins on the herbivores of wild relatives of *Bt*-crops, one must have some idea of the herbivore species that might be exposed to *Bt*-endotoxins in the tissues of their host plants. To evaluate the potential impact of herbivore mortality on those plants, raises the next logical question: Which of those herbivores might be prevented from exerting their current levels of host plant feeding should the endotoxins deter or kill them? To explore the existing data for Lepidoptera, we interrogated the Natural History Museum HOSTS database (Robinson 1999) with a short list of *Bt*-crops and wild relatives. We tallied the number of species of Lepidoptera recorded as feeding on corn, soybean, and tomato or closely related plants (Table 3.9). Hundreds of species of butterflies and moths were listed, and 376, 185, and 98 species are documented as feeding on corn,

Table 3.9   Total Number of Lepidoptera Species (subspecific categories pooled) with Larvae Recorded as Feeding on Corn, Soybean, Tomato, Potato or Closely Related Plants

Plant family	Plant species	Lepidopteran herbivores
Poaceae	*Zea diploperennis*	0
	Z. luxurians	0
	Z. mays	376
	Z. mays mays	0
	Z. mays parviglumis	0
	Z. mexicana	2
Fabaceae	*Glycine max*	185
	G. soja	8
Solanaceae	*Lycopersicon cheesmanii*	0
	L. chilense	0
	L. chmielewskii	0
	L. esculentum	98
	L. esculentum cerastiforme	1
	L. hirsutum	3
	L. hirsutum glabratum	0
	L. parviflorum	0
	L. pennellii	1
	L. peruvianum	0
	L. pimpinellifolium	0
	Solanum tuberosum	138
	S. juglandifolium	0
	S. lycopersicoides	0
	S. ochranthum	0
	S. sitiens	0

Data from the Natural History Museum's HOSTS database (see Robinson 1999).

soybean, and tomato, respectively (Table 3.9; Appendix 3.1). A list of all lepidopteran species recorded on these food plants is provided in Appendix 3.1 together with a measure of their reported host plant range and their geographical distribution. The main lessons here are (1) evaluation of fitness effects from *Bt*-transgene transfer requires some knowledge of what lepidopterans feed on the plant and might be susceptible to *Bt*-toxins; (2) queries for host plants, using the most comprehensive data set available, yields (a) at least 100 lepidopteran species feed on each crop and (b) very few records for wild relatives of those crops. Thus, if transgene transfer were expected, even rarely, evaluating the risk of increased weediness may be impeded by a lack of data. To address another issue, that of disruption of lepidopteran populations through reducing the availability of host plants on which they can complete development, a subset of the latter data was extracted to develop Table 3.10. This list addresses the level of dependence these lepidopteran species may have on a single host plant. Each of these issues will be discussed below. To our knowledge, no such lists of herbivores and host ranges have been published previously, so it has been difficult to even conceive of the number of non-target lepidopterans potentially exposed to *Bt*-toxins in crops and, if the trait is transferred, in wild relatives.

The Natural History Museum HOSTS database (described in Robinson 1999; Robinson et al., unpublished data) is a compilation of, currently, 175,000 host plant records of world Lepidoptera drawn from more than 1600 printed manuscript and electronic sources.

Table 3.10 Lepidoptera Species (subspecific categories pooled) with a Recorded Host Range of Only One Plant Genus within Which *Bt*-Manipulation Has Occurred and Which Are Potentially at Risk of Extirpation If No Alternative Host Exists

Family	Species	Distribution	Crop genus
Gelechiidae	*Bryotropha galbanella* Zeller	nt; pl	*Solanum*
	Dichomeris granivora Meyrick	nt	*Zea*
	Scrobipalpopsis solanivora Povolny	nt	*Solanum*
	Stenorumia ablunata Guenée	or	*Solanum*
Geometridae	*Zeuctoboarmia simplex* Prout	af	*Solanum*
Hesperiidae	*Bolla brennus* Godman & Salvin	nt	*Lycopersicon*
Lymantriidae	*Dasychira anophoeta* Collenette	or	*Glycine*
Noctuidae	*Agrotis subalba* Walker	af	*Solanum*
	Apamea vulgaris Grote & Robinson	na	*Zea*
	Dargida grammivora Walker	nt	*Zea*
	Eublemma aurantiaca Hampson	af	*Solanum*
	Gesonia gemma Swinhoe	or	*Glycine*
	Heterogramma circumflexalis Guenée	nt	*Zea*
	Hypena tristalis Lederer	ow	*Glycine*
	Mentaxya albifrons Geyer	af	*Solanum*
	Neuranethes angola Bethune-Baker	af	*Solanum*
	Oligia semicana Walker	na	*Zea*
	Trachea niveiplaga Walker	or	*Solanum*
Nolidae	*Gyrtothripa papuana* Hampson	au	*Solanum*
Pyralidae	*Agriphila ruricolellus* Zeller	na	*Zea*
	A. vulgivagellus Clemens	na	*Zea*
	Bradina adhaesalis Walker	or	*Glycine*
	Crambus praefectellus Zincken	na	*Zea*
	Diatraea postlineella Schaus	nt	*Zea*
	Euzophera villora Felder & Rogenh.	af; or	*Solanum*
	Mabra charonialis Walker	pl	*Glycine*
	Microcrambus elegans Clemens	na	*Zea*
	Neodactria luteolellus Clemens	na	*Zea*
	N. zeellus Fernald	na	*Zea*
	Rhodophaea nr. persicella Amsel	or	*Solanum*
Saturniidae	*Hemileuca mania* Druce	nt	*Zea*
Tortricidae	*Cryptophlebia strepsibathra* Meyrick	au	*Glycine*

Notes: Species feeding only on a *Bt*-manipulated plant genus but with a distribution outside the original distribution of the crop species and its close allies/progenitors are excluded on the grounds that they must have alternative hosts. Additional taxa with very restricted host ranges within Solanaceae occur among ithomiine butterflies (Beccaloni, personal communication). Distribution is indicated by a two-letter code; thus, af = Afrotropical region; au = Australasia; na = Nearctic region; nt = Neotropical region; or = Oriental region; ow = Old World; pl = Palaearctic.

Source: Data from the National History Museum HOSTS database (see Robinson 1999).

Although this extensive and, perhaps surprisingly, diverse list is the best available, Table 3.9 and Appendix 3.1 do not list all of the lepidopteran species that may be of interest as non-target herbivores on our sample of *Bt*-crops and wild relatives. The HOSTS database contains information culled from both the economic and non-applied entomological literature and records at least one host plant for, overall, 19% of the world's described species of Lepidoptera. Although by no means comprehensive, the HOSTS database gives broad and credible coverage that is not available from any other source.

The difficulties in knowing the potential breadth of exposure to *Bt*-crop toxins, even for a relatively well known order of insects, creates a level of uncertainty for detailed risk assessment. For example, some groups will be underrepresented because knowledge of certain groups of Lepidoptera and richness of certain categories of data vary widely (Beccaloni, unpublished data). For example, host plants of 42% of about 875 species of Nepticuloidea, and 69% of about 2000 species of Gracillarioidea (groups often collected by rearing) are known. But host plants for only 9% of the 21,900 Geometroidea and 13% of the 15,900 Pyraloidea are known. Although concerned with risk assessment for the accessible environment of a *Bt*-crop or its wild relative, documentation or access to information can vary by geographical region, as well. For the HOSTS database, abstracting of literature has been extensive and approaches practical limits for the Oriental region, North America, Western Europe, and the Afrotropical region, whereas coverage of Central and South America, Asia excluding Western Europe, and Australasia is less comprehensive. For our purposes, we expect to have disproportionately fewer host records on wild plants than on crop plants because lepidopterans commonly feeding on crop plants are probably all recorded in the literature (and in HOSTS), and herbivores on non-economic plants are recorded sporadically and eclectically. Such a bias is immediately apparent from inspection of Table 3.9. Also, it is rare to find host plant records that cite the subspecies or variety of the plant, which makes it particularly difficult to retrieve records for *Zea mays mays* and *Z. mays parviglumis* in contrast to *Z. mays* (Table 3.9). Thus, the data available to compile a list of herbivores that feed on wild relatives of *Bt*-crops or any other pest protected crops are very uneven, primarily through underreporting. In addition, misidentification, synonymy, and other potential sources of error are manifold and parallel those encountered in records of host–parasite relationships (Fitton et al. 1988).

Despite inherent inadequacies, the HOSTS databases can provide initial lists of Lepidoptera and some estimate of how dependent these herbivores are on the host plant in question. To determine the degree of this dependency, we also queried the database about diet breadth of the lepidopterans listed. At a continental or global scale, the frequency distribution of host plant usage among Lepidoptera is a hollow curve approximating to a logarithmic series (Figure 3.2) (Robinson 1998). More species (or genera) feed (or are recorded as feeding) on a single plant species (or genus) than feed on two, more feed on two than on three, and so on. At the species level, however, this observed pattern might possibly be an artifact, a product of pseudo-random incomplete sampling. Simply, the data resolution at species–species level (or at even finer resolution) is not good enough to assume conclusively that this is the actual pattern in nature.

Given the problems mentioned above, however, the following conclusions may be drawn. In the Poaceae (Table 3.11), three of four species recorded from non-crop relatives of crop plants feed also on the crop plant. In Leguminosae (Table 3.12), all species recorded from *Glycine soja* also feed on *G. max*. In Solanaceae (Table 3.13), all lepidopteran species that are recorded from the non-crop relatives of tomato also feed on *Lycopersicon esculentum* and/or *Solanum tuberosum*. However, no lepidopterans are listed as having the host plants *S. juglandifolium*, *S. lycopersicoides*, *S. sitiens*, or *S. ochranthum* (Table 3.9), so it is difficult to know for sure the level of overlap in host plant usage.

From the limited records available for these and other crop/plant queries (Letourneau, Robinson, and Hagen, unpublished data) we could speculate that 40 to 100% of the herbivores recorded from the non-economic relatives of crop plants also feed on the crop plant species. The inference seems fair that a similar proportion of the species on the crop plant might feed on the non-crop relatives (given geographical feasibility) but sufficient data have simply not been recorded.

The potential transfer of genetic modifiers that induce high herbivore mortality to close relatives of crop plants may possibly have profound implications for biodiversity. This is especially true for crops with very high acreages of *Bt*-varieties. Of 18,503 Lepidoptera species for which we have at least one host plant record, almost half (9062) are

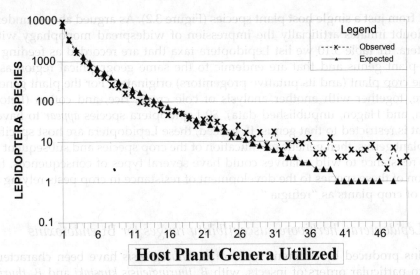

Host Plant Genera Utilized

Figure 3.2 Frequency distribution of number of genera of host plants used by 18,503 species of world Lepidoptera with hypothetical logarithmic series shown for comparison. For the logarithmic scale, only the first 50 frequency intervals are shown. The observed tail extends to one species with 172 host genera, and the predicted tail extends to one genus with 56 host genera.

Table 3.11 Pattern and Frequency of Shared Host Plants in Poaceae (species listed in and abbreviated from Appendix 3.1) among 280 Lepidoptera Species (subspecific taxa not differentiated)

Z-ma	Z-mx	Species
X		278
X	X	2

Note: 278 species are recorded from a single host plant from within this group.

Table 3.12 Pattern and Frequency of Shared Host Plants in Leguminosae (*Glycine*) (species listed in and abbreviated from Appendix 3.1) among 185 Lepidoptera Species (subspecific taxa not differentiated)

G-mx	G-sj	Species
X		177
X	X	8

Note: 177 species are recorded from a single host plant from within this group.

Table 3.13 Pattern and Frequency of Shared Host Plants in Solanaceae (*Lycopersicon* and *Solanum*) (species listed in and abbreviated from Appendix 3.1) among 176 Lepidoptera Species (subspecific taxa not differentiated)

L-es	L-hi	L-pe	S-tu	Species
			X	78
				38
X			X	57
X		X	X	2
X		X	X	1

Note: 116 species are recorded from a single hostplant from within this group.

recorded from just a single host plant species (Figure 3.2). As argued above, underrecording no doubt inflates artificially the impression of widespread monophagy within the Lepidoptera. In Table 3.10 we list Lepidoptera taxa that are recorded as feeding *only* on the crop-plant genus and that are endemic to the same geographical region as that in which the crop plant (and its putative progenitors) originated. For the plant genera dealt with here, together with another analysis of cole crops, rice, and cotton (Letourneau, Robinson, and Hagen, unpublished data), 80 Lepidoptera species *appear* to have a host range that is restricted to that genus. If, indeed, these Lepidoptera are host specific at the level of plant genus, then genetic modification of the crop species and subsequent transfer of insect resistance to wild relatives could have several types of consequences, from the extirpation of these species to the development of resistance in crop pests relying on wild relatives of crop plants as "refugia."

3.2.3   *Lepidopteran herbivore susceptibility ranges for* Bt-*endotoxins*

Endotoxins produced by different types of *B. thuringiensis* have been characterized as specific to particular orders of insects, with *B. thuringiensis kurstaki* and *B. thuringiensis aizawai* and their products formulated in commercially available sprays to control certain lepidopteran pests. *B. thuringiensis* var. *san diego* and *Bacillus thuringiensis* var. *tenebrionis* are used commercially because they act as a toxin when ingested by some coleopterans. To evaluate non-target effects of *Bt*-crops on herbivores that feed on those crops and on wild relatives that could potentially obtain the trait through introgression, more-detailed information on susceptibility is needed. This necessity for further research is especially true because herbivorous insects, even those in the target orders, vary in susceptibility not only from *cry* gene to *cry* gene, but also from species to species. The level of susceptibility also depends on the generation of pest, stage of development of the pest, and the crop line, which can determine the level of Cry protein expression (e.g., Acciarri et al. 2000, Archer et al. 2000, Ashfaq et al. 2000).

One general protocol for assessing the effects of a transgene coding for such an endotoxin in a wild relative of a crop plant would be to determine directly if any of the herbivores that feed on that plant are susceptible to the endotoxin (produced in response to the *cry* gene in question). If the herbivores are not, then neither the plant nor the herbivore is likely to experience population level changes in abundance due to transgene transfer, and risk assessment research efforts might be directed elsewhere. Unfortunately, even for the best known *Bt*-endotoxins (lepidopteran-specific), the available data on susceptibility levels among species suggest that these traits do not fall cleanly along broad taxonomic lines, such as suborders or families or even genera. Most of the research on susceptibility to *Bt*-crop endotoxins has been conducted with target pest species. For the pest Lepidoptera, variability in susceptibility exists among closely related taxa, among different instars (or sizes) of the same species, and depends on the particular protein being expressed by the plant (Macintosh et al. 1990, Frankenhuyzen et al. 1991, Ashfaq et al. 2000). As has been documented by Bourguet et al. (2000) for the ECB, if the population genetics of pest species is such that some isolation exists between individuals exploiting crop plants and those in the same species feeding on non-crop hosts, then extrapolation from the little we do know about pest susceptibility to *Bt*-crops may even be misleading.

Few data are available on the susceptibility of non-target lepidopterans to *Bt*-endotoxins. Most of the current data are based on tests with *Bt*-sprays rather than on the toxins that genetically engineered plants expressing those and related toxins produce. However, the same trends are evident as with pest insects in terms of the variable effects of *Bt*-endotoxins on non-target herbivores. *Bt*-sprays were toxic to cinnabar moth larvae *Tyria jacobaea* (L.) only when these biological control agents were exposed to the toxins as later instars

(James et al. 1993). Peacock et al. (1998) showed that, of 42 species of forest Lepidoptera tested, *B. thuringiensis* var. *kurstaki* sprays caused mortality in 27. Whereas the percentage of susceptible insects was relatively high, mortality tended to be greater when larvae in the early stages of development were tested, and susceptible species were dispersed among six of the seven lepidopteran families tested (Papilionidae, Nymphalidae, Geometridae, Lasiocampidae, Saturniidae, and Noctuidae). Nonsusceptible larvae were also found in most of the seven families tested; for example, both the Geometridae and the Noctuidae (depending on the species) contained larvae that, when exposed to *Bt*-sprays, had no mortality and larvae that experienced 100% mortality.

To determine which of the species recorded in Appendix 3.1 are susceptible to *Bt*-endotoxins, we consulted a comprehensive database of published toxicity studies for *Bt* that the Canadian Forest Service maintains (van Frankenhuyzen and Nystrom 1999). Table 3.14 lists the species in Appendix 3.1 for which *Bt*-susceptibility test results (commercially available proteins) are documented in published studies, and shows the gene that encoded endotoxin for those tests. Although, clearly, much is known about tolerance levels of some lepidopteran species that feed on *Bt*-crops or their relatives, those species constitute a very low percentage of the total species known to use *Bt*-crops or their relatives as host plants (Appendix 3.1). Not included in the database are data from private sources, which may include tests on additional species for susceptibility to proprietary (especially genetically modified) proteins. Typically, research efforts have emphasized testing of target pests or a known susceptible insect rather than a range of lepidopterans that occur either on target host plants or in geographical areas of toxin introduction. Only approximately 3% (11 of 376) of the lepidopteran species listed as feeding on *Zea mays* have undergone susceptibility studies to commercial *Bt*-toxins (Table 3.14). We are aware of no studies directed at the systematic measurement of mortality rates for the non-target lepidopterans feeding on *Bt*-crop plants or of the effects of *Bt*-sprays on the hundreds of lepidopterans recorded as feeding on *Bt*-crops and their relatives. Based on available data, then, we conclude that there is a high level of unpredictability in terms of the susceptibility of a particular lepidopteran species that may act as a significant herbivore of the wild relative of a *Bt*-crop. This conclusion is significant, because comparative information on the relative mortality of non-target herbivores feeding on traditionally bred, insect-resistant crops and their isogenic *Bt*-variety are lacking. Without experiments or empirical data, conclusions by regulatory agencies that there is no reason to expect environmental impacts from particular *Bt*-corn events "to be significantly different from those arising from the cultivation of any other variety of insect-tolerant or herbicide-tolerant corn" (USDA–APHIS 1997) are possibly, but not necessarily correct.

3.2.4   *Lepidopteran herbivores as regulators of crop yield, fitness, and invasiveness*

To assess the risk of enhanced weediness of crops or wild relatives expressing an insect-resistance transgene, the role of insect herbivory in determining plant fitness, geographical range, and competitive ability must be evaluated. Manipulative experiments designed to test the effects of naturally occurring herbivores on various aspects of plant fitness have shown that arthropods reduce productivity and individual fitness, and can represent a primary selective force for wild plants (Crawley 1983). Recent advances in herbivore–plant interactions and tests of trophic cascades theory have also shown that the success of some plant species within the community can depend upon their level of protection from herbivores (Schoener 1987, Ritchie and Tilman 1992, Moran and Hurd 1998, Letourneau and Dyer 1998, Dyer and Letourneau 1999). Indeed, the absence of herbivores on exotic plants in new habitats has allowed expansion and weediness to develop, constituting ecological release of those plant populations (Harper 1969, Hails 2000). Thus, recent and

Tables 3.14 Lepidoptera Known to Feed on *Bt*-crops (or crop relatives) That Have Been Tested for Susceptibility to Commercially Available *Bacillus thuringiensis* Cry Toxins

Genus	Species	Host[a]	Stage	Gene[b]	Author	Publication	Volume	Page	Year
Agrotis	ipsilon	c, p	N	cry01Ab10	Macintosh	J. Invertebr. Pathol.	56	258–266	1990
			N	cry01Ac1	Macintosh	J. Invertebr. Pathol.	56	258–266	1990
Chilo	suppressalis	c, t	N	cry02Aa	Sims	Southwest. Entomol.	22	395–404	1997
			L2	cry01Ac1	Fiuza	Appl. Environ. Microbiol.	62(5)	1544–1549	1996
			L1	cry01Ac1	Lee	Appl. Environ. Microbiol.	63(4)	1453–1459	1997
			L1	cry02Aa1	Lee	Appl. Environ. Microbiol.	63(4)	1453–1459	1997
Elasmolpalpus	lignosellus	c, z	N	cry01Ab6	Moar	J. Econ. Entomol.	88(3)	606–609	1995
			N	cry01Ac6	Moar	J. Econ. Entomol.	88(3)	606–609	1995
			N	cry02Aa2	Moar	J. Econ. Entomol.	88(3)	606–609	1995
Helicoverpa	armigera	c, s, t, h, p	N	cry01Ab	Padidam	J. Invertebr. Pathol.	59(1)	109–111	1992
			N	cry01Ab	Chakrabarti	J. Invertebr. Pathol.	72	336–337	1998
			N	cry01Ab	Mandaokar	World J. Microbiol. Biotechnol.	14	599–601	1998
			N	cry01Ac	Padidam	J. Invertebr. Pathol.	59(1)	109–111	1992
			N	cry01Ac	Chakrabarti	J. Invertebr. Pathol.	72	336–337	1998
			N	cry02Aa	Chakrabarti	J. Invertebr. Pathol.	72	336–337	1998
			N	cry09Ca1	Lambert	Appl. Environ. Microbiol.	62(1)	80–86	1996
	zea	c, t, s, p	N	cry01Ac1	Von Tersch	Appl. Environ. Microbiol.	57(2)	349–358	1991
			N	cry01Ac2	Von Tersch	Appl. Environ. Microbiol.	57(2)	349–358	1991
			N	cry02Ab2	Dankocsik	Mol. Microbiol.	4(12)	2087–2094	1990
			N	cry02Aa1	Dankocsik	Mol. Microbiol.	4(12)	2087–2094	1990
			N	cry01Ab	Chambers	J. Bacteriol.	173(13)	3966–3976	1991
			N	cry01Ac	Chambers	J. Bacteriol.	173(13)	3966–3976	1991
			N	cry01Aa1	Donovan	U.S. Patent 5332687			1994
			N	cry01Ab10	Macintosh	J. Invertebr. Pathol.	56	258–266	1990
			N	cry01Ac1	Macintosh	J. Invertebr. Pathol.	56	258–266	1990
			N	cry01Ac1	English	Insect Biochem. Mol. Biol.	24(10)	1025–1035	1994
			N	cry02Aa1	English	Insect Biochem. Mol. Biol.	24(10)	1025–1035	1994
			N	cry01Ac1	Bohorova	J. Econ. Entomol.	90(2)	412–415	1997
			N	cry02Aa	Sims	Southwest. Entomol.	22	395–404	1997
	zea	c, t	N	cry01Ac	Luttrell	J. Econ. Entomol.	65(5)	1849–1853	1999
			N	cry01Ab	Luttrell	J. Econ. Entomol.	65(5)	1849–1853	1999
			L4	cry1Aa[d]	Karim	Curr. Microbiol.	41(3)	214–219	2000

		s, t, p						
Heliothis	*virescens*	N, L	cry1A(b)	Wiseman	J. Entomol. Sci.	34(4)	415–425	1999
		L2	cry01Ab	Visser	J. Bacteriol.	172(12)	6783–6788	1990
		N	cry01Ab	Chambers	J. Bacteriol.	173(13)	3966–3976	1991
		L2	cry01Ab	Honee	Appl. Environ. Microbiol.	56(3)	823–825	1990
		N	cry01Ab	Luttrell	J. Econ. Entomol.	65(5)	1849–1853	1999
		N	cry01Ab10	Macintosh	J. Invertebr. Pathol.	56	258–266	1990
		N	cry01Ab2	Lee	Appl. Environ. Microbiol.	61(11)	3836–3842	1995
		N	cry01Ab2	Lee	Appl. Environ. Microbiol.	61(11)	3836–3842	1995
		L1	cry01Ab3	Lee	J. Bacteriol.	173(20)	6635–6638	1991
		N	cry01Ab5	Hofte	Microbiol. Rev.	53(2)	242–255	1989
		N	cry01Ab5	Gould	Proc. Natl. Acad. Sci. U.S.A.	89	7986–7990	1992
		N	cry01Ab5	Gould	Proc. Natl. Acad. Sci. U.S.A.	89	7986–7990	1992
		N	cry01Ab5	Schnepf	J. Biol. Chem.	265(34)	20923–20930	1990
		N	cry01Ab9	Rajamohan	J. Biol. Chem.	271(5)	2390–2396	1996
		N	cry01Ac	Chambers	J. Bacteriol.	173(13)	3966–3976	1991
		N	cry01Ac	Luttrell	J. Econ. Entomol.	65(5)	1849–1853	1999
		N	cry01Ac1	Von Tersch	Appl. Environ. Microbiol.	57(2)	349–358	1991
		N	cry01Ac1	Hofte	Microbiol. Rev.	53(2)	242–255	1989
		N	cry01Ac1	Schnepf	J. Biol. Chem.	265(34)	20923–20930	1990
		N	cry01Ac1	Lee	Appl. Environ. Microbiol.	61(11)	3836–3842	1995
		N	cry01Ac1	Macintosh	J. Invertebr. Pathol.	56	258–266	1990
		N	cry01Ac1	Lee	Appl. Environ. Microbiol.	61(11)	3836–3842	1995
		L1	cry01Ac1	Ge	J. Biol. Chem.	266(27)	17954–1795	1991
		N	cry01Ac2	Von Tersch	Appl. Environ. Microbiol.	57(2)	349–358	1991
		N	cry02Aa	Gould	Proc. Natl. Acad. Sci. U.S.A.	89	7986–7990	1992
		N	cry02Aa	Gould	Proc. Natl. Acad. Sci. U.S.A.	89	7986–7990	1992
		N	cry02Aa	Sims	Southwest. Entomol.	22	395–404	1997
		N	cry02Aa1	Donovan	J. Biol. Chem.	263(1)	561–567	1988
		N	cry02Aa1	Dankocsik	Mol. Microbiol.	4(12)	2087–2094	1990
		N	cry02Aa2	Moar	Appl. Environ. Microbiol.	60(3)	896–902	1994
		N	cry02Aa2	Kota	Proc. Natl. Acad. Sci. U.S.A.	95	1840–1845	1999
		N	cry02Ab2	Dankocsik	Mol. Microbiol.	4(12)	2087–2094	1990
		N	cry02Ac1	Wu	FEMS	81(1)	31–36	1991
		N	cry09Ca1	Lambert	Appl. Environ. Microbiol.	62(1)	80–86	1996

Tables 3.14 (continued) Lepidoptera Known to Feed on *Bt*-crops (or crop relatives) That Have Been Tested for Susceptibility to Commercially Available *Bacillus thuringiensis* Cry Toxins

Genus	Species	Host[a]	Stage	Gene[b]	Author	Publication	Volume	Page	Year
Lymantria	dispar	c	L1	cry01Ab	Wolfersberger	*Experientia*	46	475–477	1990
			L4	cry01Ab2	Liang	*J. Biol. Chem.*	270(42)	24719–24724	1995
			L4	cry01Ab2	Lee	*Appl. Environ. Microbiol.*	62(2)	583–586	1996
			N	cry01Ab2	Lee	*Biochem. Biophys. Res. Comm.*	229	139–146	1996
			L4	cry01Ab2	Lee	*Biochem. Biophys. Res. Commun.*	229	139–146	1996
			L4	cry01Ab6	Van Frankenhuyzen	*Appl. Environ. Microbiol.*	57(6)	1650–1655	1991
			L4	cry01Ab6	Van Frankenhuyzen	*Appl. Environ. Microbiol.*	63(10)	4132–4134	1997
			N	cry01Ab9	Lee	*Biochem. Biophys. Res. Commun.*	229	139–146	1996
			L4	cry01Ab9	Lee	*Biochem. Biophys. Res. Commun.*	229	139–146	1996
			N	cry01Ab9	Rajamohan	*Proc. Natl. Acad. Sci. U.S.A.*	93	14338–1434	1996
			L4	cry01Ab9	Rajamohan	*Proc. Natl. Acad. Sci. U.S.A.*	93	14338–1434	1996
			N	cry01Ac1	Von Tersch	*Appl. Environ. Microbiol.*	57(2)	349–358	1991
			L4	cry01Ac1	Liang	*J. Biol. Chem.*	270(42)	24719–24724	1995
			L4	cry01Ac1	Lee	*Appl. Environ. Microbiol.*	62(2)	583–586	1996
			L1	cry01Ac1	Wolfersberger	*Experientia*	46	475–477	1990
			N	cry01Ac2	Von Tersch	*Appl. Environ. Microbiol.*	57(2)	349–358	1991
			L4	cry01Ac6	Van Frankenhuyzen	*Appl. Environ. Microbiol.*	57(6)	1650–1655	1991
			L4	cry02Aa	Liang	*Mol. Microbiol.*	13(4)	569–575	1994
			N	cry02Aa1	Donovan	*J. Biol. Chem.*	263(1)	561–567	1988
			N	cry02Aa1	Dankocsik	*Mol. Microbiol.*	4(12)	2087–2094	1990
			L4	cry02Ab	Liang	*Mol. Microbiol.*	13(4)	569–575	1994
			N	cry02Ab2	Dankocsik	*Mol. Microbiol.*	4(12)	2087–2094	1990
			L4	cry09Ca1	Van Frankenhuyzen	*Appl. Environ. Microbiol.*	63(10)	4132–4134	1997
Mamestra	brassicae	t,p	L2	cry01Ab	Visser	*J. Bacteriol.*	172(12)	6783–6788	1990
			L1	cry01Ab5	Hofte	*Microbiol. Rev.*	53(2)	242–255	1989
			L1	cry01Ac1	Hofte	*Microbiol. Rev.*	53(2)	242–255	1989
	configurata	s, t, p	N	cry09Ca1	Lambert	*Appl. Environ. Microbiol.*	62(1)	80–86	1996
			L2	cry01Ab	Masson	*Appl. Environ. Microbiol.*	64(12)	4782–4788	1998
			L2	cry01Ab	Masson	*Appl. Environ. Microbiol.*	64(12)	4782–4788	1998
Manduca	sexta	t, p	N	cry01Ab10	Macintosh	*J. Invertebr. Pathol.*	56	258–266	1990
			N	cry01Ab10	Fischhoff	*Bio/Technology*	5	807–813	1987
			N	cry01Ab2	Lee	*Biochem. Biophys. Res. Commun.*	229	139–146	1996

Genus	Species		Gene	Stage	Author	Journal	Vol.	Pages	Year
			cry01Ab5	L1	Hofte	Eur. J. Biochem.	161	273–280	1986
			cry01Ab5	L1	Hofte	Eur. J. Biochem.	161	273–280	1986
			cry01Ab5	L1	Hofte	Microbiol. Rev.	53(2)	242–255	1989
			cry01Ab5	N	Schnepf	J. Biol. Chem.	265(34)	20923–20930	1990
			cry01Ab5	L1	Hofmann	Proc. Natl. Acad. Sci. U.S.A.	85	7844–7848	1988
			cry01Ab9	N	Rajamohan	J. Bacteriol.	177(9)	2276–2282	1995
			cry01Ab9	N	Chen	J. Biol. Chem.	270(11)	6412–6419	1995
			cry01Ab9	N	Rajamohan	J. Biol. Chem.	271(5)	2390–2396	1996
			cry01Ab9	N	Lee	Biochem. Biophys. Res. Commun.	229	139–146	1996
			cry01Ac1	N	Adang	Gene	36	289–300	1985
			cry01Ac1	L1	Hofte	Microbiol. Rev.	53(2)	242–255	1989
			cry01Ac1	L2	Ge	Proc. Natl. Acad. Sci. U.S.A.	86	4037–4041	1989
			cry01Ac1	N	Schnepf	J. Biol. Chem.	265(34)	20923–20930	1990
			cry01Ac1	N	Macintosh	J. Invertebr. Pathol.	56	258–266	1990
			cry01Ac1	L1	Schultz	Ecology	69(4)	896–897	1988
			cry02Aa	N	Sims	Southwest. Entomol.	22	395–404	1997
			cry02Aa2	N	Widner	J. Bacteriol.	171(2)	965–974	1989
			cry02Aa2	N	Widner	J. Bacteriol.	172(6)	2826–2832	1990
			cry02Ab1	N	Widner	J. Bacteriol.	171(2)	965–974	1989
			cry02Ab1	N	Widner	J. Bacteriol.	172(6)	2826–2832	1990
			cry02Ac1	N	Wu	FEMS	81(1)	31–36	1991
			cry09Ca1	N	Lambert	Appl. Environ. Microbiol.	62(1)	80–86	1996
Platynota	stultana	c, s	cry01Ab	N	Knight	J. Agric. Entomol.	15(2)	93–103	1998
			cry01Ac	N	Knight	J. Agric. Entomol.	15(2)	93–103	1998
Plodia	interpunctella	c, s, p	cry01Ab5	L2–L3	Van Rie	Science	247	72–74	1990
			cry01Ab5	L2–L3	Van Rie	Science	247	72–74	1990
			cry01Ab5	L2–L3	Van Rie	Science	247	72–74	1990
			cry01Ab5	L2–L3	Van Rie	Science	247	72–74	1990
			cry01Ab5	L3	McGaughey	J. Econ. Entomol.	87(3)	535–540	1994
			cry01Ab5	L3	McGaughey	J. Econ. Entomol.	87(3)	535–540	1994
			cry01Ab5	L3	McGaughey	J. Econ. Entomol.	87(3)	535–540	1994
			cry01Ab5	L3	McGaughey	J. Econ. Entomol.	87(3)	535–540	1994
			cry01Ab5	L3	McGaughey	J. Econ. Entomol.	87(3)	535–540	1994
			cry01Ab5	L3	McGaughey	J. Econ. Entomol.	87(3)	535–540	1994
			cry02Aa2	L3	McGaughey	J. Econ. Entomol.	87(3)	535–540	1994
			cry02Aa2	L3	McGaughey	J. Econ. Entomol.	87(3)	535–540	1994

Tables 3.14 (continued) Lepidoptera Known to Feed on *Bt*-crops (or crop relatives) That Have Been Tested for Susceptibility to Commercially Available *Bacillus thuringiensis* Cry Toxins

Genus	Species	Host[a]	Gene[b]	Stage	Author	Publication	Volume	Page	Year
			cry02Aa2	L3	McGaughey	J. Econ. Entomol.	87(3)	535–540	1994
			cry02Aa2	L3	McGaughey	J. Econ. Entomol.	87(3)	535–540	1994
			cry02Aa2	L3	McGaughey	J. Econ. Entomol.	87(3)	535–540	1994
			cry02Aa2	L3	McGaughey	J. Econ. Entomol.	87(3)	535–540	1994
Plutella	xylostella	c	cry01Ab	L3	Tabashnik	Appl. Environ. Microbiol.	60(12)	4627–4629	1994
			cry01Ab	L3	Tabashnik	Appl. Environ. Microbiol.	60(12)	4627–4629	1994
			cry01Ab	L3	Tabashnik	J. Invertebr. Pathol.	76	81–83	2000
			cry01Ab	N	Mandaokar	World J. Microbiol. Biotechnol.	14	599–601	1998
			cry01Ab5	L3	Ferre	Proc. Natl. Acad. Sci. U.S.A.	88	5119–5123	1991
			cry01Ab5	L3	Ferre	Proc. Natl. Acad. Sci. U.S.A.	88	5119–5123	1991
			cry01Ab6	L3	Granero	Biochem. Biophys. Rev.	224(3)	779–783	1996
			cry01Ab6	L3	Tang	Appl. Environ. Microbiol.	62(2)	564–569	1996
			cry01Ab8		Oeda	Gene	53	113–119	1987
			cry01Ab8	L3	Nakamura	Agric. Biol. Chem.	54	715–724	1990
			cry01Ac	L3	Tabashnik	J. Invertebr. Pathol.	76	81–83	2000
			cry01Ac1	L3	Von Tersch	Appl. Environ. Microbiol.	57(2)	349–358	1991
			cry01Ac2	L3	Von Tersch	Appl. Environ. Microbiol.	57(2)	349–358	1991
			cry01Ac6	L3	Tang	Appl. Environ. Microbiol.	62(2)	564–569	1996
			cry02Aa		Tabashnik	J. Invertebr. Pathol.	76	81–83	2000
			cry02Aa2	L3	Tang	Appl. Environ. Microbiol.	62(2)	564–569	1996
			cry02Ab1	L3	Tabashnik	Appl. Environ. Microbiol.	62(8)	2839–2844	1996
			cry02Ab1	L3	Tabashnik	Appl. Environ. Microbiol.	62(8)	2839–2844	1996
			cry09Ca	L3	Tabashnik	J. Invertebr. Pathol.	76	81–83	2000
			cry09Ca1	L3	Lambert	Appl. Environ. Microbiol.	62(1)	80–86	1996
			cry09Ca1	L3	Lambert	Appl. Environ. Microbiol.	62(1)	80–86	1996
Spodoptera	frugiperda	c, z, s, t	cry01Ab	N	Luttrell	J. Econ. Entomol.	65(5)	1849–1853	1999
			cry01Ab5	N	Aranda	J. Invertebr. Pathol.	68	203–212	1996
			cry01Ac	N	Luttrell	J. Econ. Entomol.	65(5)	1849–1853	1999
			cry01Ac1	N	Aranda	J. Invertebr. Pathol.	68	203–212	1996
			cry01Ac1	N	Luo	Appl. Environ. Microbiol.	65(2)	457–464	1999
			cry02Aa	N	Sims	Southwest. Entomol.	22	395–404	1997
			cry09Ca1	N	Lambert	Appl. Environ. Microbiol.	62(1)	80–86	1996

Species	[a]	cry gene	Stage[c]	Author	Journal	Vol(issue)	Pages	Year
Trichoplusia ni	c, s, t, p	cry01Ab	N	Moar	*Appl. Environ. Microbiol.*	56(8)	2232	1990
		cry01Ab	N	Iracheta	*J. Invertebr. Pathol.*	76	70–75	2000
		cry01Ab10	N	Macintosh	*J. Invertebr. Pathol.*	56	258–266	1990
		cry01Ab2		Thorne	*J. Bacteriol.*	166(3)	801–811	1986
		cry01Ab3	L1	Lee	*J. Bacteriol.*	173(20)	6635–6638	1991
		cry01Ab6	N	Moar	*Appl. Environ. Microbiol.*	56(8)	2232	1990
		cry01Ac	N	Moar	*Appl. Environ. Microbiol.*	56(8)	2232	1990
		cry01Ac	N	Moar	*Appl. Environ. Microbiol.*	56(8)	2232	1990
		cry01Ac	N	Iracheta	*J. Invertebr. Pathol.*	76	70–75	2000
		cry01Ac1	N	Von Tersch	*Appl. Environ. Microbiol.*	57(2)	349–358	1991
		cry01Ac1	N	Ge	*Proc. Natl. Acad. Sci. U.S.A.*	86	4037–4041	1989
		cry01Ac1	N	Macintosh	*J. Invertebr. Pathol.*	56	258–266	1990
		cry01Ac1	N	Caramori	*Gene (Amsterdam)*	98(1)	37–44	1991
		cry01Ac1	L1	Ge	*J. Biol. Chem.*	266(27)	17954–1795	1991
		cry01Ac2	N	Von Tersch	*Appl. Environ. Microbiol.*	57(2)	349–358	1991
		cry01Ac6	N	Moar	*Appl. Environ. Microbiol.*	56(8)	2232	1990
		cry02Aa	N	Sims	*Southwest. Entomol.*	22	395–404	1997
		cry02Aa1	N	Dankocsik	*Mol. Microbiol.*	4(12)	2087–2094	1990
		cry02Aa2	N	Moar	*Appl. Environ. Microbiol.*	60(3)	896–902	1994
		cry02Aa2	N	Iracheta	*J. Invertebr. Pathol.*	76	70–75	2000
		cry02Ab2	N	Dankocsik	*Mol. Microbiol.*	4(12)	2087–2094	1990
		cry02Ac1	N	Wu	*FEMS*	81(1)	31–36	1991
		cry03Aa1	L1	Herrnstadt	*Bio/Technology*	4	305–308	1986

[a] c = this species feeds on corn (*Zea mays*); z = this species feeds teosinte (*Z. mexicana* or *Z. mays* spp. *mexicana*); s = this species feeds on soybean (*Glycine max*); t = this species feeds on tomato (*Lycopersium esculentum*); p = this species feeds on potato (*Solanum tuberosum*); h = this species feeds on *L. hirsutum*.

[b] Associated *cry* genes are listed according to nomenclature by Crickmore et al. 1998.

[c] Stage tested is designated as N for neonate, L1 for first instar (larva), L3 for third instar (larva), etc.

[d] Average concentration of insecticidal protein needed to cause mortality of 50% of the larvae exposed.

Source: The *Bacillus thuringiensis* Toxin Specificity Database, available at http://www.glfc.forestry.ca/Bacillus/Bt_HomePage/netintro99.htm.

future deployment of transgenic crops with resistance traits against herbivores, and the possible movement of these traits to relatives of crop plants, could shift the balance of plant traits in a particular environment toward invasiveness. However, herbivory is only one of many factors that may act to regulate plant populations. Its level of importance is likely to vary temporally, spatially, and depending on the particular plants and insects involved. Even when herbivore pressure is evident, its effects may not extend to plant reproductive output. Gange and Brown (1989) showed that size variation in eight species of plants, including grasses and both annual and perennial forbs, was greater in communities with no disruption of herbivores compared with patches where levels of herbivory were reduced with insecticide applications. Greater size variation was due to direct effects of herbivory. Whereas for some species, herbivory resulted in reduced numbers of seed heads per plant, reduced seed weight, and fewer offspring (*Trifolium pratense*) (Gange et al. 1989), fitness changes in others were not detected (including four species of annual crucifers) (Rees and Brown 1992). Even the pattern of herbivory on individual plants can determine the extent to which it affects seed set (Mauricio et al. 1993).

We are particularly interested in lepidopteran-resistant crops, the most common type of *Bt*-crop. To assess risk associated with a lepidopteran-resistance trait in plants (that is, having *less* than the normal level of caterpillar pressure), we turn the question around to ask if this subset of herbivores could be important enough to be regulating non-crop plants or wild relatives of *Bt*-crops. Specifically, we discuss the evidence that caterpillars cause significant damage to crop plants and wild plants, and assess the value of these data for evaluating the potential for lepidopterans to alter plant fitness. Three bodies of literature contribute to an assessment of herbivore damage on plant productivity, fitness, and invasiveness: agronomic studies of pest pressure on crop yields, ecological studies of herbivore pressure on non-crop plants, and studies of biological control of weeds.

3.2.4.1   *Lepidopteran herbivores and crop yield loss*

The larvae of some moths and butterflies are clearly important as agricultural pests, responsible for reducing yields on standing crops through decreases in productivity and reduction in quality, and for causing outright crop plant mortality. The agronomic literature is mostly silent about possible yield increases due to herbivores. However, cases of yield compensation (zero net loss) are common (e.g., Brook et al. 1992), cases of net yield increases are documented occasionally (e.g., lepidopterans on alfalfa: Summers 1989), and yield increases are implied in some cases of induced resistance. We consider, briefly, the notion of induced resistance and increased attraction of natural enemies to plants after initial low levels of herbivory. However, even when herbivory is beneficial as a cue for mobilizing plant defense, the phenomenon still implies an overall negative relationship between insect damage and crop yield.

Defoliation is by far the most common form of crop damage by insect pests, and Lepidoptera are the primary defoliators in agroecosystems (Barbosa 1993). Over 60% of Lepidoptera reported by Hill (1983, 1987) as major or minor pests on food crops damage crop foliage, over 20% damage flowers, seeds, and seed pods of fruits, approximately15% damage shoots and buds, and less than 10% feed primarily on roots, tubers, phloem, or sapwood (Barbosa 1993). The actual importance of caterpillars in reducing crop yield in a particular field depends, however, on an array of factors, including the type and phenology of the crop, timing (within and between seasons), soil quality, and geographical location. For example, substantial yield losses due to pests in canola in British Columbia, Canada, were from lepidopterans — diamondback moths *Plutella xylostella* (Ypnomeutidae) and Bertha armyworms (Noctuidae) — but in the State of Georgia, aphids and seed pod weevils were primarily responsible. Nevertheless, annual listings of the 20 most-damaging insect pests or pest complexes in Georgia agriculture feature three species of noctuid caterpillars, which caused $17 to $47 million in crop losses despite $30 to $65 million worth of materials applied to control them (University of Georgia 1995, 1996, 1997). Crop yield losses can correlate directly with defoliation rates. For example, a 30% decrease in total leaf area after feeding by three different species of noctuid caterpillars — green cloverworm, *Plathypena scabra* (F.); velvetbean caterpillar, *Anticarsia gemmatalis* (Hubner); and soybean looper, *Pseudoplusia includens* (Walker) — led to a 2.5 bushel/acre decrease in Alabama soybean yields (Herbert et al. 1992).

Relevant measures of the cost in yield due to lepidopteran herbivores also can be drawn from Bt-crop efficacy trials. Field experiments using isolines with and without Cry toxins show that in the presence of lepidopteran herbivores, the latter show reduced crop yields compared with the former. For example, in efficacy field trials of transgenic tomato plants expressing Bt-endotoxin proteins, lepidopterans were added to transgenic plants and to nontransgenic controls with and without topical Bt-sprays or insecticides. Plants protected from lepidopteran defoliators and fruit-feeding lepidopterans suffered little damage compared with unprotected control lines, which were defoliated and suffered losses of marketable tomatoes (Delannay 1989).

Thus, lepidopterans can and do cause substantial damage to their host plants in agroecosystems. If lepidopterans also cause substantial damage in wild relatives of crops, then they might reduce the fitness of these plants and regulate their population densities. If the damage were to be reduced via transgene transfer of resistance traits, would populations of crop wild relatives, including volunteer offspring of the crop itself, increase? It depends. The extent to which yield loss on crops can be extrapolated to reflect fitness depression in crop descendants (crop–crop or crop–wild crosses) is not clear. An overestimation of risk for wild relatives would result if pest density on the crop were much greater than peak herbivore densities on wild relatives. An overestimation of risk might also result from comparisons in which the crop plant is inherently more susceptible to damage by the same level of pest pressure than is the wild relative. On the other hand, risk could be underestimated if, in comparison to the Bt-crop, the wild relative's fitness is not constrained by domestication. That is, the relative fitness enhancement that results from backcrossing to a phenotype more like the wild relative but with lepidopteran resistance may be much greater than can be estimated by mere yield loss in a crop plant. In many cases, however, yield loss may have little to do with parameters used for fitness measures in natural plant populations, so may not be particularly helpful in assessing effects of that herbivore on wild relatives. Thus, the ability for an herbivore to inflict crop damage and yield loss is not necessarily predictive of fitness loss in wild relatives due to the same or similar herbivores, especially if crop descendants are different in habit and form. The next logical questions would consider what evidence is available that Lepidoptera can cause fitness reduction in wild plants.

3.2.4.2 *Lepidopteran herbivores and wild plant fitness*

Because lepidopterans often feed on plant leaves, and compensatory production of foliage occurs in many plants, there could be a taxonomic bias against the capacity of these herbivores for regulating plant fitness or population densities. Although lepidopterans certainly damage marketable products in agriculture, are caterpillars a good candidate for controlling populations of wild plants? In some cases, lepidopterans have been shown to reduce plant fitness, as measured by various indicators of reproductive output (Table 3.15). Caterpillar defoliation, root damage, or vascular tissue damage sometimes work in concert with other stress factors such as plant–plant competition or herbivory by other arthropods (Table 3.15).

Perhaps more systematic data are available from biological control of weeds programs, where wild plant populations have indeed been regulated and maintained at low densities

Table 3.15 Examples from Ecological and Biological Control of Weeds Studies in Which Lepidopterans Were Shown to Reduce Fitness or Invasiveness of Their Host Plants

Plant (family or common name)	Insect herbivores	Effects of herbivory	Ref.
Lupine (Fabaceae)	Defoliators	Reduced seed set and reproductive allocation	Maron 1998
Carex laxiflora (Cyperaceae)	Defoliators	Reduced seed set and reproductive allocation	Koptur 1990
Rudbeckia hirta (Compositae)	Caterpillar defoliation	Reduced seed set	Paulissen 1987
Centaurea maculosa (spotted knapweed)	Root-mining moth	Reduced bolting, number of rosettes, and seedlings (with grass competition)	Story et al. 2000
Phragmites australis (common reed)	Stem-boring moth Achanara geminipuncta	Shoot death	Tcharntke 1999
Linaria vulgaris (yellow toadflax)	Root-mining moth Eteobalea serratella	Contributed to reduction in the competitive ability of weed in peppermint crop	Volenberg et al. 1999
Mimosa pigra (Fabaceae)	Stem-boring moth Neurostrota gunniella	Negative correlation between moth densities and seed output	Lonsdale and Farrell 1998
Cardamine cordifolia (bittercress)	Specialists including Pieris rapae	25–30% leaf area damage results in significant decrease in plant growth and seed set	Louda 1984
Parthenium hysterophorus (ragweed parthenium)	Stem-boring moth Epiblema strenuana	Reduced viable seed by 74%, and with grass competition, by 98%	Navie et al. 1998
Solidago altissima (Asteraceae)	Insect herbivores	Reduced ramet height, length and biomass of rhizomes, number of rhizomes produced	Cain et al. 1991
Solidago altissima (Asteraceae)	Gall-makers	Reduced plant fitness	Abrahamson and McCrea 1986
Helianthus annuus (sunflower)	Lepidoterans	Reduced plant fitness	Pilson 2000
Trifolium pratense (Fabaceae)	Insect herbivores	Reduced number of seed heads and individual seed weight	Gange et al. 1989
Piper arietinum (Piperaceae)	Defoliators	Reduced growth, seed production, and seed viability	Marquis 1984, 1992
Senecio jacobaea (ragwort)	Cinnabar moth and ragwort flee beetle	Reduced vegetative and sexual reproduction	James et al. 1993
Brassica rapa (field mustard)	Gall-midge	Reduced fecundity	Nakamura et al. 1995
Raphanus sativus and R. raphanistrum (wild radish)	Insect herbivores	Reduced fruit and seed production in the absence of induced resistance	Agrawal 1999
R. raphanistrum (wild radish)	Pieris rapae caterpillars	Reduced pollen number and size	Lehtila and Strauss 1999
Ipomopsis aggregata (scarlet gilia)	Seed fly and caterpillar	Reduced plant fitness	Juenger and Bergelson 1998
Solanum carolinense (horse nettle)	Beetles	75% reduction in fruit production	Wise and Sacchi 1996

by herbivores (Table 3.15). Indeed, these cases provide some of the strongest evidence that herbivory plays a critical role in reducing wild plant fitness and invasiveness. Of course, these herbivores are selected specifically to control plants, so may, in the aggregate, imply greater levels of control than would be expected for the suite of lepidopterans that could be excluded from *Bt*-crops and resistant wild relatives. Selection criteria for weed control agents could perhaps play a part in predicting which are the key herbivores of interest in risk assessment. A rich body of knowledge is certainly developing as critical control points in the life cycle of weeds (e.g., juvenile survival or seed development or overwinter survival) are identified as strong determining factors for the population dynamics of the plant. In some weed species, seedlings and juveniles appear to be the most susceptible life stages to target for control. Individuals are relatively easy to kill, and significant reductions in their numbers will reduce the overall weed status of the plant (Kriticos et al. 1999). If, on the other hand, these same controls were already functioning in a population of wild plants, removal of the controlling factors through plant resistance would release the plant, resulting in higher population densities and/or increased invasiveness. Therefore, researchers must pay some attention to whether or not the lepidopteran herbivores affected by *Bt*-based resistance are indeed those species that attack the plant at control points in the life cycle — if, indeed, something is known about control points in the wild species of interest.

Caterpillars can also have indirect effects on wild plants in their habitats, and some of these effects might be relevant to risk assessments. Feeding by caterpillars can induce plant resistance against other herbivores (Agrawal 1999, Anurag et al. 1999) and/or can increase headspace volatiles, which in turn can increase predation or parasitism by natural enemies of herbivores (Geervliet et al. 1997) (Table 3.16). Caterpillar feeding can also cause unpredicted effects on other plant species of interest in the same habitat (e.g., Callaway et al. 1999), serve as a potential selective factor in the evolution of flowering phenology (Pilson 2000), determine the level of fitness depression caused by unrelated insects (Pilson 1996, Juenger and Bergelson 1998, Naber and Aarssen 1998), reduce pollinator visitation (Strauss et al. 1999), vary in its effect depending upon the timing or pattern of damage (Marquis 1992, Mauricio et al. 1993, Table 3.16), and cause increased biodiversity in the endophagous herbivore community (Tscharntke 1999).

3.2.4.3 *Manipulative experiments on fitness effects of herbivory in wild relatives*

Manipulative experiments on the effects of *Bt*-based resistance on wild relatives of crops are limited. To our knowledge, no comparative studies of fitness impacts of lepidopterans susceptible to *Bt*-endotoxins have been carried out for wild relatives of corn, maize, potato, cotton, or tomato (Appendix 3.1). Several laboratories have concentrated recently on the effects of herbivory on sunflower and canola, cases where the same species is both a crop plant and weedy plant (Stewart et al. 1997, Cummings et al. 1999). Stewart et al. (1997) developed rapeseed (*B. napus*) lines transgenic for a *Bt*-endotoxin gene to simulate an escape of the transgene from cultivation. They showed that *B. napus* plants containing the CryIAc gene, which were initially cultivated and subsequently allowed to naturalize, experienced increased fitness compared with a nontransgenic isoline that lepidopterans defoliated. Snow and colleagues (unpublished data) are investigating the effects of lepidopteran seed predators on the fitness of wild sunflower and crop–wild hybrids. They have already shown that crop genes persist in wild sunflower populations (Linder et al. 1998). However, crop–wild hybrids were at a disadvantage because of much higher rates of pre-dispersal seed predation compared with wild sunflower (Cummings et al. 1999). If *Bt*-transgenes transferred from cultivated sunflower to wild plants, these transgenes could confer resistance to specific seed predators, and possibly reverse this trend.

Table 3.16   Examples of Insect Injury That Results in Increased Attraction of Natural Enemies or Increased Pest Resistance in the Host Plant

Plant	Herbivore	Response	Ref.
Corn seedlings	Caterpillars	Volatiles released after herbivory attract *Cotesia marginiventris* parasitoids	Turlings and Tumlinson 1992
	Spodoptera exigua	Volicitin in caterpillar saliva cues volatiles that attract the parasitoid *Microplitis croceipes*	Turlings et al. 2000
Maize	Caterpillars	*Rhopalosiphum maidis* aphids repelled by plant volatiles induced by leaf damage or caterpillar regurgitant	Bernasconi et al. 1998
	Spodoptera littoralis, Ostrinia nubilalis, aphid	Volatile signals differ among herbivores both quantitatively and qualitatively, and may attract unique sets of natural enemies	Turlings et al. 1998
Cassava	Cassava mealybug *Phenacoccus herreni*	Mealybug infested leaves emit attractants for two encyrtid parasitoids	Bertschy et al. 1997
Cotton	Corn earworm, *Helicoverpa zea*	Feeding induced compounds that may serve an immediate defensive function and attract natural enemies	McCall et al. 1994
Wild mustard and wild radish	Caterpillars	Increased density of trichomes on newly formed leaves, lower subsequent herbivore damage	Agrawal 1998
Apple	European apple sawfly larvae	Caused emissions of volatiles that may attract parasitoid	Boeve et al. 1996

3.2.4.3.1   Brassicaceae — Habitat mosaics, gene flow, and fitness effects.   Some of the crops currently being tested as lepidopteran-resistant varieties, including rice, tomato, and cole crops, are particularly interesting in the context of transgene transfer possibilities and potential introgression of lepidopteran resistance traits in populations of wild relatives. Our preliminary research involves wild *Brassica*, because the central coast of California supports a regional mosaic of land use, including the nation's richest production of cole crops and abundant coastal natural reserves for biodiversity conservation. Wild relatives of these crops are common, with coastal floras that include naturalized *B. oleracea* crops, naturalized weedy relatives, and native members of the California flora (including *Raphanus raphanistrum*, *R. sativus*, *Hirschfeldia incana*, *Erucastrum gallicum*, *B. rapa*, *Sinapis arvensis*, *Cakile maritima*, and the California natives *Cakile edentula* ssp. *californica* and *Cardamine californica*). Exotic wild relatives of cole crops are prominent weeds in agricultural and natural vegetation in the region, and native relatives are common elements of coastal scrub and understory communities. Thus, any substantial change in fitness or competitiveness of these wild relatives, through gene flow from transgenic *B. oleracea* or *B. napus* crops, could potentially affect

Table 3.17 Average Numbers of Leaves, Lepidopteran Eggs or Larvae, and Flea Beetles per Plant per Plot (*n* = 4 plots with 10 to 15 plants each) for Wild Mustard (*Brassica rapa*) and Collards (*B. oleracea*) in Experimental Plots Fertilized at Different Rates at the University of California, Santa Cruz Research and Market Gardens 1986

Date	*Brassica rapa* (mustard)			*Brassica oleracea* (collards)		
	Leaves/plant	Leps	Beetles	Leaves/plant	Leps	Beetles
July 13	2	0.0	0.0	2	0.0	0.0
July 27	7	0.05	2.0	6	0.1	1.8
Aug. 3	12	2.8	3.5	9	0.3	3.0
Aug. 20	19	No data	No data	8	1.3	No data

Source: L. R. Fox and D. K. Letourneau, unpublished data.

biodiversity and ecosystem or agroecosystem function (Randall 1996). Fitness effects of herbivory and other stresses on wild plants have been shown to be extremely sensitive to field conditions, including poor soil quality and interspecific plant competition. Any effects of lepidopteran resistance would likely be very different in cultivated vs. uncultivated habitats (Bergelson 1993, Meyer 2000). Indeed, Stewart et al. (1997) showed that individual plants experienced high levels of variability in amount of herbivory when exposed to *Bt*-susceptible lepidopterans, and low survivorship of both *Bt*-protected and non-protected *B. napus* sown into plots dominated with grasses. In cultivated plots, however, medium to high levels of defoliation increased mortality of non-transgenic plants relative to *Bt*-transgenic plants and resulted in differential reproduction favoring *Bt*-plants.

3.2.4.3.2 Brassicaceae — Preliminary studies in California. Two of the authors have begun initial laboratory and field studies to test for potential plant fitness effects of *Bt*-susceptible herbivores on wild relatives of cole crops in our central coast region (Letourneau and Hagen, unpublished data). Our first question was, do susceptible herbivores feed on local wild relatives of cole crops? To record the incidence of naturally occurring herbivores on field mustard (*B. rapa*), we monitored plants in experimental plots at the University of California Santa Cruz (UCSC) campus at weekly intervals from the time of transplanting through initial seed production (site described in Letourneau et al. 1989). Common lepidopterans on field mustard in our plots included *Pieris rapae* L. (Lepidoptera: Pieridae), imported cabbage butterfly; *Plutella xylostella* L. (Lepidoptera: Plutellidae), diamondback moth; and *Trichoplusia ni* (Lepidoptera: Noctuidae), cabbage looper. The cumulative number of lepidopteran eggs per plant increased from initial transplants in mid-July to approximately three per wild mustard plant by mid-August (Table 3.17), and flea beetle *Epitrix hirtipennis* (Coleoptera: Chrysomelidae) densities reached an average of three per plant within 2 weeks after transplanting *B. rapa* seedlings. These levels were at least as high as lepidopteran egg and flea beetle densities on the related crop plant *B. oleracea* var. *acephala* (collards) (Table 3.17).

Would a reduction of normal herbivory levels increase the fitness of wild relatives of cole crops? To answer this question, we needed a way to exclude *Bt*-endotoxin-susceptible herbivores and test for an effect of this exclusion on plant fitness parameters. To simulate the kind of decrease in lepidopteran herbivory that would be expected if lepidopteran resistance trait were transferred from cole crops to their relatives, we applied topical *Bt*-sprays to treatment plants every 3 or 4 days. This method effectively reduced average herbivory levels on wild relatives when compared with control plants sprayed with water only. Such a method may not accurately mimic the effects of actual transgene transfer. Constituative production of *Bt* endotoxins in plant tissues is likely to differ in at least two ways. Compared with topical applications of *Bt*, *Bt*-plants may have relatively higher levels

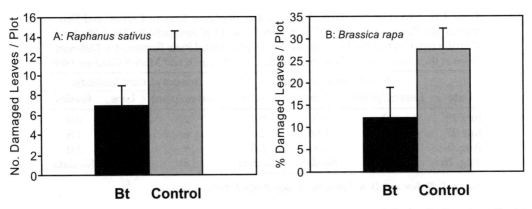

Figure 3.3   In naturally occurring *Raphanus/Brassica* stands, herbivore inhibition by *Bt*-sprays (comparison of mean damage ± SE) was detectable for both *Raphanus sativus* (A) and *Brassica rapa* (B). Compared to control plots, the mean number of leaves with lepidopteran leaf damage was significantly lower on *Bt*-treated *R. sativus* in midseason (November, 1996; Kruskal–Wallis test, p = 0.02) and later in the season, the mean percent lepidopteran leaf damage was significantly lower on *Bt*-treated *B. rapa* (December 1996; Kruskal–Wallis test, p = 0.03). For this field experiment, six replicate plots (0.5 × 0.5 m) were irrigated to allow growth of a mixed stand of *B. rapa* and *R. sativus* from the existing weed seed bank at the UCSC research farm. *Bacillus thuringiensis* topical spray treatments effective against Lepidoptera (aqueous suspension of *B. t. kurstaki* and *B. t. aizawai*) and controls (water only) were assigned in a randomized complete block design. To standardize for any low herbivore loads in plots, three individuals (first and second instars of *Pieris rapa* L.) were added to each mixed plot of *Brassica rapa* and *R. sativus* on three dates during the season.

of toxin and more widespread expression in plant parts (roots and new flowers emerging between sprays, for example). Thus, the herbivore damage reduction in our experiments could be considered a conservative amount. Nevertheless, herbivory levels on both *B. rapa* and *R. sativus* showed a significant reduction with the addition of *Bt*-endotoxin treatments on one of the two sampling dates (Figure 3.3). This experiment tested specifically for herbivory reduction with *Bt*-spray, and was not carried on past the peak flowering period. We observed, however, that the flowering phenology for *B. rapa* and *R. sativus* was shifted in treatment plants compared with controls, with the *Bt* treatment plots flowering at an earlier date. These observations are consistent with results from manipulative experiments by Agrawal et al. (1999), who showed that flowering times were delayed in response to induced resistance in wild radish after feeding by *Pieris rapae* caterpillars.

In the next series of greenhouse and field experiments, we focused on herbivory levels and seed set of black mustard *B. nigra*, another wild relative of cole crops. To measure any possible fitness effects of *Bt*-sprays themselves, we compared seed set of plants with and without *Bt*-spray applications in the absence of herbivores. Plants with *P. rapae* eggs removed (no herbivores, n = 12 plants) and sprayed with water and *Bt* produced, on average, 252 ± 30 SE seeds, and plants sprayed with water produced 225 ± 28 SE seeds per plant on average (Wilcoxin 2-sample test, $Z_{1,11} = -0.7$, $p = 0.5$). Thus, *Bt*-sprays did not themselves affect *B. nigra* seed output.

To test for potential effects of lepidopteran herbivores on plant fitness, we challenged young black mustard plants (*B. nigra*) with and without *Bt*-sprays to herbivore pressure (one or two cabbage butterfly eggs/larvae, *P. rapae*) in the laboratory for 2 weeks, and measured subsequent seed set in the greenhouse. In this experiment, juvenile *B. nigra* plants showed a significant decrease in herbivory and a significant increase in subsequent seed production when sprayed with *Bt* for only 2 weeks compared with control plants with water applications (Table 3.18).

Table 3.18 Comparative Measures of Herbivory and Seed Output of Greenhouse-Grown *Brassica nigra*[a]

Mean ± SE/plant (date)	Bt	Control	Wilcoxon 2-sample
No. *P. rapae* eggs (8/98)	1.6 ± 0.24	1.6 ± 0.24	NS
% Leaves eaten (9/98)	7 ± 3	40 ± 9	$Z_{1,16} = 3, p < 0.0085$
% Leaves eaten (10/98)	15 ± 7	25 ± 8	$Z_{1,16} = 1, p < 0.1693$
Total seeds set (12/98)	18 ± 4	8 ± 3	$Z_{1,16} = -2, p < 0.0294$

[a] *Brassica nigra* plants were caged with *Pieris rapae* for 48 h (standardized to 1–2 eggs per plant) and treated with two topical sprays per week for 2 weeks of either water (control) or lepidopteran-specific *Bt* in water (safer brand Caterpillar Killer™ containing *Bacillus thuringiensis* var. *kurstaki* (DIPEL®) with 8,800 IU of potency per mg) to simulate, in terms of lepidopteran mortality, the effects of *Bt*-gene transfer into wild *Brassica*. Values show the average number of eggs per plant at the initiation of the treatments, leaves per plant after 2 weeks, average percent of leaves per plant with caterpillar damage after 2 weeks, and average number of seeds produced per plant after 5 months.

Table 3.19 Comparative Measures of Herbivory and Seed Output of Field-grown *Brassica nigra*[a]

Mean ± SE/plant (date)	Bt	Control	Wilcoxin 2-sample
% Leaves eaten 9/00	11 ± 3	61 ± 7	$Z_{1,26} = -4, p < 0.0001$
% Leaves eaten 10/00	10 ± 2	16 ± 2	$Z_{1,26} = -3, p < 0.0094$
Total seeds set 12/00	449 ± 152	382 ± 80	$Z_{1,26} = -0.3, p < 0.7477$

[a] *Brassica nigra* plants were transplanted as seedlings to field plots (two plants per plot). One *Pieris rapae* larva was added, and plants were either sprayed with water (control) or with *Bt* and water (treatment). Values show the average percent of leaves per plant with caterpillar damage after 2 weeks and average number of seeds produced per plant after 5 months.

A field trial to replicate the laboratory experiment with *B. nigra* showed a significant reduction of herbivory on young plants sprayed with *Bt* for 6 weeks after transplanting as compared with controls. The addition of one first instar *P. rapae* larva per plant (plus rather low, ~1 egg/plant, natural levels of oviposition in the field) resulted in significantly lower herbivory after 2 weeks to plants sprayed with *Bt*-solution as compared to plants with water sprays (Table 3.19). The proportion of *B. nigra* leaves with herbivore damage on control plants in the field experiment was comparable to laboratory levels on control plants. After 18 weeks (12 weeks with no treatments of *Bt*), however, no significant treatment effect due to early differences in herbivory was manifested in seed output (Table 3.19). Thus, *B. nigra* control plants in the field experiment may have compensated for the increase in herbivory. The experiment was conducted in fertile soil and with low plant competition (regular spacing of test plants into bare soil). These results suggest, like those of Stewart et al. (1997) for *B. napus*, that in resource-rich conditions, wild *Brassica* can be plastic and may not show an increase in fitness when herbivory is reduced.

In areas where gene introgression from *Bt*-broccoli and *Bt*-cabbage into wild and weedy species can occur, reliable assessments of risk are difficult at present because the natural levels of herbivory and herbivore composition of only a few wild species is known. Even less information is currently available on normal lifetime levels of plant tissue damage by herbivores and the relationship of herbivores to individual plant fitness. Demographically, little is known about the role of *Bt*-susceptible herbivores in limiting populations of wild *Brassica* relatives, and critical control points (*sensu* Kriticos et al. 1999) are unknown in these plants. Clearly, wild relatives of cole crops experience herbivory under field conditions (Letourneau and Hagen, preliminary data), herbivory can affect plant fitness (Strauss et al. 1996), Cry toxins in wild *Brassica* can enhance fitness (Stewart et al. 1997), but fitness effects may vary as field conditions influence the level of compensatory plant responses to herbivore pressure (Bergelson 1994, Meyer 2000).

3.2.5   Foci for assessment: Plant species, individuals, or populations

A multilevel approach can make risk assessment more rigorous in terms of predicting the likelihood and severity of weediness and invasiveness of plants with *Bt*-transgenes. Different indications might come from a consideration of species characteristics, tests of individual fitness parameters, and demographic studies (in the field or using models). Commonly, however, only species characteristics are used as criteria for risk assessment. Ecological characteristics of the plant are used in invasion theory to predict which species will become major competitors (Baker 1965, 1974, Mack 1996), and habitat characteristics such as current species compositions (Lodge 1993a,b, Tilman 1997) or disturbance regimes (Ramakrishnan and Vitousek 1989, Huenneke et al. 1990) are used to estimate the likelihood of invasion.

In most cases, plant species criteria are based on Baker's (1965, 1974) list of discrete traits proposed to account for "weediness." The crop or wild relatives with which it can interbreed are evaluated for the presence of weediness traits including, among others, high overall seed production, plasticity of response, high dispersal capacity, vegetative reproduction, and rapid growth capacity. Indeed, Baker's (1965, 1974) listing of traits for an "ideal weed" has been helpful heuristically in science-based advice on risk assessment for the introduction of genetically engineered plants into the environment (Tiedje et al. 1989) and has been foundational in biosafety determinations by the USDA–APHIS. Unfortunately, a reliance on these simple criteria, especially in the absence of comparative measurements of plant performance, are no longer acceptable in all but the most straightforward cases. The reliability of using the scarceness of weediness characteristics to argue for biosafety was weakened by studies like those of Perrins et al. (1992), who failed to find any set of traits that could reliably distinguish weedy from non-weedy plant species. Thus, a distinct set of characteristics may allow accurate predictions in certain circumstances (e.g., Baker 1965, Rejmánek and Richardson 1996), but external abiotic and biotic factors can influence the success of an invader and its new associated plant species enough to obscure any predictability with respect to future invasions. Therefore, one of the most-promising methods for predicting future potential invaders is direct comparison of closely related plants (e.g., native vs. introduced congenerics) with respect to fitness parameters and demographic patterns in "common garden" experiments (Mack 1996).

Comparative studies of individual plant fitness provide another level of scrutiny for assessing the likelihood of increased weediness due to transgene transfer. Traditional measures of fecundity provide estimates of potential invasiveness, with limiting factors other than seed set determining fitness when present (Bergelson 1994). Especially if sown under field conditions, a significant increase in plant fitness parameters such as survival and reproductive output, would suggest that Cry toxins in wild relatives of crop plants pose a risk of increased weediness or invasiveness. Comparative data on individual plant performance are more difficult to obtain than the current tallying of weediness traits, but are still tractable. However, individual plant fitness measures do not always reflect population growth or range expansion. For example, even though high densities of a stem-boring moth were associated with a decrease in weed fitness, Lonsdale and Farrell (1998) found no reduction in the weediness of *Mimosa pigra*, its host plant over time (see Table 3.15). The dynamics of transgenic escape and conferred lepidopteran resistance can only be measured with long-term demographic studies of plant populations in a variety of locations (e.g., Crawley et al. 1993, Linder and Schmitt 1995, Hails et al. 1997) or estimated with demographic models. Well-designed experiments on individual fitness can provide many of the data necessary for constructing predictive models (Parker and Kareiva 1996).

Deterministic analyses of measurements made throughout the life cycle can be used to calculate differences in average population growth caused by conferred resistance to herbivores. Sensitivities and elasticities of population growth rates to different demographic traits affected by *Bt* can then be estimated. The contribution of *Bt* effects on each stage of the life cycle to the overall increase in population growth, if any, can be teased apart using Life Table Response Analysis (Kareiva et al. 1996). Finally, stochastic demographic analyses based on the field data can be used to determine whether *Bt* effects on population growth are significant in the face of environmental variation (Caswell 1989, Doak and Mills 1994) and in particular whether the effect of *Bt* can convert stable or decreasing population trajectories into increasing ones, thereby leading to significant invasiveness.

Clearly, many factors interact to determine the success of a novel plant at a given site and time of introduction, including abiotic factors and the abundance and identities of mutualists, competitors, herbivores, etc. (Crawley et al. 1996). Indirect effects of herbivory or the lack of herbivory on wild plants are also complex. For example, Callaway et al. (1999) showed how a biological control agent (knapweed root moth) introduced to reduce the invasiveness of a weed and encourage the growth of native species actually caused a reduction in the native plant's fitness via root exudate–mediated effects triggered by the lepidopteran. As more traits are added to crop plants, risk assessment of the consequences of gene flow will be more complex. However, predicting the potential invasiveness of plant species in different habitats is extremely important, given the fact that invasions can lead to a reduction in biological diversity and threaten ecosystem integrity (Daehler and Strong 1996).

3.3 Knowledge gaps in environmental impact assessment of Bt-transgenes

Hoffman (1990) used the incorporation of *Bt*-endotoxins in crop plants as an example to pose a series of ecological questions relevant to ecosystem integrity and biodiversity. She speculated about plant community effects of increased survival, reproduction, or invasiveness of wild crop relatives, the consequences of increased competitive ability on plant biodiversity, and long-distance effects of susceptible pollinators whose populations have been reduced by toxic host plants. A decade later, the speculations are still valid, the complexity of insect–plant and plant–plant interactions is more appreciated, and we have a very weak scientific database on which to base regulatory decisions. Despite the expanded cultivation of *Bt*-crops in the United States and worldwide, very basic ecological questions remain unanswered, creating a scientific vacuum instead of a scientific foundation for the assessment of environmental effects. If barriers to the incorporation and expression of the trait (and its promoters) do not exist, then given enough time and expansion of *Bt*-crops, *cry* genes will likely enter wild plant populations. The overwhelming conclusion of recent studies designed to analyze containment strategies for small-scale field tests with flowering transgenic or otherwise detectable crops is that gene flow will occur if there are wild relatives in the region that can hybridize with the crop (Kareiva et al. 1994, Klinger and Ellstrand 1994). Kareiva et al. (1994) urged that ecologists begin to fill information gaps about the consequences of such gene flow to wild gene pools in order to have the tools to make regulatory decisions and to mitigate harmful consequences. These information gaps include basic questions about the forces that control the distribution, abundance, and diversity of wild plant species in terrestrial communities (e.g., McNaughton 1986, Doak 1991, Price 1992, Letourneau and Dyer 1997), the influence of *Bt*-based toxins on naturally occurring Lepidoptera (Smith and Couche 1991), and the actual role of particular traits to affect substantially these forces in natural and agricultural habitats.

Theoretically, a herbivore resistance trait could indeed increase fitness of weeds in natural populations and its invasiveness in the community. Yet, we are faced with scientific controversies about the role of consumer dynamics as forces controlling plant population levels and community structure. For predicting the outcome of a particular crop, manipulative experiments become necessary but are not available (Fretwell 1977, Strong 1984, 1992, McNaughton 1986, Turkington et al. 1993, Kareiva et al. 1994, Royal Society of Canada 2000). As shown in our chapter, current predictions about the behavior of non-target lepidopterans on crop wild relatives must be made in the absence of any comprehensive species list of these herbivores. If none of these species is susceptible to *Bt*-based endotoxins, or none is currently regulating populations of their host plant, then increased weediness is not likely. If susceptible non-target herbivores are not rare and endangered, or being relied upon as genetic stock in refugia against resistance development, or acting as important pollinators in the ecosystem, then the environmental price for transgene tranfer from commercialized *Bt*-crops to their wild relatives will be insignificant. But we are not much better at answering these questions now than we were when they were first posed.

Acknowledgments

We are grateful to Pedro Barbosa, Lee Dyer, Norm Ellstrand, Roslyn McKendry, Tom Miller, John Sulzycki, and Fequiere Vilsaint for helpful comments on initial drafts. Cathy Carlson, Rachel O'Malley, and Angie Shelton assisted with laboratory, fieldwork, and supervision of undergraduate student crews. We thank Cheryl VanDeVeer, University of California, Santa Cruz, Document Publishing and Editing Center, for excellent technical improvements; Angie Shelton for providing early instar larvae for field experiments; and Carl Nystrom for compiling lepidopteran susceptibility test information for the species in question.

Literature cited

Abbott, R. J., Ecological risks of transgenic crops, *Trends Ecol. Evol.*, 9, 280, 1994.

Abrahamson, W. G. and McCrea, K. D., Nutrient and biomass allocation in *Solidago altissima*: effects of two stem gallmakers, fertilization, and ramet isolation, *Oecologia* (Berlin), 68, 174, 1986.

Acciarri, N., Vitelli, G., Arpaia, S., Mennella, G., Sunseri, F., and Rotino, G. L., Transgenic resistance to the Colorado potato beetle in *Bt*-expressing eggplant fields, *HortScience*, 35, 722, 2000.

Agrawal, A. A., Induced responses to herbivory in wild radish: effects on several herbivores and plant fitness, *Ecology* (Washington, D.C.), 80, 1713, 1999.

Agrawal, A. A., Strauss, S. Y., and Stout, M. J., Costs of induced responses and tolerance to herbivory in male and female fitness components of wild radish, *Evolution*, 53, 1093, 1999.

Archer, T. L., Schuster, G., Patrick, C., Cronholm, G., Bynum, E. D., and Morrison, W. P., Whorl and stalk damage by European and Southwestern corn borers to four events of *Bacillus thuringiensis* transgenic maize, *Crop Prot.*, 19, 181, 2000.

Ashfaq, M., Young, S. Y., and McNew, R. W., Development of *Spodoptera exigua* and *Helicoverpa zea* (Lepidoptera: Noctuidae) on transgenic cotton containing CrylAc insecticidal protein, *J. Entomol. Sci.*, 35, 360, 2000.

Bachelor, J. S., Bollgard cotton performance expectations for North Carolina producers, *Carol. Cotton Notes*, 99-5-A, May 1999.

Bachelor, J. S., Bollgard cotton performance expectations for North Carolina producers, *Carol. Cotton Notes*, CCN-00-3D, March 23, 2000.

Baker, H. G., Characteristics and modes of origin of weeds, in *Genetics of Colonizing Species*, Baker, H. G. and Stebbins, G. L., Eds., Academic Press, New York, 1965, 147.

Baker, H. G., The evolution of weeds, *Annu. Rev. Ecol. Syst.*, 5, 1, 1974.

Barbosa, P., Lepidopteran foraging on plants in agroecosystems: constraints and consequences, in *Ecological and Evolutionary Constraints on Foraging Caterpillars*, Stamp, N. E. and Casey, T. M., Eds., Chapman & Hall, New York, 1993, 29.

Bergelson, J., Changes in fecundity do not predict invasiveness: a model study of transgenic plants, *Ecology*, 75, 249, 1994.

Bernasconi, M. L., Turlings, T. C. J., Ambrosetti, L., Bassetti, P., and Dorn, S., Herbivore-induced emissions of maize volatiles repel the corn leaf aphid, *Rhopalosiphum maidis*, *Entomol. Exp. Appl.*, 87, 133, 1998.

Bertschy, C., Turlings, T. C. J., Bellotti, A. C., and Dorn, S., Chemically-mediated attraction of three parasitoid species to mealybug-infested cassava leaves, *Fla. Entomol.*, 80, 383, 1997.

Betz, F. S., Hammond, B. G., and Fuchs, R. L., Safety and advantages of *Bacillus thuringiensis*-protected plants to control insect pests, *Regul. Toxicol. Pharmacol.*, 32, 156, 2000.

Bing, D. J., Downey, R. K., and Rakow, G. F. W., Hybridizations among *Brassica napus*, *B. rapa* and *B. juncea* and their two weedy relatives *B. nigra* and *Sinapis arvensis* under open pollination conditions in the field, *Plant Breed.*, 115, 470, 1996.

Brunoehler, R., Bt-hybrids score high for performance, *Soybean Dig.*, Nov. 1, 1998.

Boeve, J.-L., Lengwiler, U., Tollsten, L., Dorn, S., and Turlings, T. C. J., Volatiles emitted by apple fruitlets infested by larvae of the European apple sawfly, *Phytochemistry* (Oxford), 42, 373, 1996.

Bourguet, D., Bethenod, M. T., Trouve, C., and Frederique, V., Host-plant diversity of the European corn borer *Ostrinia nubilalis*: What value for sustainable transgenic insecticidal Bt maize? *Proc. R. Soc. Biol. Sci. Ser. B*, 267, 1177, 2000.

Bryant, K. J., Robertson, W. C., and Lorenz III, G. M., Economic evaluation of Bollgard cotton in Arkansas, in *Proceedings of Beltwide Cotton Conferences*, Cotton Economics and Marketing Conference, National Cotton Council of America, Memphis, TN, 1998.

Cain, M. L., Carson, W. P., and Root, R. B. Long-term suppression of insect herbivores increases the production and growth of *Solidago altissima* rhizomes, *Oecologia*, 88, 251, 1991.

Callaway, R. M., DeLuca, T. H., and Belliveau, W. M., Biological-control herbivores may increase competitive ability of the noxious weed *Centaurea maculosa*, *Ecology*, 80, 1196, 1999.

Carpenter, J. E., Case Studies in Benefits and Risks of Agricultural Biotechnology: Roundup Ready Soybeans and Bt Field Corn, National Center for Food and Agricultural Policy, Washington, D.C., 2001, available at http://www.ncfap.org/pup/biotech/benefitsandrisks.pdf.

Carpenter, J. E. and Gianessi, L. P., Agricultural Biotechnology: Updated Benefit Estimates, National Center for Food and Agricultural Policy, Washington, D.C., 2001, available at http://www.ncfap.org/pup/biotech/updatedbenefits.pdf.

Caswell, H., *Matrix Population Models*, Sinauer Associates, Sunderland, MA, 1989.

Colwell, R. K., Norse, E. A., Pimentel, D., Sharples, F. E., and Simberloff, D., Genetic engineering in agriculture, *Science*, 229, 111, 1985.

Crawley, M. J., *Herbivory, the Dynamics of Animal-Plant Interactions*, Studies in Ecology, vol. 10, University of California Press, Berkeley, 1983, 437 pp.

Crawley, M. J., Hails, R. S., Rees, M., Kohn, D., and Buxton, J., Ecology of transgenic oilseed rape in natural habitats, *Nature* (London), 363, 620, 1993.

Crawley, M. J., Harvey, P. H., and Purvis, A., Comparative ecology of the native and alien floras of the British Isles, *Philos. Trans. R. Soc. London B Biol. Sci.*, 351, 1251, 1996.

Crickmore, N., Zeigler, D. R., Feitelson, J., Schnept, E., Van Rie, J., Lereclus, D., Baum, J., and Dean, D. H., Revision of the nomenclature for the *Bacillus thuringiensis* pestical crystal proteins, *Microbiol. Mol. Biol. Rev.*, 62, 807, 1998.

Cummings, C. L., Alexander, H. M., and Snow, A. A., Increased pre-dispersal seed predation in sunflower crop-wild hybrids, *Oecologia* (Berlin), 121, 330, 1999.

Daehler, C. C. and Strong, D. R., Status, prediction and prevention of introduced cordgrass *Spartina* spp. invasions in Pacific estuaries, USA, *Biol. Conserv.*, 78, 51, 1996.

Darmency, H., The impact of hybrids between genetically modified crop plants and their related species: introgression and weediness, *Mol. Ecol.*, 3, 37, 1994.

De Cosa, B., Moar, W., Lee, S.-B., Miller, M., and Daniell, H., Overexpression of the Bt cry2Aa2 operon in chloroplasts leads to formation of insecticidal crystals, *Nat. Biotechnol.*, 19, 71, 2001.

DeJoode, D. R. and Wendel, F. F., Genetic diversity and origin of the Hawaiian island cotton, *Gossypium tomentosum*, *Am. J. Bot.*, 79, 1311, 1992.

de la Fuente, J. M., Ramirez-Rodriguez, V., Cabrera-Ponce, J. L., and Herrera-Estrella, L., Aluminum tolerance in transgenic plants by alteration of citrate synthesis, *Science*, 276, 1566, 1997.

Delannay, X., La Vallee, B. J., Proksch, R. K., Fuchs, R. L., Sims, S. R. et al., Field performance of transgenic tomato plants expressing the *Bacillus thuringiensis* var. *kurstaki* insect control protein, *Biotechnology*, 7, 61, 1989.

Doak, D. F., The consequences of herbivory for dwarf fireweed: different time scales, different morphological scales, *Ecology*, 72, 1397, 1991.

Doak, D. F. and Mills, L. S., A useful role for theory in conservation, *Ecology* (Tempe), 75, 615, 1994.

Dornbos, D., *Bt* Value Story, Novartis Seeds, Inc., Golden Valley, MN, 1998.

Drinkwater, L. E., Letourneau, D. K., Workneh, F., van Bruggen, A. H. C., and Shennan, C., Fundamental differences between conventional and organic tomato agroecosystems in California, *Ecol. Appl.*, 5, 1098, 1995.

Dyer, L. A. and Letourneau, D. K., Trophic cascades in a complex terrestrial community, *Proc. Natl. Acad. Sci. U.S.A.*, 96, 5072, 1999.

Ellstrand, N. C., Pollen as a vehicle for the escape of engineered genes, *Trends Biotechnol.*, 6, S30, 1988.

Ellstrand, N. C., Prentice, H. C., and Hancock, J. F., Gene flow and introgression from domesticated plants into their wild relatives, *Annu. Rev. Ecol. Syst.*, 30, 539, 1999.

ERS–USDA (U.S. Economic Research Service), Genetically engineered crops for pest management, ERS-USDA, Washington, D.C., 1999, available at http://www.econ.ag.gov/whatsnew/issues/biotech/index.htm.

ERS–USDA (U.S. Economic Research Service), Impacts of Adopting Genetically Engineered Crops in the U.S. — Preliminary Results, 2000, available at http://www.econ.ag.gov/whatsnew/ issues/gmo/.

Falck-Zepeda, J. B., Traxler, G., and Nelson, R. G., Rent creation and distribution from the first three years of planting *Bt* cotton, Briefs No. 14, Report, International Service for the Acquisition of Agri-biotech Applications (ISAAA), Ithaca, NY, 1999.

Farnham, D. E. and Pilcher, C., *Bt*-corn hybrid evaluation: year 2, *Iowa State Univ. Integrated Crop Manage. Newsl.*, December 7, 1998.

Fawcett, R. S., Analysis of the YieldGard corn technology, Monsanto Company, 01 August 1998

Fernandez-Cornejo, J. and McBride, W., Genetically engineered crops for pest management in U.S. agriculture: farm-level effects, AER-786, Economic Research Service, U.S. Department of Agriculture, April, 2000, available at http://www.ers.usda.gov/epubs/pdf/aer786/.

Fitton, M. G., Shaw, M. R., and Gauld, I. D., Pimpline ichneumon-flies, Hymenoptera, Ichneumonidae (Pimplinae), *Handbooks for the Identification of British Insects*, 7, 1, 1988.

Fretwell, S. D., The regulation of plant communities by food chains exploiting them, *Perspect. Biol. Med.*, 20, 169, 1977.

Fryxell, P. A., *The Natural History of the Cotton Tribe* (*Malvaceae, Tribe Gossypieae*), Texas A&M University Press, College Station, 1979, 245.

Gange, A. C. and Brown, V. K., Insect herbivory affects size variability in plant populations, *Oikos*, 56, 351, 1989.

Gange, A. C., Brown, V. K., Evans, I. M., and Storr, A. L., Variation in the impact of insect herbivory on *Trifolium pratense* through early plant succession, *J. Ecol.*, 77, 537, 1989.

Geervliet, J. B. F., Posthumus, M. A., Vet, L. E. M., and Dicke, M., Comparative analysis of headspace volatiles from different caterpillar-infested or uninfested food plants of *Pieris* species, *J. Chem. Ecol.*, 23, 2935, 1997.

Gianessi, L. P. and Carpenter, J. E., Agricultural biotechnology: insect control benefits, National Center for Food and Agricultural Policy, Washington, D.C., July 1999, available at http://www.ncfap.org/pup/biotech/insectcontrolbenefits.pdf.

Gibson, J. W., IV, Laughlin, D., Luttrell, R. G., Parker, D., Reed, J., and Harris, A., Comparison of costs and returns associated with *Heliothis* resistant *Bt* cotton to non-resistant varieties, *Proceedings of Beltwide Cotton Conference*, Cotton Economics and Marketing Conference, 1, 244, 1997.

Goodman, R. M. and Newell, N., Genetic engineering of plants for herbicide resistance: status and prospects, in *Engineered Organisms in the Environment: Scientific Issues*, Halvorson, H. O., Pramer, D., and Rogul, M., Eds., American Society for Microbiology, Washington, D.C., 1985, 47.

Goto, F., Yoshihara, T., Shigemoto, N., Toki, S., and Takaiwa, F., Iron fortification of rice seed by the soybean ferritin gene, *Nat. Biotechnol.*, 17, 282, 1999.

Gould, F., Evolutionary biology and genetically engineered crops, *BioScience*, 38, 26, 1988.

Hails, R. S., Genetically modified plants — the debate continues, *TREE*, 15, 14, 2000.

Hails, R. S., Rees, M., Kohn, D. D., and Crawley, M. J., Burial and seed survival in *Brassica napus* subsp. oleifera and *Sinapis arvensis* including a comparison of transgenic and non-transgenic lines of the crop, *Proc. R. Soc. London Ser. B Biol. Sci.*, 264, 1, 1997.

Harper, J. L., The role of predation in vegetational diversity, presented at Brookhaven Symposium on Diversity and Stability in Ecological Systems, Brookhaven National Laboratory, Brookhaven, NY, 1969.

Hauser, T. P., Jorgensen, R. B., and Ostergard, H., Fitness of backcross and F2 hybrids between weedy *Brassica rapa* and oilseed rape (*B. napus*), *Heredity,* 81, 436, 1998b.

Hauser, T. P., Shaw, R. G., and Østergård, H., Fitness of F1 hybrids between weedy *Brassica rapa* and oilseed rape (*B. napus*), *Heredity,* 81, 429, 1998b.

Herbert, D. A., Jr., Mack, T. P., Backman, P. A., and Rodriguez-Kabana, R., Validation of a model for estimating leaf-feeding by insects in soybean, *Crop Prot.,* 11, 27, 1992.

Hill, D. S., *Agricultural Insect Pests of the Tropics and Their Control,* 2nd ed., Cambridge University Press, Cambridge, U.K., 1983.

Hill, D. S., *Agricultural Insect Pests of Temperate Regions and Their Control,* Cambridge University Press, Cambridge, U.K., 1987.

Hoffman, C. A., Ecological risks of genetic engineering of crop plants: scientific and social analyses are critical to realize benefits of the new techniques, *BioScience,* 40, 434, 1990.

Höfte, H. and Whiteley, H. R., Insecticidal crystal proteins in *Bacillus thuringiensis, Microbiol. Rev.,* 53, 242, 1989.

Houseman, J. G., Campos, F., Thie, N. M. R., Philogene, B. J. R., Atkinson, J., Morand, P., and Arnason, J. T., Effects of the maize-derived compounds DIMBOA and MBOA on growth and digestive processes of European corn borer (Lepidoptera: Pyralidae), *J. Econ. Entomol.,* 85, 669, 1992.

Hoy, C. W., Feldman, J., Gould, F., Kennedy, G. G., Reed, G., and Wyman, J. A., Naturally occurring biological control in genetically engineered crops, in *Conservation Biological Control,* Barbosa, P., Ed., Academic Press, New York, 1998, 1985.

Huenneke, L. F., Hamburg, S. P., Koide, R., Mooney, H. A., and Vitousek, P. M., Effects of soil resources on plant invasion and community structure in Californian serpentine grassland, *Ecology,* 71, 478, 1990.

James, C., Global Status of Transgenic Crops in 1997, ISAAA Briefs No. 5, International Service for the Acquisition of Agri-biotech Applications, Ithaca, NY, 1997, available at http://www.isaaa.org/publications/briefs/Brief_5.htm.

James, C., Global Review of Commercialized Transgenic Crops: 1998, ISAAA Briefs No. 8, International Service for the Acquisition of Agri-biotech Applications, Ithaca, NY, 1998, available at http://www.isaaa.org/publications/briefs/Brief_8.htm.

James, C., Global Review of Commercialized Transgenic Crops: 1999, ISAAA Briefs No. 12: Preview, International Service for the Acquisition of Agri-biotech Applications, Ithaca, NY, 1999, available at http://www.isaaa.org/publications/briefs/Brief_12.htm.

James, C., Global Status of Commercialized Transgenic Crops: 2000, ISAAA Briefs No. 21: Preview, International Service for the Acquisition of Agri-biotech Applications, Ithaca, NY, 2000, available at http://www.isaaa.org/publications/briefs/Brief_21.htm.

James, C. and Krattiger, A. F., Global Review of Field Testing and Commercialization of Transgenic Plants, 1986–1995: The First Decade of Crop Biotechnology, ISAAA Briefs No. 1, International Service for the Acquisition of Agri-biotech Applications, Ithaca, NY, 1996, available at http://www.isaa.org/publications/briefs/Brief_1.htm.

James, R. R., Miller, J. C., and Lighthart, B., *Bacillus thuringiensis* var. *kurstaki* affects a beneficial insect, the cinnabar moth (Lepidoptera: Arctiidae), *J. Econ. Entomol.,* 86, 334, 1993.

Jørgensen, R. B., Andersen, B., Landbo, L., Mikkelsen, T. R., Dias, J. S., Crute, I., and Monteiro, A. A., Spontaneous hybridization between oilseed rape (*Brassica napus*) and weedy relatives, *Acta Hortic.,* 407, 193, 1996.

Juenger, T. and Bergelson, J., Pairwise versus diffuse natural selection and the multiple herbivores of scarlet gilia, *Ipomopsis aggregata, Evolution,* 52, 1583, 1998.

Kareiva, P., Morris, W., and Jacobi, C. M., Studying and managing the risks of cross-fertilization between transgenic crops and wild relatives, *Mol. Ecol.,* 3, 15, 1994.

Kareiva, P., Parker, I. M., and Pascual, M., Can we use experiments and models in predicting the invasiveness of genetically engineered organisms, *Ecology* (Washington, D.C.), 77, 1670, 1996.

Klinger, T. and Ellstrand, N. C., Engineered genes in wild populations — fitness of weed-crop hybrids of *Raphanus sativus, Ecol. Appl.,* 4, 117, 1994.

Koptur, S., Early season defoliation can affect *Carex laxiflora* Cyperaceae seed set in same year or reproductive allocation in subsequent year, *Bull. Ecol. Soc. Am.,* 71(2 Suppl.), 217, 1990.

Krattiger, A. F., Insect Resistance in Crops: A Case Study of *Bacillus thuringiensis* (*Bt*) and Its Transfer to Developing Countries, ISAAA Briefs No. 2, International Service for the Acquisition of Agri-biotech Applications, Ithaca, NY, 1997, available at http://www.isaaa.org/publications/briefs/Brief_2.htm.

Kriticos, D., Brown, J., Radford, I., and Nicholas, M., Plant population ecology and biological control: *Acacia nilotica* as a case study, *Biol. Control*, 16, 230, 1999.

Lacey, L. A., Horton, D. R., Chauvin, R. L., and Stocker, J. M., Comparative efficacy of *Beauveria bassiana*, *Bacillus thuringiensis*, and aldicarb for control of Colorado potato beetle in an irrigated desert agroecosystem and their effects on biodiversity, *Entomol. Exp. Appl.*, 93, 189, 1999.

Lehtila, K. and Strauss, S. Y., Effects of foliar herbivory on male and female reproductive traits of wild radish, *Raphanus raphanistrum*, *Ecology* (Washington, D.C.), 80, 116, 1999.

Leroy, T., Henry, A. M., Royer, M., Altosaar, I., Frutos, R., Duris, D., and Philippe, R., Genetically modified coffee plants expressing the *Bacillus thuringiensis* cry1Ac gene for resistance to leaf miner, *Plant Cell Rep.*, 19, 382, 2000.

Letourneau, D. K. and Dyer, L. A., Density patterns of Piper ant-plants and associated arthropods: top-predator trophic cascades in a terrestrial system? *Biotropica*, 30, 162, 1998.

Letourneau, D. K. and Fox, L. R., Effects of experimental design and nitrogen on cabbage butterfly oviposition, *Oecologia*, 80, 211, 1989.

Linder, C. R. and Schmitt, J., Potential persistence of escaped transgenes: performance of transgenic, oil-modified *Brassica* seeds and seedlings, *Ecol. Appl.*, 5, 1056, 1995.

Linder, C. R., Taha, I., Seiler, G. J., Snow, A. A., and Rieseberg, L. H., Long-term introgression of crop genes into wild sunflower populations, *Theor. Appl. Genet.*, 96, 339, 1998.

Llewellyn, D., Cousins, Y., Mathews, A., Hartweck, L., and Lyon, B., Expression of *Bacillus thuringiensis* insecticidal protein genes in transgenic crop plants, *Agric. Ecosyst. Environ.*, 49, 85, 1994.

Lodge, D. M., Biological invasions — lessons for ecology, *Trends Ecol. Evol.*, 8, 133, 1993a.

Lodge, D. M., Prediction and biological invasions — reply, *Trends Ecol. Evol.*, 8, 380, 1993b.

Lonsdale, W. M. and Farrell, G. S., Testing the effects on *Mimosa pigra* of a biological control agent *Neurostrata gunniella* (Lepidoptera: Gracillariidae), plant competition and fungi under field conditions, *Biocontrol Sci. Technol.*, 8, 485, 1998.

Louda, S. M., 1984, Herbivore effect on stature, fruiting, and leaf dynamics of a native crucifer, *Ecology*, 65, 1379, 1984.

Lozzia, G. C., Biodiversity and structure of ground beetle assemblages (*Coleoptera Carabidae*) in *Bt* corn and its effects on non-target insects, *Boll. Zool. Agrar. Bachic.*, 31, 37, 1999.

Macintosh, S. C., Stone, T. B., Sims, S. R., Hunst, P. L., Greenplate, J. T., Marrone, P. G., Perlak, J., Fischhoff, D. A., and Fuchs, R. L., Specificity and efficacy of purified *Bacillus thuringiensis* proteins against agronomically important insects, *J. Invertebr. Pathol.*, 56, 258, 1990

Mack, R. N., Predicting the identity and fate of plant invaders — emergent and emerging approaches, *Biol. Conserv.*, 78, 107, 1996.

Malvar, T. and Gilmer, A. J., Hybrid *Bacillus thuringiensis* delta-endotoxins with novel broad-spectrum insecticidal activity, Official Gazette of the U.S. Patent and Trademark Office Patents, Washington, D.C., Jan. 25, 2000.

Maron, J. L., Insect herbivory above- and belowground: individual and joint effects on plant fitness, *Ecology*, 79, 1281, 1998.

Marquis, R. J., Leaf herbivores decrease fitness of a tropical plant, *Science* (Washington, D.C.), 226, 537, 1984.

Marquis, R. J., A bite is a bite is a bite? Constraints on response to folivory in *Piper arietinum* (Piperaceae), *Ecology* (Tempe), 73, 143, 1992.

Marra, M. G. and Hubbell, B., Economic impacts of the first crop biotechnologies, 1998, available at http://www.ag-econ.ncsu.edu/faculty/marra/firstcrop/img001.gif.

Mauricio, R., Bowers, M. D., and Bazzaz, F. A., Pattern of leaf damage affects fitness of the annual plant *Raphanus sativus* (Brassicaceae), *Ecology* (Washington, D.C.), 74, 2066, 1993.

May, R., Genetically modified foods: facts, worries, policies and public confidence, Office of Science and Technology, Great Britain, 1999, available at http://www.dti.gov.uk/ost/ostbusiness/index.htm.

McCall, P. J., Turlings, T. C. J., Loughrin, J., Proveaux, A. T., and Tumlinson, J. H., Herbivore-induced volatile emissions from cotton (*Gossypium hirsutum* L.) seedlings, *J. Chem. Ecol.*, 20, 3039, 1994.

McNaughton, S. J., On plants and herbivores, *Am. Nat.*, 128, 765, 1986.

Metz, P. L. J., Jacobsen, E., Nap, J. P., Pereira, A., and Stiekema, W. J., The impact on biosafety of the phosphinothricin-tolerance transgene in inter-specific *B. rapa* × *B. napus* hybrids and their successive backcrosses, *Theor. Appl. Genet.*, 95, 442, 1997.

Metz, T. D., Dixit, R., and Earle, E. D., Agrobacterium tumefaciens-mediated transformation of broccoli (*Brassica oleracea* var *italica*) and cabbage (*B. oleracea* var. *capitata*), *Plant Cell Rep.*, 15, 287, 1995.

Meyer, G. A., Interactive effects of soil fertility and herbivory on *Brassica nigra*, *Oikos*, 88, 433, 2000.

Mikkelsen, T. R., Andersen, B., and Jorgensen, R. B., The risk of crop transgene spread, *Nature* (*London*), 380, 31, 1996.

Monsanto, 2001, available at http://www.NatureMark.com.

Monsanto, YieldGard: the whole plant the whole season, 1997 *Bt*-yield studies, Monsanto 1/15/98, 1999.

Moran, M. D. and Hurd, L. E., A trophic cascade in a diverse arthropod community caused by a generalist arthropod predator, *Oecologia* (Berlin), 113, 126, 1998.

Mullins, J. W. and Mills, J. M., Economics of Bollgard versus non-Bollgard cotton in 1998, in *Proceedings of the Beltwide Cotton Conferences*, Vol. 2, National Cotton Council of America, Memphis, TN, 1999, 958.

Naber, A. C. and Aarssen, L. W., Effects of shoot apex removal and fruit herbivory on branching, biomass and reproduction in *Verbascum thapsus* (Scrophulariaceae), *Am. Midl. Nat.*, 140, 42, 1998.

Nakamura, R. R., Mitchell-Olds, T., Manasse, R. S., and Lello, D., Seed predation, pathogen infection and life-history traits in *Brassica rapa*, *Oecologia* (Berlin), 102, 324, 1995.

NASS–USDA, "Crop Production," October 8, 1999.

NASS–USDA, "Acreage," June 30, 2000.

National Agricultural Statistics Service, Agricultural Chemical Usage: Field Crops Summary, NASS–USDA, Washington, D.C., 1991–2000.

Navie, S. C., Priest, T. E., McFadyen, R. E., and Adkins, S. W., Efficacy of the stem-galling moth *Epiblema strenuana* Walk (Lepidoptera: Torricidae) as a biological control agent for ragweed *Parthenium* (*Parthenium hysterophorus* L.), *Biol. Control*, 13, 1, 1998.

NRC (National Research Council), Field Testing Genetically Modified Organisms: Framework for Decision, National Academy Press, Washington, D.C., 1989.

NRC (National Research Council), Genetically Modified Pest-Protected Plants: Science and Regulation, National Academy Press, Washington, D.C., 2000.

Parker, I. M. and Kareiva, P., Assessing the risks of invasion for genetically engineered plants: acceptable evidence and reasonable doubt, *Biol. Conserv.*, 78, 193, 1996.

Paulissen, M. A., Exploitation by, and the effects of, caterpillar grazers on the annual, *Rudbeckia hirta* (Compositae), *Am. Midl. Nat.*, 117, 439, 1987.

Peacock, J. W., Schweitzer, D. F., Carter, J. L., and Dubois, N. R., Laboratory assessment of the effects of *Bacillus thuringiensis* on native Lepidoptera, *Environ. Entomol.*, 27, 450, 1998.

Perrins, J., Williamson, M., and Fitter, A., Do annual weeds have predictable characters, *Acta Oecol. Int. J. Ecol.*, 13, 517, 1992.

Pilson, D., Two herbivores and constraints on selection for resistance in *Brassica rapa*, *Evolution*, 50, 1492, 1996.

Pilson, D., Herbivory and natural selelction on flowering phenology in wild sunflower, *Helianthus annuus*, *Oecologia*, 122, 72, 2000.

Price, P. W., Plant resources as the mechanistic basis for insect herbivore population dynamics, in *Effects of Resource Distribution on Animal-Plant Interactions*, Hunter, M. D., Ohgushi, T., and Price, P. W., Eds., Academic Press, San Diego, CA, 1992, 139.

Randall, J. M., Weed control for the preservation of biological diversity, *Weed Technol.*, 10, 370, 1996.

Rees, M. and Brown, V. K., Interactions between invertebrate herbivores and plant competition, *J. Ecol.*, 80, 353, 1992.

Regal, P. J., The adaptive potential of genetically engineered organisms in nature, *Trends Biotechnol.*, 6, S36, 1988.

ReJesus, R. M., Greene, J. K., Hamming, M. D., and Curtis, C. E., Economic analysis of insect management strategies for transgenic *Bt*-cotton production in South Carolina, in *1997 Proceedings Beltwide Cotton Conferences*, Cotton Economics and Marketing Conference, Vol. 1, Natural Cotton Council of America, Memphis, TN, 1997, 247.

Rejmanek, M. and Richardson, D. M., What attributes make some plant species more invasive, *Ecology,* 77, 1655, 1996.

Rice, M., Yield performance of *Bt*-corn, *Iowa State Univ. Integrated Crop Manage. Newsl.,* January 19, 6, 1998.

Rice, M. E. and Ostlie, K., European corn borer management in field corn: a survey of perceptions and practices in Iowa and Minnesota, *J. Prod. Agric.,* 10, 628, 1997.

Ritchie, M. E. and Tilman, D., Interspecific competition among grasshoppers and their effect on plant abundance in experimental field environments, *Oecologia,* 89, 524, 1992.

Robinson, G. S., Bugs, hollow curves and species-diversity indexes, *STATS,* American Statistical Association, 21, 8, 1998.

Robinson, G. S., HOSTS — a database of the hostplants of the world's Lepidoptera, *Nota Lepidopterol.,* 22, 35, 1999.

Robinson, G. S., Ackery, P. R., Kitching, I. J., Beccaloni, G. W., and Hernández, L. M., Hostplants of the moth and butterfly caterpillars of America north of Mexico, unpublished data.

Robinson, G. S., Ackery, P. R., Kitching, I. J., Beccaloni, G. W., and Hernández, L. M., HOSTS — The Natural History Museum's Database of Caterpillar Hostplants and Foods, Public intranet database, 2000, available at http://flood.nhm.ac.uk/cgi-bin/perth/hosts/index.dsml.

Robinson, G. S., Kitching, I. J., Ackery, P. R., Beccaloni, G. W., and Hernández, L. M., Hostplants of the moth and butterfly caterpillars of the Oriental Region, unpublished data.

Rosset, P., personal communication, 1989.

Royal Society of Canada, An Expert Panel Report on the Future of Food Biotechnology, The Royal Society of Canada, Ottawa, Canada, 2001.

Schoener, T. W., Leaf pubescence in buttonwood: community variation in a putative defense against defoliation, *Proc. Natl. Acad. Sci. U.S.A.,* 84, 7992, 1987.

Scientists' Working Group on Biosafety, *Manual for Assessing Ecological and Human Health Effects of Genetically Engineered Organisms,* The Edmonds Institute, Edmonds, WA, 1998.

Settle, W. H., Ariawan, H., Astuti, E. T., Cahyana, W., Hakim, A. L., Hindayana, D., Lestari, A. S., and Pajarningsih, S., Managing tropical rice pests through conservation of generalist natural enemies and alternative prey, *Ecology* (Washington, D.C.), 77, 1975, 1996.

Sharma, H. C. and Ortiz, R., Transgenics, pest management, and the environment, *Curr. Sci.* (Bangalore), 79, 421, 2000.

Smith, R. A. and Couche, G. A., The phylloplane as a source of *Bacillus thuringiensis* variants, *Appl. Environ. Microbiol.,* 57, 311, 1991.

Snow, A. A. and Moran-Palma, P., Commercialization of transgenic plants: potential ecological risks, *BioScience,* 47, 86, 1997.

Snow, A. A., Andersen, B., and Jorgensen, R. B., Costs of transgenic herbicide resistance introgressed from *Brassica napus* into weedy *B. rapa, Mol. Ecol.,* 8, 605, 1999.

Stark, C. R., Jr., Economics of transgenic cotton: some indications based on Georgia producers, in *Proceedings of the Beltwide Cotton Conferences,* Cotton Economics and Marketing Conference, Vol. 1, National Cotton Council of America, Memphis, TN, 1997, 251.

Stark, J. D., Tanigoshi, L., Bounfour, M., and Antonelli, A., Reproductive potential: its influence on the susceptibility of a species to pesticides, *Ecotoxicol. Environ. Saf.,* 37, 273, 1997.

Stewart, C. N. and Prakash, C. S., Chloroplast-transgenic plants are not a gene flow panacea, *Nat. Biotechnol.,* 16, 401, 1998.

Stewart, C. N., All, J. N., Raymer, P. L., and Ramachandran, S., Increased fitness of transgenic insecticidal rapeseed under insect selection pressure, *Mol. Ecol.,* 6, 773, 1997.

Story, J. M., Good, W. R., White, L. J., and Smith, L., Effects of the interaction of the biocontrol agent *Agapeta zoegana* L. (Lepidoptera: Cochylidae) and grass competition on spotted knapweed, *Biol. Control,* 17, 182, 2000.

Strauss, S. Y., Siemens, D. H., Decher, M. B., and Mitchell-Olds, T., Ecological costs of plant resistance to herbivores in the currency of pollination, *Evolution,* 53, 1105, 1999.

Strong, D. R., Density-vague ecology and liberal population regulation in insects, in *A New Ecology: Novel Approaches to Interactive Systems,* Price, P. W., Slobodchikoff, C. N., and Gaud, W. S., Eds., Wiley, New York, 1984, 313.

Strong, D. R., Are trophic cascades all wet — differentiation and donor-control in speciose ecosystems, *Ecology,* 73, 747, 1992.

Summers, C. G., Effect of selected pests and multiple pest complexes on alfalfa productivity and stand persistence, *J. Econ. Entomol.*, 82, 1782, 1989.

Tiedje, J. M., Colwell, R. K., Grossman, Y. L., Hodson, R. E., Lenski, R. E., Mack, R. N., and Regal, P. J., The planned introduction of genetically engineered organisms: ecological considerations and recommendations, *Ecology*, 70, 298, 1989.

Tilman, D., Community invasibility, recruitment limitation, and grassland biodiversity, *Ecology*, 78, 81, 1997.

Tomiuk, J., Hauser, T. P., and Bagger-Jorgensen, R., A- or C-chromosomes, does it matter for the transfer of transgenes from *Brassica napus*, *Theor. Appl. Genet.*, 100, 750, 2000.

Tscharntke, T., Insects on common reed (*Phragmites australis*): community structure and the impact of herbivory on shoot growth, *Aquat. Bot.*, 64, 399, 1999.

Tu, J., Zhang, G., Datta, K., Xu, C. He, Y., Zhang, Q., Khush, G. S., and Datta, S. K., Field performance of transgenic elite commercial hybrid rice expressing *Bacillus thuringiensis* delta-endotoxin, *Nat. Biotechnol.*, 18, 1101, 2000.

Turkington, R., Klein, E., and Chanway, C. P., Interactive effects of nutrients and disturbance — an experimental test of plant strategy theory, *Ecology*, 74, 863, 1993.

Turlings, T. C. J. and Tumlinson, J. H., Systemic release of chemical signals by herbivore-injured corn, *Proc. Natl. Acad. Sci. U.S.A.*, 89, 8399, 1992.

Turlings, T. C. J., Bernasconi, M., Bertossa, R., Bigler, F., Caloz, G., and Dorn, S., The induction of volatile emissions in maize by three herbivore species with different feeding habits: possible consequences for their natural enemies, *Biol. Control*, 11, 122, 1998.

Turlings, T. C. J., Alborn, H. T., Loughrin, J. H., and Tumlinson, J. H., Volicitin, an elicitor of maize volatiles in oral secretion of *Spodoptera exigua*: isolation and bioactivity, *J. Chem. Ecol.*, 26, 189, 2000.

University of Georgia, Summary of Losses from Insect Damage and Costs of Control in Georgia, http://www.bugwood.caes.uga.edu, 1995, 1996, 1997.

USDA–APHIS, Response to Monsanto Petition 94-308-01p for Determination of Nonregulated Status for Lepidopteran-Resistant Cotton Lines, U.S. Department of Agriculture, Animal and Plant Health Inspection Service, Biotechnology, Biologics, and Environmental Protection, 1995, available at ftp://www.aphis.usda.gov/pub/bbep/Determinations/ascii/9430801p_det.txt.

USDA–APHIS, Determination of Nonregulated Status for Bt Cry9C Insect Resistant and Glufosinate Tolerant Corn Transformation Event CBH-351, U.S. Department of Agriculture, Animal and Plant Health Inspection Service Plant Protection and Quarantine Scientific Services, Riverdale, Maryland, 1997, available at ftp://www.aphis.usda.gov/pub/bbep/Determinations/ascii/9726501p_det.txt.

USDA–APHIS Biotechnology Permits Database, available at http://www.nbiap.vt.edu/index.html, March 2001.

U.S. EPA, Biopesticides Registration Action Document: *Bt*-Plant-Pesticides, 1999, available at http://www.epa.gov/pesticides/biopesticides/news/news-bt-crops-sap-oct.htm.

van Frankenhuyzen, K. and Nystrom, C., The *Bacillus thuringiensis* Toxin Specificity Database, 1999, available at http://www.glfc.forestry.ca/Bacillus/Bt_HomePage/netintro99.htm (date site accessed).

Van Hamburg, H. and Guest, P. J., The impact of insecticides on beneficial arthropods in cotton agro-ecosystems in South Africa, *Arch. Environ. Contam. Toxicol.*, 32, 63, 1997.

Volenberg, D. S., Hopen, H. J., and Campobasso, G., Biological control of yellow toadflax (*Linaria vulgaris*) by *Eteobalea serratella* in peppermint (*Mentha piperita*), *Weed Sci.*, 47, 226, 1999.

Wiebold, W. J., Morris, C. G., Mason, H. L., Knerr, D. R., Hasty, R. W., Fritts, T. G., and Adams, E., Missouri Crop Performance 2000, Corn, Crop performance testing, available at http://agebb.missouri.edu/cropperf/corn/.

Wise, M. J. and Sacchi, C. F., Impact of two specialist insect herbivores on reproduction of horse nettle, *Solanum carolinense*, *Oecologia*, 108, 328, 1996.

Wolfenbarger, L. L. and Phifer, P. R., The ecological risks and benefits of genetically engineered plants, *Science*, 290, 2088, 2000.

Zehnder, G. W., Yao, C., Wei, G., and Kloepper, J. W., Influence of methyl bromide fumigation on microbe-induced resistance in cucumber, *Biocontrol Sci. Technol.*, 10, 687, 2000.

Zhu, Y., Chen, H., Fan, J., Wang, Y., Li, Y., Chen, J., Fan, J., Yang, S., Hu, L., Leung, H., Mew, T. W., Teng, P. S., Wang, Z., and Mundt, C. C., Genetic diversity and disease control in rice, *Nature* (London), 406, 718, 2000.

Appendix 3.1: Lepidoptera that feed on corn, soybean, tomato, or closely related plants

Host plant specificity is indicated by number of plant families and genera recorded as the food plants of the species; thus "10/24" indicates that 10 families and 24 genera of plants are recorded as hosts. Where the number of host plant families exceeds that of the number of genera, excess families are each for a record in which no plant genus is specified. Distribution is indicated by a two-letter code; thus, af = Afrotropical region; au = Australasia and Pacific; cm = cosmopolitan; hl = Holarctic region (Nearctic + Neotropical); hw = Hawaii; ia = Indo-Australian region (Oriental + Australasia); na = Nearctic region; nt = Neotropical region; nw = New World (Nearctic + Neotropical regions); nz = New Zealand; or = Oriental region; ow = Old World; pl = Palearctic region. Distribution is simplified from individual records and may involve overlap and/or duplication. Data from The Natural History Museum HOSTS database.

Poaceae

Zea mays

Acrolophidae.   *Acrolophus popeanella* Clemens [2/2] (na; nt).

Arctiidae.   *Alpenus investigatorum* Karsch [15/24] (af; pl); *Alpenus maculosa* Stoll [14/24] (af); *Alpenus pardalina* Rothschild [1/1] (af); *Amsacta lactinea* Cramer [26/45] (or); *Amsacta lineola* Fabricius [8/15] (or); *Amsacta moorei* Butler [7/14] (or); *Apantesis phalerata* Harris [4/5] (na); *Cisseps fulvicollis* Hübner [3/5] (na); *Creatonotos gangis* Linnaeus [8/19] (au; or); *Creatonotos transiens* Walker [20/28] (or); *Estigmene acrea* Drury [23/46] (na; nt; nw); *Grammia arge* Drury [8/10] (na); *Grammia phyllira* Drury [5/5] (na); *Grammia virgo* Linnaeus [6/7] (na); *Halysidota tessellaris* Smith [19/36] (na; nt; nw); *Holomelina aurantiaca* Hübner [4/4] (na); *Hypercompe indecisa* Walker [13/19] (nt); *Hyphantria cunea* Drury [49/100] (hl; na; pl); *Metarctia inconspicua* Holland [3/6] (af); *Pericallia ricini* Fabricius [31/45] (or); *Pyrrharctia isabella* Smith [25/37] (na); *Spilosoma curvilinea* Walker [3/3] (af); *Spilosoma lutescens* Walker [14/19] (af); *Spilosoma lutescens lutescens* Walker [14/21] (af); *Spilosoma lutescens screabilis* Wallengren [10/14] (af); *Spilosoma nigricosta* Holland [1/2] (af); *Spilosoma obliqua* Walker [27/54] (or); *Spilosoma virginica* Fabricius [44/86] (cm; na).

Cosmopterigidae.   *Pyroderces amphisaris* Meyrick [2/3] (or); *Pyroderces badia* Hodges [16/18] (hw; na); *Pyroderces falcatella* Stainton [8/10] (au; or); *Pyroderces hemizopha* Meyrick [2/3] (af; ow); *Pyroderces rileyi* Walsingham [13/18] (au; hl; hw; na; nt; or); *Pyroderces simplex* Walsingham [11/17] (af; cm; or; ow; pl).

Eupterotidae.   *Eupterote petola* Moore [1/5] (or).

Gelechiidae.   *Dichomeris granivora* Meyrick [1/1] (nt); *Phthorimaea operculella* Zeller [6/14] (af; au; cm; hw; na; nt; nw; or; ow; pl); *Sitotroga cerealella* Olivier [2/8] (af; cm; hw; na; nt; or).

Geometridae.   *Cleora nigrisparsalis* Janse [3/3] (af); *Eupithecia abbreviata* Stephens [3/3] (pl); *Eupithecia scopariata* Rambur [1/1] (pl); *Eupithecia unedonata* Mabille [1/1] (pl); *Gymnoscelis rufifasciata* Haworth [13/16] (pl); *Pleuroprucha insulsaria* Guenée [8/14] (na; nt); *Scopula emissaria* Walker [6/9] (or); *Thalassodes quadraria* Guenée [14/22] (af; au; or).

Hesperiidae.   *Ancyloxypha numitor* Fabricius [1/8] (na); *Borbo borbonica* Boisduval [1/5] (af); *Borbo gemella* Mabille [1/4] (af); *Lerema accius* Smith [1/10] (na; nt; nw); *Lerodea eufala* Edwards [1/9] (na; nt; nw); *Nyctelius nyctelius* Latreille [1/6] (nt); *Parnara guttatus*

Bremer & Grey [2/9] (or; ow; pl); *Parnara guttatus apostata* Snellen [2/5] (or); *Parnara naso bada* Moore [2/6] (au; or); *Pelopidas conjuncta* Herrich-Schäffer [1/7] (or; ow); *Pelopidas conjuncta conjuncta* Herrich-Schäffer [1/6] (or); *Pelopidas mathias* Fabricius [3/11] (af; ia; or; ow); *Perichares philetes* Gmelin [2/12] (nt); *Telicota bambusae* Moore [2/6] (au; or); *Zenonia zeno* Trimen [1/2] (af).

Limacodidae. *Acharia stimulea* Clemens [20/22] (hl; na); *Darna ochracea* Moore [4/4] (nt; or); *Darna pallivitta* Moore [5/7] (or); *Latoia vivida* Walker [15/20] (af; or); *Parasa semperi* Holloway [6/5] (or).

Lycaenidae. *Strymon melinus* Hübner [31/67] (na; nw).

Lymantriidae. *Bembina isabellina* Heylaerts [2/2] (or); *Cifuna locuples* Walker [6/12] (or; pl); *Clethrogyna turbata* Butler [20/35] (or); *Euproctis lunata* Walker [14/24] (or); *Euproctis pseudoconspersa* Strand [6/8] (or; ow); *Euproctis varians* Walker [9/17] (or); *Laelia suffusa* Walker [2/6] (or); *Lymantria dispar* Linnaeus [43/80] (hl; na; pl); *Olene inclusa* Walker [22/32] (au; or); *Olene mendosa* Hübner [50/97] (af; au; ia; or); *Orgyia leucostigma* Smith [51/101] (na); *Orgyia mixta* Snellen [20/32] (af); *Orgyia osseata* Walker [21/31] (or); *Orvasca subnotata* Walker [13/22] (or; ow); *Psalis pennatula* Fabricius [11/18] (af; or; ow); *Somena scintillans* Walker [38/76] (or); *Sphrageidus producta* Walker [12/15] (af); *Sphrageidus virguncula* Walker [12/25] (or).

Noctuidae. *Achatodes zeae* Harris [4/5] (na); *Acronicta oblinita* Smith [24/34] (na); *Acronicta rumicis* Linnaeus [26/46] (pl); *Actebia praecox* Linnaeus [5/6] (ow; pl); *Aegoceropsis rectilinea* Boisduval [7/9] (af); *Agrochola helvola* Linnaeus [10/14] (na; pl); *Agrotis clavis* Hufnagel [5/6] (or; pl); *Agrotis crinigera* Butler [6/10] (hw; pl); *Agrotis exclamationis* Linnaeus [11/15] (pl); *Agrotis gladiaria* Morrison [11/17] (na); *Agrotis interjectionis* Guenée [8/10] (au; or); *Agrotis ipsilon* Hufnagel [30/74] (af; au; cm; hw; ia; na; nt; nz; or; ow; pl); *Agrotis longidentifera* Hampson [7/9] (af); *Agrotis malefida* Guenée [11/19] (na; nt; nw); *Agrotis orthogonia* Morrison [11/27] (na); *Agrotis puta* Hübner [3/3] (pl); *Agrotis radians* Guenée [7/10] (au); *Agrotis repleta* Walker [11/20] (nt); *Agrotis segetum* Denis & Schiff. [21/45] (af; or; ow; pl); *Agrotis spinifera* Hübner [3/3] (af); *Agrotis subterranea* Fabricius [17/39] (na; nt; nw); *Agrotis tokionis* Butler [6/8] (ow; pl); *Agrotis venerabilis* Walker [5/8] (na); *Agrotis vetusta* Walker [13/20] (na); *Amphipoea americana* Speyer [2/2] (na); *Amphipoea burrowsi* Chapman [1/1] (or); *Amphipoea fucosa* Freyer [1/3] (pl); *Amphipoea oculea* Linnaeus [2/5] (hl; pl); *Amphipoea ussuriensis* Peterson [1/1] (pl); *Amphipoea velata* Walker [3/3] (na); *Anagrapha falcifera* Kirby [11/17] (na); *Anicla ignicans* Guenée [4/10] (nt; pl); *Anicla infecta mahalpa* Schaus [1/3] (nt); *Anomis texana* Riley [2/2] (na; nt); *Apamea amputatrix* Fitch [7/9] (na); *Apamea devastator* Brace [9/19] (na); *Apamea finitima* Guenée [2/6] (na); *Apamea lateritia* Hufnagel [1/3] (pl); *Apamea ophiogramma* Esper [2/6] (pl); *Apamea sordens* Hufnagel [1/11] (na; ow; pl); *Apamea vulgaris* Grote & Robinson [1/1] (na); *Argyrogramma verruca* Fabricius [13/17] (na; nt; nw); *Autoba brachygonia* Hampson [5/12] (af; or); *Autoba nr. versicolor* Walker [1/1] (af); *Autographa californica* Speyer [26/44] (na); *Autographa gamma* Linnaeus [25/75] (cm; na; nt; pl); *Autoplusia egena* Guenée [5/7] (na; nt; nw); *Bathytricha aethalion* Turner [1/1] (au); *Bathytricha truncata* Walker [4/7] (au); *Brithys pancratii* Cyrillo [6/11] (af; ow; pl); *Busseola fusca* Fuller [3/8] (af); *Celaena leucostigma* Hübner [4/5] (or; ow; pl); *Celaena reniformis* Grote [3/3] (na); *Chrysodeixis acuta* Walker [18/26] (af; or; ow); *Chrysodeixis chalcites* Esper [28/75] (af; au; cm; or; ow; pl); *Chrysodeixis eriosoma* Doubleday [18/29] (au; ia; or; pl); *Dargida grammivora* Walker [1/1] (nt); *Elaphria grata* Hübner [6/6] (na; nt); *Eublemma dimidialis* Fabricius [2/5] (au; or); *Eublemma gayneri* Rothschild [2/3] (af; or; ow; pl); *Euxoa albipennis* Grote [4/5] (na); *Euxoa auxiliaris* Grote [16/42] (na); *Euxoa cymograpta* Hampson [4/4] (af); *Euxoa dargo* Strecker [2/2] (na); *Euxoa declarata* Walker [4/6] (na); *Euxoa detersa* Walker [8/14] (na); *Euxoa messoria* Harris [16/27] (na); *Euxoa niveilinea* Grote

[2/3] (na); *Euxoa ochrogaster* Guenée [10/18] (na; pl); *Euxoa olivia* Morrison [2/2] (na); *Euxoa tessellata* Harris [15/28] (na); *Euxoa tritici* Linnaeus [8/13] (pl); *Faronta albilinea* Hübner [5/23] (na; nt; nw); *Faronta diffusa* Walker [1/17] (na); *Feltia jaculifera* Guenée [16/30] (na); *Feltia subgothica* Haworth [9/18] (cm; na; nw; pl); *Gortyna basalipunctata* Graeser [1/1] (or); *Gortyna fortis* Butler [1/1] (pl); *Grammodes geometrica* Fabricius [7/9] (af; or; ow); *Helicoverpa armigera* Hübner [40/102] (af; au; cm; ia; na; nt; nz; or; ow; pl); *Helicoverpa assulta* Guenée [8/17] (af; au; or; ow); *Helicoverpa gelotopoeon* Dyar [7/12] (nt); *Helicoverpa zea* Boddie [38/111] (af; hw; na; nt; nw; or); *Heliothis peltigera* Denis & Schiff. [12/27] (af; or; ow; pl); *Heterogramma circumflexalis* Guenée [1/1] (nt); *Hydraecia amurensis* Staudinger [1/1] (pl); *Hydraecia immanis* Guenée [4/10] (na); *Hydraecia micacea* Esper [15/27] (hl; na; pl); *Hypena scabra* Fabricius [12/31] (na); *Lacanobia subjuncta* Grote & Robinson [12/14] (na); *Lacinipolia renigera* Stephens [12/18] (na; pl); *Leucania humidicola* Guenée [1/6] (nt); *Leucania latiuscula* Herrich-Schäffer [1/12] (na; nt); *Leucania loreyi* Duponchel [5/17] (af; au; or; ow; pl); *Leucania phaea* Hampson [1/4] (af; pl); *Leucania secta* Herrich-Schäffer [1/2] (nt); *Leucania venalba* Moore [1/2] (au; or); *Lophoruza semiscripta* Mabille [2/2] (af); *Luperina dumerilii* Duponchel [1/2] (pl); *Luperina testacea* Denis & Schiff. [1/6] (pl); *Lycophotia porphyrea* Denis & Schiff. [7/12] (hw; pl); *Macronoctua onusta* Grote [3/5] (na); *Mamestra configurata* Walker [16/33] (na; nt); *Melanchra picta* Harris [26/48] (hl; na); *Meropleon cosmion* Dyar [1/4] (na); *Mesapamea concinnata* Heinicke [1/1] (pl); *Mesapamea secalis* Linnaeus [2/10] (ow; pl); *Mesapamea stipata* Morrison [4/11] (na); *Mocis frugalis* Fabricius [5/15] (af; au; ia; or); *Mocis latipes* Guenée [10/37] (na; nt; nw); *Mythimna convecta* Walker [3/11] (au); *Mythimna l-album* Linnaeus [1/2] (pl); *Mythimna scirpi* Duponchel [1/1] (pl); *Mythimna zeae* Duponchel [1/3] (ow); *Nephelodes minians* Guenée [7/12] (na); *Ochropleura herculea* Corti and Draudt [1/2] (or); *Oligia fractilinea* Grote [1/2] (na); *Oligia semicana* Walker [1/1] (na); *Oruza divisa* Walker [3/3] (au; or); *Palthis asopialis* Guenée [5/6] (na); *Papaipema cataphracta* Grote [15/26] (na); *Papaipema nebris* Guenée [31/86] (na); *Papaipema sauzalitae* Grote [4/5] (na); *Peridroma saucia* Hübner [40/86] (cm; na; nt; pl); *Persectania aversa* Walker [1/2] (nz); *Phragmatiphila truncata* Walker [1/3] (au); *Pseudaletia adultera* Schaus [4/14] (nt); *Pseudaletia separata* Walker [12/32] (au; ia; nz; or); *Pseudaletia unipuncta* Haworth [18/53] (au; cm; hw; na; nt; or; pl); *Pseudoleucania minna* Butler [1/3] (nt); *Richia albicosta* Smith [3/5] (na); *Sciomesa biluma* Nye [1/3] (af); *Sesamia botanephaga* Tams & Bowden [3/10] (af); *Sesamia calamistis* Hampson [3/13] (af); *Sesamia cretica* Lederer [2/7] (af; or; ow; pl); *Sesamia inferens* Walker [3/24] (or; ow); *Sesamia nonagrioides* Lefèbvre [4/6] (af; ow; pl); *Sesamia penniseti* Tams & Bowden [1/8] (af); *Sesamia poebora* Tams & Bowden [1/3] (af); *Sesamia poephaga* Tams & Bowden [1/2] (af); *Sesamia uniformis* Dudgeon [1/8] (ia; or); *Simyra henrici* Grote [6/10] (na); *Spaelotis clandestina* Harris [18/25] (na); *Speia vuteria* Stoll [3/8] (nt; ow); *Spiramater lutra* Guenée [16/26] (na); *Spodoptera albula* Walker [11/13] (nt); *Spodoptera cilium* Guenée [4/8] (af; or; ow; pl); *Spodoptera compta* Walker [1/5] (or); *Spodoptera dolichos* Fabricius [14/22] (na; nt); *Spodoptera eridania* Stoll [31/63] (na; nt; nw); *Spodoptera exempta* Walker [7/29] (af; au; hw; or; ow); *Spodoptera exigua* Hübner [35/83] (af; au; cm; hw; na; or; pl); *Spodoptera frugiperda* Smith [34/91] (af; cm; na; nt; nw); *Spodoptera littoralis* Boisduval [25/43] (af; na; or; ow; pl); *Spodoptera litura* Fabricius [69/166] (af; au; cm; ia; nt; or; ow); *Spodoptera mauritia* Boisduval [7/20] (af; au; cm; hw; ia; or; ow); *Spodoptera ornithogalli* Guenée [25/53] (cm; na; nt); *Spodoptera pecten* Guenée [7/15] (ia; or); *Spodoptera triturata* Walker [1/3] (af); *Spragueia onagrus* Guenée [3/3] (na); *Trichoplusia ni* Hübner [27/65] (af; cm; hw; na; nt; or; pl); *Trichoplusia orichalcea* Fabricius [24/57] (af; au; or; ow); *Tripseuxoa strigata* Hampson [4/7] (nt); *Tycomarptes inferior* Guenée [3/3] (af); *Vietteania torrentium* Guenée [2/3] (af); *Xestia c-nigrum* Linnaeus [31/51] (hl; na; or; pl); *Xestia dolosa* Franclemont [5/6] (na).

Nolidae.   *Earias biplaga* Walker [5/20] (af; au); *Earias insulana* Boisduval [9/22] (af; au; or; ow; pl); *Nola fovifeferoides* Poole [1/2] (af); *Nola sorghiella* Riley [2/5] (na; nt; nw); *Nola spermophaga* Fletcher [1/1] (af).

Notodontidae. *Phalera combusta* Walker [1/4] (or); *Phalera lydenburgi* Distant [1/1] (af).

Nymphalidae. *Acraea acerata* Hewitson [5/7] (af); *Euptoieta claudia* Cramer [15/24] (na; nt; nw); *Melanitis leda* Linnaeus [2/33] (af; au; or; ow; pl); *Vanessa cardui* Linnaeus [25/104] (af; au; cm; hw; na; or; pl).

Oecophoridae. *Autosticha solita* Meyrick [1/1] (au); *Borkhausenia minutella* Linnaeus [1/1] (pl); *Hofmannophila pseudospretella* Stainton [6/6] (cm; na; pl).

Psychidae. *Thyridopteryx ephemeraeformis* Haworth [50/92] (hl; na; nt).

Pyralidae. *Achyra affinitalis* Lederer [7/10] (au); *Achyra coelatalis* Walker [1/3] (af; or); *Achyra massalis* Walker [1/2] (af; or); *Achyra rantalis* Guenée [8/14] (na; nt; nw); *Achyra similalis* Guenée [6/8] (na; nt); *Agriphila ruricolellus* Zeller [1/1] (na); *Agriphila vulgivagellus* Clemens [1/1] (na); *Amyelois transitella* Walker [18/33] (na; nt; nw); *Ancylolomia inornata* Staudinger [1/3] (af); *Bleszynskia malacellus* Duponchel [2/5] (af; au; ow); *Cadra cautella* Walker [35/61] (af; au; cm; na; nt; or); *Chilo agamemnon* Bleszynski [1/5] (af; ow); *Chilo auricilia* Dudgeon [1/9] (au; or); *Chilo diffusilinea* Joannis [1/4] (af); *Chilo luteellus* Motschulsky [1/2] (pl); *Chilo orichalcociliella* Strand [1/6] (af); *Chilo partellus* Swinhoe [1/7] (af; or; ow); *Chilo polychrysa* Meyrick [2/11] (au; or; pl); *Chilo sacchariphagus* Bojer [1/7] (af; or; ow); *Chilo suppressalis* Walker [3/19] (af; au; hw; ia; na; or; ow); *Chilo zacconius* Bleszynski [1/5] (af); *Cnaphalocrocis medinalis* Guenée [4/14] (au; or; ow); *Cnaphalocrocis suspicalis* Walker [1/3] (au; or); *Cnaphalocrocis trapezalis* Guenée [1/14] (af; au; cm; na; nt; or); *Coniesta ignefusalis* Hampson [1/4] (af); *Conogethes punctiferalis* Guenée [31/55] (au; ia; or; pl); *Corcyra cephalonica* Stainton [19/39] (af; cm; hw; na; nt; or); *Crambus laqueatellus* Clemens [1/1] (na); *Crambus praefectellus* Zincken [1/1] (na); *Crypsiptya coclesalis* Walker [4/12] (or; ow); *Cryptoblabes gnidiella* Millière [30/49] (af; cm; hw; na; nt; or; pl); *Diatraea crambidoides* Grote [1/3] (na; nw); *Diatraea grandiosella* Dyar [1/3] (nt; nw); *Diatraea lineolata* Walker [1/5] (nt; nw); *Diatraea muellerella* Dyar & Heinrich [1/2] (nt); *Diatraea postlineella* Schaus [1/1] (nt); *Diatraea saccharalis* Fabricius [2/19] (na; nt; nw; or); *Doloessa viridis* Zeller [10/12] (or); *Duponchelia fovealis* Zeller [3/3] (af; pl); *Ectomyelois ceratoniae* Zeller [23/42] (af; cm; hw; na; nt; or; pl); *Elasmopalpus lignosellus* Zeller [23/38] (cm; na; nt; nw); *Eldana saccharina* Walker [3/8] (af); *Eoreuma loftini* Dyar [1/6] (na); *Ephestia kuehniella* Zeller [12/21] (hl; na; nt; pl); *Etiella zinckenella* Treitschke [9/40] (af; au; cm; na; nt; or; pl); *Euzophera semifuneralis* Walker [13/18] (na; nw); *Fissicrambus mutabilis* Clemens [1/9] (na); *Hellula rogatalis* Hulst [2/5] (na); *Herpetogramma bipunctalis* Fabricius [17/25] (cm; na; nt; or); *Herpetogramma licarsisalis* Walker [6/15] (af; au; or; ow); *Loxostege sticticalis* Linnaeus [17/35] (hl; na; pl); *Marasmia cochrusalis* Walker [1/2] (nt); *Marasmia venilialis* Walker [1/8] (af; au; ia; or); *Microcrambus elegans* Clemens [1/1] (na); *Monoctenocera brachiella* Hampson [2/6] (or); *Monoctenocera leucania* Felder & Rogenh. [1/2] (or); *Moodna bisinuella* Hampson [2/3] (na; nw); *Mussidia nigrivenella* Ragonot [8/15] (af; na); *Nacoleia octasema* Meyrick [6/6] (au; or); *Neodactria caliginosellus* Clemens [3/4] (na); *Neodactria luteolellus* Clemens [1/1] (na); *Neodactria zeellus* Fernald [1/1] (na); *Nomophila noctuella* Denis & Schiff. [6/11] (af; hw; nt; or; ow; pl); *Omiodes indicata* Fabricius [12/29] (af; cm; nt; or); *Oncocera semirubella* Scopoli [4/9] (or; ow; pl); *Ostrinia furnacalis* Guenée [10/27] (au; ia; or); *Ostrinia kasmirica* Moore [2/3] (or); *Ostrinia kurentzovi* Mutuura & Munroe [2/2] (ow); *Ostrinia narynensis* Mutuura & Munroe [2/2] (ow); *Ostrinia nubilalis* Hübner [16/39] (au; cm; hl; na; or; pl); *Ostrinia obumbratalis* Lederer [5/7] (na); *Ostrinia palustralis* Hübner [1/1] (or); *Ostrinia quadripunctalis* Denis & Schiff. [1/1] (pl); *Ostrinia scapulalis* Walker [5/6] (pl); *Ostrinia zealis* Guenée [2/2] (ow); *Ostrinia zealis varialis* Bremer [2/2] (ow); *Paralipsa gularis* Zeller [22/28] (hl; hw; na); *Pediasia trisecta* Walker [2/4] (na); *Phidotricha agriperda* Dyar [2/2] (nt); *Phidotricha erigens* Ragonot [9/12] (na; nt); *Pleuroptya ruralis* Scopoli [4/4] (pl); *Plodia interpunctella* Hübner [25/48] (af; cm; na; nt; or; pl); *Pococera atramentalis* Lederer [7/7] (na; nt); *Pyralis farinalis* Linnaeus [8/17] (af; cm; na; nt; or; pl); *Pyralis manihotalis* Guenée [10/13] (cm; na; nt; or); *Scirpophaga incertulas* Walker [2/10] (or; ow); *Scirpophaga xanthogastrella* Walker [1/3] (or; ow); *Spoladea recurvalis* Fabricius [23/37] (af; au; cm; hw; ia; na; nt; or); *Stenachroia elongella* Hampson [1/5] (or); *Talis quercella* Denis & Schiff. [4/4] (pl).

Saturniidae.   *Anisota stigma* Fabricius [5/6] (na); *Automeris io* Fabricius [29/59] (na; nw); *Automeris io io* Fabricius [35/85] (na; nw); *Eacles imperialis imperialis* Drury [24/40] (na; nt); *Eacles imperialis magnifica* Walker [15/26] (nt); *Eacles penelope penelope* Cramer [3/4] (nt); *Hemileuca mania* Druce [1/1] (nt); *Hemileuca oliviae* Cockerell [1/15] (na; nw).

Scythrididae.   *Eretmocera impactella* Walker [8/11] (or).

Sphingidae.   *Hippotion celerio* Linnaeus [18/43] (af; au; cm; or; ow; pl); *Leucophlebia lineata* Westwood [1/3] (ia; or); *Xylophanes tersa* Linnaeus [8/14] (na; nt; nw).

Tineidae.   *Opogona sacchari* Bojer [19/25] (au; cm; hw; na; nt; pl); *Opogona subcervinella* Walker [2/4] (pl); *Setomorpha rutella* Zeller [19/23] (af; au; cm; ia; na; nt; or; ow); *Tinea fictrix* Meyrick [1/1] (ow).

Tortricidae.   *Amorbia emigratella* Busck [22/30] (hw; na; nt); *Archips micaceana* Walker [43/87] (ia; or); *Cnephasia longana* Haworth [20/37] (cm; na; pl); *Lobesia aeolopa* Meyrick [23/41] (af; au; or; ow; pl); *Platynota nigrocervina* Walsingham [8/11] (na); *Platynota stultana* Walsingham [25/39] (na; nt; nw); *Sparganothis sulfureana* Clemens [21/41] (na); *Thaumatotibia leucotreta* Meyrick [23/28] (af; na; or; pl).

Yponomeutidae.   *Plutella xylostella* Linnaeus [11/22] (af; au; cm; hw; ia; na; nt; or; pl).

Zea mexicana

Noctuidae.   *Spodoptera frugiperda* Smith [34/91] (af; cm; na; nt; nw).

Pyralidae.   *Elasmopalpus lignosellus* Zeller [23/38] (cm; na; nt; nw).

Fabaceae

Glycine max

Arctiidae.   *Amsacta lactinea* Cramer [26/45] (or); *Estigmene acrea* Drury [23/46] (na; nt; nw); *Pericallia ricini* Fabricius [31/45] (or); *Spilarctia casigneta* Kollar [4/7] (or); *Spilosoma jacksoni* Rothschild [5/8] (af); *Spilosoma obliqua* Walker [27/54] (or); *Spilosoma sumatrana* Swinhoe [11/15] (or); *Spilosoma virginica* Fabricius [44/86] (cm; na); *Trichaeta bivittata* Walker [2/2] (af).

Choreutidae.   *Brenthia pavonacella* Clemens [1/5] (na; nt).

Gelechiidae.   *Anarsia ephippias* Meyrick [2/10] (or); *Anarsia nigricana* Park [1/1] (pl); *Aproaerema anthyllidella* Hübner [1/5] (pl); *Aproaerema modicella* Deventer [2/11] (or); *Bilobata subsecivella* Zeller [2/5] (ia; or; ow); *Dichomeris acuminatus* Staudinger [1/13] (af; hw; ia; na; nt; or; ow; pl); *Mesophleps palpigera* Walsingham [3/14] (af; cm; ia; or); *Stomopteryx simplexella* Walker [1/3] (au).

Geometridae.   *Anacamptodes fragilaria* Grossbeck [15/24] (na); *Anavitrinella pampinaria* Guenée [25/42] (na; nt); *Ascotis selenaria* Denis & Schiff. [35/74] (af; or; ow; pl); *Ectropis bistortata* Goeze [18/23] (pl); *Ectropis obliqua* Prout [12/16] (or; pl); *Hyposidra talaca* Walker [28/68] (au; ia; or); *Iridopsis syrniaria* Guenée [1/1] (nt); *Macaria abydata* Guenée [2/8] (cm; hw; ia; na; nt); *Oxydia nimbata* Guenée [1/1] (nt); *Scopula remotata* Guenée [4/4] (or);

Semiothisa regulata Fabricius [2/2] (nt); *Synchlora frondaria* Guenée [5/14] (na; nt); *Synchlora frondaria frondaria* Guenée [4/8] (na); *Tephrina murinaria* Denis & Schiff. [1/1] (pl).

Gracillariidae.   *Caloptilia soyella* Deventer [1/7] (au; or; pl); *Phodoryctis caerulea* Meyrick [5/20] (af; au; or; ow; pl).

Hesperiidae.   *Chioides catillus* Cramer [1/5] (na; nt); *Spialia galba* Fabricius [3/5] (or); *Urbanus proteus* Linnaeus [4/17] (na; nt; nw).

Lycaenidae.   *Lampides boeticus* Linnaeus [6/41] (af; au; hw; or; ow; pl); *Leptotes pirithous* Linnaeus [10/18] (af; ia; pl); *Virachola antalus* Hopffer [1/4] (af); *Zizeeria knysna* Trimen [7/13] (af; or; pl); *Zizina labradus* Godart [1/5] (au; nz); *Zizina labradus labradus* Godart [1/16] (au); *Zizina otis* Fabricius [2/13] (au; or; pl).

Lymantriidae.   *Cifuna locuples* Walker [6/12] (or; pl); *Dasychira anophoeta* Collenette [1/1] (or); *Dasychira georgiana* Fawcett [13/22] (af); *Euproctis taiwana* Shiraki [5/6] (or); *Orgyia basalis* Walker [16/23] (af); *Orgyia mixta* Snellen [20/32] (af); *Orgyia postica* Walker [34/72] (au; ia; or; pl); *Somena scintillans* Walker [38/76] (or).

Lyonetiidae.   *Microthauma glycinella* Kuroko [1/2] (pl).

Noctuidae.   *Achaea mormoides* Walker [3/3] (af); *Achaea sordida* Walker [2/3] (af); *Acontia dacia* Druce [2/2] (na); *Acronicta lutea* Bremer and Grey [1/1] (pl); *Actebia praecox* Linnaeus [5/6] (ow; pl); *Agrapha agnata* Staudinger [2/2] (ow); *Agrotis ipsilon* Hufnagel [30/74] (af; au; cm; hw; ia; na; nt; nz; or; ow; pl); *Agrotis repleta* Walker [11/20] (nt); *Agrotis segetum* Denis & Schiff. [21/45] (af; or; ow; pl); *Allagrapha aerea* Hübner [3/3] (na); *Amyna octo* Guenée [11/22] (af; au; na; or; ow); *Anagrapha falcifera* Kirby [11/17] (na); *Anticarsia gemmatalis* Hübner [4/21] (na; nt); *Anticarsia irrorata* Fabricius [4/16] (af; au; or; ow); *Argyrogramma verruca* Fabricius [13/17] (na; nt; nw); *Autographa gamma* Linnaeus [25/75] (cm; na; nt; pl); *Autographa precationis* Guenée [11/17] (na); *Autoplusia egena* Guenée [5/7] (na; nt; nw); *Caenurgina erechtea* Cramer [4/7] (na); *Chrysodeixis acuta* Walker [18/26] (af; or; ow); *Chrysodeixis chalcites* Esper [28/75] (af; au; cm; or; ow; pl); *Chrysodeixis eriosoma* Doubleday [18/29] (au; ia; or; pl); *Ctenoplusia oxygramma* Geyer [3/6] (cm; na); *Discestra trifolii* Hufnagel [17/34] (cm; na; or; pl); *Egira natalensis* Butler [1/1] (af); *Euxoa detersa* Walker [8/14] (na); *Gesonia gemma* Swinhoe [1/1] (or); *Helicoverpa armigera* Hübner [40/102] (af; au; cm; ia; na; nt; nz; or; ow; pl); *Helicoverpa gelotopoeon* Dyar [7/12] (nt); *Helicoverpa zea* Boddie [38/111] (af; hw; na; nt; nw; or); *Heliothis peltigera* Denis & Schiff. [12/27] (af; or; ow; pl); *Heliothis virescens* Fabricius [19/50] (na; nt; nw); *Heliothis viriplaca* Hufnagel [7/13] (or; pl); *Hypena rostralis* Linnaeus [3/3] (ow; pl); *Hypena scabra* Fabricius [12/31] (na); *Hypena tristalis* Lederer [1/1] (ow); *Lacanobia suasa* Denis & Schiff. [15/25] (pl); *Mocis frugalis* Fabricius [5/15] (af; au; ia; or); *Mocis latipes* Guenée [10/37] (na; nt; nw); *Mocis mayeri* Boisduval [1/6] (af); *Mocis undata* Fabricius [7/24] (af; or); *Peridroma saucia* Hübner [40/86] (cm; na; nt; pl); *Pseudaletia adultera* Schaus [4/14] (nt); *Pseudoplusia includens* Walker [22/46] (na; nt); *Pyrrhia umbra* Hufnagel [11/22] (hl; na; or; pl); *Rachiplusia nu* Guenée [9/18] (nt); *Rachiplusia ou* Guenée [8/13] (na); *Richia albicosta* Smith [3/5] (na); *Sarcopolia illoba* Butler [3/4] (pl); *Spodoptera albula* Walker [11/13] (nt); *Spodoptera eridania* Stoll [31/63] (na; nt; nw); *Spodoptera exigua* Hübner [35/83] (af; au; cm; hw; na; or; pl); *Spodoptera frugiperda* Smith [34/91] (af; cm; na; nt; nw); *Spodoptera latifascia* Walker [16/24] (na; nt; nw); *Spodoptera littoralis* Boisduval [25/43] (af; na; or; ow; pl); *Spodoptera litura* Fabricius [69/166] (af; au; cm; ia; nt; or; ow); *Spodoptera ornithogalli* Guenée [25/53] (cm; na; nt); *Syngrapha rectangula* Kirby [3/7] (na); *Trichoplusia ni* Hübner [27/65] (af; cm; hw; na; nt; or; pl); *Trichoplusia orichalcea* Fabricius [24/57] (af; au; or; ow).

Nolidae.   *Giaura sceptica* Swinhoe [2/4] (or).

Notodontidae.   *Antheua simplex* Walker [1/9] (af).

Nymphalidae.   *Euptoieta claudia* Cramer [15/24] (na; nt; nw); *Neptis hylas* Linnaeus [10/32] (or; pl); *Vanessa cardui* Linnaeus [25/104] (af; au; cm; hw; na; or; pl).

Pantheidae.   *Panthea acronyctoides* Walker [3/8] (af; na).

Pieridae.   *Colias erate* Esper [1/6] (af; or; ow; pl); *Colias eurytheme* Boisduval [2/21] (na; nt); *Colias lesbia* Fabricius [1/3] (nt); *Colias lesbia pyrrhothea* Hübner [1/8] (nt); *Colias philodice* Godart [1/16] (na; nw); *Eurema deva* Doubleday [1/4] (nt); *Eurema elathea elathea* Cramer [1/5] (nt); *Eurema lisa euterpe* Ménétriés [1/2] (nt).

Pyralidae.   *Achyra rantalis* Guenée [8/14] (na; nt; nw); *Argyria divisella* Walker [1/1] (nt); *Bradina adhaesalis* Walker [1/1] (or); *Cadra cautella* Walker [35/61] (af; au; cm; na; nt; or); *Conogethes punctiferalis* Guenée [31/55] (au; ia; or; pl); *Corcyra cephalonica* Stainton [19/39] (af; cm; hw; na; nt; or); *Diaphania indica* Saunders [9/19] (af; au; cm; or; pl); *Elasmopalpus lignosellus* Zeller [23/38] (cm; na; nt; nw); *Ephestia kuehniella* Zeller [12/21] (hl; na; nt; pl); *Etiella behrii* Zeller [1/10] (af; au; or); *Etiella hobsoni* Butler [1/2] (ia; or); *Etiella zinckenella* Treitschke [9/40] (af; au; cm; na; nt; or; pl); *Herpetogramma bipunctalis* Fabricius [17/25] (cm; na; nt; or); *Loxostege sticticalis* Linnaeus [17/35] (hl; na; pl); *Mabra charonialis* Walker [1/1] (pl); *Maruca testulalis* Geyer [5/26] (af; au; hw; nt; or; ow; pl); *Nomophila nearctica* Munroe [7/12] (na); *Nomophila noctuella* Denis & Schiff. [6/11] (af; hw; nt; or; ow; pl); *Ocrasa nostralis* Guenée [4/4] (af; na; nt; nw); *Omiodes diemenalis* Guenée [6/29] (au; ia; or); *Omiodes indicata* Fabricius [12/29] (af; cm; nt; or); *Omiodes poeonalis* Walker [2/6] (af; or); *Omiodes simialis* Guenée [11/19] (na); *Pleuroptya ruralis* Scopoli [4/4] (pl); *Plodia interpunctella* Hübner [25/48] (af; cm; na; nt; or; pl); *Spoladea recurvalis* Fabricius [23/37] (af; au; cm; hw; ia; nt; or).

Sphingidae.   *Agrius convolvuli* Linnaeus [12/24] (af; au; cm; ia; or; ow; pl); *Clanis bilineata* Walker [1/4] (cm; or); *Platysphinx phyllis* Roths. & Jordan [1/5] (af).

Tortricidae.   *Adoxophyes moderatana* Walker [7/8] (au; or); *Adoxophyes orana* Fischer v. Rösl. [18/29] (or; ow; pl); *Adoxophyes privatana* Walker [31/53] (ia; or; ow; pl); *Archips breviplicanus* Walsingham [12/19] (pl); *Archips fraterna* Tuck [2/2] (or); *Archips fuscocupreanus* Walsingham [12/22] (na; pl); *Archips micaceana* Walker [43/87] (ia; or); *Archips mimicus* Walsingham [14/20] (or); *Archips occidentalis* Walsingham [10/12] (af); *Archips seditiosa* Meyrick [7/10] (ia; or); *Choristoneura lafauryana* Ragonot [8/13] (pl); *Cnephasia asseclana* Denis & Schiff. [9/15] (na; pl); *Cryptophlebia ombrodelta* Lower [10/32] (af; au; hw; ia; or); *Cryptophlebia strepsibathra* Meyrick [1/1] (au); *Cydia fabivora* Meyrick [1/2] (nt); *Epinotia aporema* Walsingham [1/4] (nw); *Eucosma aporema* Walsingham [1/5] (na; nt); *Grapholita compositella* Fabricius [1/5] (pl); *Gynnidomorpha vectisana* Humphreys & Wstw [2/2] (pl); *Homona coffearia* Nietner [34/60] (au; ia; or; ow; pl); *Homona magnanima* Diakonoff [24/34] (ow; pl); *Homona mermerodes* Meyrick [3/3] (or); *Leguminivora glycinivorella* Matsumura [2/5] (af; or; ow; pl); *Leguminivora ptychora* Meyrick [1/11] (af; or; ow); *Matsumuraeses falcana* Walsingham [1/5] (or); *Matsumuraeses phaseoli* Matsumura [1/7] (or; ow; pl); *Matsumuraeses ussuriensis* Caradja [1/3] (pl); *Matsumuraeses vicina* Kuznetsov [1/2] (pl); *Olethreutes transversana* Christoph [3/3] (pl); *Phalonidia curvistrigana* Stainton [2/5] (pl); *Platynota stultana* Walsingham [25/39] (na; nt; nw); *Sparganothis sulfureana* Clemens [21/41] (na).

Glycine soja

Gelechiidae.   *Aproaerema modicella* Deventer [2/11] (or).

Noctuidae.   *Agrotis ipsilon* Hufnagel [30/74] (af; au; cm; hw; ia; na; nt; nz; or; ow; pl); *Amyna octo* Guenée [11/22] (af; au; na; or; ow); *Spodoptera litura* Fabricius [69/166] (af; au; cm; ia; nt; or; ow).

Pyralidae. *Omiodes indicata* Fabricius [12/29] (af; cm; nt; or).

Tortricidae. *Homona coffearia* Nietner [34/60] (au; ia; or; ow; pl); *Leguminivora ptychora* Meyrick [1/11] (af; or; ow); *Matsumuraeses phaseoli* Matsumura [1/7] (or; ow; pl).

Solanaceae

Lycopersicon esculentum

Arctiidae. *Hypercompe icasia* Cramer [11/13] (nt); *Hypercompe indecisa* Walker [13/19] (nt); *Utetheisa pulchella* Linnaeus [7/12] (af; nt; or; ow; pl).

Gelechiidae. *Keiferia lycopersicella* Walsingham [2/5] (cm; hw; na; nt); *Phthorimaea operculella* Zeller [6/14] (af; au; cm; hw; na; nt; nw; or; ow; pl); *Scrobipalpa ergasima* Meyrick [1/2] (af; or; ow); *Scrobipalpula absoluta* Meyrick [2/5] (nt); *Symmetrischema tangolias* Gyen [1/3] (au; cm; na; nt; nz).

Hesperiidae. *Bolla brennus* Godman & Salvin [1/1] (nt).

Noctuidae. *Abagrotis alternata* Grote [8/12] (na); *Acontia porphyrea* Butler [1/1] (af); *Agrotis gladiaria* Morrison [11/17] (na); *Agrotis ipsilon* Hufnagel [30/74] (af; au; cm; hw; ia; na; nt; nz; or; ow; pl); *Agrotis malefida* Guenée [11/19] (na; nt; nw); *Agrotis orthogonia* Morrison [11/27] (na); *Agrotis radians* Guenée [7/10] (au); *Agrotis repleta* Walker [11/20] (nt); *Agrotis subterranea* Fabricius [17/39] (na; nt; nw); *Agrotis vetusta* Walker [13/20] (na); *Alabama argillacea* Hübner [3/5] (na; nt; nw); *Anomis flava* Fabricius [9/22] (af; au; cm; na; nt; or; ow); *Asota caricae* Fabricius [11/12] (au; ia; or); *Autographa californica* Speyer [26/44] (na); *Chrysodeixis acuta* Walker [18/26] (af; or; ow); *Chrysodeixis chalcites* Esper [28/75] (af; au; cm; or; ow; pl); *Chrysodeixis eriosoma* Doubleday [18/29] (au; ia; or; pl); *Euplexia lucipara* Linnaeus [16/18] (ow; pl); *Euxoa auxiliaris* Grote [16/42] (na); *Euxoa messoria* Harris [16/27] (na); *Euxoa tessellata* Harris [15/28] (na); *Euxoa tritici* Linnaeus [8/13] (pl); *Feltia jaculifera* Guenée [16/30] (na); *Feltia subgothica* Haworth [9/18] (cm; na; nw; pl); *Helicoverpa armigera* Hübner [40/102] (af; au; cm; ia; na; nt; nz; or; ow; pl); *Helicoverpa assulta* Guenée [8/17] (af; au; or; ow); *Helicoverpa gelotopoeon* Dyar [7/12] (nt); *Helicoverpa zea* Boddie [38/111] (af; hw; na; nt; nw; or); *Heliothis peltigera* Denis & Schiff. [12/27] (af; or; ow; pl); *Heliothis phloxiphaga* Grote & Robinson [10/22] (na); *Heliothis tergemina* Fabricius [1/3] (nt); *Heliothis virescens* Fabricius [19/50] (na; nt; nw); *Hydraecia micacea* Esper [15/27] (hl; na; pl); *Lacanobia oleracea* Linnaeus [22/41] (pl); *Lascoria orneodalis* Guenée [4/4] (nt); *Lycophotia porphyrea* Denis & Schiff. [7/12] (hw; pl); *Mamestra brassicae* Linnaeus [14/27] (or; pl); *Mamestra configurata* Walker [16/33] (na; nt); *Noctua pronuba* Linnaeus [15/22] (hl; na; or; pl); *Papaipema cataphracta* Grote [15/26] (na); *Papaipema nebris* Guenée [31/86] (na); *Peridroma saucia* Hübner [40/86] (cm; na; nt; pl); *Pseudaletia adultera* Schaus [4/14] (nt); *Pseudoplusia includens* Walker [22/46] (na; nt); *Rachiplusia nu* Guenée [9/18] (nt); *Spodoptera dolichos* Fabricius [14/22] (na; nt); *Spodoptera eridania* Stoll [31/63] (na; nt; nw); *Spodoptera exempta* Walker [7/29] (af; au; hw; or; ow); *Spodoptera exigua* Hübner [35/83] (af; au; cm; hw; na; or; pl); *Spodoptera frugiperda* Smith [34/91] (af; cm; na; nt; nw); *Spodoptera latifascia* Walker [16/24] (na; nt; nw); *Spodoptera littoralis* Boisduval [25/43] (af; na; or; ow; pl); *Spodoptera litura* Fabricius [69/166] (af; au; cm; ia; nt; or; ow); *Spodoptera ornithogalli* Guenée [25/53] (cm; na; nt); *Spodoptera praefica* Grote [16/35] (na); *Trichoplusia ni* Hübner [27/65] (af; cm; hw; na; nt; or; pl); *Xestia c-nigrum* Linnaeus [31/51] (hl; na; or; pl).

Nymphalidae. *Mechanitis lysimnia* Fabricius [2/6] (nt); *Mechanitis polymnia* Linnaeus [2/5] (nt).

Psychidae. *Apterona crenulella helix* Siebold [19/39] (pl).

Pyralidae. *Chilo suppressalis* Walker [3/19] (af; au; hw; ia; na; or; ow); *Duponchelia fovealis* Zeller [3/3] (af; pl); *Euzophera osseatella* Treitschke [9/15] (cm; pl); *Euzophera perticella* Ragonot [2/5] (or); *Herpetogramma infuscalis* Guenée [1/2] (nt); *Leucinodes orbonalis* Guenée [7/9] (af; cm; na; or; ow; pl); *Lineodes integra* Zeller [1/5] (na; nt; nw); *Neoleucinodes elegantalis* Guenée [2/4] (na; nt); *Ostrinia nubilalis* Hübner [16/39] (au; cm; hl; na; or; pl); *Phycita clientella* Zeller [2/3] (or); *Rhectocraspeda periusalis* Walker [8/11] (na; nt; nw); *Sceliodes laisalis* Walker [2/3] (af; cm; pl); *Spoladea recurvalis* Fabricius [23/37] (af; au; cm; hw; ia; na; nt; or).

Sphingidae. *Acherontia atropos* Linnaeus [16/38] (af; ow; pl); *Acherontia lachesis* Fabricius [19/45] (ia; or); *Acherontia styx* Westwood [12/30] (ia; or; pl); *Coelonia fulvinotata* Butler [14/29] (af); *Erinnyis ello* Linnaeus [10/18] (na; nt; nw); *Hyles lineata* Fabricius [36/61] (af; au; na; nt; nw; pl); *Manduca diffissa petuniae* Boisduval [1/5] (nt); *Manduca lichenea* Burmeister [1/2] (nt); *Manduca lucetius* Stoll [1/5] (nt); *Manduca quinquemaculatus* Haworth [2/9] (na; nt; nw); *Manduca quinquemaculatus blackburni* Butler [1/3] (hw); *Manduca sexta* Linnaeus [7/16] (na; nt; nw); *Manduca sexta jamaicensis* Butler [1/4] (nt); *Manduca sexta paphus* Cramer [3/7] (nt); *Manduca sexta sexta* Linnaeus [2/16] (na; nt); *Sphinx separatus* Neumoegen [1/1] (na).

Tortricidae. *Acleris variegana* Denis & Schiff. [7/14] (cm; hl; na; pl); *Amorbia emigratella* Busck [22/30] (hw; na; nt); *Cacoecimorpha pronubana* Hübner [24/36] (hl; pl); *Platynota stultana* Walsingham [25/39] (na; nt; nw).

Lycopersicon esculentum cerastiforme

Noctuidae. *Spodoptera litura* Fabricius [69/166] (af; au; cm; ia; nt; or; ow).

Lycopersicon hirsutum

Gelechiidae. *Keiferia lycopersicella* Walsingham [2/5] (cm; hw; na; nt).

Noctuidae. *Chrysodeixis chalcites* Esper [28/75] (af; au; cm; or; ow; pl); *Heliothis virescens* Fabricius [19/50] (na; nt; nw).

Lycopersicon pennellii

Solanum tuberosum

Arctiidae. *Estigmene acrea* Drury [23/46] (na; nt); *Hypercompe indecisa* Walker [13/19] (nt); *Spilosoma virginica* Fabricius [45/87] (na; or).

Cosmopterigidae. *Stagmatophora serratella* Treitschke [2/4] (pl).

Gelechiidae. *Brachmia insulsa* Meyrick [2/2] (or); *Bryotropha galbanella* Zeller [2/1] (nt; pl); *Gnorimoschema epithymella* Staudinger [1/3] (pl); *Helcystogramma convolvuli* Walsingham [2/4] (af; na; or); *Keiferia lycopersicella* Walsingham [2/5] (hw; na; nt; or); *Phthorimaea operculella* Zeller [6/14] (af; au; hw; na; nt; or; pl); *Scrobipalpopsis solanivora* Povolny [1/1] (nt); *Scrobipalpula absoluta* Meyrick [2/5] (nt); *Symmetrischema tangolias* Gyen [1/3] (au; na; nt; nz; or); *Tildenia glochinella* Zeller [1/2] (na; nt).

Geometridae. *Cleora alienaria gelidaria* Walker [5/5] (or); *Stenorumia ablunata* Guenée [1/1] (or); *Zeuctoboarmia simplex* Prout [1/1] (af).

Hepialidae. *Hepialus humuli* Linnaeus [13/19] (pl); *Korscheltellus lupulina* Linnaeus [10/18] (pl); *Pharmacis fusconebulosa* de Geer [3/3] (pl).

Lycaenidae.   *Loxura atymnus* Stoll [3/3] (or); *Tmolus echion* Linnaeus [14/12] (hw; nt; or).

Lymantriidae.   *Olene mendosa* Hübner [50/97] (af; au; or).

Noctuidae.   *Abagrotis alternata* Grote [8/12] (na); *Achaea finita* Guenée [5/7] (af); *Agrapha limbirena* Guenée [10/20] (af; or); *Agrotis exclamationis* Linnaeus [11/15] (pl); *Agrotis gladiaria* Morrison [11/17] (na); *Agrotis ipsilon* Hufnagel [30/74] (af; au; hw; na; nt; or; pl); *Agrotis longidentifera* Hampson [7/9] (af); *Agrotis malefida* Guenée [11/19] (na; nt); *Agrotis orthogonia* Morrison [11/27] (na); *Agrotis repleta* Walker [11/20] (nt); *Agrotis segetum* Denis & Schiff. [21/45] (af; or; pl); *Agrotis subalba* Walker [1/1] (af); *Agrotis subterranea* Fabricius [17/39] (na; nt); *Agrotis tokionis* Butler [6/8] (or; pl); *Amyna octo* Guenée [11/22] (af; au; na; or); *Anicla ignicans* Guenée [4/10] (nt; pl); *Ariathisa abyssinia* Guenée [4/5] (af; or); *Autographa californica* Speyer [26/44] (na); *Autographa gamma* Linnaeus [25/75] (na; nt; or; pl); *Chrysodeixis acuta* Walker [18/26] (af; or); *Chrysodeixis argentifera* Guenée [6/9] (au); *Chrysodeixis chalcites* Esper [28/75] (af; au; or; pl); *Chrysodeixis eriosoma* Doubleday [18/29] (au; or; pl); *Copitarsia incommoda* Walker [4/5] (nt); *Copitarsia turbata* Herrich-Schaeffe [3/3] (nt); *Episteme adulatrix* Kollar [2/2] (or); *Eublemma aurantiaca* Hampson [1/1] (af); *Euxoa albipennis* Grote [4/5] (na); *Euxoa auxiliaris* Grote [16/42] (na); *Euxoa cymograpta* Hampson [4/4] (af); *Euxoa declarata* Walker [4/6] (na); *Euxoa detersa* Walker [8/14] (na); *Euxoa messoria* Harris [16/27] (na); *Euxoa nigricans* Linnaeus [11/12] (pl); *Euxoa obelisca* Denis & Schiff. [5/6] (pl); *Euxoa ochrogaster* Guenée [10/18] (na; pl); *Euxoa tessellata* Harris [15/28] (na); *Euxoa tritici* Linnaeus [8/13] (pl); *Feltia jaculifera* Guenée [16/30] (na); *Gortyna flavago* Denis & Schiff. [9/18] (or; pl); *Helicoverpa armigera* Hübner [40/102] (af; au; na; nt; nz; or; pl); *Helicoverpa zea* Boddie [38/111] (af; hw; na; nt; or); *Heliothis virescens* Fabricius [19/50] (na; nt); *Hemieuxoa interrupta* Maassen [2/2] (nt); *Hydraecia medialis* Smith [2/2] (na); *Hydraecia micacea* Esper [15/27] (na; pl); *Lacanobia oleracea* Linnaeus [22/41] (pl); *Lacanobia subjuncta* Grote & Robinson [12/14] (na); *Lycophotia porphyrea* Denis & Schiff. [7/12] (hw; pl); *Mamestra brassicae* Linnaeus [14/27] (or; pl); *Mamestra configurata* Walker [16/33] (na; nt); *Melanchra picta* Harris [26/48] (na; pl); *Mentaxya albifrons* Geyer [1/1] (af); *Mentaxya ignicollis* Walker [2/2] (af); *Mocis undata* Fabricius [7/24] (af; or); *Neogalea sunia* Guenée [8/11] (au; hw; na; nt; or); *Neuranethes angola* Bethune-Baker [1/1] (af); *Noctua pronuba* Linnaeus [15/22] (na; or; pl); *Ochropleura flammatra* Denis & Schiff. [13/19] (or; pl); *Papaipema cataphracta* Grote [15/26] (na); *Papaipema nebris* Guenée [31/86] (na); *Peridroma saucia* Hübner [40/86] (na; nt; or; pl); *Phlogophora meticulosa* Linnaeus [12/23] (pl); *Pseudoleucania bilitura* Guenée [3/4] (nt); *Pseudoplusia includens* Walker [22/46] (na; nt); *Rachiplusia nu* Guenée [9/18] (nt); *Simplicia caeneusalis* Walker [8/10] (au; hw; or); *Spodoptera dolichos* Fabricius [14/22] (na; nt); *Spodoptera eridania* Stoll [31/63] (na; nt); *Spodoptera exempta* Walker [7/29] (af; au; hw; or); *Spodoptera exigua* Hübner [34/84] (af; au; hw; na; or; pl); *Spodoptera frugiperda* Smith [34/91] (af; na; nt; or); *Spodoptera littoralis* Boisduval [25/43] (af; na; or; pl); *Spodoptera litura* Fabricius [69/166] (af; au; nt; or); *Spodoptera ornithogalli* Guenée [25/53] (na; nt; or); *Spodoptera praefica* Grote [16/35] (na); *Syngrapha circumflexa* Linnaeus [7/8] (af; or; pl); *Trachea niveiplaga* Walker [1/1] (or); *Trichoplusia ni* Hübner [27/65] (af; hw; na; nt; or; pl); *Trichoplusia orichalcea* Fabricius [24/58] (af; au; or); *Tripseuxoa strigata* Hampson [4/7] (nt); *Xestia c-nigrum* Linnaeus [31/51] (na; or; pl).

Nolidae.   *Gyrtothripa papuana* Hampson [1/1] (au).

Nymphalidae.   *Mechanitis lysimnia* Fabricius [2/6] (nt); *Vanessa cardui* Linnaeus [25/104] (af; au; hw; na; or; pl).

Pieridae.   *Anaphaeis aurota* Fabricius [3/7] (af; or).

Psychidae.   *Pteroma pendula* Joannis [18/33] (au; or).

Pyralidae.   *Achyra affinitalis* Lederer [7/10] (au); *Ephestia elutella* Hübner [15/17] (na; nt; or; pl); *Ephestia kuehniella* Zeller [12/21] (na; nt; pl); *Euzophera osseatella* Treitschke [9/15] (or; pl); *Euzophera ostricolorella* Hulst [2/3] (na); *Euzophera perticella* Ragonot [2/5] (or); *Euzophera villora* Felder & Rogenh. [1/1] (af; or); *Hymenia perspectalis* Hübner [9/14] (na; nt; or); *Leucinodes orbonalis* Guenée [7/9] (af; na; or; pl); *Lineodes integra* Zeller [1/5] (na; nt); *Loxostege sticticalis* Linnaeus [17/35] (na; pl); *Ostrinia nubilalis* Hübner [16/39] (au; na; or; pl); *Plodia interpunctella* Hübner [25/48] (af; na; nt; or; pl); *Pyralis farinalis* Linnaeus [9/18] (af; na; nt; or; pl); *Pyralis manihotalis* Guenée [10/13] (na; nt; or); *Rhectocraspeda periusalis* Walker [8/11] (na; nt); *Rhodophaea nr. persicella* Amsel [1/1] (or); *Udea profundalis* Packard [10/12] (na).

Saturniidae.   *Leucanella memusae* Walker [10/13] (nt).

Sesiidae.   *Synanthedon rileyana* Edwards [1/2] (na).

Sphingidae.   *Acherontia atropos* Linnaeus [16/38] (af; or; pl); *Acherontia lachesis* Fabricius [19/45] (or); *Acherontia styx* Westwood [12/30] (or; pl); *Manduca diffissa petuniae* Boisduval [1/5] (nt); *Manduca quinquemaculatus* Haworth [2/9] (na; nt); *Manduca sexta* Linnaeus [7/16] (na; nt); *Manduca sexta jamaicensis* Butler [1/4] (nt); *Manduca sexta paphus* Cramer [3/7] (nt); *Manduca sexta sexta* Linnaeus [2/16] (na; nt).

Tortricidae.   *Amorbia emigratella* Busck [22/30] (hw; na; nt); *Cnephasia asseclana* Denis & Schiff. [9/15] (na; pl).

chapter four

Resisting resistance to Bt-corn

David A. Andow

Contents

4.1   Bt-*corn and pest resistance*

In 1995, Ciba Seeds (now Novartis) and Mycogen Seeds together introduced the first commercial *Bt*-corn hybrids in the United States. This was followed in 1996 by *Bt*-hybrids introduced by Northrup-King (now Novartis) and Monsanto. *Bt*-corn is made by transferring genes from the soil-dwelling bacterium *Bacillus thuringiensis* Berliner to corn, *Zea mays* L. These genes, which are called *cry* genes, code for crystal protein toxins (Cry toxins) that are toxic to some insects but not to mammals.

Table 4.1   Area of Transgenic Crops in the World from 1996 to 1999 (in millions of hectares)

	1996	1997	1998	1999
A. Country				
United States	1.5	8.1	20.5	28.7
China	1.1	1.8	+	0.3
Canada	0.1	1.3	2.8	4.0
Argentina	0.1	1.4	4.3	6.7
Australia	+	+	0.1	0.1
Mexico	+	+	+	+
B. Crop				
Soybean	0.5	5.1	14.5	21.6
Corn	0.3	3.2	8.3	11.1
Cotton	0.8	1.4	2.5	3.7
Canola	0.1	1.3	2.5	3.4
Tobacco	1.0	1.7	+	+
Tomato	0.1	0.1	+	+
Potato	+	+	+	+
C. Trait				
Herbicide tolerance	0.7	6.9	20.1	31.0
Insect resistance	1.0	4.7	8.0	11.8
Virus resistance	1.1	1.8	+	+
Quality traits	+	+	+	+
D. Total	2.8	12.8	27.8	39.9

Note: + indicates that <100,000 ha were grown. The first commercial crops were planted in China during the early 1990s. The first commercial production in the United States was tomatoes during 1994. Several crops were first commercialized during 1995, including *Bt* corn. (A) Area by country. By 1999, small areas of transgenic crops were grown in Spain, France, South Africa, Portugal, Romania, and the Ukraine. (B) Area by crop. Several minor crops are not listed. (C) Area by transgenic trait. This does not always sum to the worldwide total because some crops have more than one transgenic trait. (D) Total area worldwide.

Data from James (1997, 1998, 1999).

The present generation of *Bt*-corn protects the plant from damage by the larvae of European corn borer, *Ostrinia nubilalis* (Hübner) [Lepidoptera: Crambidae] (Ostlie et al. 1997, Rice and Pilcher 1998), and southwestern corn borer, *Diatraea grandiosella* Dyar [Lepidoptera: Crambidae] (Porter et al. 2000). Because these species are two of the most significant pests of corn in the U.S. corn belt, *Bt*-corn can in principle provide farmers with significant economic benefits. These perceived benefits have resulted in a rapid adoption by U.S. farmers (Table 4.1). Other moth pests, especially the noctuids fall armyworm *Spodoptera frugiperda* (J. E. Smith), corn earworm *Helicoverpa zea* (Boddie), black cutworm *Agrotis ipsilon* (Hufnagel), armyworm *Pseudaletia unipunctata* (Haworth), and stalk borer *Papaipema nebris* (Guenée), are not effectively controlled by *Bt*-field corn. Moreover, this generation of *Bt*-corn has no effect on non-moth pests, such as mites and beetles. Consequently, *Bt*-corn does not eliminate the need for farmers to remain vigilant in monitoring insect pests, and in some cases may not affect insecticide use patterns.

Bt-corn has been the focus of considerable controversy worldwide. Non-target species effects, antibiotic resistance marker genes, movement of transgenes to wild relatives of corn, and a large number of cultural factors have led to public resistance of *Bt*-corn in

Table 4.2 Area of Bt-Corn, Bt-Cotton, and Bt-Potato

	1995	1996	1997	1998	1999	2000
Bt-corn	+	0.3	2.8	6.7	9.6	6.3
Bt-cotton	+	0.7	0.8	na	2.2	2.2
Bt-potato	+	3,650	10,000	20,000	23,000	na[a]

[a] na = data not available.

Note: 1995–1997 and 2000 for the United States only; 1998–1999 for the world, although nearly all was planted in the United States. Corn and cotton data are in million hectares. Potato data are in hectares.

Data from James (1997, 1998, 1999) and NASS (2000).

Europe and Japan. This chapter focuses on a different scientific issue related to resistance. By killing pest insects, Bt-corn creates selection pressure on pests to evolve resistance to Bt-corn. Resisting the evolution of resistance is the complex scientific and management problem that is the central focus of this chapter.

4.2 Use of transgenic crops worldwide

The first commercial transgenic crops were planted in China during the early 1990s, primarily virus-resistant tobacco and tomato. In the United States, the first commercialized crop was Calgene's FLVR SAVR tomato in 1994. This product never achieved large acreage in part because it did not pack well for shipping. During the following year, numerous transgenic crops were commercialized, and by 1996, the United States was planting more transgenic crops than any other country in the world (Table 4.1A). The large decrease in planting area in China during 1998 (Table 4.1A) was caused by a collapse of the international markets for transgenic tobacco and tomato. During 1999, 12 countries planted some transgenic crops but except for the United States, Canada, and Argentina, the areas were small. Indeed, about 72% of the world's transgenic crops was planted in the United States alone during 1999.

Initially, a large variety of transgenic crops were planted (Table 4.1B). By 1999, however, four crops dominated: soybean, corn, cotton, and canola. The primary traits are herbicide tolerance and insect resistance (Table 4.1C). In 1999, herbicide-tolerant soybeans, Bt-corn, herbicide-tolerant corn, Bt-cotton, herbicide-tolerant cotton, and herbicide tolerant canola accounted for over 99% of the commercial transgenic crops grown worldwide (James 1999).

Although Bt-genes have been incorporated into broccoli, cabbage, canola, cotton, corn, eggplant, poplar, potato, soybean, tobacco, and tomato, the only crops planted on significant commercial acreage have been Bt-corn, Bt-cotton, and Bt-potato. Since their introduction during 1995, the cropping area of all of these transgenic crops has grown substantially (Table 4.2). By 1999, Bt-corn was grown on 9.6 million ha, fully one quarter of the total transgenic crop area worldwide. Bt-cotton lagged behind substantially in total area because about five times more corn than cotton is grown in the United States. Bt-potatoes failed to penetrate the market because of perceived detrimental agronomic properties and a resistant market. Clearly, Bt-corn in the United States is one of the dominant transgenic crops in the world today. Interestingly, Bt-corn area has decreased during 2000 (NASS 2000), probably in response to market uncertainty.

4.3 Types of Bt-corn

Not all Bt-corn hybrids are the same, and not all Bt-cry genes are the same. Indeed, Bt-Cry toxin in Bt-corn is not exactly the same as Bt-Cry toxin in B. *thuringiensis*, the source bacterium for *cry* genes. *Bacillus thuringiensis* is a soil-dwelling bacterium that produces large amounts of insecticidal delta-endotoxin when it sporulates. These endotoxins are biologically inactive protein toxins that crystallize into characteristic shapes. In bacteria, the endotoxins are mixtures of several specific crystalline protein toxins (Cry toxins) that are classified into several numbered major classes, which are themselves subdivided into many minor classes. The microbial *Bt*-insecticides targeting larvae of moths and butterflies typically contain toxins in the Cry1A class. Other strains of *B. thuringiensis* have different mixtures of Cry toxins. In addition, the bacteria also produce other toxins with insecticidal properties. The present generation of *Bt*-corn contains Cry1Ab, Cry1Ac, or Cry9C toxin, although most of the *Bt*-corn contains only Cry1Ab. Future versions of *Bt*-corn that contain various combinations of these toxins and Cry1F are under development.

The delta-endotoxins kill insects by a complex process. After ingestion, the crystals must dissolve in the insect midgut. This occurs readily when the pH of the midgut is alkaline, but occurs hardly at all under acidic conditions. In the presence of certain enzymes, the crystal releases a 130- to 135-kDa biologically inactive protoxin of Cry1Ab or Cry1Ac. In a series of poorly understood reactions, this protoxin is processed by proteolytic enzymes to yield a 65-kDa activated toxin that can bind to receptors on the midgut epithelium. This receptor–toxin complex somehow induces pore formation in the midgut wall, lysis of the midgut, septicemia, and rapid death of the insect. In short, the insect dies of stomach ulcers. The bacterium reproduces in the insect as a saprovore feeding on the cadaver, not as a pathogen feeding on living tissue. Cry1Ac is similar to Cry1Ab, but Cry9C is produced naturally as a 70.46-kD activated protein toxin.

In *Bt*-corn, the *Bt-cry* gene has been modified in at least two ways. First the gene has been truncated so that the resultant protein is not identical to the inactive protoxin from the bacterium. Different companies have truncated the genes differently, resulting in several different truncated *cry* genes in *Bt*-corn. Bacterial genes do not express well in plants, in part because bacterial genomes are enriched with the nucleotides A and T, whereas plant genomes are enriched in G and C. Consequently, the truncated *cry* genes are modified further to enrich their G-C content, allowing high levels of expression in plants. As discussed elsewhere in this volume (Chapter 7 by Hilbeck), these alterations may change the ecological effects of these toxins.

To make *Bt*-corn, it has been necessary to make a genetic construct containing the truncated *cry* gene, a marker gene, promoters for each, and other elements to facilitate expression and integration into the corn genome. These genetic constructs are used to transform corn cells, usually using a biolistics method. From the many transformation events associated with each construct, one is selected for commercial development. In the United States, five events have been commercialized in field corn, sweet corn, and popcorn (Table 4.3). These five form the present generation of *Bt*-corn.

The most complete information about the present generation of products is available for Event 176, BT-11, and CBH 351. Event 176 is commercialized by Novartis, which acquired Ciba Seeds, and Mycogen. It has two copies of a truncated *cry1Ab* gene, which has been truncated so that the gene product is quite similar to the activated Cry1Ab toxin. The promoters for these genes are the corn phosphoenol pyruvate carboxylase (PEPC) and a pollen-specific promoter, and the markers are ampicillin resistance and phosphinthricin herbicide resistance both of which are regulated by the cauliflower mosaic virus 35S promoter. In addition, the event has an intron of the corn PEPC gene to facilitate expression in corn. BT-11 is commercialized by Novartis, which acquired Northrup King. It has one copy of a truncated *cry1Ab* gene with the cauliflower mosaic virus (CaMV) 35S promoter; this gene is not truncated down to the active Cry1Ab toxin, but is shortened from the original bacterial gene. The marker is a phosphinothricin (= glufosinate ammonium) herbicide resistance gene, which is regulated by the CaMV 35S promoter, and the event

Table 4.3 Characteristics of the Five *Bt*-Corn Constructs[a]

Event	Truncated *cry* gene	Promoter	Marker	Expression of marker	Other elements
Event 176 (Ciba Seeds and Mycogen Seeds) pCIB4431[b]	*cry1Ab*, two copies[c]	PEPC promoter and a pollen-specific promoter	*Amp[r]* and *PPT[R]* (*bar*) with 35S (CaMV)[d]	*amp[r]* and PAT	Corn PEPC intron
BT-11 (Northrup-King) pZO1502[e]	*cry1Ab*, one copy[f]	35S (CaMV)	*PPT[R]* (*bar*) and *amp[r]* with 35S (CaMV)	PAT, *amp[r]* gene not present in final construct	Intron of corn ADH 1S gene
Mon 810 (Monsanto) PV-ZMCT01	*cry1Ab*[g]	35S (CaMV)	EPSPS NptII, *promoter?*	No detectable gene products of *nptII*	?
CBH 351 (Plant Genetic Systems)	*cry9C*, one copy[h]	35S (CaMV)[b]	*PPT[R]* (*bar*), ~four copies, promoter?	PAT	?
DBT 418 (DeKalb Seeds) three plasmids	*cry1Ac*, two copies[i]	35S (CaMV)[j]	*PPT[R]* (*bar*) two copies,[k] promoter?	PAT	?

Note: As of February 22, 2000, DBT-418 has not been marketed in the United States. Registrations for Event 176 will expire in 2001. Bt-11 is also in sweet corn and popcorn.

[a] PEPC = phosphoenol pyruvate carboxylase; CaMV = cauliflower mosaic virus; *nptII* codes for neomycin phosphotransferase, which confers resistance to several antibiotics; *amp[r]* codes for ampicillin resistance, another kind of antibiotic; *PPT[R]* = phosphinthricin herbicide resistant; PAT = phosphinothricin acetyltransferase, which is the product of *PPT[R]*; for Event 176, BT-11, and DBT-418 it is from *Streptomyces viridochromogenes* (*bar*); EPSPS = 5-enolpyruvyl shikimate-3-phosphate synthase (bacterial origin) provides glyphosate herbicide resistance; *bar* PAT gene from *Streptomyces hybroscopicus*; ADH = alcohol dehydrogenase.

[b] Supplemental information for this event from Koziel et al. 1993.

[c] Truncated toxin with 648 of the 1155 amino acids in the non-truncated version; Koziel et al. 1996.

[d] USDA permit 92-042-01.

[e] Additional information on this event from USDA permit 92-017-03.

[f] Truncated *cry* gene (EPA 2000f). Characterized by SDS-PAGE. The gene comes from the same source as that used in Mon 810, but introduced on plasmid pZO1502. The degree of truncation is not specified, but the core proteins of both plant and microbe after digestion with typsin have the same molecular weight.

[g] Truncated *cry* gene (EPA 2000f). Characterized by Western blots. Introduced on plasmid PV-ZMCT01, which itself is two plasmids, PV-ZMBK07 and PV-ZMGT10, ballistically introduced together into corn. The original gene was further truncated during process of making Mon 810. The number of introduced *cry* genes is not specified, but it could be more than one. Western blots of Mon 810 not treated with trypsin reveal many bands, with one band comigrating with a full-length cry1Ac standard. After trypsin digestion, an intensified band comigrated with the Cry1Ab tryptic core standard. Thus, Cry1Ab toxin is produced by Mon 810, but other related molecules may also be produced.

[h] One amino acid substitution in internal sequence, two additional amino acids at N-terminus (EPA 2000e). Characterized by SDS-PAGE, Western, Northern, and Southern blots, terminal amino acid sequencing, and glycosylation tests.

[i] First 613 amino acids of the protoxin.

[j] Not available in public literature.

[k] One copy of *bar* is rearranged. Also includes a partial copy of *pinII*. Products of *pinII* not detectable.

Source: EPA 2000a–e, except as noted.

has an intron of the corn alcohol dehydrogenase 1S gene to facilitate expression in corn. Novartis should be recognized for publishing the linkage maps and details of the structure of the toxins in its events (although more detail would be welcome for BT-11), and all others should be encouraged to do likewise.

Two of the events have had the structure of their toxins characterized in detail, but the linkage map is not available. CBH 351 is commercialized by Aventis, which acquired Plant

Table 4.4   Expression of Cry Toxin in *Bt*-Corn Plants (mg/g)

Event	Grain	Leaf	Pollen	Pith	Root	Whole plant
Event 176	<5	4.4	7.1			0.6
BT-11	1.4 (kernel)	3.3	<0.09 (pollen dry weight)		2.2–37.0 (protein)	6.3
Mon 810[a]	0.19–0.39 (grain)	10.34	<0.09 (pollen dry weight)			4.65
CBH 351	18.6 (kernel)	44	0.24	2.8	25.87	250
DBT 418	43	1.2	Not detectable			0.15–1.0

Note: CBH 351 is Cry9C, DBT 418 is Cry1Ac, and the others are Cry1Ab. All values are expressed per fresh tissue weight unless otherwise noted.

[a] Based on 1994 field data. Data from other years provide slightly different values for grain, leaf, and whole plant concentrations.

Data from U.S. EPA Fact Sheets (EPA 2000a–e) and EPA (2000f).

Genetic Systems. It has one copy of a *cry9C* gene with the CaMV 35S promoter, which produces a 73.79-kD toxin, 3.42 kD larger than the toxin produced by the bacterium. The plant toxin appears to lose three to seven amino acid residues from the N-terminus from protease degradation to produce a protein similar to the bacterial protein. This protein toxin is characterized in considerable more detail than the Cry1A toxins in order to evaluate its potential for allergenicity to humans. The marker used in this event is phosphinthricin herbicide resistance with an unspecified promoter. Most events contain an intron to facilitate expression in corn, but this is not specified for CBH 351. DBT 418 was commercialized by DeKalb, which was later acquired by Monsanto. It is the only event expressing Cry1Ac toxin and contains two copies of a gene truncated to express activated toxin. The plant Cry toxin has the first 613 amino acid residues of the bacterial protoxin, and both genes use the CaMV promoter. The marker system is quite complicated, but in the end, it appears that phosphinthricin herbicide resistance is the only one expressed. Its promoter is not specified. These details about the structure of the gene product from the *cry* gene are important for understanding potential effects on non-target species, as these subtle changes may influence the range of species affected by the gene product.

The remaining event has not been adequately described in the public literature, lacking both detailed characterization of the toxin and a published linkage map. Mon 810 is commercialized by Monsanto and was formed from two different constructs. It contains at least one copy of a truncated *cry1Ab* gene with the cauliflower mosaic virus (CaMV) 35S promoter; this gene is not truncated down to the active Cry1Ab toxin, but is shortened from the original bacterial gene. Although the original gene is the same truncated gene used to make BT-11, in Mon 810 it is further reduced in size. The number of gene inserts in Mon 810 is not specified, and the diversity of expression products may indicate that there is more than one. The markers are *nptII*, an antibiotic-resistance gene, and a glyphosate herbicide-resistance gene, with unspecified promoters. Expressed products of *nptII* are not detected in corn plants. All events should be adequately described in the scientific literature.

The different events can result in phenotypic differences in expression of activated Cry toxin (Table 4.4). BT-11 and Mon 810 have similar but not identical levels of expression in the whole plant. The similarity might have been predicted because they share a similar truncated *cry* gene and use the same promoter, but the differences suggest real differences in expression. Event 176 uses two different promoters, one expressing in green leaf tissue and one that is pollen specific. Consequently, expression of Cry toxin is lower in grain and the whole plant but higher in pollen than BT-11 and Mon 810. DBT 418 expresses Cry1Ac toxin, which is not as toxic as Cry1Ab toxin to the major pests of corn in the

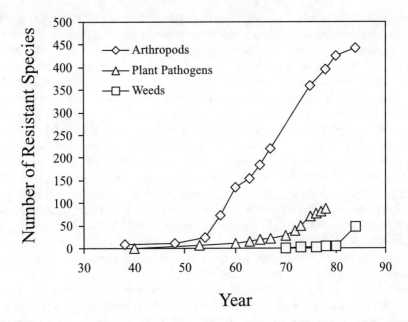

Figure 4.1   Number of species resistant to insecticides and acaricides (arthropods), fungicides (plant pathogens), and herbicides (weeds). (Data originally from Georghiou 1986.)

United States, and also expresses this toxin at a low concentration. CBH 351 expresses Cry9C, which is more toxic than Cry1Ab, at high concentrations. These variations have important implications for resistance management, as will be discussed below.

4.4   Need for resistance management

Resistance in insects to pest control is a serious problem worldwide. Although resistance problems have been known for nearly 100 years, resistance became a significant agricultural problem after World War II (Figure 4.1), when modern, intensive agricultural technologies proliferated. Resistance occurs quickly when there is strong, uniform selection on a pest population for long periods of time over spatially extensive areas. Modern intensive agriculture, with its reliance on pesticides, monoculture, and uniform production practices provides these conditions, and resistance has proliferated. In the United States alone, the cost of resistance to insecticides and acaricides has been about $133 million annually in extra insecticide applications, measured in 1980 dollars (Pimentel et al. 1980). For some pests, such as Colorado potato beetle and diamondback moth, resistance is so extensive that few effective pest control alternatives remain. In northeastern Mexico and the lower Rio Grande of Texas, resistance to insecticides evolved in tobacco budworm, a pest of cotton, in early 1970. This caused the loss of about 700,000 acres of cotton (Adkisson 1971, 1972, NAS 1975), devastating the local communities. Resistance to *Bt*-toxins has been documented in 17 insect species (Tabashnik 1994, Huang et al. 1997), so it is now widely assumed that resistance to *Bt*-corn can occur.

The goal of *Bt*-corn resistance management is to delay or prevent the evolution of resistance in European corn borer, southwestern corn borer, and corn earworm. Effective resistance management will allow farmers to use *Bt*-corn for a long period of time. Because *Bt*-corn can provide yield benefits to farmers averaging between 7 and 18 bushels/acre in the northern corn belt (Rice and Pilcher 1998), if it costs an additional $10/acre, *Bt*-corn can net a farmer $4 to $26/acre even at very low corn prices. Loss of this income because of resistance evolution could have significant detrimental effects on struggling farm families. *Bt*-corn, however, is not without risk to farmers. If there is little insect pest damage, there may be no yield increase, and the farmer can lose the $10/acre paid for the *Bt*-corn seed. Most U.S. corn receives no insecticide applications, but on the 5% that does, the potential benefits from using *Bt*-corn are uncertain. In much of this "high-insecticide" use area, spider mites, a leaf-feeding pest, are a problem and miticides are commonly applied. *Bt*-corn does not control these mites, but the miticides do control European corn borer and southwestern corn borer. Consequently, it is not yet clear if farmers in these regions will receive benefits from the use of *Bt*-corn and if insecticide use will decline.

Needless to say, resistance management is also beneficial to the companies that sell *Bt*-corn. The corn seed market is a highly competitive, $4 billion/year market in the United States, and now that most major seed companies are selling *Bt*-corn hybrids, prolonging the life of this product will enable the companies to make additional profits. If profit were the only reason for resistance management, there would be little need for society to intervene to ensure that effective resistance management occurs. The major users of *Bt*-corn would be the major beneficiaries of resistance management, and they would pay the costs of poor stewardship and resistance failures.

Two other reasons have compelled society to take an active role in ensuring that effective resistance management is implemented. First, there are other farmers who depend on *Bt*-based insecticides and do not use *Bt*-corn. For example, under present guidelines, *Bt*-sprays, but not *Bt*-corn, can be used as a part of organic agricultural production. If pests evolved resistance to *Bt*-corn, the other *Bt*-based insecticides would likely become ineffective against those pests. In other words, farmers who experience no benefit from *Bt*-corn might have to pay the costs of poor stewardship by others. Resistance management helps protect the interests of those farmers who do not use *Bt*-corn but do rely on *Bt*-insecticides.

Second, resistance management of *Bt*-corn is important because it preserves a pest control method that results in less harm to the environment and human health than many other insecticides. *Bt*-based pest control has several significant advantages over traditional synthetic insecticides. Cry toxins have a narrow range of non-target species effects, very low mammalian toxicity, no record of carcinogenicity, and no documented adverse effects in aquatic environments. Loss of *Bt*-based controls because of the evolution of resistance would probably increase use of insecticides that are more harmful to the environment or human health. In addition, effective resistance management can help stabilize pest control in the future. Although the U.S. Environmental Protection Agency (EPA) registers pesticides only after lengthy reviews, unregistered pesticides can be used under emergency exemptions. Use of unregistered pesticides under emergency exemptions may cause unanticipated environmental or human health risks. During 1991–1994, about 30% of all emergency exemptions requests were made, at least in part, because of resistance (Matten et al. 1996). With effective resistance management, the need for emergency exemptions could be significantly reduced.

4.5   *The static resistance management strategy*

Insect resistance management (IRM) can be characterized as either responsive or preemptive. Responsive strategies react to the widespread occurrence of field resistance, while preemptive strategies attempt to avoid or delay resistance before it occurs in the field (Brown 1981, Dennehy 1987, Sawicki and Denholm 1987). Most IRM strategies have been reactive (Denholm and Rowland 1992), yet for transgenic insecticidal crops preemptive strategies have predominated (Gould 1998). IRM for transgenic *Bt*-crops, and *Bt*-corn in particular, is based on the preemptive high-dose plus refuge strategy (EPA 2000a–e).

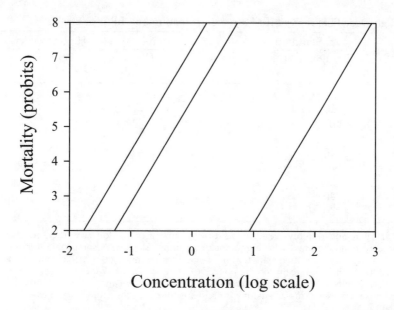

Figure 4.2 Hypothetical mortality of *SS*, *SR*, and *RR* genotypes as a function of *Bt*-crystal protein concentration. (From Georghiou and Taylor 1977, Tabashnik et al. 1992.)

4.5.1 Scientific assumptions

The high-dose plus refuge strategy requires that *Bt*-corn produce a high toxin concentration and that non-*Bt*-corn is planted nearby as a refuge for pests. This preemptive strategy relies on three essential assumptions (Andow and Hutchison 1998). Computer simulation models show that if the first three assumptions discussed below hold, the evolution of resistance will be substantially delayed (Comins 1977, Tabashnik and Croft 1982, Gould 1986, Roush 1994, 1996, Alstad and Andow 1995). The fourth assumption has practical implications for resistance monitoring.

4.5.1.1 High dose

Plant tissue must be very toxic (Figure 4.2, concentration *h* or higher), so that resistance is functionally recessive in *Bt*-corn (Tabashnik and Croft 1982). If resistance is determined by allelic variation at a single locus, let *R* designate the resistance allele and *S* designate the susceptible allele. At toxin concentration *h*, nearly all resistance heterozygotes will be killed, providing selection against the *R* allele to counter the selective advantage of the *RR* homozygotes. In non-*Bt*-corn there is no selection on resistance.

4.5.1.2 Resistance rare

The resistance alleles must be sufficiently rare — the frequency should be $<10^{-3}$ (Roush 1994); even lower is better — so that nearly all resistance alleles will be in heterozygote genotypes. If nearly all *R* alleles are heterozygotes, they can be eliminated by the *Bt*-corn. Of course, both *R* and *S* alleles survive in the non-*Bt*-corn. For example, suppose the frequency of resistance alleles is 0.001. Under random mating, about 2×10^{-3} individuals will be *RS* heterozygotes and 1×10^{-6} will be *RR* homozygotes. If the frequency of resistance alleles is 0.0001, about 2×10^{-4} individuals will be *RS* heterozygotes and 1×10^{-8} will be *RR* homozygotes. Because a 100-acre corn field can have between 300,000 and 15,000,000 European corn borers, the number of *RR* homozygotes will be increasingly rare as the frequency of resistance alleles is lower.

Table 4.5   High-Dose Plus Refuge IRM Strategies Required by EPA for 2000

	Standard		Southern region		High insecticide use region	
	Refuge, %	Distance, mile	Refuge, %	Distance, mile	Refuge, %	Distance, mile
Event 176						
Field corn	20	$1/2$	50	$1/2$	X	X
Popcorn	20/40	$1/2$	20/40	$1/2$	X	X
BT-11						
Field corn	20	$1/2$	50	$1/2$	20	$1/4$
Sweet corn	N/R		N/R		N/R	
Mon 810	20	$1/2$	50	$1/2$	20	$1/4$
CBH 351	20	$1/2$	20	$1/2$	20	$1/4$
DBT 418	20	$1/2$	X	X	X	X

Note: The southern region roughly corresponds to areas that grow cotton. The high-insecticide-use region roughly corresponds to areas with high attack by southwestern corn borer that are not cotton-growing areas. The standard requirements are for all other corn-growing regions. All 20 and 50% refuges may be treated with insecticides, except for the popcorn refuge for Event 176, for which, if treated, the refuge must be 40%. X indicates that the event cannot be planted. N/R means nothing required for that criterion. Sweet corn has additional management requirements to reduce the probability of overwintering survival of resistant insects.

Source: EPA 2000a–e.

4.5.1.3   Random mating

The non-*Bt*-corn refuges must be interspersed sufficiently among the *Bt*-corn fields. The assumption is that insect pests emerging from the non-*Bt*-refuges will mate randomly with insects surviving the *Bt*-fields (Comins 1977, Tabashnik and Croft 1982). If this occurs, then nearly all resistant homozygotes will mate with susceptible homozygotes, producing heterozygous progeny that cannot survive on the *Bt*-crop. For example, suppose that the initial frequencies of resistant and susceptible alleles are $p = 0.001$ and $q = 0.999$, respectively. If resistant *RR* insects from *Bt*-plots mate randomly with insects from refuges, then desirable $RR \times SS$ matings can be a million times more common than undesirable $RR \times RR$ matings (depending on refuge size and selection differential).

4.5.1.4   Spatial structure

Pest populations must be structured so that there is minimal spatial structure (Caprio and Tabashnik 1992). When the population is spatially structured, there is likely to be considerable variance in resistance allele frequency among populations at different locations. Consequently, the time to failure would be highly variable, and more populations would have to be monitored.

4.5.2   IRM requirements for the year 2000

The EPA (2000a–e) has used the high-dose plus refuge strategy as its core strategy for IRM in *Bt*-corn (Table 4.5). The standard refuge configuration is for 20% of the corn to be non-*Bt*-corn refuge that is located within $1/2$ mile of the *Bt*-corn, which is referred to as the 20% structured refuge strategy. This is generally interpreted to mean that the distance from the closest field edges of the *Bt*-corn and refuge should be within $1/2$ mile. It is also possible that it means that all *Bt*-corn plants should be within $1/2$ mile of a refuge; that is, the farthest field edge of the *Bt*-corn field must be within $1/2$ mile of the refuge. The EPA (2000b) does not require any refuge for *Bt*-sweet corn, because there are deemed to be sufficient other refuges available on farms and because it is projected to reach only a small acreage (at most 125,000 acres by 2002). The EPA (2000c) is maintaining previous IRM requirements for *Bt*-popcorn, and because no popcorn is grown in the southern region of the United States, only the standard requirements are needed (Table 4.5).

The EPA (2000a–d) has designated special IRM requirements in the southern region where cotton is grown (Table 4.5). In this region, corn earworm is common and feeds on both corn and cotton. Because *Bt*-cotton (expressing Cry1Ac) is grown in this region, and cotton is the major crop for most farmers, the EPA believes it is necessary to protect against evolution of resistance in corn earworm in this region in order to preserve both *Bt*-cotton and *Bt*-corn. Because all corn earworm populations pass a generation on corn during the annual cycle, a larger corn refuge of 50% non-*Bt*-corn will relax selection on corn prior to selection on *Bt*-cotton. The corn refuges are expected to delay resistance to both crops. CBH 351 is not required to have a larger than 20% refuge in the southern region because Cry9C is not toxic to corn earworm (EPA 2000e).

For the 2000 growing season, EPA identified an additional region for special IRM requirements, called the high-insecticide-use region. In this region (parts of Colorado, Kansas, Texas, and Oklahoma), insecticides are commonly used, primarily against southwestern corn borer and mites. Consequently, it is likely that refuges will be sprayed with insecticides, reducing their effectiveness at delaying resistance. In part to compensate for this, the EPA requires that refuges are planted closer (within 1/4 mile) to *Bt*-corn, so that survivors from the sprayed refuge do not have as far to go to find the resistant individuals emerging from the *Bt*-corn.

4.5.3 *Validity of the scientific assumptions*

The validity of the scientific assumptions underlying present IRM strategies can be evaluated for European corn borer, southwestern corn borer, and corn earworm, the primary pests affected by *Bt*-corn. This analysis is summarized in Table 4.6.

4.5.3.1 *High dose*

Unfortunately, the high-dose assumption cannot be scientifically verified for any of the key pest species at the present time (Andow and Hutchison 1998). As illustrated in Figure 4.2, the high dose assumption requires that *RS* heterozygotes are functionally recessive. This can only be determined for actual resistance alleles. Because resistance to *Bt*-corn has not yet been recovered in any of the species, this assumption simply cannot be tested. A concerted effort must be made to find resistance in all three species so that the high-dose assumption can be properly evaluated.

There have been some attempts to designate a high dose as some arbitrary toxin concentration, such as 25 times the toxin concentration needed to kill susceptible larvae (EPA 1998). This concept relies on previous data from the insecticide resistance literature that 25 times the LC_{50} may be a functional high dose, which means that such a concentration would kill nearly all *RS* heterozygotes. The LC_{50} is the concentration of toxin at which 50% of the individuals in the tested population die (*lethal concentration 50*). Using this line of argument, Caprio et al. (2000) argue that a better definition of high dose is 50 times the LC_{50}. Huang et al. (1999) found resistance in European corn borer, in which the LC_{50} of heterozygotes to *B. t. kurstaki* was about 65 times that of susceptible insects. This resistance was inherited as a single, autosomal incompletely dominant trait. This example shows that a definition of 25 times the LC_{50} would be insufficient to guarantee a high dose for European corn borer.

The value of the EPA standard is that it is readily operationalized for transgenic *Bt*-crops. If the genetics and physiology of resistance to *Bt*-crops were the same as the genetics and physiology of resistance to insecticides, this would be a reasonable approach. The

Table 4.6   Status of Assumptions of the High-Dose plus Refuge Strategy for *Bt*-Corn Resistance Management for the Key Target Pests

Criterion	Event 176	BT-11	MON 810	CBH 351	DBT 418
European corn borer					
High dose					
Overall	–	+?	+?	+?	–
1st Gen.					
a	0+	0+	0+	0+	0+
b	0+	0+	0+	0+	0+
2nd Gen.					
a	–	0+	0+	0+	–
b	–	0+	0+	0+	–
Resistance rare	(+)/?	(+)	(+)	?	?
Random mating	+/–	+/–	+/–	+/–	+/–
Southwestern corn borer					
High dose					
Overall	–	+?	+?	+?	–
1st Gen.					
a	0+	0+	0+	0+	0+
b	0+	0+	0+	0+	(–)
2nd Gen.					
a	–	0+	0+	0+	–
b	–	0+	0+	0+	–
Resistance rare	?	?	?	?	?
Random mating	?	?	?	?	?
Corn earworm					
High dose					
Overall	–	–	–	na	–
a	–	–	–	na	–
b	–	–	–	na	–
Resistance rare	?	?	?	na	?
Random mating	(+)	(+)	(+)	na	(+)

Note: ++ indicates the assumption has been verified; – indicates that the assumption does not hold; 0/+ indicates that the assumption has not been falsified; na indicates the assumption is not relevant; ? indicates no scientific data are available; ( ) indicates that the assumption is nearly verified. The criteria to falsify the high-dose assumption are (1) toxin concentration in the plants is uniform and high in space and time during the period of pest attack, and (2) the plants kill >99% of susceptible individuals. See text for additional details.

fundamental problem with this approach, however, is that it does not allow one to anticipate whatever novel resistance mechanisms might occur. This approach creates a false sense of security about the validity of the high-dose assumption. The actual data suggest that, although the veracity of the high-dose assumption is probable for some *Bt*-corn events (see below), it is not yet possible to verify that assumption.

Present data on resistance to *Bt* in several insects provide cautious optimism that the high-dose assumption is valid for some *Bt*-corn events. The genetics and physiology of *Bt*-toxin resistance have been investigated in several insects. Resistance is characterized by the resistance ratio, which is the ratio of the LC_{50} of the resistant line to the LC_{50} of the susceptible line. The resistance ratios for various insects to *Bt*-toxin vary from 12 to >1000. For high resistance ratios, resistance to *Bt*-toxin is inherited as a single, autosomal fully or incompletely recessive trait. In Indian meal moth, the resistance ratio reached 253 to *B. t. kurstaki* and resistance was inherited as a single, autosomal recessive trait (McGaughey and Beeman 1988). In tobacco budworm the resistance ratio reached >2100 for Cry1Ab toxin, and resistance was inherited as a single autosomal incompletely recessive trait (Gould et al. 1995). Diamondback moth has evolved resistance to *Bt*-toxin several times. In a resistant population from the Philippines, resistance to Cry1Ab toxin reached a ratio >200 (Ferré et al. 1991), and was inherited as a single, autosomal incompletely recessive trait with a paternal effect (Martinez-Ramirez et al. 1995). A resistant population from Osaka, Japan had a resistance ratio of 704 to a commercial formulation of *Bt*, and resistance was inherited as a single, incompletely recessive autosomal trait (Hama et al. 1992). A Hawaiian population had a resistance ratio of 720 to *B. t. kurstaki*, and resistance was inherited as a single, autosomal recessive trait (Tabashnik et al. 1992).

The mechanism of resistance for these high-resistance-ratio traits appears to center on the receptors in the midgut epithelium and associated reactions, and for most examples is consistent with a recessive inheritance model. Reduced binding of Cry toxin to receptors in the brush border membrane of the midgut epithelium is the primary mechanism of resistance in Indian meal moth (Van Rie et al. 1990) and diamondback moth (Ferré et al. 1991). However, resistance in a Hawaiian diamondback moth population is not associated with reduced binding (Masson et al. 1995, Escriche et al. 1995), suggesting that another resistance mechanism can provide high levels of resistance. Resistance based on reduced binding is likely to be recessive. If *SS* susceptible individuals produce receptors that bind the toxins, then *RS* individuals may produce half as many receptors, which will still bind toxin resulting in death. Only *RR* individuals will produce no receptors and be resistant.

For lower resistance ratios, the genetics of resistance is more complex. Gould et al. (1992) selected a resistant strain of tobacco budworm (CP 73) that had a resistance ratio of only 50. This resistant strain showed no differences in binding characteristics when compared with a susceptible strain, and the resistance was inherited either as an incomplete recessive or an additive trait. Sims and Stone (1991) selected another resistant strain of tobacco budworm with a resistance ratio of 12. Resistance was multigenic and inherited as a dominant trait in the F_1 generation. Huang et al. (1999) selected a colony of European corn borer to have a resistance ratio of 65 to *B. t. kurstaki*, and showed that resistance was inherited as a single, autosomal incompletely dominant trait. The mechanistic basis of this resistance is not known, but the result raises the concern that a high-level dominant resistance might be possible in European corn borer.

Significantly, none of these resistance populations with low resistance ratios can survive on *Bt*-crop plants (either *Bt*-corn or *Bt*-cotton). This suggests that resistance to the present generation of *Bt*-crops will require a high resistance ratio (>>100). Because the empirical data suggest that these high levels of resistance are likely to be recessive, *Bt*-corn may meet the high-dose assumption for European corn borer. Recent work on the mechanisms of resistance (Masson et al. 1995, Escriche et al. 1995) and the inheritance of resistance (Huang et al. 1999), however, raises questions about the validity of this assumption. These questions can only be fully addressed when resistance to *Bt*-corn is recovered.

Although verification of the high-dose assumption is at present fraught with difficulties, in some cases the assumption can be scientifically rejected, because a less exacting test can be used to reject an assumption than to verify it. For example, a corollary to the assumption that the *RS* heterozygotes are functionally recessive is that survival of *SS* homozygotes is extremely low. At the high-dose concentration, *h* (Figure 4.2), survival of *SS* homozygotes is <1%. If survival of *SS* homozygotes were higher, then many more *RS* heterozygotes would survive, and the advantage of the high-dose plus refuge strategy would be lost. Three criteria have been developed and used to evaluate this possibility (Andow and Hutchison 1998, ILSI 1999), and two are particularly relevant. A high dose is unlikely if toxin concentration in the plants is (1) not uniform among plant parts and during the period of pest attack, or if (2) the plants kill less than 99% of susceptible individuals. Data on the concentrations of toxin in corn plants for the different transformation events are given in Table 4.4. During vegetative growth, toxin concentrations are low in DBT 418, but high in all of the other events.

European corn borer. In the central corn belt, European corn borer has two generations/ year on corn. The first occurs during the vegetative stage of corn and is usually complete before tassel elongation. The second begins during the pollen shedding and silking stages. Larvae from the second generation usually overwinter as last instars in the cornstalks. In some areas, a univoltine population attacks corn between the two generations of the bivoltine populations. Because this generation of the univoltine population begins during the vegetative stage of corn, efficacy is similar to that for the first generation of the bivoltine population. Survival of unselected European corn borer populations on the different *Bt*-corn events is quite variable among varieties and between the generations of the bivoltine population. The extensive efficacy data for the five transformation events indicate that BT-11 and Mon 810 both have excellent control for both first and second generation. Event 176 and DBT 418 have excellent control during the first generation. After anthesis, toxin concentrations drop in Event 176 and there is good survival of susceptible larvae that are associated with the young corn ear during the early instars. The declining levels of toxin may kill larvae in other locations on the plant. DBT 418 provides economically significant control during the second generation by causing enough mortality to susceptible larvae. Less extensive data on CBH 351 indicate that CBH 351 has excellent control in both the first and second generation. Consequently, while BT-11, MON 810, and CBH 351 may be high-dose events, Event 176 and DBT 418 are not.

Southwestern corn borer. Similar to European corn borer, southwestern corn borer has two generations/year, both of which overlap substantially with the European corn borer generations. Efficacy data are not as extensive for this species, but in all trials efficacy for southwestern corn borer has been similar to that for European corn borer. First-generation survival on DBT 418 is higher than for any of the other events. Consequently, BT-11, MON 810, and CBH 351 may be high-dose events for southwestern corn borer, but Event 176 and DBT 418 probably are not.

Corn earworm. Corn earworm has several generations per year, but only one generation is associated with corn at any location. During this generation, larvae complete development in the ear. None of the events provides excellent control of corn earworm, and although BT-11 and MON 810 provide uniform toxin concentrations, corn earworm tolerates higher concentrations of Cry1A toxin than either European or southwestern corn borer. Thus, none of the events should be considered high-dose events against corn earworm. This provides the essential justification for the EPA requirement (Table 4.4) to have 50% refuges in the cotton-growing areas where corn earworm is most prevalent. CBH 351 probably does not control this species at all, creates little risk of resistance, and therefore there is little need to alter the requirements for this event.

4.5.3.2   Resistance rare

Resistance must be rare enough to ensure that most resistance alleles occur in heterozygote genotypes, thereby allowing the high-dose plus refuge strategy to work. In a series of simulations, Roush (1994) suggested that recessive resistance gene frequency should be $<1 \times 10^{-3}$ for the high-dose plus refuge strategy to delay resistance effectively. Although estimating allele frequencies at that low level can be challenging, two methods have been developed to estimate rare resistance frequencies, an F_2 screen (Andow and Alstad 1998) d an in-field screen (Andow and Hutchison 1998, Venette et al. 2000).

European corn borer. The frequency of resistance in European corn borer has been ated using an F_2 screen (Andow and Alstad 1998, 1999). This method screens for any

Table 4.7 Estimated Recessive Resistance Allele Frequency Using an F_2 Screen in Several Populations of European Corn Borer

Population	n	Bt-corn resistance Expected frequency	95% CI	Partial resistance Number recovered	Expected frequency	95% CI
LeSueur, MN, 1996	81	0.0030	≤0.009	1	0.0054	(0.001, 0.015)
Ames, IA, 1997	188	0.0013	≤0.0039	2	0.0039	(0.0008, 0.0094)
Lamberton, MN, 1999	153	0.0016	≤0.0057	2	0.0048	(0.001, 0.012)

Sources: Andow et al. 1998, 2000, Andow and Alstad 1999, Andow and Ives, in press.

and all resistance alleles in a population by inbreeding isofemale lines so that resistance is expressed in some of the F_2 larvae. In three populations sampled, no resistant isofemale line was found, and the 95% credibility interval for recessive resistance to Bt-corn is almost as low as 1×10^{-3} (Table 4.7). Unpublished F_2 screen data extend this result for a French population of European corn borer where exposure to Cry toxin has been slight (Bourguet et al. unpublished). The in-field screen has been used extensively with Minnesota populations as well. These unpublished results corroborate the results from the F_2 screen. Partial resistance to Bt-corn is prevalent in populations of European corn borer (Table 4.7). Together, these results suggest that resistance in European corn borer may be low enough to enable the high-dose plus refuge strategy to be effective.

Southwestern corn borer. Resistance has not been systematically evaluated in this species. Because of the resistance risk associated with this species, research on the frequency of resistance in this species is needed.

Corn earworm. Similarly, the frequency of resistance to Bt-corn has not been systematically evaluated in this species. Because a significant number of susceptible corn earworms survive on all of the different Bt-corn hybrids, it may be difficult to distinguish the true resistant insects from the susceptible individuals.

4.5.3.3 *Random mating*

Random or near-random mating is needed to dilute out the resistance alleles in the Bt-corn fields, creating heterozygote offspring that can be killed by Bt-corn. Random mating is difficult to observe directly because moths mate at night using pheromone cues. The effects of non-random mating in a population can be observed indirectly by measuring the degree of inbreeding in the local population. This can be done by genotyping many individuals in the local population using co-dominant genetic markers, such as selectively neutral electrophoretic allozymes. Under random mating, the frequency of heterozygotes is predicted to be $2p(1 - p)$, where p is the frequency of one allelic variant. By comparing the observed heterozygosity with the predicted heterozygosity, an inbreeding index, F_{IS}, can be calculated to measure the deviation from random mating in the local population (Wright 1978). A non-zero F_{IS} indicates the population is not mating randomly, and clearly requires rejection of the random-mating hypothesis.

That random mating occurs among individuals in a local population does not necessarily imply that random mating will occur between resistant and susceptible moths. For example, resistant individuals feeding on Bt-corn may develop more slowly than susceptible ones feeding on non-Bt-corn, thereby creating temporal asynchrony and non-random mating (Gould 1998, Liu et al. 1999). This possibility cannot be investigated until resistant insects are recovered.

As discussed below, the estimation of F_{IS} has not been done for two of the three pest species, so weaker, alternative lines of argument have been developed to suggest that random mating is possible. These lines of argument are based on measures of the move rates and distances of adult moths. These dispersal arguments are flawed for several reasons. First, they do not distinguish between pre-mating and post-mating movement. If mating occurs before dispersal, the possibility of non-random mating is enhanced. If it occurs after dispersal, then the possibility of random mating is enhanced. It is not known on a population-wide basis how much mating occurs before or after dispersal for any of the species of concern for *Bt*-corn resistance management, but it is likely that these measures of dispersal will lead to overestimation of gene flow. Second, a dispersing insect does not necessarily contribute offspring to the population in its new habitat, and such non-reproductive dispersal will likely lead to artificially high estimates of gene flow.

European corn borer. Although allozyme variation in European corn borer has been examined (Harrison and Vawter 1977, Cardé et al. 1978, Glover et al. 1991, Ni 1995, Wang et al. 1995), only Bourguet et al. (2000) and Alstad et al. (unpublished) have evaluated local population structure and measured F_{IS}. Bourguet et al. (2000) sampled 29 sites (cornfields) in France, collecting about 40 individuals per site. They evaluated six polymorphic allozyme loci per corn borer, and found no evidence of non-random mating. Alstad et al. (unpublished) sampled 45 sites covering the range of European corn borer in North America, collecting about 70 individuals per site. Four polymorphic allozyme loci were examined per corn borer, and they found in 16 of these local populations evidence of non-random mating, all with heterozygote deficiencies. This suggests that in North America, local non-random mating may be common, violating a core assumption of the high-dose plus refuge strategy.

Research on dispersal in European corn borer provides somewhat conflicting evidence. Chiang (1972) showed that when European corn borer first invaded Minnesota, it spread across the state at a rate of 100 km/year, implying a high rate of dispersal. Showers (1993) examined the spread of an introduced deleterious genotype, and found that it occurred as far away as 32 km from the release area. Chiang et al. (1965) found that during the spring large numbers of adults migrate from central Iowa to southern Minnesota, a distance of 150 km. Each of these dispersal estimates is likely to overestimate the rate of gene flow. Spread into unoccupied habitats can happen significantly faster than into occupied habitats (Shigesada and Kawasaki 1997), and the distance an introduced deleterious genotype can be found from a point of release is largely a function of how deleterious the genotype is. The large spring dispersal may land the dispersing insects in habitats unsuitable for reproduction, because they arrive too early for the host plants to have reached a suitable growth stage.

Mark–release–recapture experiments have produced more modest dispersal values. Marking insects in the field with rubidium, Legg (1983) found that second-generation moths can disperse 3 km. Showers marked laboratory insects with vital dyes, and recaptured males at pheromone traps at four distances (200, 800, 3200, and 6400 m) during the second flight period. About 60% were recaptured at the closest traps (200 m) and only 6% were caught at the two more-distant traps. Hunt also used laboratory insects marked with vital dyes, but recaptured insects at light traps so both male and female moths were recaptured. Nearly all of the recaptured unmated females were within 500 m of the release site during the second flight period (Hunt, unpublished data). Field experiments examining populations in 20-acre refuge plots adjacent to 60 acres of *Bt*-corn revealed considerable suppression of populations of second-generation corn borer within 100 m of the *Bt*-corn (Andow and Alstad, unpublished). This was caused by reduced oviposition, and implies that most movement of females is restricted to within 100 m of their natal site.

Together, these data suggest that in some locations local populations do not mate randomly and that the majority of adults may disperse significantly less than ¹/₂ mile in a generation. Thus, the assumption that refuges planted within ¹/₂ mile of *Bt*-corn will ensure random mating is probably true only some of the time. In many cases there will be significant non-random mating at this spatial scale for European corn borer. The consequences of this observed non-random mating are discussed below.

Resistant European corn borer larvae could have significant developmental delays, which could contribute to greater levels of nonrandom mating (Gould 1994, Liu et al. 1999). Larvae resistant to low concentrations of Cry1Ab toxin develop more slowly on diets containing toxins than do susceptible larvae feeding on diets without toxins (Andow, unpublished data). The effect of a developmental delay on resistance evolution, however, is likely to vary geographically. For most regions with more than one generation, developmental delays during the final European corn borer generation may be unlikely to increase non-random mating. For example, emergence of adults during the second flight in southern Minnesota is prolonged over 3 to 4 weeks, with the last adults frequently emerging too late in the season to successfully parent offspring that can mature and overwinter. Consequently, resistant adults emerging late during the final flight may either encounter many susceptible mates or be too late to parent offspring successfully. In all regions, possible developmental delays could affect the timing and nonrandom mating of the first flight by delaying post-diapause development. Relatively little information is known about post-diapause development in European corn borer. Post-diapause development does not depend on feeding on corn tissue, so the effect of resistance alleles on development rate might not be great. In three-generation areas, however, developmental delays might increase non-random mating during the second flight.

Southwestern corn borer. The local population genetic structure of this species has not been examined, although between population allozyme variation has been studied (McCauley et al. 1990, 1995). Dispersal distances of this insect have not been recorded quantitatively. This species invaded the United States and spread rapidly, at rates up to 50 km/year (Chippendale 1979). Its rate of dispersal appears to be on a similar order of magnitude as that of European corn borer. The potential effects of developmental delays have not been evaluated, although the response of this species might be similar to that of European corn borer.

Corn earworm. Very little work on population genetic structure has been done on corn earworm. Sell et al. (1975) examined one electrophoretic locus (esterase 2) in 22 samples of larvae from a total of seven sites. With an average of 45 individuals/sample, they found no evidence of non-random mating. Generally, it is not sufficient to base conclusions about population genetic structure on results from only one locus.

Several studies have been conducted on the population genetic structure of relatives of corn earworm, *Helicoverpa armigera* (Hübner), *Heliothis punctigera* Wallengren, and *H. virescens* (F.). There has been no evidence of non-random mating in any of these species (Sluss and Graham 1979, Daly and Gregg 1985, Korman et al. 1993, Nibouche et al. 1998).

Dispersal of heliothine moths is extensive (Haile et al. 1975, Sparks et al. 1986, Farrow and Daly 1987, Daly 1989), and mark–release–recapture experiments indicate dispersal is on the order of 10 km/generation (Schneider et al. 1989). Even if corn earworm disperses only 10 km/generation, if some of these dispersing individuals successfully mate, the $1/2$ mile refuge requirement might ensure sufficiently random mating for this species.

The effects of developmental delays for this species may be complex. In the southeastern United States, developmental delays during the first generation on *Bt*-corn could increase non-random mating and delay the timing of attack of resistant larvae on *Bt*-cotton. Delaying the timing of attack may lead to spatial segregation of resistant and susceptible individuals as cotton fields become more or less attractive to ovipositing moths during the season. Spatial segregation and additional developmental delays associated with *Bt*-cotton might further exacerbate non-random mating. In the northern United States, corn earworm does not overwinter and migrates from the southern United States every year. Developmental delays in the south will probably delay the time of migration and the ultimate destination in the north. These, in turn, will probably lead to spatial segregation of resistant and susceptible individuals because corn is attractive to ovipositing moths for only a limited period during silking. If these individuals successfully migrate back to the south at the end of the season, a complex spatially inhomogeneous distribution of resistance allele frequencies may ensue.

4.5.3.4 Regional spatial structure

Spatial population structure is most readily estimated by evaluating the degree of genetic differentiation of neutral genetic markers among spatially separated populations. This is measured by the inbreeding statistic F_{ST} (Wright 1978), which is different from the previously discussed F_{IS}. An $F_{ST} = 0$ implies no population subdivision, while $F_{ST} > 0$ implies population subdivision. The larger the F_{ST}, the more subdivided the population. The spatial scale of the subdivision will depend on the distance separating the originally sampled populations.

European corn borer. F_{ST} has been estimated in European corn borer in France and North America, and a related species, Asian corn borer, has been examined in China. Bourguet et al. (2000) found that European corn borer in southwestern France formed one panmictic population in an area of about 350×150 km. There was small, but significant differentiation between populations in northern and southern France, among the two northern regions, and within each of the northern regions ($F_{ST} = 0.011$ to 0.015). This implies between 16.4 and 22.4 reproductively successful migrants move between the north and south and among the northern populations each generation, which is a relatively high level of gene flow. Wang et al. (1995) provided data from which F_{ST} can be calculated for nine populations of Asian corn borer in China spanning a 2000×4000 km area. An $F_{ST} = 0.008$ was found, a very high level of gene flow. In contrast, North American populations appear to exhibit considerable population subdivision (Alstad et al., unpublished). The average F_{ST} was over 0.175, an extremely high value, which could imply that only one migrant/generation moves between populations separated by about 300 km. This would imply that panmictic populations occur on less than 300×300 km areas, giving approximately 50 panmictic populations in North America. An alternative interpretation is that the genetic structure of European corn borer in North America has not reached equilibrium, and the high F_{ST} reflects in part the historical pattern of invasion and geographical spread of the species.

Southwestern corn borer. Interestingly, F_{ST} estimates for southwestern corn borer are also extremely high. Five populations sampled from within a 2000-km equilateral triangle had an F_{ST} of 0.33, implying about 0.5 migrants/generation (McCauley et al. 1990, 1995). Insufficient samples have been examined in this species to estimate the geographical scale of spatial subdivision, but southwestern corn borer may have a spatial genetic structure similar to European corn borer. Southwestern corn borer is also a recent invader of North America, so historical factors might contribute to the high F_{ST}.

Corn earworm. There are too few data on corn earworm to estimate its population structure (Sell et al. 1975). The related *H. virescens* has been studied twice, with somewhat variable results. Sluss and Graham (1979) examined allozyme variability from 11 populations from North Carolina to California. Their data were analyzed by Daly (1989), who calculated $F_{ST} = 0.048$, a migration rate of five individuals/generation. Subsequent work by Korman et al. (1993), who examined 60 sites from Georgia to Texas, found $F_{ST} = 0.002$, a migration rate of 125 individuals/generation. Although there are significant unresolved differences between these studies, both suggest relatively low population subdivision in *H. virescens* in North America. Because the population structure of corn earworm may be similar to *H. virescens*, these data might indicate that panmictic populations of corn earworm are significantly larger than those of European corn borer and southwestern corn borer in North America.

4.5.4 An assessment of the present IRM plans

The original 20 to 40% structured refuge and recently proposed 20% refuge were initially developed to incorporate a safety margin that would hedge against faulty assumptions in the high-dose plus refuge strategy (Andow and Hutchison 1998). It was expected that, as data accumulated, this margin could be adjusted as indicated by the empirical data. The present status of data, however, suggests that there is no remaining safety margin. In other words, a 20% unsprayed refuge should suffice to delay the evolution of resistance in European corn borer and southwestern corn borer if all of the assumptions of the high-dose plus refuge strategy hold, but if any of them do not, the refuge may be insufficient. Consequently, management alternatives should be sought to restore some safety margin in the IRM strategy for *Bt*-corn.

Of the five *Bt*-corn events, two are not high-dose events (Event 176 and DBT 418). The remaining three appear to be high-dose events for European and southwestern corn borer, but it is not be possible to know this for certain because resistance has not yet been found. Recessivity of resistance to these three events has been assumed on theoretical grounds, based on the mechanisms of resistance and extrapolation about resistance in other insects to Cry toxins. This assumption has been challenged, and some evidence has been found to indicate that resistance may be dominant (Huang et al. 1999). The weight of evidence, however, indicates that resistance is more likely to be recessive than dominant. Thus, while this assumption has not been confirmed, it also has not been disproved. The geographical restrictions on use of Event 176 and DBT 418 (see Table 4.4) appear to be scientifically justifiable, because neither event is a high-dose event. For these events, a pure refuge strategy might be effective, leading to increased refuge requirements for Event 176 and DBT 418.

Although BT-11 provides high levels of control of corn earworm in sweet corn (Lynch et al. 1999), none of the events in field corn appears to provide consistent high-dose expression for corn earworm. For these events, the 50% corn refuge requirement in cotton-growing regions where corn earworm is prevalent appears to be a reasonable IRM requirement. CBH 351 appears to be an exception because it does not appear to provide any control of corn earworm. Special consideration for corn earworm resistance management for this event may not be necessary.

Resistance alleles to Cry1Ab appear to be sufficiently rare in European corn borer in southern Minnesota and central Iowa, although additional research is necessary to prove this (Andow et al. 1998, 2000, Andow and Alstad 1999). If these regions are representative of other parts of the range of European corn borer, the assumption of rare resistance could be confirmed throughout the range with additional work. Resistance frequencies in European corn borer to Cry9C or Cry1Ac have not been examined. The frequency of resistance in southwestern corn borer and corn earworm has not been studied for any of the *Bt*-events. Research should be conducted to fill these data gaps.

Local random mating in European corn borer does not occur in some localities. When this assumption is violated, a 20% unsprayed refuge may barely preserve corn borer susceptibility for 15 years if the remaining assumptions of the high-dose plus refuge strategy are valid (ILSI 1999, NC205 1998). Local mating structure in southwestern corn borer has not been evaluated. The weak evidence for corn earworm suggests that local random mating may be common, but convincing evidence is needed before any conclusions can be reached.

European corn borer and southwestern corn borer both appear to have significant regional genetic structure, while corn earworm does not. If true, this implies that monitoring of resistance must be performed at a finer spatial scale for European corn borer and southwestern corn borer than for corn earworm.

The hypothesis that an insecticide-sprayed refuge provides effective resistance management in any corn-growing region lacks scientific support at present. In general, the rate of resistance evolution is proportional to the selective advantage of the R allele over the S allele (May and Dobson 1986), and spraying the refuge with any insecticide gives the R allele an additional selective advantage. Killing susceptible insects increases the relative fitness of the R allele compared to the remaining S alleles. Consequently, spraying the refuge will accelerate the rate of evolution of resistance. If sprays are sufficiently infrequent, isolated, and inefficacious, it is possible that spraying the refuge with insecticide has little effect on the rate of resistance evolution. Given that European corn borer populations show periodic outbreaks over extensive spatial regions that could merit control, it is uncertain whether realistic patterns of insecticide use will be sufficiently isolated, infrequent, and inefficacious.

Together, the data to date suggest the following:

1. The IRM strategy for *Bt*-corn aimed at corn earworm may be adequate.
2. A higher refuge proportion is needed for Event 176 and DBT 418.
3. The 20% structured refuge may provide 15 years of susceptibility in European corn borer for BT-11 and Mon 810, but the margin of safety appears to be limited.
4. The frequency of resistance to CBH 351 is not known.
5. There are major data gaps for all events for southwestern corn borer, although in many respects the relevant information for southwestern corn borer is similar to that for European corn borer.

In summary, even for the best characterized events (BT-11 and Mon 810), there is no safety margin to protect from additional failures in the assumptions of the high-dose plus refuge strategy or in implementation. In what follows, I suggest that adaptive IRM may restore some safety margin.

4.6   Implementation and compliance

All models of resistance management assume full compliance with the resistance management strategy. While the vast majority of farmers do comply with most required farm practices, a small minority generally do not. Moreover, because *Bt*-corn resistance management requires the planting of a refuge, where pest losses might occur, farmers may be reluctant to fully implement an effective refuge. Refuges might be planted on poorer soils or later in the planting period, or receive lower levels of management or higher levels of insecticide. Any of these practices would decrease the effectiveness of the refuge and, if practiced widely, could jeopardize resistance management. The consequences of incomplete compliance should be investigated.

The EPA now requires that compliance be monitored by the registrants, i.e., those companies that own the *Bt*-corn products. Registrants ensure compliance by requiring farmers to sign legal agreements binding them to plant a refuge. Growers must sign the agreements to be able to purchase *Bt*-corn. The proportion of growers signing the agreements is used to measure compliance. It is believed that the legal penalties for breaking the agreement are sufficient incentive to ensure that the farmer will actually plant the refuge as agreed. Potentials for abuse of such a system of reporting are obvious, and an independent means of measuring compliance is needed.

4.7   An adaptive resistance management strategy

Although it has been emphasized that IRM strategies must be dynamic and adaptive (Forrester 1990), static strategies are more common in practice and change is usually forced by events that make it necessary to respond. Based on the Australian experience in cotton, Forrester (1990, Forrester et al. 1993, Forrester and Bird 1996) has made increasingly pointed pleas for adaptive resistance management, echoed by Denholm and Rowland (1992). An adaptive strategy relies on an effective and sensitive resistance-monitoring system. The lack of an effective and sensitive monitoring system remains a significant constraint on developing and implementing adaptive IRM strategies (NRC 1986, Denholm 1990, Sawicki 1996).

Given the absence of a safety margin in the present high-dose plus refuge strategy for Bt-corn and concerns about implementation and compliance, it would be risky indeed to rely solely on this preemptive strategy. A more robust, adaptive resistance management strategy would include proactive monitoring that would allow management to change as the evolution of resistance proceeds but before there were control failures. Such a monitoring and response system would provide the safety margin necessary to hedge against failures in the scientific assumptions of the preemptive high-dose plus refuge strategy and failures in implementation.

Andow and Ives (in press) outlined the structure of such an adaptive strategy for Bt-corn. It involved estimating a time frame for monitoring and responses, evaluating potential monitoring methods, and assessing potential management responses. The central conclusion was that given realistic time frames, there were monitoring methods that appeared to be cost-effective and adaptive management responses that could be effective at delaying the evolution of resistance to Bt-corn. These methods and responses are outlined below.

Based on a theoretical model adapted from Comins (1977), resistance alleles must be detected at frequencies of $\leq 5 \times 10^{-3}$ to provide enough time to adapt management. Previous work on the costs of monitoring indicates that an F_2 screen (Andow and Alstad 1998) is the most cost-effective method for monitoring recessive resistance to Bt-corn and an in-field screen (Venette et al. 2000) is the most cost-effective method for monitoring dominant resistance. Both methods can detect and measure resistance at frequencies of $\leq 5 \times 10^{-3}$ for ~$5000 in variable costs per sample.

Two types of adaptive responses can be taken. The first would reduce the selective advantage of the resistance allele by increasing mortality or reducing fecundity of resistant types in the Bt-field, or reducing mortality or increasing fecundity of insects associated with the refuge. The second type of response would modify the mating system so that fewer resistance alleles are passed on to future generations. Because of logistical issues associated with confirming measurements and organizing a response, there is likely to be a 2-year time delay between detecting and measuring the resistance frequency and taking an adaptive response. Under these conditions, increasing refuge size to 66% (from 20%) can prolong susceptibility ten generations, and decreasing survival and reproduction of moths from Bt-corn fields by 90% can prolong susceptibility by ten generations. Modification of the mating system by changing movement rates and attracting susceptible males into Bt-corn fields could prolong susceptibility for >20 generations. Although these results suggest that adaptive resistance management could increase the durability of Bt-corn, much work remains to prove that such strategies are feasible and effective management interventions.

4.8   Conclusions

IRM for Bt-corn involves at present five different transformation events for three different pests: European corn borer, southwestern corn borer, and corn earworm. Three events express Cry1Ab toxin (Event 176, BT-11, and MON 810), one expresses Cry1Ac toxin (DBT 418), and one expresses Cry9C toxin (CBH 351). The structures of Event 176 and BT-11 are well described, but there is considerable uncertainty about the structure of the other three events. Recently, the owners of Event 176 and DBT 418 have indicated that these events will be phased out of commercial production in the United States, so future IRM strategies may be able to ignore them. New events are under continual development, such as one based on Cry1F toxin, so the present IRM plans will need to evolve in the future.

The high-dose plus 20% structured refuge strategy is currently required for all events and pests in the United States, with some exceptions. The corn refuge must be within $^{1}/_{2}$ mile of *Bt*-corn and it may be treated with a non-*Bt*-insecticide. In southern regions, where corn earworm can feed on cotton, the corn refuge must be 50%, and in regions where insecticide use has been high, the corn refuge must be within $^{1}/_{4}$ mile of *Bt*-corn.

Because none of the events is a true high-dose event for corn earworm, the larger refuge in the southern regions would appear to be justified scientifically. Requiring the refuge to be closer in the high-insecticide-use area is based on a sound scientific principle, but whether that requirement makes for sufficiently effective resistance management is scientifically uncertain. Spraying the refuge with insecticides certainly will accelerate resistance evolution. In the majority of the U.S. corn belt, however, the magnitude of change in resistance evolution from realistic insecticide spray regimes against European corn borer is not known scientifically.

Two events are not true high-dose events for either European corn borer or southwestern corn borer. These are Event 176 and DBT 418. Both are prohibited from being grown in the high-insecticide-use region, and DBT 418 cannot be grown in the southern region. Event 176 may be grown in the southern region with a 50% refuge. These restrictions and requirements appear to be scientifically justified. Because these events are not high-dose events, a larger 50% refuge in the main corn belt of the United States could be scientifically justified.

For the remaining three events, most of the scientific evidence suggests the assumptions of the high-dose plus refuge strategy will be met for both European and southwestern corn borer. Nevertheless, significant uncertainty remains. These three events may be high-dose events for European and southwestern corn borer, but this cannot be confirmed until resistance is recovered in each species. Increased effort is needed to find resistance to Cry1Ab and Cry9C in the target pest species. Resistance in European corn borer to the Cry1Ab events may be rare enough, but the frequency of resistance to Cry9C and the frequency of resistance to either toxin in southwestern corn borer are not known. Several populations of European corn borer exhibit significant local non-random mating. Because of this, the high-dose plus 20% structured refuge may be just barely adequate, assuming that all other scientific assumptions of the strategy are confirmed. The mating structure of local populations of southwestern corn borer is not known.

It is important to develop an adaptive resistance management strategy to hedge against potential failures in any of the other assumptions of the present resistance management strategy or any failures in implementation. Theoretical analysis suggests that this may be possible, but more research on intervention tactics is needed. Specifically, it may be possible to manage population structure and increase the frequency of mating between resistant and susceptible individuals. If this proves feasible, significant improvements to resistance management may be possible.

Acknowledgments

I thank Deborah Letourneau and Beth Burrows for providing the opportunity to write this chapter, and L. Buschmann, J. Harmon, G. Heimpel, W. Hutchison, R. Venette, and

J. White for their comments, which greatly improved the chapter. The work was supported in part by USDA Regional Research Project NC205 and USDA-NRI 99-35302-7937.

Literature cited

Adkisson, P. L., Objective uses of insecticides in agriculture, in *Agricultural Chemicals: Harmony or Discord for Food, People, and the Environment,* Swift, J. E., Ed., Division of Agricultural Science, University of California, Berkeley, 1971, 43–51.

Adkisson, P. L., The integrated control of the insect pests of cotton, *Proc. Tall Timbers Conf. Ecol. Anim. Control Habitat Manage.,* 4, 175, 1972.

Alstad, D. N. and Andow, D. A., Managing the evolution of resistance to transgenic plants, *Science,* 268, 1894, 1995.

Andow, D. A. and Alstad, D. N., The F_2 screen for rare resistance alleles, *J. Econ. Entomol.,* 91, 572, 1998.

Andow, D. A. and Alstad, D. N., Credibility interval for rare resistance allele frequencies, *J. Econ. Entomol.,* 92, 755, 1999.

Andow, D. A. and Hutchison, W. D., Corn resistance management, in *Now or Never: Serious Plans to Save a Natural Pest Control,* Mellon, M. and Rissler, J., Eds., Union of Concerned Scientists, Washington, D.C., 1998, chap. 2.

Andow, D. A. and Ives, A., Adaptive resistance management, *Ecol. Appl.,* in press.

Andow, D. A. et al., Using an F_2 screen to search for resistance alleles to *Bacillus thuringiensis* toxin in European corn borer, *J. Econ. Entomol.,* 91, 579, 1998.

Andow, D. A. et al., Frequency of resistance to *Bacillus thuringiensis* toxin Cry1Ab in an Iowa population of European corn borer (Lepidoptera: Crambidae), *J. Econ. Entomol.,* 93, 26, 2000.

Bourguet, D. et al., Gene flow in the European corn borer *Ostrinia nubilalis*: implications of the sustainability of transgenic insecticidal maize, *Proc. R. Soc. London,* B 267, 117, 2000.

Brown, T. M., Countermeasures for insecticide resistance, *Bull. Entomol. Soc. Am.,* 27, 198, 1981.

Caprio, M. A. and Tabashnik, B. E., Gene flow accelerates local adaptation among finite populations: simulating the evolution of insecticide resistance, *J. Econ. Entomol.,* 85, 611, 1992.

Cardé, R. T. et al., European corn borer: pheromone polymorphism or sibling species? *Science,* 199, 555, 1978.

Chiang, H. C., Dispersion of the European corn borer (Lepidoptera: Pyralidae) in Minnesota and South Dakota, 1945 to 1970, *Environ. Entomol.,* 1, 157, 1972.

Chiang, H. C., Sisson, V., and Ewert, M. A., Northerly movement of corn borer moth, *Ostrinia nubilalis,* in southern Minnesota, *Proc. Minn. Acad. Sci.,* 33, 17, 1965.

Chippendale, G. M., The southwestern corn borer, *Diatraea grandiosella*: a case history of an invading insect, *Mo. Agric. Exp. Sta. Res. Bull.,* 1031, 1979.

Comins, H. N., The development of insecticide resistance in the presence of migration, *J. Theor. Biol.,* 64, 177, 1977.

Daly, J. C., The use of electrophoretic data in a study of gene flow in the pest species *Heliothis armigera* (Hübner) and *H. punctigera* Wallengren (Lepidoptera: Noctuidae), in *Electrophoretic Studies on Agricultural Pests,* Loxdale, H. D. and den Hollander, J., Eds., Oxford University Press, New York, 1989, 115–141.

Daly, J. C. and Gregg, P., Genetic variation in *Heliothis* in Australia: species identification and gene flow in two pest species *H. armigera* (Hübner) and *H. punctigera* Wallengren (Lepidoptera: Noctuidae), *Bull. Entomol. Res.,* 75, 169, 1985.

Denholm, I., Monitoring and interpreting changes in insecticide resistance, *Funct. Ecol.,* 4, 601, 1990.

Denholm, I. and Rowland, M. W., Tactics for managing pesticide resistance in arthropods: theory and practice, *Annu. Rev. Entomol.,* 37, 91, 1992.

Dennehy, T. J., Decision-making for managing pest resistance to pesticides, in *Combating Resistance to Xenobiotics: Biological and Chemical Approaches,* Ford, M. G. et al., Eds., Ellis Horwood, Chichester, U.K., 1987, 118–126.

EPA, Report, FIFRA Scientific Advisory Panel, Subpanel on *Bacillus thuringiensis (Bt)* plant-pesticides and resistance management, February 9–10, 1998. U.S. Environmental Protection Agency, Washington, D.C., 1998.

EPA, Biopesticide Fact Sheet: *Bacillus thuringiensis* Cry1Ab delta-endotoxin and the genetic material necessary for its production in corn [MON 810] (006430), EPA 730-F-00-006, U.S. Environmental Protection Agency, Washington, D.C., 2000a.

EPA, Biopesticide Fact Sheet: *Bacillus thuringiensis* Cry1Ab delta-endotoxin and the genetic material necessary for its production (Plasmid vector pZO1502) in corn [Bt11] (006444), EPA 730-F-00-002, U.S. Environmental Protection Agency, Washington, D.C., 2000b.

EPA, Biopesticide Fact Sheet: *Bacillus thuringiensis* Cry1Ab delta-endotoxin and the genetic material necessary for its production (Plasmid vector pCIB4431) in corn [Event 176] (006458), EPA 730-F-00-003, U.S. Environmental Protection Agency, Washington, D.C., 2000c.

EPA, Biopesticide Fact Sheet: *Bacillus thuringiensis* subspecies *kurstaki* Cry1Ac delta-endotoxin and the genetic material necessary for its production in corn [Dekalb DBT-418] (006463), EPA 730-F-00-004, U.S. Environmental Protection Agency, Washington, D.C., 2000d.

EPA, Biopesticide Fact Sheet: *Bacillus thuringiensis* subspecies *tolworthii* Cry9C protein and the genetic material necessary for its production in corn (006466), EPA 730-F-00-005, U.S. Environmental Protection Agency, Washington, D.C., 2000e.

EPA, Preliminary Risk and Benefit Assessments for *Bt* Plant-Pesticides, Briefing Document for the October 18–20, 2000 Science Advisory Panel Meeting, U.S. Environmental Protection Agency, Washington, D.C., 2000f.

Escriche, B. et al., Immunohistochemical detection of binding of Cry1A crystal proteins of *Bacillus thuringiensis* in highly resistant strains of *Plutella xylostella* (L.) from Hawaii, *Biochem. Biophys. Res. Commun.*, 212, 388, 1995.

Farrow, R. A. and Daly, J. C., Long-range movement as an adaptive strategy in the genus *Heliothis* (Lepidoptera: Noctuidae): a review of its occurrence and detection in four pest species, *Aust. J. Zool.*, 35, 1, 1987.

Ferré, J. et al., Resistance to the *Bacillus thuringiensis* bioinsecticide in a field population of *Plutella xylostella* is due to a change in a midgut membrane receptor, *Proc. Natl. Acad. Sci. U.S.A.*, 88, 5119, 1991.

Forrester, N. W., Designing, implementing and servicing an insecticide resistance management strategy, *Pest. Sci.*, 28, 167, 1990.

Forrester, N. W. and Bird, L. J., The need for adaptation to change in insecticide resistance management strategies: the Australian experience, in *Molecular Genetics and Evolution of Pesticide Resistance*, Brown, T. M., Ed., American Chemical Society, Washington, D.C., 1996, 160–168.

Forrester, N. W. et al., Management of pyrethroid and endosulfan resistance in *Helicoverpa armigera* (Lepidoptera: Noctuidae) in Australia, *Bull. Entomol. Res.*, Suppl. 1, 1, 1993.

Georghiou, G. P., The magnitude of the resistance problem, in Pesticide Resistance: Strategies and Tactics for Management, National Academy Press, Washington, D.C., 1986, 14–43.

Georghiou, G. P. and Taylor, C. E., Operational influences in the evolution of insecticide resistance, *J. Econ. Entomol.*, 70, 653, 1977.

Glover, T. J. et al., Gene flow among three races of European corn borers (Lepidoptera: Pyralidae) in New York State, *Environ. Entomol.*, 20, 1356, 1991.

Gould, F., Simulation models for predicting durability of insect-resistant germ plasm: Hessian fly (Diptera: Cecidomyiidae)-resistant winter wheat, *Environ. Entomol.*, 15, 11, 1986.

Gould, F., Potential problems with high-dose strategies for pesticidal engineered crops, *Biocontrol Sci. Technol.*, 4, 451, 1994.

Gould, F., Sustainability of transgenic insecticidal cultivars: integrating pest genetics and ecology, *Annu. Rev. Entomol.*, 43, 701, 1998.

Gould, F. et al., Broad-spectrum resistance to *Bacillus thuringiensis* toxins in *Heliothis virescens*, *Proc. Natl. Acad. Sci. U.S.A.*, 89, 7986, 1992.

Gould, F. et al., Selection and genetic analysis of a *Heliothis virescens* (Lepidoptera: Noctuidae) strain with high levels of resistance to *Bacillus thuringiensis* toxins, *J. Econ. Entomol.*, 88, 1545, 1995.

Haile, D. G., Snow, J. W., and Young, J. R., Movement by adult *Heliothis* released on St. Croix to other islands, *Environ. Entomol.*, 4, 225, 1975.

Hama, H., Suzuki, K., and Tanaka, H., Inheritance and stability of resistance to *Bacillus thuringiensis* formulations of the diamondback moth, *Plutella xylostella* (Linnaeus) (Lepidoptera: Yponomeutidae), *Appl. Entomol. Zool.*, 27, 355, 1992.

Harrison, R. G. and Vawter, A. T., Allozyme differentiation between pheromone strains of the European corn borer, *Ostrinia nubilalis, Ann. Entomol. Soc. Am.,* 70, 717, 1977.

Huang, F. et al., Inheritance of resistance to *Bacillus thuringiensis* toxin (Dipel ES) in the European corn borer, *Science,* 284, 965, 1999.

ILSI (International Life Sciences Institute–Health and Environmental Sciences Institute), *An Evaluation of Insect Resistance Management in* Bt *Field Corn: A Science-Based Framework for Risk Assessment and Risk Management, Report of an Expert Panel,* ILSI Press, Washington, D.C., 1999.

James, C., *The Global Review of Commercialized Transgenic Crops: 1997,* ISAAA Briefs No. 5, International Service for the Acquisition of Agri-biotech Applications, Ithaca, NY, 1997.

James, C., *The Global Review of Commercialized Transgenic Crops: 1998,* ISAAA Briefs No. 8, International Service for the Acquisition of Agri-biotech Applications, Ithaca, NY, 1998.

James, C., Preview in *Global Status of Commercialized Transgenic Crops: 1999,* ISAAA Briefs No. 12, International Service for the Acquisition of Agri-biotech Applications, Ithaca, NY, 1999.

Korman, A. K. et al., Population structure in *Heliothis virescens* (Lepidoptera: Noctuidae): an estimate of gene flow, *Ann. Entomol. Soc. Am.,* 86, 182, 1993.

Koziel, M. G. et al., Field performance of elite transgenic maize plants expressing an insecticidal protein derived from *Bacillus thuringiensis, Nature Bio/Technol.,* 11, 194, 1993.

Koziel, M. G. et al., Transgenic maize for the control of European corn borer and other maize pests, *Ann. N.Y. Acad. Sci.,* 792, 164, 1996.

Legg, D. E., Dispersion, Distribution, and Subsequent Infestations of the European Corn Borer, *Ostrinia nubilalis* (Hübner), among Southern Minnesota Corn Fields, Ph.D. thesis, University of Minnesota, St. Paul, 1983.

Liu, Y. B. et al., Development time and resistance to *Bt* crops, *Nature,* 400, 519, 1999.

Lynch, R. E. et al., Evaluation of transgenic sweet corn hybrids expressing CryIA(b) toxin for resistance to corn earworm and fall armyworm (Lepidoptera: Noctuidae), *J. Econ. Entomol.,* 92, 246, 1999.

Martinez-Ramirez, A. C. et al., Inheritance of resistance to a *Bacillus thuringiensis* toxin in a field population of diamondback moth (*Plutella xylostella*), *Pest. Sci.,* 43, 115, 1995.

Masson, L. et al., Kinetics of *Bacillus thuringiensis* toxin binding with brush border membrane vesicles from susceptible and resistant larvae of *Plutella xylostella, J. Biol. Chem.,* 270, 11887, 1995.

Matten, S. R. et al., The U.S. Environmental Protection Agency's role in pesticide resistance management, in *Molecular Genetics and Evolution of Pesticide Resistance,* Brown, T. M., Ed., American Chemical Society, Washington, D.C., 1996, 243–253.

May, R. M. and Dobson, A. P., Population dynamics and the rate of evolution of pesticide resistance, in Pesticide Resistance: Strategies and Tactics for Management, National Academy Press, Washington, D.C., 1986, 170–193.

McCauley, D. E. et al., Genetic differentiation of populations of the southwestern corn borer (Lepidoptera: Pyralidae) from the United States and Mexico, *Ann. Entomol. Soc. Am.,* 83, 586, 1990.

McCauley, D. E. et al., Genetic differentiation accompanying range expansion by the southwestern corn borer (Lepidoptera: Pyralidae), *Ann. Entomol. Soc. Am.,* 88, 357, 1995.

McGaughey, W. H. and Beeman, R. W., Resistance to *Bacillus thuringiensis* in colonies of Indianmeal moth and almond moth (Lepidoptera: Pyralidae), *J. Econ. Entomol.,* 81, 28, 1988.

NASS, Prospective Plantings, March 31, 2000, available at http://usda.mannlib.cornell.edu/reports/nassr/field/pcp-bbp/psp10300.pdf, National Agricultural Statistical Service, Agricultural Statistics Board, USDA, Washington, D.C., 2000.

NC205, North Central regional research on ecology and management of European corn borer and other stalk-boring Lepidoptera, Supplement to NCR-602, Bt Corn & European Corn Borer: Long-Term Success through Resistance Management, available at http://ent.agri.umn.edu/ecb/nc205doc.htm, 1998.

Ni, Y.-S., Genetic Variation of European Corn Borer in Delaware and Genetic Comparison with Asian Corn Borer, M.S. thesis, University of Delaware, Newark, 1995.

Nibouche, S. et al., Allozyme polymorphism in the cotton bollworm *Helicoverpa armigera* (Lepidoptera: Noctuidae): comparison of African and European populations, *Heredity,* 80, 438, 1998.

NRC, Pesticide Resistance: Strategies and Tactics for Management, National Academy Press, Washington, D.C., 1986.

Ostlie, K. R., Hutchison, W. D., and Hellmich, R. L., Eds., Bt *Corn and European Corn Borer: Long-Term Success through Resistance Management*, NCR Publication No. 602, University of Minnesota, St. Paul, 1997.

Pimentel, D. et al., Environmental and social costs of pesticides: a preliminary assessment, *Oikos*, 34, 126, 1980.

Porter, P. et al., Bt Corn Technology in Texas: A Practical View, Texas Agricultural Extension Service, B-6090, 1-00, 2000.

Rice, M. E. and Pilcher, C. D., Potential benefits and limitations of transgenic Bt corn for management of the European corn borer (Lepidoptera: Crambidae), *Am. Entomol.*, 44, 75, 1998.

Roush, R. T., Managing pests and their resistance to *Bacillus thuringiensis*: can transgenic crops be better than sprays? *Biocontrol Sci. Technol.*, 4, 501, 1994.

Roush, R. T., Can we slow adaptation by pests to insect-resistant transgenic crops? in *Biotechnology and Integrated Pest Management*, Persley, G., Ed., CAB International, Wallingford, U.K., 1996, 242–263.

Sawicki, R. M., Definition, detection and documentation of insecticide resistance, in *Combating Resistance to Xenobiotics: Biological and Chemical Approaches*, Ford, M. G. et al., Eds., Ellis Horwood, Chichester, U.K., 1996, 105–117.

Sawicki, R. M. and Denholm, I., Management of resistance to pesticides in cotton pests, *Trop. Pest Manage.*, 33, 262, 1987.

Schneider, J. C. et al., Movement of *Heliothis virescens* (Lepidoptera: Noctuidae) in Mississippi in the spring: implications of area-wide management, *Environ. Entomol.*, 18, 868, 1989.

Sell, D. K., Whitt, G. S., and Luckmann, W. H., Esterase polymorphism in the corn earworm, *Heliothis zea* (Boddie): a survey of temporal and spatial allelic variation in natural populations, *Biochem. Genet.*, 13, 885, 1975.

Shigesada, N. and Kawasaki, K., *Biological Invasions: Theory and Practice*, Oxford University Press, Oxford, U.K., 1997.

Showers, W. B., Diversity and variation of European corn borer populations, in *Evolution of Insect Pests*, Kim, K. C. and McPheron, B. A., Eds., John Wiley & Sons, New York, 1993, 287–309.

Sims, S. R. and Stone, T. B., Genetic basis of tobacco budworm resistance to an engineered *Pseudomonas fluorescens* expressing the δ-endotoxin of *Bacillus thuringiensis kurstaki*, *J. Invertebr. Pathol.*, 57, 206, 1991.

Sluss, T. P. and Graham, H. M., Allozyme variation in natural populations of *Heliothis virescens*, *Ann. Entomol. Soc. Am.*, 72, 317, 1979.

Sparks, A. N. et al., Insects captured in light traps in the Gulf of Mexico, *Ann. Entomol. Soc. Am.*, 79, 132, 1986.

Tabashnik, B. E., Evolution of resistance to *Bacillus thuringiensis*, *Annu. Rev. Entomol.*, 39, 47, 1994.

Tabashnik, B. E. and Croft, B. A., Managing pesticide resistance in crop-arthropod complexes: interactions between biological and operational factors, *Environ. Entomol.*, 11, 137, 1982.

Tabashnik, B. E. et al., Inheritance of resistance to *Bacillus thuringiensis* in diamondback moth (Lepidoptera: Plutellidae), *J. Econ. Entomol.*, 85, 1046, 1992.

Van Rie, J. et al., Mechanism of insect resistance to the microbial insecticide *Bacillus thuringiensis*, *Science*, 247, 72, 1990.

Venette, R. C., Hutchison, W. D., and Andow, D. A., An in-field screen for early detection and monitoring of insect resistance to *Bacillus thuringiensis* in transgenic crops, *J. Econ. Entomol.*, 93, 1055, 2000.

Wang, R. et al., Allozyme differentiation among nine populations of the corn borer (*Ostrinia*) in China, *Biochem. Genet.*, 33, 413, 1995.

Wright, S., *Evolution and the Genetics of Populations*, Vol. 4, *Variability within and among Natural Populations*, University of Chicago Press, Chicago, 1978.

chapter five

Ecological risks of transgenic virus-resistant crops

Alison G. Power

Contents

5.1 Transgenic virus resistance

Diseases caused by plant viruses result in significant yield losses in crops worldwide (Thresh 1980). Since most plant viruses are transmitted to crops by insect vectors, strategies for virus control may target either the virus itself or its insect vector. Many of the most common management strategies rely on vector control because natural genetic resistance to viruses has been elusive for many important crops despite the efforts of plant breeders to discover and employ such resistance. With the advent of plant genetic engineering, new opportunities have arisen for developing virus resistance in crops.

Most major food crops worldwide have been genetically engineered for virus resistance via the insertion of viral genes into the plant genome. Most of these genes encode viral coat proteins (Beachy et al. 1990) although other genes, such as those encoding viral replicase, have also been successfully inserted into plants. This type of engineered resistance has been successful in disrupting replication of a range of viruses in a number of crops (Gadani et al. 1990). Although the mechanisms that result in protection are incompletely understood, plants expressing virus coat protein genes typically require greater inoculum for infection, show infection delays, and exhibit milder disease symptoms if inoculation is successful (Miller and Young 1995). Recent studies suggest that protection may be effective even if the viral proteins themselves are not produced, due to a mechanism known as RNA-mediated resistance (for reviews, see van den Boogaart et al. 1998, Waterhouse et al. 1999). RNA-mediated resistance may allow us to develop strategies for engineered virus resistance that are less ecologically risky than the strategies currently in use. Possibilities for reducing the ecological risks of transgenic virus resistance through transgene design are discussed at the end of this chapter.

Ecological risk has two components: probability and hazard. Following hazard identification, determining the probability of a particular hazard occurring is the next step of risk assessment. It is also essential to evaluate the magnitude of the damage that results from such an event. In the case of transgenic virus resistance, we are only beginning to assess probabilities of potential risks and we are far from a clear understanding of the extent of the hazards that may result from these risks. However, a framework for categorizing ecological risks has emerged over the past few years (see Miller et al. 1997, Tepfer and Balazs 1997, Hammond et al. 1999). Potential ecological risks associated with the widespread adoption of engineered virus resistance fall into three major categories: (1) recombination between viral transgenes and invading viruses; (2) interactions between transgene products and invading viruses, such as synergies or heterologous encapsidation (see below); and (3) transgene movement from transgenic crops to wild relatives via hybridization. In all of these categories, both the probability of the phenomenon occurring and the extent of the damage associated with the phenomenon should be assessed.

5.2   Risks of virus–transgene recombination

The hazards associated with virus–transgene recombination include the formation of novel viruses with modified virulence, host range, or transmission characteristics. Increased virulence could lead to greater damage to hosts of the virus, including any wild hosts in natural habitats. Such increased plant damage could lead to changes in competitive relations between plants, which in turn could have profound effects on natural plant communities. If recombination were to lead to increased host range, similar changes in plant competitive relations would be expected. Moreover, any changes in transmission characteristics resulting from recombination could allow the recombinant virus to colonize hosts that were previously unavailable to the parental virus. Again, this could lead to significant changes in species interactions within natural plant communities.

How likely is virus–transgene recombination? Accumulating evidence suggests that recombination between an invading virus and the viral RNA encoded by the transgenic plant is highly probable (de Zoeten 1991, Miller et al. 1997, Aaziz and Tepfer 1999b, Hammond et al. 1999). We know that recombination between two distinct viruses can occur when they simultaneously infect a plant. Evidence for virus–virus recombination in a broad range of viruses comes from both laboratory experiments and virus molecular phylogenies (for recent reviews, see Aaziz and Tepfer 1999b, Hammond et al. 1999). However, few studies to date have examined recombination among viruses under natural conditions of mixed infections with no artificially imposed selection pressure (but see Aaziz and Tepfer 1999a). Although recombinant viruses often have lower fitness than parental types, the recombinant virus sometimes has higher fitness than its parents (Fernández-Cuartero et al. 1994). This higher fitness could allow the recombinant to outcompete the parental viruses and dominate the population.

Recombination between viral transgenes and invading viruses that infect the plant naturally also has been demonstrated numerous times (see Aaziz and Tepfer 1999b, Hammond et al. 1999). Although most of the studies demonstrating such recombination have designed the experimental conditions to select for recombinants, a few studies have shown significant amounts of virus–transgene recombination even under conditions of moderate to weak selection (Wintermantel and Schoelz 1996, Borja et al. 1999). Falk and Bruening (1994) and Hammond et al. (1999) have argued that (1) recombination between virus RNA and transgene RNA is unlikely to occur with any greater frequency than recombination between viruses in natural infections; and (2) any new recombinant virus is unlikely to have higher fitness than naturally occurring viruses. Although the data are scarce, the conclusion that the recombination rate will not be higher for virus–transgene recombinants seems reasonable. However, as de Zoeten (1991) points out, in a monoculture of transgenic plants expressing viral genes, every infection with an invading virus is a "mixed" infection. Thus, the total number of recombinants is likely to rise, even if the rate of recombination is equal to or less than the rate of natural recombination among viruses.

The fitness of the recombinant virus could be higher than that of parental viruses if recombination resulted in changes in insect transmission, increased host range, or altered virulence. Any of these modifications could provide a selective advantage that would allow the recombinant to spread and thereby could increase the risk of a recombinant virus invading natural plant communities. To date, there is no evidence for changes in insect transmission due to recombination. However, Lecoq et al. (1993) reported that a virus that could not normally be transmitted by insects acquired aphid transmissibility through heterologous encapsidation (see below). This suggests that there is potential for changes in insect transmission as a result of recombination. Further, as mentioned above, changes in transmission could allow the recombinant virus to colonize hosts that were previously unavailable to the virus. Since recent studies suggest that viruses are more constrained by the specificity of their relations with vectors than with hosts (Power 2000), changes in insect transmission are likely to have significant impacts on viral host range and virus epidemiology.

Studies that compare the fitness of transgene–virus recombinants to parental viruses are rare, but the number of such studies is increasing. Research on cauliflower mosaic caulimovirus in transgenic tobacco has demonstrated expansions of viral host range and changes in symptom severity, an indicator of viral virulence, resulting from transgene–virus recombination (Schoelz and Wintermantel 1993). Such recombination occurred even under selection pressure of only weak to moderate intensity (Wintermantel and Schoelz 1996). In their 1996 study, Wintermantel and Schoelz found that recombinants were quite common under strong selection (i.e., recombinants were detected in 36% of transgenic plants) and reasonably common under weak to moderate selection (13% of transgenic plants). They also found that, in direct competition with its parental virus, a recombinant was significantly more aggressive than the parent, outcompeting it as the infection proceeded (Kiraly et al. 1998). In another study, Borja et al. (1999) detected high levels of recombination between tomato bushy stunt tombusvirus and host transgenes under moderate selection. Again, recombinants, which required two recombination steps to become viable, rapidly dominated in infected plants. These studies illustrate the potential for transgenic virus resistance to have significant impact on the population biology of viruses in nature.

One recent study attempted to examine the probability of recombinant viruses arising under field conditions by screening for changes in the characteristics of viruses found in infected transgenic potato plants. Over the course of 6 years, Thomas et al. (1998) exposed 65,000 transgenic potatoes expressing potato leafroll virus coat protein or replicase genes to field infections of wild-type viruses. Viruses found to be infecting transgenic plants were examined for changes in symptoms, serology, host range, and transmission characteristics, and no unusual variants were detected. Although this does not rule out the possibility that recombination occurred, it does suggest that the probability is rather low that recombinants with significant effects on virus fitness will persist in this system. This type of study is clearly a useful first step in assessing risk in the field. Unfortunately, low-probability events are difficult to detect, but still may result in significant ecological impacts, such as those described above.

Another significant obstacle to predicting the likelihood of spread and persistence of transgene–virus recombinants is the paucity of data about competition between viruses in the field (Power 1996) and about the spread of viruses from crops to wild plant hosts. Although there have been numerous surveys for particular viruses in wild or weedy plants that are suspected of being reservoirs of viruses (Thresh 1980), apparently no studies have monitored the movement of viruses from crops to wild hosts.

5.3   Other interactions between viruses and transgenes

Another hazard of releasing crops with engineered virus resistance is the enhancement of virus spread through interactions between transgene products and invading viruses (de Zoeten 1991, Miller and Young 1995, Miller et al. 1997). One type of interaction that probably poses relatively low ecological risks is heterologous encapsidation or transcapsidation. When two viruses replicate simultaneously in mixed infections, encapsidation of viral RNA of one virus by the coat protein of another virus can occur (Rochow 1970, 1977, Creamer and Falk 1990, Wen and Lister 1991). Transcapsidation is known to alter vector specificity and the efficiency of virus transmission (Rochow 1977), and it occurs naturally in mixed infections, which can be common for some viruses in some hosts (Hammond et al. 1999). However, if transgenic plants are widely adopted, the probability of "mixed" infections would be much higher than is now common (de Zoeten 1991). Under those conditions, transgenic plants expressing the coat protein of one virus might allow the transmission of a second virus by an herbivorous insect that would not normally be a vector. Since each species of insect vector has a unique plant host range, this increase in vector capability could allow movement of a virus into a plant that would not normally be exposed to that particular virus.

A number of authors have reported transcapsidation of invading viruses with the coat protein expressed by transgenic plants. This phenomenon has been detected in transgenic plants producing coat proteins of a wide range of viruses, including tobamoviruses (Osbourn et al. 1990, Holt and Beachy 1991), potyviruses (Farinelli et al. 1992, Lecoq et al. 1993, Hammond and Dienelt 1997), and bromoviruses (Candelier-Harvey and Hull 1993). Lecoq et al. (1993) have shown that even a virus not normally transmitted by insects can acquire aphid transmissibility through transcapsidation in a transgenic plant. The probability of such an event occurring would depend, in part, on the level of coat protein expression in the transgenic plants, a level that is already highly variable. The probability would also depend on the particular combination of viruses or virus strains, since encapsidation is typically limited to closely related viruses. Despite the real possibility that such events might occur in transgenic plants, the outcome may still pose few hazards to natural habitats. Once the transcapsidated virus has been transmitted to a plant, virus replication would produce the original parental virus since the RNA of the transcapsidated virus would still encode the original coat protein. Thus, although such transmission events might result in the colonization of new hosts, those events may be unlikely to lead to long-term changes in transmission patterns.

A few recent studies have tried to examine the probability of a change in vector transmission arising under field conditions. Fuchs et al. (1999) demonstrated the transmission of an aphid-nontransmissible potyvirus (a strain of zucchini yellow mosaic virus) in fields of transgenic squash expressing the coat protein of an aphid-transmissible strain of watermelon mosaic virus. Under experimental conditions that should have allowed epidemics to occur, only about 3% of uninoculated plants became infected with a virus that was normally aphid nontransmissible, apparently through aphid transmission. Although the mechanism underlying this anomalous transmission was not investigated directly, indirect evidence suggested that it was caused by transcapsidation rather than by virus–transgene recombination. A similar study with transgenic melon and squash expressing coat proteins of an aphid-transmissible strain of cucumber mosaic virus did not detect any transmission of nontransmissible strains (Fuchs et al. 1998). A study of transgenic potatoes expressing coat protein and replicase genes of potato leafroll virus, which is aphid transmitted, also did not detect changes in aphid transmission of several different potato viruses (Thomas et al. 1998). Thus, the risk of changes in aphid transmission due to transcapsidation clearly exists, but evidence to date suggests that such changes do not occur at high rates, and their occurrence is not a certainty. However, the inherent challenges of effective monitoring and detection of low-probability events makes it difficult to evaluate the true risk of these interactions.

A second type of interaction between transgene products and invading virus may present more significant risks, because it can result in increased disease severity in the crop (Miller et al. 1997). Synergistic interactions between viruses in mixed infections can result in disease that is much more severe than that caused by either virus alone. Such effects are common among some viruses such as the luteoviruses (Wen et al. 1991), where synergies appear to be due to the polymerase gene (Miller et al. 1997). Because of this, transgenic plants expressing replicase genes may present the highest risk. This is unfortunate because evidence to date suggests that transgenic plants expressing luteovirus polymerase show greater virus resistance than those expressing coat protein (Thomas et al. 1995). However, if further research indicates that this risk is substantial, the use of coat protein genes might offer a lower-risk strategy for developing transgenic resistance to luteoviruses.

Synergy is primarily an agronomic risk, rather than a risk to natural plant communities, because the increase in disease severity would be limited to the transgenic plant. However, synergism could result in higher rates of viral replication, which in turn could lead to a general increase in virus population levels, and that would negatively affect nontransgenic crops. To the extent that viruses move from crops to natural plant communities, they could contribute to greater disease pressure in those communities as well. Moreover, if transgenes for resistance escape into wild hosts (see below), increased disease severity due to transgene–virus synergies in wild hosts could lead to changes in competitive outcomes among plants in natural habitats.

5.4 Risks due to transgene escape

There is considerable concern that transgenic crops or transgenes might escape. In the course of space and time (via seed banks), they might access wild plant populations (e.g.,

Ellstrand and Hoffman 1990, Crawley et al. 1993, Kareiva et al. 1994, Linder and Schmitt 1994, Raybould and Gray 1994, Schmitt and Linder 1994, Snow and Moran-Palma 1997, Raybould 1999). In the case of escaped transgenes, the probability of transgene introgression would vary with the potential for crop hybridization with wild relatives; the consequences would vary with the general fitness of the resulting hybrids and the particular effect of the transgene on plant fitness and performance. Hybridization between crops and their wild relatives has been documented for 12 of the 13 most important world food crops (Ellstrand et al. 1999) and for a wide range of less important crops (Snow and Moran-Palma 1997). In addition, several studies have examined the fitness of first-generation hybrids as an indication of the potential for transgene establishment in noncrop hosts. Although two of the studies showed reduced fitness in hybrids (Jørgensen et al. 1996, Snow et al. 1998), most found hybrid fitness to be similar to that of the wild parent (Langevin et al. 1990, Klinger and Ellstrand 1994, Frello et al. 1995, Linder and Schmitt 1995, Mikkelsen et al. 1996, Arriola and Ellstrand 1997). These studies taken together suggest that in many cases, transgene movement through hybridization is quite likely.

Few studies have addressed the next steps in transgene establishment, i.e., the persistence of crop genes in wild host populations following initial hybridization and the selective advantage of the transgene resulting from its effects on host biology. Lengthy persistence of genes from cultivated sunflowers has been detected in populations of wild sunflowers (Whitton et al. 1997, Linder et al. 1998), although a separate study found that hybrids of cultivated and wild sunflowers had higher seed predation than their parents (Cummings et al. 1999). Mikkelsen et al. (1996) demonstrated the movement of a transgene for herbicide tolerance from oilseed rape, *Brassica napus,* into the weedy relative *B. campestris.* The transgene would confer a selective advantage in environments where the herbicide was being used for weed control, thus potentially contributing to increased agronomic problems. In addition, no cost to carrying herbicide resistance was detected in hybrids of oilseed rape and the wild relative *B. rapa* (Snow et al. 1999). Therefore, transgenes for herbicide resistance would be expected to persist in wild *B. rapa* populations even if there were no selection via the application of herbicides.

Transgenes for pest and pathogen resistance may confer a selective advantage to wild relatives under an even broader range of conditions. The fitness consequences for plants damaged by insect herbivores are well documented (see Letourneau et al., Chapter 3 of this volume). More recently, biologists have also begun to measure the fitness impacts of pathogens on natural plant populations and communities (for review, see Jarosz and Davelos 1995). Although relatively few studies have examined the fitness consequences of viruses in natural plant populations, several recent studies have demonstrated significant negative impacts of virus infection on the growth, survivorship, and reproduction of purslane, *Portulaca oleracea* (Friess and Maillet 1996), the wild composite *Eupatorium makinoi* (Funayama et al. 1997), and wild cabbage, *Brassica oleracea* (Maskell et al. 1999).

As Raybould (1999) points out, it is much easier to measure and model transgene dispersal than it is to predict its consequences. Experimental studies that investigate the impacts of transgene flow are sorely needed.

5.5   Ecological risks of transgenic resistance to barley yellow dwarf virus

In our laboratory, we have been studying the impacts of the barley yellow dwarf viruses (BYDVs) on wild oats and on its competitive relations with cultivated oats and other grasses, in order to assess the potential hazards associated with movement of transgenic virus resistance from cultivated oats to wild oats. A brief summary of the BYDV system and some of our research is described below.

5.5.1 Transgenic resistance to barley yellow dwarf virus

The aphid-transmitted barley yellow dwarf viruses are useful for addressing ecological risks of transgenic virus resistance. This is so for several reasons: the economic importance of the BYDVs, their exceptionally broad host range among wild grasses, the extensive acreage of their crop hosts, and the high likelihood that viral transgenes will be used in the near future. Barley yellow dwarf is one of the most economically important diseases of grasses worldwide (Irwin and Thresh 1990, D'Arcy and Burnett 1995), and it is among the most prevalent of all viral diseases (Thresh 1980). The barley yellow dwarf viruses (BYDVs) are phloem-limited luteoviruses that are obligately transmitted in a persistent manner by several species of grass-feeding aphids to cultivated and wild grasses (D'Arcy and Burnett 1995, Power and Gray 1995, Miller and Rasochová 1997). The viruses typically cause stunting and yellowing of the host, often resulting in total loss of seed production. In spite of the economic importance of barley yellow dwarf in a wide range of crops, such as wheat, oats, barley, and rice, few natural resistance genes have been identified in any of the major crops. Because conventional breeding has not achieved satisfactory levels of resistance, researchers are using genetic engineering to construct artificial resistance genes targeted against BYDVs (Miller and Young 1995).

Transgenic cereals with genetically engineered resistance to BYDVs are currently being developed and field-tested. Recently, maize (Wyatt et al. 1993), barley (Wyatt et al. 1993, Lister et al. 1994a, McGrath et al. 1997), and oats (Lister et al. 1994b, McGrath et al. 1997) have been transformed with BYDV coat protein genes. Laboratory tests suggest that concentrations of BYDVs are suppressed in at least some of the transgenic lines, and promising lines are undergoing further testing (Miller and Young 1995). Transgenic oats expressing a BYDV replicase gene have also been developed (Koev et al. 1996, 1998). Some lines show reduced viral concentrations (i.e., true resistance), whereas others show tolerance (i.e., limited yield reduction, despite high viral concentrations). Other strategies for artificial resistance to BYDVs, which utilize satellite RNAs, antisense RNA, ribozymes, and defective interfering RNAs, are currently under development or consideration (Miller and Young 1995). Once stable, effective virus resistance has been achieved, cultivars with these resistance genes are likely to be attractive to farmers. Such cultivars might be expected to spread rapidly throughout the cereal growing regions of the United States.

5.5.2 Ecological risks of transgene–virus recombination

The potential for both recombination and interactions between viruses and transgene products is probably higher for luteoviruses such as the BYDVs than for most other plant viruses (Miller et al. 1997). This is suggested by evidence that recombination has occurred relatively recently in the evolution of these viruses (Gibbs 1995). Given the expectation that transgene–virus recombination is likely to occur, at least at low rates, it is important to assess the probability that a recombinant virus would move from a transgenic crop to a wild host. Although the movement of BYDVs from crops to wild hosts has never been studied directly in the field, surveys and laboratory experiments suggest that it probably happens readily. Several surveys have detected high prevalence of BYDVs in wild grasses in various locations (Fargette et al. 1982, Grafton et al. 1982, Guy et al. 1986, Griesbach et al. 1990, D'Arcy 1995, Power and Remold 1996). Based on virus distribution, we can infer that some strains probably do move from crops to wild hosts, since these strains are much more prevalent in wild hosts on the periphery of fields containing crop hosts than in similar areas without crop hosts (Power and Remold 1996).

We have carried out a number of laboratory transmission studies in which isolates of BYDVs were collected from field crops and wild grasses and transferred to new host species using three common species of vector aphids, *Rhopalosiphum maidis, R. padi,* and *Sitobion avenae* (S. Remold and A. Power, unpublished data). Our data suggest that most isolates are transferred more readily between some pairs of plant species than between others. In general, we find a significantly higher probability of successful virus transmission when isolates are transferred between plants of the same host species or closely related species than when they are transferred between distantly related hosts, such as grass species in different subfamilies (S. Remold and A. Power, unpublished data). For example, isolates collected from domesticated oats (*Avena sativa*) are most easily transferred back to oats, but are also readily transferred to wild oats (*A. fatua*). Given the spatial proximity of domesticated and wild oats in oat fields throughout the western United States, this suggests a high likelihood of virus isolates in cultivated oats moving to wild oats. To the extent that any recombination among viral transgenes and invading viruses occurs in cultivated oats, these new recombinant viruses are quite likely to move into wild oats.

5.5.3   Ecological risks of transgene movement

Ecological risks resulting from the movement of transgenes from genetically engineered cereals to wild grasses must be assessed in terms of the probablity of the event, the probability of transgene fixation, and the potential hazards that might result. Because of the presence of interfertile wild relatives in and around crop fields, the likelihood of gene flow is very high. Because transgenes can confer resistance to pathogens, higher fitness is likely to result. Potential hazards associated with resistant weeds can be identified for both agricultural and natural habitats.

5.5.3.1   Probability of transgene movement

Transgene escape may pose particularly high risks for cereal crops expressing transgenic resistance to BYDVs because of the paucity of natural resistance to BYDVs in some wild relatives such as wild oats. Accumulating evidence suggests that both the probability of transgene transfer to wild relatives and the probability that the transgene conveys fitness advantages are likely to be high for some cereals targeted for transgenic BYDV resistance. Of the crops likely to be engineered for resistance to BYDVs, oats (*A. sativa*), barley (*Hordeum vulgare*), and wheat all hybridize with wild relatives in the United States and other parts of their range (Snow and Moran-Palma 1997, Seefeldt et al. 1998). In particular, wild oats (*A. fatua*) and squirreltail grass (*H. jubatum*) are extremely troublesome weeds in the western United States (Holm 1977). Both species, along with other wild species of *Avena* and *Hordeum*, become infected with BYDVs and show typical disease symptoms (Griesbach et al. 1990, C. Malmström and A. Power, unpublished data), indicating that fitness may be reduced in infected plants. Although both oats and barley are largely self-pollinating and their outcrossing rates are considered to be low, forced hybridization with wild relatives has been common in breeding programs for both crops (e.g., Brown 1980). Under field conditions designed to maximize outcrossing, Shorter et al. (1978) found average outcrossing rates of 5.8% across four matings of *A. sativa–A. sterilis* hybrids, but outcrossing rates as high as high as 23.8% were detected for some generations of particular matings. In natural mixed populations of *A. fatua* and *A. sterilis*, outcrossing rates of 4.8% have been reported. Given the success of forced hybridization between cultivated oats and various species of wild oats (Brown 1980), there is significant potential for gene flow from cultivated oats to wild oats and for further gene flow from these hybrids. Despite the expectation that gene flow is likely to fall off significantly with increasing distance between a crop and its wild relatives, low rates of gene flow can occur even at substantial distances for barley (Wagner and Allard 1991) and other crops (see Klinger, Chapter 1 of this volume). Even extremely low rates of gene flow can have significant effects on the evolution of wild relatives, if the genes confer a selective advantage on the recipient (Kareiva et al. 1994).

5.5.3.2   Fitness consequences of transgene movement to wild oats

Would the transfer of BYDV resistance transgenes to wild relatives of oats or barley confer a fitness advantage to these species? Recent work in our laboratory suggests that BYDVs can cause a very significant reduction in the fitness of wild oats (A. Power, unpublished data). In several greenhouse and field experiments, we found significant reductions of growth and reproduction in wild oats inoculated with BYDVs. For example, Californian wild oats inoculated at the two-leaf stage with a Californian strain (PAV) of BYDV produced significantly reduced aboveground (Figure 5.1A) and belowground (Figure 5.1B) biomass. It was striking that the effects of BYDV were much more severe for wild oats than for domesticated oats in this experiment. Given the prevalence of BYDVs in wild oats, the spread of genetically engineered BYDV resistance to wild oats via hybridization likely would enhance the fitness of those weeds. Although some resistance or tolerance to BYDV infection has been detected in collections of wild oats from the north central United States (Rines et al. 1980), we have encountered no evidence of resistance or tolerance in any of the populations we have examined (A. Power, unpublished data). Since effective BYDV resistance has not been developed through conventional plant breeding of oats, there has been no previous opportunity for resistance genes to move to wild oats. In this case, genetically engineered resistance would constitute a novel trait in wild oats.

5.5.3.3   Ecological hazards associated with transgene movement

The movement of transgenes for BYDV resistance into wild oat populations poses two types of risk, one agronomic and the other ecological. In terms of agronomic risk, acquisition of BYDV resistance by wild oats may make this weedy species a more significant competitor with cultivated cereals. The competitive effects of wild oats on cultivated oats, barley, and wheat are well documented (e.g., Cousens et al. 1991, Satorre and Snaydon 1992), but until our recent work, no studies have examined the impact of BYDVs on competitive relations among these species (see below).

Increased weed fitness would be agronomically disadvantageous, but perhaps more importantly, it could have significant impacts on natural plant communities. Introduced wild oats already compete heavily with native grasses in natural California grasslands, causing declines in native grass populations and impeding current efforts at grassland restoration (Barbour et al. 1993, Dyer and Rice 1997). Increased fitness of wild oats through the acquisition of transgenic resistance could result in the release of these species from ecological constraints normally imposed by infection with BYDVs (Schmitt and Linder 1994). This is turn could result in significant negative impacts on native grassland ecosystems. Populations of wild oats commonly show dramatic symptoms of BYDV infection, and our data indicate high levels of infection in Californian populations of *A. fatua* (C. Malmström and A. Power, unpublished data). Understanding the influence of BYDVs on the competitive relations between wild oats and co-occurring native grasses, as well on their weediness in crop fields, would allow us to better predict the potential consequences of any escape of transgenic BYDV resistance to wild relatives.

5.5.3.4   Virus impacts on competitive pressure from wild oats

We have begun to explore the potential consequences of transgenic BYDV resistance movement to wild oats by investigating the effects of BYDVs on competition between

A. Aboveground Biomass

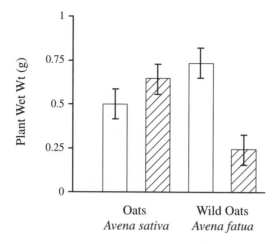

B. Belowground Biomass

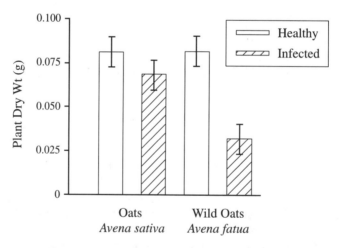

Figure 5.1   Biomass of healthy and BYDV-infected domesticated oats (*Avena sativa*) and wild oats (*A. fatua*) at 28 days after planting. (A) Aboveground biomass. (B) Belowground biomass. Error bars designate standard error of the mean.

wild oats, *A. fatua*, and domesticated oats, *A. sativa* (A. Power, unpublished data). We used a replacement series design with total plant density held constant, and manipulated two factors, plant species and virus infection, following an approach similar to that described in Marks et al. (1991) and Clay et al. (1993). Treatments included *A. sativa* monocultures with or without BYDV, *A. fatua* monocultures with or without BYDV, and mixtures of *A. sativa* and *A. fatua* with or without BYDV. This design allowed us to compare intraspecific vs. interspecific competition for each plant species, the impact of BYDV on each species, and the impact of BYDV on both intraspecific and interspecific competition for each species. All plants designated as infected were inoculated with viruliferous aphids (*R. padi*) at the four-leaf stage, to avoid excessive mortality due to early infection. Plants

A. Vegetative Biomass

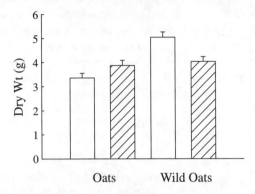

B. Seed Yields

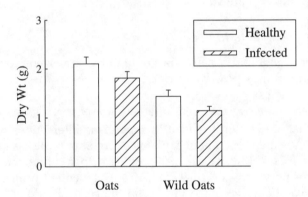

Figure 5.2 Yield of healthy and BYDV-infected domesticated oats (*Avena sativa*) and wild oats (*A. fatua*) when grown together in a two-species mixture. (A) Vegetative biomass. (B) Seed yields. Error bars designate standard error of the mean.

designated as uninfected were mock-inoculated with an equal number of virus-free aphids to control for any possible effects of aphid feeding on plant performance.

For the purposes of risk assessment, we were most interested in comparing domesticated oat–wild oat competition when both species were infected with BYDV (i.e., current ambient conditions) with competition under conditions of no infection (i.e., the predicted scenario when transgenic resistance moves from oats to wild oats). Our data suggest that infection with BYDV more or less "levels the playing field" in the competitive relations between oats and wild oats. Under ambient conditions of BYDV infection, there is no difference between the two species in vegetative biomass, whereas when no BYDV is present, wild oats produce significantly more vegetative biomass than domesticated oats (Figure 5.2A). The seed yields of both wild oats and domesticated oats are lower when infected, but the decrease in seed production of wild oats is proportionately greater than the decrease in domesticated oats (21 vs. 14%; Figure 5.2B). Thus, under ambient conditions of infection with BYDV, seed production is more suppressed in wild oats than in domesticated oats. Release from this suppression through transgenic resistance is likely to benefit wild oats more than domesticated oats.

Our experiments indicate that if transgenic resistance to BYDVs moves from cultivated oats to wild oats, wild oats might become more severe weeds in an agronomic context. Although yields of domesticated oats are reduced by competition whether or not BYDV is present, wild oats exert greater competitive pressure and have much greater seed production when they are not infected by BYDV. Greater seed production may lead to increased population sizes in later generations. These responses to the reduction in disease pressure on wild oats would also be expected in natural grassland ecosystems, where wild oats are already exerting significant competitive pressure on native grasses.

5.6   Prospects for risk reduction

A number of recent articles have suggested strategies for lowering the ecological risks of transgenic virus resistance, while still taking advantage of the transgenic resistance approach. In their comprehensive review of risks of crops with transgenic virus resistance, Hammond et al. (1999) describe a number of different strategies, mostly aimed at minimizing risks of recombination or other transgene–virus interactions, which might mitigate ecological risk under some circumstances. The most-promising recommendations from these authors and others (Hull 1994, Allison et al. 1996, Aaziz and Tepfer 1999b, Hammond et al. 1999, Rubio et al. 1999) are summarized below.

5.6.1   Transgene constructs to avoid

There is general agreement that certain types of high-risk transgene constructs should be avoided, whereas others may pose lesser risks.

5.6.1.1   Functional genes that may interact with other viruses

In general, genes that are known to result in synergistic interactions with other viruses should not be used. As mentioned above, plants expressing replicase genes may present high risks for luteoviruses (Miller et al. 1997), while helper component proteins may be particularly risky for potyviruses (Pruss et al. 1997). Movement proteins from one virus may permit the spread of an unrelated virus, and therefore should also be avoided.

5.6.1.2   Replicase recognition sequences

Subgenomic promoters and some noncoding regions can contain replicase recognition sequences, which can result in transcription of complementary transcripts from transgenes. These sequences are also best avoided (Allison et al. 1996).

5.6.2   Transgene constructs that may be less risky

5.6.2.1   Transgenes with low expression levels

Low transgene expression levels can reduce the risk of recombination, because the fewer the transcripts produced, the lower the potential for recombination with invading viruses (Allison et al. 1996). Use of these transgenes might be most successful for risk reduction when RNA-mediated resistance is involved, to the extent to which both transgene transcript and invading virus are rapidly degraded (Hammond et al. 1999).

5.6.2.2   Genes from mild endemic viral isolates

The use of transgenes from endemic isolates, rather than from exotic isolates, is preferable for two reasons. First, these endemic isolates are most likely to confer resistance to local virus strains. Second, endemic isolates should be less likely than exotic isolates to form recombinant viruses with markedly modified pathogenicity or host range (Hammond et al. 1999).

5.6.2.3 *Untranslatable or antisense sequences*

These sequences cannot give rise to functional protein products and therefore they cannot lead to interactions between invading viruses and transgene products, such as transcapsidation. In addition, multiple mutations that distribute nonsense codons along the transcript may be particularly effective in preventing the incorporation of functional fragments during recombination (Hammond et al. 1999).

5.6.2.4 *Defective genes*

Deleting or mutating domains responsible for vector specificity in a coat protein gene could reduce the risk of transcapsidation (Hammond et al. 1999). Disruption or deletion of sequences known to be replication complex initiation sites is recommended (Allison et al. 1996), along with the use of nonfunctional replicase or movement protein.

5.6.3 *Other approaches to risk reduction*

Strategies to avoid recombination and transgene–virus interactions are useful, but cannot address other categories of risk such as transgene escape. As described above, transgene escape is a significant hazard when transgenes are likely to confer a selective advantage to hybrids of crops and their relatives. Although our experiments have demonstrated the potential for a selective advantage of transgenic BYDV resistance in wild oats, the general paucity of data about the effects of viruses on wild plants in natural habitats makes it difficult to evaluate the generality of our results. However, recent studies of other viruses in natural plant communities do suggest that transgenic virus resistance should be advantageous to noncultivated species in at least some systems (e.g., Friess and Maillet 1996, Funayama et al. 1997, Maskell et al. 1999). Given these studies and the scarcity of information about the impacts of viruses in natural plant communities, the cautious approach would be to avoid deployment of transgenic virus resistance where crops have wild relatives with which they interbreed.

5.7 *Conclusions and policy implications*

Regulatory policies surrounding the use of transgenic virus resistance are still under review at this time. The U.S. Environmental Protection Agency (EPA) asserted jurisdiction over transgenic virus resistance in plants in its 1994 proposed rule on pesticidal substances in transgenic plants. In that rule, the EPA proposed an exemption for viral coat proteins expressed in transgenic plants, on the basis of a lack of evidence of significant negative impacts on human health or the environment. A recent report by the National Research Council on genetically modified pest-protected plants similarly concluded that many crops with transgenic virus resistance are unlikely to pose significant risks to human health or the environment (NRC 2000). However, the report questions the categorical exemption of transgenic virus resistance because of concerns about the potential for outcrossing with weedy relatives. Based on these concerns, the report recommended that EPA should not categorically exempt viral coat proteins from regulation. In addition, other types of virus transgenes, such as replicase genes, have not been addressed explicitly by the regulatory framework.

Advances in our understanding of the molecular aspects of the risks of transgenic virus resistance are proceeding faster than advances in our understanding of ecological aspects of these risks. A major impediment to accurately predicting and reducing the ecological risks of transgenic virus resistance is the difficulty of monitoring viral processes in the field. Recent attempts to screen for transgene–virus interactions have been useful, but those attempts also illustrate the problems of detecting low-probability events (Fuchs et al. 1998, 1999, Thomas et al. 1998). At the same time, we need many more studies that examine the epidemiology and fitness effects of viruses in natural communities. Only by understanding the ecological functioning of these communities in the absence of transgenic crops will we be able to predict the consequences of the deployment of transgenic virus resistance. Moreover, a better understanding of the ecological context of transgenic virus resistance will aid in developing sound regulatory policies for this new technology.

Literature cited

Aaziz, R. and Tepfer, M., Recombination between genomic RNAs of two cucumoviruses under conditions of minimal selection pressure, *Virology*, 263, 282, 1999a.

Aaziz, R. and Tepfer, M., Recombination in RNA viruses and in virus-resistant transgenic plants, *J. Gen. Virol.*, 80, 1339, 1999b.

Allison, R. F., Schneider, W. L., and Greene, A. E., Recombination in plants expressing viral transgenes, *Semin. Virol.*, 7, 417, 1996.

Arriola, P. E. and Ellstrand, N. C., Fitness of interspecific hybrids in the genus *Sorghum*: persistence of crop genes in wild populations, *Ecol. Appl.*, 7, 512, 1997.

Barbour, M., Pavlik, B., Drysdale, F., and Lindstrom, S., *California's Changing Landscapes: Diversity and Conservation of California Vegetation*, California Native Plant Society, Sacramento, 1993.

Beachy, R. N., Loesch-Fries, S., and Tumer, N. E., Coat protein-mediated resistance against virus infection, *Annu. Rev. Phytopathol.*, 28, 451, 1990.

Borja, M., Rubio, T., Scholthof, H. B., and Jackson, A. O., Restoration of wild-type virus by double recombination of tombusvirus mutants with a host transgene, *Mol. Plant Microb. Interact.*, 12, 153, 1999.

Brown, C. M., Oat, in *Hybridization of Crop Plants*, Fehr, W. R. and Hadley, H. H., Eds., American Society of Agronomy, Madison, WI, 1980, 427.

Candelier-Harvey, P. and Hull, R., Cucumber mosaic virus genome is encapsidated in alfalfa mosaic virus coat protein expressed in transgenic tobacco plants, *Transgenic Res.*, 2, 277, 1993.

Clay, K., Marks, S., and Cheplick, G. P., Effects of insect herbivory and fungal endophyte infection on competitive interactions among grasses, *Ecology*, 74, 1767, 1993.

Cousens, R. D., Weaver, S. E., Martin, T. D., Blair, A. M., and Wilson, J., Dynamics of competition between wild oats (*Avena fatua* L.) and winter cereals, *Weed Res.*, 31, 203, 1991.

Crawley, M. J., Hails, R. S., Rees, M., Kohn, D., and Buxton, J., Ecology of transgenic oilseed rape in natural habitats, *Nature*, 363, 620, 1993.

Creamer, R. and Falk, B. W., Direct detection of transcapsidated barley yellow dwarf luteoviruses in doubly infected plants, *J. Gen. Virol.*, 71, 211, 1990.

Cummings, C. L., Alexander, H. M., and Snow, A. A., Increased pre-dispersal seed predation in sunflower crop-wild hybrids, *Oecologia*, 121, 330, 1999.

D'Arcy, C. J., Symptomatology and host range of barley yellow dwarf, in *Barley Yellow Dwarf: 40 Years of Progress*, D'Arcy, C. J. and Burnett, P. A., Eds., American Phytopathological Society Press, St. Paul, MN, 1995, 9.

D'Arcy, C. J. and Burnett, P. A., *Barley Yellow Dwarf: 40 Years of Progress*, American Phytopathological Society Press, St. Paul, MN, 1995.

de Zoeten, G. A., Risk assessment: do we let history repeat itself? *Phytopathology*, 81, 585, 1991.

Dyer, A. R. and Rice, K. J., Intraspecific and diffuse competition: the response of *Nassella pulchra* in a California grassland, *Ecol. Appl.*, 7, 484, 1997.

Ellstrand, N. and Hoffman, C., Hybridization as an avenue of escape for engineered genes, *BioScience*, 40, 438, 1990.

Ellstrand, N. C., Prentice, H. C., and Hancock, J. F., Gene flow and introgression from domesticated plants into their wild relatives, *Annu. Rev. Ecol. Syst.*, 30, 539, 1999.

Falk, B. W. and Bruening, F., Will transgenic crops generate new viruses and new diseases? *Science*, 263, 1395, 1994.

Fargette, D., Lister, R. M., and Hood, E. L., Grasses as a reservoir of barley yellow dwarf virus in Indiana, *Plant Dis.*, 66, 1041, 1982.

Farinelli, L., Malnoe, P., and Collet, G. F., Heterologous encapsidation of potato virus Y strain O (PVYo) with the transgenic coat protein of PVY strain N (PVYn) in *Solanum tuberosum* cv. Bintje, *Bio-Technology*, 10, 1020, 1992.

Fernández-Cuartero, B., Burgyán, J., Aranda, M. A., Salánki, K., Moriones, E. and García-Arenal, F., Increase in the relative fitness of a plant virus RNA associated with its recombinant nature, *Virology*, 203, 373, 1994.

Frello, S., Hansen, K. R., Jensen, J., and Jørgensen, R. B., Inheritance of rapeseed (*Brassica napus*)-specific RAPD markers and a transgene in the cross *B. juncea* × (*B. juncea* × *B. napus*), *Theor. Appl. Genet.*, 91, 236, 1995.

Friess, N. and Maillet, J., Influence of cucumber mosaic virus infection on the intraspecific competitive ability and fitness of purslane (*Portulaca oleracea*), *New Phytol.*, 132, 103, 1996.

Fuchs, M., Klas, F. E., McFerson, J. R., and Gonsalves, D., Transgenic melon and squash expressing coat protein genes of aphid-borne viruses do not assist the spread of an aphid nontransmissible strain of cucumber mosaic virus in the field, *Transgenic Res.*, 7, 449, 1998.

Fuchs, M., Gal-On, A., Raccah, B., and Gonsalves, D., Epidemiology of an aphid non-transmissible potyvirus in fields of nontransgenic and coat protein transgenic squash, *Transgenic Res.*, 8, 429, 1999.

Funayama, S., Hikosaka, K., and Yahara, T., Effects of virus infection and growth irradiance on fitness components and photosynthetic properties of *Eupatorium makinoi* (Compositae), *Am. J. Bot.*, 84, 823, 1997.

Gadani, F., Mansky, L. M., Medici, R., Miller, W. A., and Hill, J. H., Genetic engineering of plants for virus resistance, *Arch. Virol.*, 115, 1, 1990.

Gibbs, M., The luteovirus supergroup: rampant recombination and persistent partnerships, in *Molecular Basis of Virus Evolution*, Gibbs, A. J., Calisher, C. H., and Garcia-Arenal, F., Eds., Cambridge University Press, Cambridge, U.K., 1995, 351.

Grafton, K. F., Poehlman, J. M., Sehgal, O. P., and Sechler, D. T., Tall fescue as a natural host and aphid vectors of barley yellow dwarf virus in Missouri (*Rhopalosiphum padi*), *Plant Dis.*, 66, 318, 1982.

Griesbach, J. A., Steffenson, B. J., Brown, M. P., Falk, B. W., and Webster, R. K., Infection of grasses by barley yellow dwarf viruses in California, *Crop Sci.*, 30, 1173, 1990.

Guy, P. L., Johnstone, G. R., and Duffus, J. E., Occurrence and identity of barley yellow dwarf viruses in Tasmanian pasture grasses, *Aust. J. Agric. Res.*, 37, 43, 1986.

Hammond, J. and Dienelt, M. M., Encapsidation of potyviral RNA in various forms of transgene coat protein is not correlated with resistance in transgenic plants, *Mol. Plant Microb. Interact.*, 10, 1023, 1997.

Hammond, J., Lecoq, H., and Raccah, B., Epidemiological risks from mixed virus infections and transgenic plants expressing viral genes, *Adv. Virus Res.*, 54, 189, 1999.

Holm, L. G., *The World's Worst Weeds: Distribution and Biology*, University of Hawaii Press, Honolulu, 1977.

Holt, C. A. and Beachy, R. N., *In vivo* complementation of infectious transcripts from mutant tobacco mosaic virus cDNAs in transgenic plants, *Virology*, 181, 109, 1991.

Hull, R., Resistance to plant viruses: obtaining genes by non-conventional approaches, *Euphytica*, 75, 195, 1994.

Irwin, M. E. and Thresh, J. M., Epidemiology of barley yellow dwarf virus: a study in ecological complexity, *Annu. Rev. Phytopathol.*, 28, 393, 1990.

Jarosz, A. M. and Davelos, A. L., Effects of disease in wild plant populations and the evolution of pathogen aggressiveness, *New Phytol.*, 129, 371, 1995.

Jørgensen, R. B., Andersen, B., and Mikkelsen, T. R., Spontaneous hybridization between oilseed rape (*Brassica napus*) and weedy relatives, *Acta Hortic.*, 407, 193, 1996.

Kareiva, P., Morris, W., and Jacobi, C. M., Studying and managing the risk of cross-fertilization between transgenic crops and wild relatives, *Mol. Ecol.*, 3, 15, 1994.

Kiraly, L., Bourque, J. E., and Schoelz, J. E., Temporal and spatial appearance of recombinant viruses formed between cauliflower mosaic virus (CaMV) and CaMV sequences present in transgenic *Nicotiana bigelovii*, *Mol. Plant Microb. Interact.*, 11, 309, 1998.

Klinger, T. and Ellstrand, N. C., Engineered genes in wild populations: fitness of weed-crop hybrids of *Raphanus sativus*, *Ecol. Appl.*, 4, 117, 1994.

Koev, G., Mohan, B. R., Beckett, R., and Miller, W. A., Transgenic resistance to barley yellow dwarf virus: replication, recombination, and risks, in *Proceedings of the Biotechnology Risk Assessment Symposium*, Levin, M., Grim, C., and Angle, S., Eds., University of Maryland, Gaithersburg, 1996, chap. 7.

Koev, G., Mohan, B. R., Dinesh-Kumar, S. P., Torbert, K. A., Somers, D. A., and Allen, W. A., Extreme reduction of disease in oats transformed with the 5' half of the barley yellow dwarf virus-PAV genome, *Phytopathology*, 88, 1013, 1998.

Langevin, S. A., Clay, K., and Grace, J. B., The incidence and effects of hybridization between cultivated rice and its related weed red rice (*Oryza sativa* L.), *Evolution*, 44, 1000, 1990.

Lecoq, H., Ravelonandro, M., Wipf-Scheibel, C., Monsion, M., Raccah, B., and Dunez, J., Aphid transmission of a non-aphid-transmissible strain of zucchini yellow mosaic potyvirus from transgenic plants expressing the coat protein of plum pox potyvirus, *Mol. Plant Microb. Interact.*, 6, 403, 1993.

Linder, C. R. and Schmitt, J., Assessing the risks of transgene escape through time and crop-wild hybrid persistence, *Mol. Ecol.*, 3, 23, 1994.

Linder, C. R. and Schmitt, J., Potential persistence of escaped transgenes: performance of transgenic oil-modified *Brassica* seeds and seedlings, *Ecol. Appl.*, 5, 1056, 1995.

Linder, C. R., Taha, I., Seiler, G. J., Snow, A. A., and Rieseberg, L. H., Long-term introgression of crop genes into wild sunflower populations, *Theor. Appl. Genet.*, 96, 339, 1998.

Lister, R. M., Vincent, J. R., and McGrath, P. F., Coat protein mediated resistance to barley yellow dwarf virus in barley, *Phytopathology*, 84, 1117, 1994a.

Lister, R. M., Vincent, J. R., Lei, C.-H., and McGrath, P. F., Coat protein mediated resistance to barley yellow dwarf virus in oats, *Phytopathology*, 84, 1117, 1994b.

Marks, S., Clay, K., and Cheplick, G. P., Effects of fungal endophytes on interspecific and intraspecific competition in the grasses *Festuca arundinacea* and *Lolium perenne*, *J. Appl. Ecol.*, 28, 194, 1991.

Maskell, L. C., Raybould, A. F., Cooper, J. I., Edwards, M. L., and Gray, A. J., Effects of turnip mosaic virus and turnip yellow mosaic virus on the survival, growth and reproduction of wild cabbage (*Brassica oleracea*), *Ann. Appl. Biol.*, 135, 401, 1999.

McGrath, P. F., Vincent, J. R., Lei, C. H., Pawlowski, W. P., Torbet, K. A., Gu, W., Kaeppler, H. F., Wan, Y., Lemaux, P. G., and Rines, H. R., Coat protein-mediated resistance to isolates of barley yellow dwarf in oats and barley, *Eur. J. Plant Pathol.*, 103, 695, 1997.

Mikkelsen, T. R., Andersen, B., and Jørgensen, R. B., The risk of crop transgene spread, *Nature*, 380, 31, 1996.

Miller, W. A. and Rasochová, L., Barley yellow dwarf viruses, *Annu. Rev. Phytopathol.*, 35, 167, 1997.

Miller, W. A. and Young, M. J., Prospects for genetically engineered resistance to barley yellow dwarf viruses, in *Barley Yellow Dwarf: 40 Years of Progress*, D'Arcy, C. J. and Burnett, P. A., Eds., American Phytopathological Society, St. Paul, MN, 1995, 345.

Miller, W. A., Koev, G., and Mohan, B. R., Are there risks associated with transgenic resistance to luteoviruses? *Plant Dis.*, 81, 700, 1997.

NRC (National Research Council), Genetically Modified Pest-Protected Plants: Science and Regulation, National Academy Press, Washington, D.C., 2000.

Osbourn, J. K., Sarkar, S., and Wilson, T. M. A., Complementation of coat protein-defective TMV mutants in transgenic tobacco plants expressing TMV coat protein, *Virology*, 179, 921, 1990.

Power, A. G., Competition between viruses in a complex plant-pathogen system, *Ecology*, 77, 1004, 1996.

Power, A. G., Insect transmission of plant viruses: a constraint on virus variability, *Curr. Opin. Plant Biol.*, 3, 335, 2000.

Power, A. G. and Gray, S. M., Virus-vector interactions and the transmission of barley yellow dwarf viruses by aphids, in *Barley Yellow Dwarf: 40 Years of Progress*, D'Arcy, C. J. and Burnett, P. A., Eds., American Phytopathological Society, St. Paul, MN, 1995, 259.

Power, A. G. and Remold, S. K., Incidence of barley yellow dwarf virus in wild grass populations: implications for biotechnology risk assessment, in *Proceedings of the Biotechnology Risk Assessment Symposium*, Levin, M., Grim, C. and Angle, S., Eds., University of Maryland, Gaithersburg, 1996, chap. 3.

Pruss, G., Xe, G., Shi, X. M., Carrington, J. C., and Vance, V. B., Plant viral synergism: the potyviral genome encodes a broad-range pathogenicity enhancer that transactivates replication of hetero-logous viruses, *Plant Cell*, 9, 859, 1997.

Raybould, A. F., Transgenes and agriculture — going with the flow? *Trends Plant Sci.*, 4, 247, 1999.

Raybould, A. F. and Gray, A. J., Will hybrids of genetically modified crops invade natural commu-nities? *Trends Ecol. Evol.*, 9, 85, 1994.

Rines, H. W., Stuthman, D. D., Briggle, L. W., Youngs, V. L., Jedlinski, H., Smith, D. H., Webster, J. A., and Rothman, P. G., Collection and evaluation of *Avena fatua* for use in oat improvement, *Crop Sci.*, 20, 63, 1980.

Rochow, W. F., Barley yellow dwarf virus: phenotypic mixing and vector specificity, *Science*, 167, 875, 1970.

Rochow, W. F., Dependent virus transmission from mixed infections, in *Aphids as Virus Vectors*, Harris, F. and Maramorosch, K., Eds., Academic Press, New York, 1977, 253.

Rubio, T., Borja, M., Scholthof, H. B., and Jackson, A. O., Recombination with host transgenes and effects of virus evolution: an overview and opinion, *Mol. Plant Microb. Interact.*, 12, 87, 1999.

Satorre, E. H. and Snaydon, R. W., A comparison of root and shoot competition between spring cereals and *Avena fatua* L., *Weed Res.*, 32, 45, 1992.

Schmitt, J. and Linder, C. R., Will escaped transgenes lead to ecological release? *Mol. Ecol.*, 3, 71, 1994.

Schoelz, J. E. and Wintermantel, W. M., Expansion of viral host range through complementation and recombination in transgenic plants, *Plant Cell*, 5, 1669, 1993.

Seefeldt, S. S., Zemetra, R., Young, F. L., and Jones, S. S., Production of herbicide-resistant jointed goatgrass (*Aegilops cylindrica*) × wheat (*Triticum aestivum*) hybrids in the field by natural hybrid-ization, *Weed Sci.*, 46, 632, 1998.

Shorter, R., Gibson, P., and Frey, K. J., Outcrossing rates in oat species crosses (*Avena sativa* L. × *A. sterilis* L.), *Crop Sci.*, 18, 877, 1978.

Snow, A. A. and Moran-Palma, P., Commercialization of transgenic plants: potential ecological risks, *BioScience*, 47, 86, 1997.

Snow, A. A., Moran-Palma, P., Rieseberg, L. H., Wszelaki, A., and Seiler, G. J., Fecundity, phenology, and seed dormancy of F-1 wild-crop hybrids in sunflower (*Helianthus annuus*, Asteraceae), *Am. J. Bot.*, 85, 794, 1998.

Snow, A. A., Andersen, B., and Jorgensen, R. B., Costs of transgenic herbicide resistance introgressed from *Brassica napus* into weedy *B. rapa*, *Mol. Ecol.*, 8, 605, 1999.

Tepfer, M. and Balazs, E., *Virus-Resistant Transgenic Plants: Potential Ecological Impact*, Springer-Verlag, Berlin, 1997, 126.

Thomas, P. E., Kaniewski, W. K., Reed, G. L., and Lawson, E. C., Transgenic resistance to potato leafroll virus in Russet Burbank potatoes, in *Environmental Biotic Factors in Integrated Plant Disease Control*, Manka, M., Ed., Polish Phytopathogical Society, Poznan, 1995, 551.

Thomas, P. E., Hassan, S., Kaniewski, W. K., Lawson, E. C., and Zalewski, J. C., A search for evidence of virus/transgene interactions in potatoes transformed with the potato leafroll virus replicase and coat protein genes, *Mol. Breed.*, 4, 407, 1998.

Thresh, J. M., The origin and epidemiology of some important plant virus diseases, *Appl. Biol.*, 5, 1, 1980.

van den Boogaart, T., Lomonossoff, G. P., and Davies, J. W., Can we explain RNA-mediated virus resistance by homology-dependent gene silencing? *Mol. Plant Microb. Interact.*, 11, 717, 1998.

Wagner, D. B. and Allard, R. W., Pollen migration in predominantly self-fertilizing plants: barley, *J. Hered.*, 82, 302, 1991.

Waterhouse, P. M., Smith, N. A., and Wang, M. B., Virus resistance and gene silencing: killing the messenger, *Trends Plant Sci.*, 4, 452, 1999.

Wen, F. and Lister, R. M., Heterologous encapsidation in mixed infections among four isolates of barley yellow dwarf virus, *J. Gen. Virol.*, 72, 2217, 1991.

Wen, F., Lister, R. M., and Fattouh, F. A., Cross-protection among strains of barley yellow dwarf virus, *J. Gen. Virol.*, 72, 791, 1991.

Whitton, J., Wolf, D. E., Arias, D. M., Snow, A. A., and Rieseberg, L. H., The persistence of cultivar alleles in wild populations of sunflowers five generations after hybridization, *Theor. Appl. Genet.*, 95, 33, 1997.

Wintermantel, W. M. and Schoelz, J. E., Isolation of recombinant viruses between cauliflower mosaic virus and a viral gene in transgenic plants under conditions of moderate selection pressure, *Virology*, 223, 156, 1996.

Wyatt, S. D., Berger, P. H., and Lemaux, P. G., Coat protein gene-mediated resistance to barley yellow dwarf virus in maize and barley, *Phytopathology*, 83, 1374, 1993.

chapter six

Impacts of genetically engineered crops on non-target herbivores: Bt-corn and monarch butterflies as a case study

John E. Losey, John J. Obrycki, and Ruth A. Hufbauer

Contents

6.1   Introduction

A wide variety of transgenic insecticidal crops have been field-tested in the United States and three — corn, cotton, and potato — have been commercialized and planted extensively (Gould 1998, Federici 1998). These crops have been modified to produce insecticidal toxins derived from genes of the bacterium *Bacillus thuringiensis* (*Bt*). Planting of millions of hectares of these crops has raised questions about selection for resistant pest populations and ecological disruption of food webs (Gould 1998, McGaughey et al. 1998, Obrycki et al. 2001). This chapter focuses on identifying and assessing the risks to non-target herbivorous insect species resulting from transgenic insecticidal corn that was developed to control selected lepidopteran species feeding on aboveground portions of the corn plant (Koziel et al. 1993, Pilcher et al. 1997, Williams et al. 1997, 1998, Anonymous 1999). Many of the principles developed for this risk assessment will be applicable to risk assessment for other crops and toxins.

6.2   Risk identification: Determining which herbivore species are at risk

The first step in any risk assessment is developing a list of species that might be at risk. Once a complete list is generated, the relative level of risk can be estimated for each individual species. Four factors largely determine which herbivorous insect species might be affected by transgenic plants: (1) the range of activity of the toxin produced toward various herbivore taxa, (2) the phenology of toxin expression, (3) the level of toxin expression, and (4) the plant tissues where the toxin is expressed.

6.2.1   Toxin specificity

All transgenic insecticidal plants currently registered in the United States have been transformed to contain genetic material from the soil bacterium *B. thuringiensis* (*Bt*). Although *Bt* has a wide host range, individual strains are usually very specific in their range of activity. The most widely utilized *Bt*-crops are *Bt*-potatoes, which produce a toxin effective against beetles (Coleoptera), and *Bt*-corn and cotton, which produce toxins specific to the caterpillars of butterflies and moths (Lepidoptera). New hybrids of *Bt*-corn are being developed that contain the beetle toxin. The limited range of taxa affected by particular *Bt*-strains theoretically limits the direct non-target impacts of these transgenic crops to a single insect order, although there is the possibility of indirect effects on predators of the target order (Hilbeck et al. 1998). If toxins are eventually combined or "stacked" in a single hybrid, the range of pest species protected against and the range of non-target species potentially affected will be expanded.

6.2.2   Toxin expression profile

Knowing which plant tissues express the toxin can further narrow the list of potentially impacted species. For example, if toxin production is limited to the aboveground portions

Table 6.1 Host Plants of Arctiid Moths (Lepidoptera: Arctiidae) That Are Associated with Corn

Plant record	Plant species	Common name	Plant family	Occurrence in corn	Ref.
1	*Apocynum cannabinum*	Hemp dogbane	Apocynaceae	Common	Ransom and Kells 1998
2	*Asclepias syriaca*	Common milkweed	Asclepiadaceae	Common	Wax et al. 1981
3	*Helilanthus annuus*	Common sunflower	Asteraceae	Common	Wax et al. 1981
4	*Campsis radicans*	Trumpetcreeper	Bignoniaceae	Rare	Wax et al. 1981
5	*Chenopodium album*	Common lambsquarter	Chenopodiaceae	Abundant	Glenn et al. 1997
6	*Salsola iberica*	Russian thistle	Chenopodiaceae	Common	Schweizer et al. 1998
7	*Ipomoea purpurea*	Purple morning glory	Convolvulaceae	Common	Webster and Cobble 1997
8	*Malva neglecta*	Common mallow	Malvaceae	Rare	Schweizer et al. 1998
9	*Polygonum convolvulus*	Wild buckwheat	Polygonaceae	Rare	Schweizer et al. 1998
10	*Portulaca oleracea*	Common purslane	Portulaceae	Abundant	Schweizer et al. 1998
11	*Rubus allegheniensis*	Wild blackberry	Rosaceae	Rare	Glenn et al. 1997
12	*Solanum rostratum*	Buffalo bur	Solanaceae	Common	Wax et al. 1981

of the plant, root-feeding herbivores are unlikely to be affected. Although decaying plant tissues may become mobile in the soil and impact the detritus cycle (Saxena et al. 1999), most living plant tissues are stationary and the herbivores that attack cultivated plants are fairly well documented. Herbivores that feed extensively on stationary plant tissues are potential pests, so confining toxins to these tissues serves to maximize exposure of target pests and minimize exposure of non-target organisms.

6.2.3 Pollen: Assessing risks from a mobile tissue

Unlike most other plant tissues, pollen can be extremely mobile and *Bt*-toxin is expressed in the pollen of some transgenic plants (Fearing et al. 1997). Toxic transgenic pollen has the potential to be dispersed onto other plants where it might be consumed by a wide variety of herbivorous insects. We are compiling an initial list of lepidopteran species that could potentially be affected by pollen from *Bt*-corn plants (Losey et al., unpublished). Below we provide data from one lepidopteran family, Arctiidae, as an illustration of the process involved in generating the list. This process could be applied to a broad range of transgenic plants and toxins to assess risks. For example, if another wind-pollinated plant produced pollen toxic to Coleoptera (beetles), then a similar approach could be adapted with the focus on beetles instead of butterflies and moths.

6.2.3.1 Plants within the pollen shadow

Since plants are the framework of ecological communities, plant species within range of pollen dispersal will, to a large degree, determine which other species are most likely to be present and subject to the effects of the pollen. Thus, the first step in the initial screening process is to determine which plants occur within the pollen shadow (Table 6.1). In this example we list only those plants that occur within the corn pollen shadow that are known hosts of Arctiid moths. Our ongoing survey has, as of this writing, identified 57 plant

Table 6.2 Arctiid Moths (Lepidoptera: Arctiidae) Reported to Feed as Larva on Plants That Occur within the Corn Pollen Shadow

Lepidopteran species	Common name	Abundance	Phenology Jn	Jl	Ag	Plant utilization Ref.	Plant record
Apantesis nevadensis	—	—	—	—	—	Covell 1984	8
A. ornata	—	—	—	—	—	Covell 1984	5
A. parthenice	—	—	N	—	Y	Forbes 1923	5
Arctia caja	Great tiger moth	Uncommon/ rare	Y	Y	Y	Covell 1984	5
Cycnia tenera	Dogbane tiger moth	Common	Y	Y	Y	Forbes 1923	1
Diacrisia vagans	—	—	—	—	—	Covell 1984	5
D. virginica	Yellow bear moth	Abundant	Y	N	Y	Forbes 1923	3,5,6,7, 9,11,12
Eubaphe aurantiaca	—	—	Y	Y	Y	Forbes 1923	5
Euchaetias egle	Milkweed tussock moth	Common	Y	Y	—	Richard and Heitzman 1987	5
Haploa colona	Colona moth	Uncommon	Y	—	N	Covell 1984	2
H. lecontei	Leconte's Haploa	Common	Y	Y	—	Covell 1984	11
Hyphantria cunea	Fall webworm	Abundant	Y	Y	Y	Richard and Heitzman 1987	4,5, 10,11
Isia isabella	Isabella tiger moth; black-ended bear	Common	Y	N	Y	Forbes 1923	3,11

Note: Jn = June, Jl = July, Ag = August, Y = yes, N = no.

species in 17 families (Losey et al. unpublished). Because most non-crop plants in and around transgenic crop fields are considered weeds, the makeup of these plant communities is fairly well known. Unfortunately, what is not known for most of the plants associated with transgenic crops is the proportion of their total distribution that falls within the pollen shadow.

6.2.3.2 Herbivores on plants within the pollen shadow

The next step in our sequential approach is to determine which herbivores feed on plants within the pollen shadow. By cross-listing the plant species with the herbivore species that feed on these plants, an initial list of non-target herbivores can be generated (Table 6.2). Although our example provides the data for one family, our survey has so far identified 206 species of Lepidoptera in 20 families that feed on plants within the corn pollen shadow (Losey et al. unpublished). It should be emphasized that a record of an herbivore feeding on a plant does not provide quantitative information on the relative importance of that host for the herbivore species. Specialist herbivores feeding on plants that grow exclusively near transgenic fields are much more likely to be affected by transgenic pollen than herbivores that feed on plants growing in several habitats, in addition to transgenic field margins. This type of niche analysis is probably best reserved for those species that are identified as at risk in the initial screen.

6.2.3.3 Phenological concordance between pollen and herbivores

Once it has been determined which herbivore species feed on plant species within the pollen shadow of a given transgenic crop, the next step will be to determine which of those herbivore species are feeding in the larval stage during anthesis. The pollen shed period is well known for most crop plants. To encounter transgenic pollen, larvae must be feeding during or immediately after this period, while the pollen is still on their host plant. Although exact phenological data are not available for many of the lepidopteran species identified, it is often possible to at least determine which of the summer months the larvae of most species are known to be active (see Table 6.2). Corn pollen sheds from mid-July to mid-August and thus larvae feeding in the early or late summer are unlikely to encounter corn pollen.

6.2.3.4   Relative susceptibility to the transgenic toxin

The final step in predicting which herbivore species are likely to be affected would be to determine each species' relative susceptibility to the toxin expressed in the pollen. For example, although the toxin in *Bt*-corn is active against many lepidopteran families, there is variation in the levels of susceptibility (Peacock et al. 1998, 1993, Pilcher et al. 1997, Williams et al. 1997, 1998). This variation occurs both within and among families of Lepidoptera. For example, a recent study found no effect of *Bt*-corn pollen on the black swallowtail, *Papilio polyxenes*, under field conditions, whereas *Papilio canadensis* was found to be highly susceptible to the toxin (in a spray formulation) under field conditions (Johnson et al. 1995, Wraight et al. 2000). Thus, not only does susceptibility vary widely, but the results of susceptibility tests will be influenced by testing conditions.

Obviously it would be extremely useful to move beyond simply noting that variability exists toward predictions of susceptibility for a given non-target species. It may be possible eventually to link susceptibility with phylogeny to allow this type of prediction. Unfortunately, there are not enough data currently available to complete this analysis at even the family level. However, even with the existing data gaps, integrating the available information on distribution, phenology, host range, and susceptibility will allow an initial categorization of the risk to a given herbivore species. Species at potentially high risk could then be identified for further testing.

6.3   Risk assessment: Determining the severity of risk

Once at-risk species are identified, the next step is to begin a detailed assessment of the extent of the risk to those herbivore populations. To do this, we propose a two-tiered approach that combines the level of exposure to the toxin and the susceptibility of the herbivore species to the toxin. Although this approach applies directly to the determination of risk from toxic pollen, it could be adapted for exposure to other plant tissues.

6.3.1   Exposure to the toxin

For a given herbivore population, the first step in risk assessment is to determine the level of exposure to the toxin from a transgenic plant. A simple mean exposure level (e.g., pollen grains/leaf area) is not sufficient information for an accurate determination of risk. A population exposure distribution must be developed for each species potentially at risk (Figure 6.1A). This exposure distribution reflects the fact that a proportion of the herbivore population, those on plants within fields of transgenic crops, will be exposed to the maximum toxin concentrations whereas those farther from the fields will be exposed to progressively lower toxin concentrations.

6.3.2   Susceptibility to the toxin

Once the exposure distribution is calculated, the mortality caused by a range of toxin concentrations as a standard dose–response relationship within that range can be quantified (Figure 6.1B). The dose–response for an individual species will provide more detailed

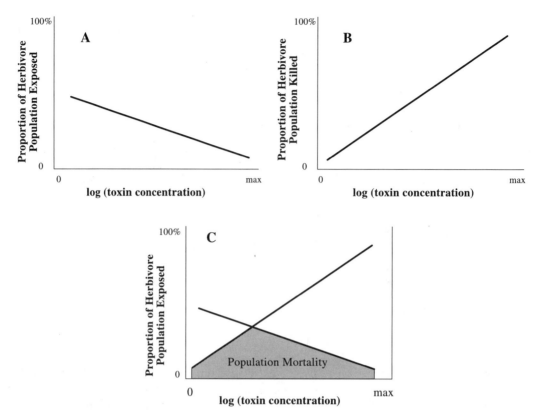

Figure 6.1   (A) Herbivore population exposure distribution. (B) Dose–response for herbivore mortality due to transgenic toxin. (C) Population level mortality for herbivore species calculated from A and B.

information than the general classification of susceptibility for higher taxonomic levels outlined in Section 6.2.3. To understand fully the impact of a given toxin, sublethal effects on growth, development, and behavior must be assessed in addition to direct mortality.

6.3.3   Combining exposure and susceptibility to predict risk

By combining the data on exposure and susceptibility for a given population the population-level mortality attributable to that particular toxin or plant tissue (Figure 6.1C) can be calculated by the following formula:

$$\int_{\text{toxin (concentration)} = 0}^{\text{toxin (concentration)} = \text{max}} \text{Exposure (toxin)} \times \text{Mortality (toxin)} \, d \, \text{toxin} = \text{Population mortality}$$

Several environmental and biological factors can affect the level of toxin exposure and the mortality caused at each exposure level. We discuss these factors in some detail for the monarch butterfly in Section 6.5. Determining the magnitude of variation in these factors will allow a prediction of the range of the impact of a given transgenic toxin on a herbivore species. The variance of the predicted mean mortality will be contingent on the accuracy of the data available and the inherent variability in these factors. Clearly, as more data become available on susceptibility and the factors that affect exposure, the accuracy of predictions regarding the magnitude of non-target effects will be enhanced.

6.4 Pollen from Bt-*corn: A potential risk vector*

Genetically engineered resistance to insects, derived by inserting genetic material from the bacterium *Bt* into plants, has been one of the first commercial successes of agricultural biotechnology. Plants successfully transformed with the *Bt*-gene produce a crystalline protein that is normally produced by the bacterium, and is selectively toxic to certain insects. At least 18 different *Bt*-crops have been approved for field testing in the United States and three — corn, cotton, and potato — have been approved for commercial use (Andow and Hutchison 1998).

Bt-transformed plants are generally considered to have little or no impact on non-target organisms (Ostlie et al. 1997). However, the impact of pollen from transgenic plants on non-target organisms has not been fully examined. Most commercial *Bt*-corn hybrids are known to express the *Bt*-toxin in pollen tissue (Fearing et al. 1997), and corn pollen is dispersed at least 60 m (Raynor et al. 1972), and possibly more than 200 m, by the wind (Louette 1995). Over 7 million acres of *Bt*-corn were planted in the United States in 1998, limited only by seed availability, and the amount of *Bt*-corn production is projected to increase substantially over the next several years (Andow and Hutchison 1998). This means that large quantities of insufficiently tested, unreglated, transgenic pollen with insecticidal properties will drift out of *Bt*-corn fields each year. The toxin in *Bt*-pollen is active primarily against species of Lepidoptera. Potentially, many non-target Lepidoptera could consume *Bt*-pollen that has blown on to their host plants from nearby *Bt*-corn fields. The amount of toxin originating from *Bt*-corn could represent a direct risk for non-target Lepidoptera and an indirect risk for other insect species.

6.5 Monarch butterflies: A species at risk

Several factors combine to place monarch butterflies, *Danaus plexippus*, at particularly high risk from *Bt*-corn pollen. Monarchs migrate annually from overwintering sites in Mexico to spring and summer breeding areas across eastern North America producing up to five generations (Brower 1996). Overwintering sites are extremely small areas (approximately 800 km^2) and they are threatened by habitat destruction (Brower 1996). The threat to monarch populations in their seemingly vast northern breeding grounds is more insidious. Monarch larvae feed exclusively on milkweed leaves (Malcom et al. 1993) and the primary host plant of monarch butterflies in the northern United States and southern Canada is the common milkweed, *Asclepias syriaca* (Malcom et al. 1989). The common milkweed is a secondary successional plant that frequently occurs in and around the edges of corn fields (Bhowmik 1994, Yenish et al. 1997, Hartzler and Buhler 2000) where they are colonized and fed on by monarchs (Yenish et al. 1997). This general overlap between corn and milkweed and the existence of other serious threats already facing monarch populations were early indications that the potential risk to monarchs from *Bt*-corn pollen needed to be investigated further.

6.5.1 Monarchs as a model for studying non-target effects

The monarch butterfly is a uniquely suitable lepidopteran model for studying the potential effects of *Bt*-corn pollen on non-target organisms. Monarch butterflies are a common, large, attractive species found throughout much of North America. Because all the eastern monarch population migrates from summer breeding grounds from the central states through southeastern Canada to small, isolated overwintering grounds in the mountains of central Mexico, their populations are extremely vulnerable to environmental or human-induced pressures in both regions. The monarch migratory system and eastern populations are considered to be an "endangered biological phenomenon" (Brower 1994). Few other Lepidoptera species are so seriously threatened, or have attracted such widespread concern of the American public.

Because of public attention, there is no other non-pest, non-target lepidopteran species whose long-term population cycles are as closely monitored. Currently, there are groups that monitor adult monarch overwintering populations in Mexico (Brower, unpublished data), spring remigration populations in different parts of the United States (Malcolm et al. 1993, Borkin and Brower, unpublished data, Brower et al., in progress), midsummer populations nationwide (NABA-Xerces 4th of July Butterfly Count), and fall migratory populations (Walton and Brower 1996, Monarch Watch – fall migration counts). Larval monarch abundances are also currently being monitored (Prysby and Oberhauser, in progress). Further, new technology has been developed that allows identification of the natal region from which adult monarchs originated based on hydrogen and carbon isotope ratios from larval host plants (Wassenaar and Hobson 1998). The resulting data indicate that 50% of the monarch population that makes it to the overwintering grounds in Mexico (or the breeding pool for all future generations) originates from the "corn belt" region of the midwestern United States (Wassenaar and Hobson 1998).

With the combination of these new techniques and ongoing monitoring efforts, it will be possible to determine if the proportion of monarchs in Mexico originating in the corn belt changes over time. Monarchs are an excellent model system for correlating levels of larval mortality from transgenic *Bt*-pollen with the size of adult populations in Mexico.

6.5.2   Initial studies on Bt-*corn and monarchs*

Although there has been a substantial amount of material written on the effects of *Bt*-corn on monarch butterflies, when this chapter was prepared there were only two peer-reviewed articles presenting new data on this topic (Losey et al. 1999, Jesse and Obrycki 2000). We will focus primarily on these articles and cite unpublished works where they are particularly illuminating. The Losey et al. (1999) study was initiated in the summer of 1998 following the observation that leaves of many plants, including milkweeds (*Asclepias* spp.), received a "dusting" of pollen from nearby cornfields during the period of pollen-shed. Based on this observation, Losey et al. (1999) examined the survival as well as the consumption and growth rates of monarch larvae feeding on milkweed leaves (*A. curassavica*) dusted with untransformed corn pollen, *Bt*-pollen, and on control leaves with no pollen (Losey et al. 1999; and see Table 6.3).

Table 6.3   Characteristics of Transgenic Corn Hybrids Used in Tests of Non-Target Effects on the Monarch Butterfly

Location	Hybrid	Characteristic	Level of *Bt*-toxin in pollen	Ref.
Cornell University	Novartis N4640 *Bt*	Bt11	0.09 µg/g	Losey et al. 1999
	—	Control — unrelated, untransformed hybrid	0	
Iowa State University	Novartis MAX 454	Event 176	7.1 µg/g	Jesse and Obrycki 2000
	Novartis 4494	Isogenic control	0	
	Novartis 7333*Bt*	Event Bt11	0.09 µg/g	
	Novartis 7333	Isogenic control	0	

The *Bt*-corn pollen was toxic to the monarch larvae. The survival rate of larvae feeding on leaves dusted with *Bt*-pollen (56%) was significantly lower than the survival rates of larvae feeding either on leaves dusted with untransformed pollen or on control leaves with no pollen (both 100%, *p* = 0.008) (Losey et al. 1999). Because there was no mortality on leaves dusted with untransformed pollen, all of the mortality on leaves dusted with *Bt*-pollen appeared to be due to the effects of the *Bt*-toxin.

Losey et al. (1999) also demonstrated a significant effect of corn pollen on monarch feeding behavior (*p* = 0.0001). The mean cumulative number of leaves consumed per larva was significantly lower on leaves dusted with *Bt*-pollen (0.57 ± 0.14 SEM, *p* = 0.001) and on leaves dusted with untransformed pollen (1.12 ± 0.09 SEM, *p* = 0.007) than consumption on control leaves without pollen (1.61 ± 0.09 SEM). The reduced rates of larval feeding on pollen-encrusted leaves may represent a gustatory response of this highly specific herbivore to the presence of a "non-host" stimulus. However, such a hypothesized feeding deterrence alone cannot explain the nearly twofold decrease in consumption rate on leaves with *Bt*-pollen compared with leaves with untransformed pollen (*p* = 0.004). Lacking any evidence that monarch larvae are able to "taste" the *Bt*-protein, we suggest that the lower consumption rate is due to toxic effects of *Bt*-pollen following ingestion.

The low consumption rates of larvae fed on leaves with *Bt*-pollen translated into slower growth rates. This trend is illustrated by the low average weight of the surviving larvae that fed on *Bt*-pollen-dusted leaves (0.16 ± 0.03 SEM g) compared with the average weight of larvae that fed on control leaves (0.38 ± 0.02 SEM g, *p* = 0.0001) at the end of the experiment. In other studies involving lepidopteran larvae that have ingested the *Bt*-toxin, low larval weights were correlated with lower survival rates (Halcomb et al. 1996) and with lower adult fecundity (Abo et al. 1995). Thus it is reasonable to predict that the overall impact of larvae feeding on *Bt*-pollen would be substantially greater than the 44% mortality we measured for the second to third instars. This effect seems to be due to the presence of the *Bt*-toxin and not due to some other factor in corn pollen since larvae feeding on untransformed pollen actually grew significantly larger (0.49 ± 0.03 SEM g) than those fed on control leaves (*p* = 0.0021). Presumably this is due to the high protein content of the pollen (Haydak 1970).

Jesse and Obrycki (2000) corroborated the results of Losey et al. (1999) and extended them by providing the first evidence that pollen from *Bt*-corn naturally deposited on milkweed in a cornfield causes significant mortality to monarch larvae. Larvae feeding for 48 h on common milkweed, *A. syriaca*, dusted with pollen from *Bt*-corn plants suffered significantly higher rates of mortality (20 ± 3 SEM%) compared with larvae feeding on leaves with no pollen (3 ± 3 SEM%), or feeding on leaves with non-*Bt*-pollen (0%). Based on these and other feeding assays and their quantification of pollen dispersal, Jesse and Obrycki (2000) conclude that effects of *Bt*-pollen on monarchs may be observed at least 10 m from transgenic field borders, with the highest mortality rates occurring on milkweed plants within 3 m.

6.5.3   *Using behavioral and ecological methods to assess risk to monarch populations*

Several environmental and biological factors may have substantial effects on the level of pollen exposure and the mortality caused at each exposure level. These factors include the phenological overlap of pollen on milkweed with monarch larval feeding, the deposition of pollen as a function of distance from corn, the distribution of milkweed relative to corn, and the behavior of monarch adults and larvae.

In general, the monarch's phenology leads to a very high probability of encounter with pollen from *Bt*-corn. Pollen shed in corn can occur any time from late June to mid-August

(Anonymous 1982), directly overlapping the period when monarch larvae feed and develop on milkweeds (Brower 1996). A model is currently being developed to provide more-detailed predictions of the overlap of corn pollen with each monarch larval instar (Calvin and Taylor, unpublished data).

Field deposition of transgenic corn pollen on *A. syriaca* in Iowa was assessed during the summers of 1998 and 1999 (Jesse and Obrycki, 2000). Deposition of pollen was quantified by placing potted milkweed plants within a cornfield, 0.2, 1, and 3 m from the field edge in 1998, and within the field, 0.2, 1, 3, 5, and 10 m from the edge in 1999. The number of pollen grains on 0.79 cm^2 leaf disks taken from the milkweed plants was counted.

The cumulative deposition of transgenic pollen in 1998 was highest within the cornfield (74 to 217 pollen grains/cm^2) and decreased to between 6 and 20 pollen grains/cm^2 at 3 m from the edge of the field. Similarly, in 1999, the deposition within the field was highest within the cornfield (80 to 115 pollen grains/cm^2) and decreased to 5 to 7 pollen gains/cm^2 at 3 m and 1 pollen grain/cm^2 at 10 m from the field edge (Jesse and Obrycki, 2000). These pollen concentrations on milkweeds are lower than those from previous studies in which greased glass was used as a mechanical pollen trap (Raynor et al. 1972). As noted above, substantial monarch mortality following exposure to *Bt*-pollen from MAX 454 hybrid corn (event 176) was recorded at these pollen levels (Jesse and Obrycki 2000).

A survey of milkweed distribution conducted in Iowa found that milkweeds commonly are found in and near cornfields (Hartzler and Buhler, 2000). This confirms the findings of earlier studies that the common milkweed is a secondary successional plant that frequently occurs in and around the edges of corn fields (Bhowmik 1994, Yenish et al. 1997).

6.5.3.1   The effect of monarch behavior on pollen exposure

One major area affecting pollen exposure for which very few data exist is the importance of adult and larval monarch behavior. The foraging behavior of monarch larvae or the oviposition behavior of monarch adults may be influenced by the presence of corn pollen or the proximity of corn plants. In the virtual absence of any behavioral data, it is impossible to predict if adult or larval monarchs will seek or avoid contact with corn pollen. Specifically, if monarchs actively seek out corn pollen, then their exposure will increase and, conversely, it would decrease if they avoid pollen.

6.5.3.1.1   The role of larval feeding behavior.   If monarch larvae demonstrate an extreme aversion to pollen, then consumption of pollen in the field may be negligible and this species may be "behaviorally immune" to this form of the toxin. Conversely, if larvae perceive corn pollen as a nutritious addition to their normal diet, they may actually seek it out or remain longer on pollen-dusted leaves. There is preliminary evidence for a nutritional benefit from corn pollen in the initial Cornell study (Losey et al. 1999).

6.5.3.1.2   The role of adult oviposition behavior.   If female monarchs avoid ovipositing on plants that are dusted with pollen, exposure of monarchs to *Bt*-pollen would decrease. Exposure to pollen would also decrease if monarchs avoid ovipositing near cornfields or if they have difficulty locating milkweed plants within cornfields. Several recent experiments have assessed the effect of corn plants and corn pollen on oviposition behavior. These experiments were conducted at different spatial scales (cage, flight chamber, field) and the results varied widely. The results of field studies probably provide the best measure of behavior for risk assessment but the differences in results between spatial scales may be important for understanding underlying behavioral mechanisms.

The effect of corn plant proximity and corn pollen presence on the oviposition behavior of the monarch butterfly, *Danaus plexippus* (Danaidae), was assessed in cage and flight chamber studies (Tschenn et al. 2001). The proportions of monarch eggs oviposited on milkweed plants dusted with a *Bt*-corn hybrid, an untransformed hybrid, gravel dust, and undusted control plants were recorded from a cage study. None of the treatments differed significantly in the relative proportion of eggs found. The effect of *Bt* and untransformed corn plant proximity and corn pollen presence were also assessed in a flight chamber. A significantly higher proportion of eggs (96%) were recovered from patches of milkweed plants not surrounded by corn plants and a significantly higher proportion of eggs (nearly 70%) were recovered from patches of milkweed plants not dusted with corn pollen. There were no significant differences in the effects of *Bt*-corn plants compared with untransformed corn plants or in the effects of *Bt*-corn pollen compared with untransformed pollen. These flight chamber data suggest that, if given a choice, monarchs may avoid ovipositing on milkweeds in cornfields or on the edge of fields if they are dusted with corn pollen. This could potentially reduce overall exposure to *Bt*-corn pollen.

Field studies of monarch oviposition and feeding on milkweeds in and around cornfields were initiated at both Cornell and Iowa State in 1999. Initial surveys for monarch eggs and larvae at Cornell focused on five habitat types; cornfield edges, alfalfa fields, road sides, open areas, and old fields. Monarch eggs or larvae were found only on milkweed plants on the edges of cornfields and in alfalfa fields (Losey, unpublished data). Because of the frequency of mowing of alfalfa, none of the eggs or larvae found there would have survived to become adults.

At Iowa State, four sites with high concentrations of milkweeds near cornfields were selected from a survey of milkweeds in Iowa (Hartzler and Buhler 2000). For 7 weeks, beginning on July 12, the number of monarchs on milkweeds located within 2 m of the road, within the road ditch, on either side of the first row of corn, and within the field, was counted at each site. Monarch larvae were found feeding on milkweeds in all field locations (Jesse and Obrycki, unpublished data). This confirms earlier reports that milkweed plants in cornfields are colonized and fed on by monarchs (Yenish et al. 1997).

6.5.4 Beyond direct mortality: assessing sublethal effects

Most research programs aimed at determining the risk to monarch populations from *Bt*-corn pollen have focused on direct mortality following consumption. There has been very little work done on the potential impact of larval consuming sublethal concentrations of *Bt*-corn pollen. By compromising the integrity of the gut lining, even low doses of the *Bt*-toxin leave lepidopteran larvae vulnerable to secondary infections that in turn may lower their fitness. Fitness-lowering effects could include slower growth rates, behavior changes, reduced size or vigor in immatures, and reduced size, vigor, or fecundity in adult butterflies. While sublethal effects are more difficult to measure because they may only become apparent after a long period of time, their overall impact on monarch populations may equal or exceed the impact of direct mortality.

Losey et al. (1999) found evidence for sublethal effects in addition to mortality. The rate of leaf consumption by larvae feeding on leaves dusted with *Bt*-corn pollen was significantly lower than the rate of consumption on control leaves. The low consumption rates of larvae fed on leaves with *Bt*-pollen translated into slower growth rates. This trend is illustrated by the low average weight of the surviving larvae that fed on *Bt*-pollen dusted leaves (0.16 ± 0.03 SEM g) compared with the average weight of larvae that fed on control leaves (0.38 ± 0.02 SEM g, $p = 0.0001$) at the end of the experiment. Jesse and Obrycki (2000) found no effect on adult weight, size, or lipid content following exposure to a sublethal dose of *Bt*-corn pollen. In other studies involving lepidopteran larvae that have ingested the *Bt*-toxin, low larval weights were correlated with lower survival rates (Halcomb et al. 1996) and with lower adult fecundity (Abo et al. 1995).

Table 6.4   Causes of Monarch Mortality

Human-Related Causes
Loss of habitat
— Overwintering sites
— summer breeding areas
Pest management
— Insecticide sprays
— Transgenic insecticidal plants (e.g., *Bt*-corn)
Natural Causes
Predators, pathogens, parasitoids
Harsh climatic conditions (e.g., drought, snowstorms)
Age/physiological stress

Adapted from Monsanto Company, 2000.

These sublethal effects could have serious implications for monarch fitness. One implication is that short-term studies of the effects of *Bt*-corn pollen may seriously underestimate the level of mortality. Larvae that exhibit no mortality after exposure to low doses of *Bt*-corn pollen may exhibit very high levels at later stages of their life cycle. Further, if the larval development is delayed and the phenology of individuals feeding on *Bt*-pollen is shifted, then those adult butterflies might experience reduced mating opportunities. An even more serious problem would arise if the migratory generation of monarchs emerged as adults after the relatively small window of warm weather that allows them to make the long flight to Mexico.

Potential reductions in size and vigor could have equally important fitness implications. Similar to shifts in phenology, reductions in size or vigor could reduce the ability of adult monarchs to mate or disperse to find suitable oviposition sites. For the migratory generation, smaller or less vigorous individuals might be less likely to complete the flight to Mexico and back. In addition, small larvae may be more susceptible to natural enemies such as predators, parasitoids, and pathogens. Many parasitoids and predators are known to preferentially attack smaller-sized individuals. By reducing larval size or even delaying larval growth, *Bt*-corn pollen may be extending the "window of vulnerability" of those larvae and thereby increasing the mortality due to predation (Benrey and Denno 1997).

All the research to date on the effects of *Bt*-corn pollen on monarch butterflies has involved relatively short-term exposures of the larvae to the pollen. These short-term studies will almost certainly underestimate the magnitude of the sublethal effects. Clearly, longer-term laboratory studies and replicated life-table studies in the field are needed to determine the nature and magnitude of any sublethal effects.

6.5.5   *The relative importance of mortality from* Bt-corn

Ingestion of lethal levels of *Bt*-corn pollen is just one of many sources of monarch mortality (Table 6.4). In general, causes of monarch mortality can be divided into two categories, human-related and natural causes. This categorization is useful because it separates threats to monarchs into those that may be mitigated from those that are beyond human control.

Although data from large temporal or spatial scales are inadequate to rank human-related causes quantitatively in terms of their importance, there is general agreement that loss of overwintering habitat is the single largest source of monarch mortality. The relative importance of other human-related mortality sources is less clear. Certainly some summer breeding area is lost due to urbanization and the use of herbicides that eliminate milkweed. However, since milkweed is a secondary succession plant and the edges of agricultural fields provide exactly the type of "disturbed" habitat in which milkweed thrives, it seems possible that there is more potential monarch habitat than would have been available two centuries ago, before North America was so widely cultivated. Similarly, the importance of mortality from insecticide sprays is also difficult to rank. Many insecticides have the potential to kill monarch larvae and adults but, as with *Bt*-corn pollen, it is difficult to determine the level of monarch exposure to these chemicals.

The mortality caused by *Bt*-corn pollen and that caused by insecticide sprays are potentially linked since both represent pest management options in corn production. If increased use of *Bt*-corn leads to decreased use of more toxic insecticides, then *Bt*-corn could actually lower overall monarch mortality. However, there is substantial evidence to indicate very little, if any, reduction in insecticide use for the main target of *Bt*-corn, the European corn borer, over the past 3 years and, during that time, the proportion of corn acres planted to *Bt*-corn rose from 0 to almost 25% (Obrycki et al. 2000).

It is important to note that, with the possible exception of insecticide sprays, the mortality from *Bt*-corn pollen is additive with the mortality caused by each of these other sources. There are very few data on the level of additional mortality that an insect population can sustain without becoming unstable and dropping into long-term decline. Thus, even if the additional mortality caused by *Bt*-corn is only a small fraction of the total mortality caused by the other sources, it could theoretically be enough to send monarch populations into a downward spiral. For example, if monarch mortality from egg to adult was 95% before the advent of *Bt*-corn (not an unreasonable estimate) and *Bt*-corn caused an additional 3% mortality, the extra 3% might be enough to have a negative impact on monarch populations over a long timescale. Quantification of the level of additional mortality necessary to disrupt long-term lepidopteran population cycles is a high priority for future research. It may be possible to conduct this type of analysis with existing life history data. The results would have implications far beyond monarchs and *Bt*-corn.

6.6   Who cares? Why are non-target impacts important?

Why should we be concerned about the potential non-target effects of transgenic crops or any other pest management tactic? The answer lies in the ecological roles or "services" provided by these species.

6.6.1   Ecological services of non-target insects

The importance of predators such as ladybeetles and pollinators such as honeybees is clear to most people and there is general agreement that they should be protected. In some places, there is even legislation to ensure that they are protected from pest management practices. Unfortunately, the ecological role of other non-target species, such as the butterflies and moths, is less obvious. Many people find the only value of many such insects is their aesthetic value. Nevertheless, these non-target herbivores play vital roles in agroecosytems. If transgenic crops or any other factor has a negative impact on this guild, those ecological functions may be threatened. As an example of the importance of a group of non-target herbivores to an ecosystem, we will outline the ecological roles of non-target Lepidoptera.

6.6.2   The ecological value of butterflies and moths

Butterflies and moths play several pivotal ecological roles that are often overlooked. Among the most important of these roles are their contributions as (1) pollinators, (2) regulators of weed species, and (3) prey/hosts for predators, parasitoids, and pathogens. Although pollination is generally attributed to domesticated and wild bees, certain flowers have unique shapes that allow pollination only by the specialized mouth parts of butterflies and moths. It has been estimated that 8% of the 240,000 flowering plant species worldwide are primarily pollinated by butterflies or moths (Buchmann and Nabhan 1996). Further, the larvae of butterflies and moths play an important role in the suppression of many potential weed species. To date, 58 species of Lepidoptera have been imported into the United States from other countries specifically to suppress populations of introduced weeds (Julien and Griffiths 1998). Native herbivores are also useful in suppressing native weed populations (Julien and Griffiths 1998). Finally, many larval and adult Lepidoptera are food sources for both other insects (Price 1997) and vertebrate predators such as bats, birds, fish, salamanders, frogs, toads, lizards, turtles, and snakes (Daly et al. 1998). Some of these predators and parasitoids are specialized on their lepidopteran prey (Godfray 1993) and would suffer if prey densities declined precipitously. For example, declines in populations of partridges, *Perdix perdix*, have been linked to decreased densities of insects caused by insecticide treatments in cereal crops (de Snoo and de Leeuw 1996). The loss or even decrease in the magnitude of these ecological services could destabilize ecosystems. Thus, any factor with the potential to cause large-scale negative impacts on lepidopteran species deserves serious consideration.

6.7   Minimizing non-target impacts

Recent investigations on potential adverse effects of *Bt*-crops on non-target herbivores have led to increased efforts in risk assessment. There are several approaches to reducing toxin exposure of susceptible lepidopteran herbivores that are not pests of corn.

6.7.1   Minimizing the effects of genetically engineered pollen

If current research determines that there are substantial negative impacts of pollen from *Bt*-corn on monarchs or other Lepidoptera, it may be possible to modify the utilization of *Bt*-corn to minimize these impacts. The simplest strategy would be to utilize only those *Bt*-corn hybrids that do not express the *Bt*-toxin in the pollen. Expression in pollen is known to be controlled by a single gene promoter (Fearing et al. 1997) and there are several commercial hybrids that do not express detectable levels of *Bt*-toxins in pollen (Andow and Hutchison 1998, Andow, Chapter 4 of this volume). The drawback of this strategy is that *Bt*-corn hybrids that do not express the *Bt*-toxin in the pollen may not be as effective against the target pest (the European corn borer, *Ostrinia nubilalis*) and the lowered efficacy may present problems for resistance management programs built around high target population mortality (Ostlie et al. 1997).

An alternative solution is the creation of "buffer zones" of non-*Bt*-corn around *Bt*-corn fields. It has been demonstrated that very little corn pollen goes beyond three to four rows from the original plant (Louette 1995). A "buffer zone" of at least that many rows would serve to trap most of the transgenic pollen. It is generally accepted and may be legislated that a certain proportion of each field planted to *Bt*-corn must be planted to non-*Bt*-corn as a refuge for resistance management (Ostlie et al. 1997). Thus, it may be possible to make the refuge into a buffer zone at no added expense. The size and shape of the non-*Bt*-corn areas would have to be designed carefully to ensure that they effectively serve both purposes.

6.7.2   The regulatory response

Based on the initial monarch study (Losey et al. 1999) and subsequent research results reported at the Monarch Butterfly Research Symposium held in Chicago, Illinois on November 2, 1999, the U.S. Environmental Protection Agency (EPA) required registrants of *Bt*-corn hybrids to submit additional data regarding potential impacts on non-target herbivores (see Appendix 6.1). The EPA specifically requested data to address the potential impact of *Bt*-corn pollen on non-target Lepidoptera. Most of the data required generally followed the hierarchy outlined in this chapter. It is important to note that in addition to the monarch, the Karner Blue butterfly receives special emphasis because it is a listed endangered species. Completing the process outlined in Section 6.2.3 will allow identification of other rare, threatened, and endangered lepidopteran species that might also be at risk. Working with endangered and threatened species poses a unique challenge since there are restrictions against harming them in the testing process.

Data were submitted to the EPA by the registrants at or around the deadline date but, because the bulk of this research was initiated less than 1 year earlier, most studies were preliminary. The only studies on the risk to monarchs from *Bt*-corn pollen published in peer-reviewed journals before 2001 were Losey et al. (1999) and Jesse and Obrycki (2000). The data submitted will be used by the EPA in its consideration of re-registration requests for *Bt*-corn events.

The potential impact of *Bt*-corn pollen on non-target herbivores was integrated into the EPA regulations for *Bt*-corn in December 1999 (see Appendix 6.2A). In the letter to registrants, the EPA issued a directive to "implement voluntary measures, through the refugia requirements, that will protect non-target insects, particularly the Monarch butterfly." In an attachment to this letter (Appendix 6.2B) the EPA provided specific instructions for the planting of refuges so that they can also function as buffer zones to minimize non-target impacts (see Section 6.7.1).

6.8 An ecological approach for assessing risk to non-targets

In retrospect, the appropriateness of the initial decision to approve *Bt*-corn without more testing of non-target effects is debatable, but it is clear that the EPA moved quickly to incorporate new data into their regulations and to outline a program to investigate non-target effects more fully. This suggests that any weakness in the regulatory system for transgenic crops stems not from a lack of ability to respond to potential problems that arise after crops are commercialized, but from a failure to detect potential problems in the registration process. We suggest that this failure arises from the basic philosophy behind the current assessment of non-target effects, and we propose a more robust paradigm that will be more likely to detect potential problems in the registration process.

The current approach to non-target risk assessment focuses primarily on direct mortality of herbivores that consume transgenic plant tissue within the field (Figure 6.2A). Although this is an important aspect, it misses many of the more complex negative impacts that might have equal or greater impact on herbivore populations (Figure 6.2B). Specifically, a complete assessment of risk to non-target herbivores should include (1) forms of the toxin that are mobile (e.g., pollen, root exudates); (2) sublethal effects such as those outlined in Section 6.5.4, and (3) complex trophic interactions (e.g., Section 6.6.1). In addition, based on the variability among herbivores in ecological factors that affect exposure and physiological factors that affect susceptibility, a wider range of carefully chosen species should be tested to ensure that risk to non-targets as a group is being adequately characterized. Specific details will depend on each crop and toxin being registered, but these types of modifications in the risk assessment system will increase the likelihood that effects such as that of *Bt*-corn pollen on butterflies are discovered before the transgenic crop has already been widely adopted. In many instances if a potential threat is identified early in the process, it may be possible to modify the transgenic crop or the cropping system to minimize its effects on non-target species. When potential non-target risks are discovered after a transgenic crop has been widely adopted, the cost to the producer in

A.   Toxicological Approach

Herbivore

↑

Bt-corn

Parameters Measured
- **Direct Mortality**

B.   Ecological Approach

Predators   Parasitoids   Predators   Parasitoids

↖   ↗   ↖   ↗

Herbivore   Herbivore

↑   pollen   ↑

Bt-Corn   Weed

Parameters Measured
- **Direct Mortality**
- **Sublethal Effects**
- **Complex Trophic Interactions**

Figure 6.2   (A) Toxicological approach; (B) ecological approach. (From Obrycki, J. J. et al., *BioScience*, 51, 353, 2001. © 2001 American Institute of Biological Sciences. With permission.)

terms of consumer and investor rejection can be very high. Thus, while a more comprehensive program to assess the risk to non-target species will certainly be more expensive, the greater expense is justified by the benefits to both environmental safety and long-term value of the transgenic crop for pest management.

Literature cited

Abo El-Ghar, G. E. S., Radwan, H. S. A., El-Bermawy, Z. A., and Zidan, L. T. M., Sublethal effects of avermectin B_1, β-Exotoxin of *Bacillus thuringiensis* and diflubenzuron against cotton leafworm (Lepidoptera: Noctuidae), *J. Appl. Entomol.*, 119, 309, 1995.

Andow, D. A. and Hutchison, W. D., *Bt*-corn resistance management, in *Now or Never: Serious New Plans to Save Natural Pest Control*, in Mellow, M. and Rissler, J., Eds., Union of Concerned Scientists, Cambridge, MA, 1998, 19.

Anonymous, How a Corn Plant Develops, Based on Range of Planting Dates, Average Climatic Conditions and Common Corn Hybrid Phenology, Iowa State University, Special Report No. 48, 1982.

Anonymous, Biopesticide Fact Sheet, EPA Office of Pesticide Programs, available at www.epa.gov/pesticides/biopesticides/factsheets, 1999.

Benrey, B. and Denno, R. F., The slow-growth-high-mortality hypothesis: a test using the cabbage butterfly, *Ecology*, 78, 987, 1997.

Bhowmik, P. C., Biology and control of common milkweed *(Asclepias syriaca)*, *Rev. Weed Sci.*, 6, 227, 1994.

Brower, L. P., A new paradigm in conservation of biodiversity: endangered biological phenomena, in *Principles of Conservation Biology*, Meffe, G. and Carroll, C., Eds., Sinauer Associates, Sunderland, MA, 1994, 104.

Brower, L. P., Monarch butterfly orientation: missing pieces of a magnificent puzzle, *J. Exp. Biol.*, 199, 93, 1996.

Buchmann, S. L. and Nabhan, G. P., *The Forgotten Pollinators*, Island Press, Washington, D.C., 1996.

Covell, C. V., *A Field Guide to the Moths of Eastern North America*, Houghton Mifflin, Boston, 1984.

Daly, H. V., Doyen, J. T., and Purcell, A. H., *Introduction to Insect Biology and Diversity*, 2nd ed., Oxford University Press, Oxford, U.K., 1998.

de Snoo, G. R. and de Leeuw, J., Non-target insects in unsprayed cereal edges and aphid dispersal, *J. Appl. Entomol.*, 120, 501, 1996.

Fearing, P. L., Brown, D., Vlachos, D., Meghji, M., and Privalle, L., Quantitative analysis of CryIA(b) expression in *Bt* maize plants, tissues, and silage and stability of expression over successive generation, *Mol. Breed.*, 3, 169, 1997.

Federici, B. A., Broadscale use of pest-killing plants to be true test, *Calif. Agric.*, 52, 14, 1998.

Forbes, W. T. M., *The Lepidoptera of New York*, Cornell University Press, Ithaca, NY, 1923.

Glenn, S., Phillips, W. H., and Kalnay, P., Long-term control of perennial broadleaf weeds and triazine-resistant common lambsquarters (*Chenopodium album*) in no-till corn (*Zea mays*), *Weed Technol.*, 11, 436, 1997.

Godfray, H. C. J., *Parasitoids*, Princeton University Press, Princeton, NJ, 1993.

Gould, F., Sustainability of transgenic insecticidal cultivars: integrating pest genetics and ecology, *Annu. Rev. Entomol.*, 43, 701, 1998.

Halcomb, J. L., Benedict, J. H., Cook, B., and Ring, D. R., Survival and growth of bollworm and tobacco budworm on nontransgenic and transgenic cotton expressing a CryIA insecticidal protein (Lepidoptera: Noctuidae), *Environ. Entomol.*, 25, 250, 1996.

Hartzler, R. G. and Buhler, D. D., Occurrence of common milkweed (*Asclepias syriaca*) in cropland and adjacent areas, *Crop Prot.*, 19, 363, 2000.

Haydak, M. H., Honey bee nutrition, *Annu. Rev. Entomol.*, 15, 143, 1970.

Hilbeck, A., Baumgartner, M., Fried, P. M., and Bigler, F., Effects of transgenic *Bacillus thuringiensis* corn-fed prey on mortality and development time of immature *Chrysoperla carnea* (Neuroptera: Crysopidae), *Environ. Entomol.*, 27, 480, 1998.

Jesse, L. C. H. and Obrycki, J. J., Field deposition of *Bt* transgenic corn pollen: lethal effects on the monarch butterfly, *Oecologia*, 125, 241, 2000.

Johnson, K. S., Scriber, J. M., Nitao, J. K., and Smitley, D. R., Toxicity of *Bacillus thuringiensis* var. *kurtstaki* to three nontarget lepidoptera in field studies, *Environ. Entomol.*, 24, 288, 1995.

Julien, M. H. and Griffiths, M. W., *Biological Control of Weeds: A World Catalogue of Agents and Their Target Weeds*, 4th ed., CABI Publishing, Wallingford, U.K., 1998.

Koziel, M. G., Beland, G. L., Bowman, C., Carozzi, N. B., Crenshaw, R., Crossland, L., Dawson, J., Desai, N., Hill, M., Kadwell, S., Launis, K., Lewis, K., Maddox, D., McPherson, K., Meghji, M. R., Merlin, E., Rhodes, R., Warren, G. W., Wright, M., and Evola, S. V., Field performance of elite transgenic maize plants expressing an insecticidal protein derived from *Bacillus thuringiensis*, *Bio/Technology*, 11, 194, 1993.

Losey, J. E., Rayor, L. S., and Carter, M. E., Transgenic pollen harms monarch butterflies, *Nature*, 399, 214, 1999.

Louette, D., Seed exchange among farmers and gene flow among maize varieties in traditional agricultural systems, in *Gene Flow among Maize Landraces, Improved Maize Varieties, and Teosinte: Implications for Transgenic Maize*, Serratos, J. A., Wilcox, M. C., and Castillo-Gonzales, F., Eds., El Batan, Mexico, D. F., Centro Internacional de Mejoramiento de Maiz y Trigo, 1995, 56.

Malcolm, S. B., Cockrell, B. J., and Brower, J. P., Cardenolide fingerprint of monarch butterflies reared on common milkweed, *Asclepias syriaca* L., *J. Chem. Ecol.*, 15, 819, 1989.

Malcolm, S. B., Cockrell, B. J., and Brower, L. P., Spring recolonization of eastern North America by the monarch butterfly: successive brood or single sweep migration? in *Biology and Conservation of the Monarch Butterfly*, Malcolm, S. B. and Zalucki, M. P., Eds., Natural History Museum of Los Angeles County, Los Angeles, CA, 1993, 219.

McGaughey, W. H., Gould, F., and Gelernter, W., *Bt* resistance management, *Nature Biotechnol.*, 16, 144, 1998.

Monarch Watch, available at http://www.MonarchWatch.org/tagmig/index.htm.

Monsanto Company, *Butterflies and Bt Corn Pollen: Lab Research and Field Realities*, Monsanto Company, St. Louis, MO, 2000.

Obrycki, J. J., Losey, J. E., Taylor, O. R., and Jesse, L. H., Transgenic insecticidal corn: beyond insecticidal toxicity to ecological complexity, *BioScience*, 51, 353, 2001.

Ostlie, K. R., Hutchison, W. D., and Hellmich, R. L., Bt Corn and the European Corn Borer, NCR Publ. 602, University of Minnesota, St. Paul, 1997.

Peacock, J. W., Wagner, D. L., and Schweitzer, D. F., Impacts of *Bt* on non-target lepidoptera, *Newsl. Mich. Entomol. Soc.*, 38, 1, 1993.

Peacock, J. W., Schweitzer, D. F., Carter, J. L., and Dubois, N. R., Laboratory assessment of the effects of *Bacillus thuringiensis* on native lepidoptera, *Environ. Entomol.*, 27, 450, 1998.

Pilcher, C. D., Rice, M. E., Obrycki, J. J., and Lewis, L. C., Field and laboratory evaluations of transgenic *Bacillus thuringiensis* corn on secondary lepidopteran pests (Lepidoptera: Noctuidae), *J. Econ. Entomol.*, 90, 669, 1997.

Price, P. W., *Insect Ecology*, 3rd ed., John Wiley & Sons, New York, 1997.

Ransom, C. V. and Kells, J. J., Hemp dogbane (*Apocynum cannabinum*) control in corn (*Zea mays*) with selective postemergence herbicides, *Weed Technol.*, 12, 631, 1998.

Raynor, G. S., Ogden, E. C., and Hayes, J. V., Dispersion and deposition of corn pollen from experimental sources, *Agron. J.*, 64, 420, 1972.

Richard, J. and Heitzman, J. E., *Butterflies and Moths of Missouri*, Missouri Department of Conservation, Jefferson City, 1987.

Saxena, D., Flores, S., and Stotzky, G., Insecticidal toxin in root exudates from *Bt* corn, *Nature*, 402, 480, 1999.

Schweizer, E. E., Westra, P., and Lybecker, D. W., Seedbank and emerged annual weed populations in cornfields (*Zea mays*) in Colorado, *Weed Technol.*, 12, 243, 1998.

Tschenn, J., Losey, J. E., Jesse, L. H., Obrycki, J. J., and Hufbauer, R., Effect of corn plants and corn pollen on monarch butterfly (Lepidoptera: Danaidae) oviposition behavior, *Environ. Entomol.* 30, 495, 2001.

Walton, R. K. and Brower, L. P., Monitoring the fall migration of the monarch butterfly, *Danaus plexippus* L. (Nymphalidae: Danainae) in eastern North America: 1991–1994, *J. Lepid. Soc.*, 50, 1, 1996.

Wassenaar, L. I. and Hobson, K. A., Natal origins of wintering migrant monarch butterflies in Mexico: new isotopic evidence, *Proc. Natl. Acad. Sci. U.S.A.*, 95, 15436, 1998.

Wax, L. M., Fawcett, R. S., and Isely, D., *Weeds of the North Central States*, University of Illinois at Urbana-Champaign, Urbana, 1981.

Webster, T. M. and Cobble, H. D., Changes in weed species composition of the southern United States: 1974 to 1995, *Weed Technol.*, 11, 308, 1997.

Williams, W. P., Sagers, J. B., Hanten, J. A., Davis, F. M., and Buckley, P. M., Transgenic corn evaluated for resistance to fall armyworm and southwestern corn borer, *Crop Sci.*, 37, 957, 1997.

Williams, W. P., Buckley, P. M., Sagers, J. B., and Hanten, J. A., Evaluation of transgenic corn for resistance to corn earworm (Lepidoptera: Noctuidae), fall armyworm (Lepidoptera: Noctuidae), and southwestern corn borer (Lepidoptera: Crambidae) in a laboratory bioassay, *J. Agric. Entomol.*, 15, 105, 1998.

Wraight, C. L., Zangerl, A. R., Carroll, M. J., and Berenbaum, M. R., Absence of toxicity of *Bacillus thuringiensis* pollen to black swallowtails under field conditions, *Proc. Natl. Acad. Sci. U.S.A.*, 97, 7700, 2000.

Yenish, J. P., Fry, T. A., Durgan, B. R., and Wyse, D. L., Establishment of common milkweed (*Asclepias syriaca*) in corn, soybean and wheat, *Weed Sci.*, 45, 44, 1997.

Appendix 6.1: EPA "call-in" for data pertaining to the effects of Bt-corn on non-target Lepidoptera*

Bt-Corn Data Call-In

12/15/99

On 12/15/99, BPPD issued a data call-in for all *Bt*-corn plant-pesticides. The data call-in has been issued to require data on non-target lepidopteran effects. Some of the data required include toxicity testing (on the Monarch butterfly and a relative of the endangered Karner Blue butterfly), milkweed dispersal, and pollen dispersal. Protocols are due in March 2000 and the data is due in March 2001.

PROTOCOL AND DATA SUBMISSION

Prior to commencing the required studies, you must submit proposed study protocols to the Agency for approval by the Agency. The proposed study protocols must (1) describe your proposed study design, explain how it addresses each of the required study elements stated below, and include detailed rationale for your proposed design elements, (2) identify your proposed choice of representative inbreds and hybrids and provide detailed rationale for your proposed choices, and (3) explain how the study design meets generally accepted standards for statistical adequacy.

(1) Determine and report (in square miles) the total land mass in North America that contains milkweed and monarchs vs. the total amount of land at the edge of cornfields where milkweed could be exposed to *Bt*-corn pollen.
(2) Determine and report what species of milkweed monarchs feed on.
(3) Determine and report what percentage of milkweed in the corn belt is found in row crop areas vs. roadsides, pastures and other non-row crop areas.
(4) Provide surveys of cornfields in representative corn growing states to determine how much milkweed is in the fields.
   (a) Determine and report whether milkweeds are closer, farther or at random distance to corn.
   (b) Provide data on the relative abundance of milkweed in the cornfield pollen shadow vs. areas further than 60 meters away from cornfields.
   (c) Determine and report whether herbicides are effective in cornfields in eliminating milkweeds. If so, determine which herbicides are most effective.
(5) Determine and report what is the relationship between monarch colonization of milkweeds and distance to corn.
   (a) Determine and report the distribution of monarch eggs and larvae on milkweeds relative to cornfields.
   (b) Quantify and report the pollen on milkweed leaves within the pollen shadow and up to 60 meters from the edge of *Bt*-corn fields.
   (c) Provide the distances from the edge of cornfield at which LD50 concentrations of Bt pollen are found for each *Bt*-corn hybrid.
(6) Determine and report the LD50s for the Cry protein in your *Bt*-corn active ingredient(s) for a) monarch larvae and for b) larvae of a relative of the endangered Karner Blue butterfly. The Karner Blue butterfly relative tested must be from the genus *Lycaeides*, such as the

---

* Available at www.epa.gov/pesticides/biopesticides/otherdocs/bt_dci.htm.

Northern Blue butterfly (*Lycaeides idas*). If it is not feasible to test a butterfly from the genus *Lycaeides*, then you must provide justification regarding why such testing is not feasible and test a butterfly from a genus within the family Lycaenidae. The Karner Blue butterfly must not be tested.

(7) Determine the monarch larvae LD50s for pollen, for representative inbreds and hybrids from your transformation event(s), containing your *Bt*-corn plant-pesticide and report the results on both a weight and a number of pollen grains basis.

(8) Determine and report each instar larval survival and developmental effects in the presence of *Bt*-pollen, for representative inbreds and hybrids from your transformation event(s), containing your *Bt*-corn plant-pesticide.

(9) The Cornell data (Losey, J., L. Rayor, and M. Carter. 1999. Transgenic pollen harms monarch larvae. Nature 399:214.) show that less larval feeding took place on pollinated milkweed leaves than on non-pollinated leaves. Therefore:

   (a) Determine and report what is the probability of corn pollen consumption by monarch larvae on Milkweed leaves;

   (b) Determine and report whether foraging larvae actively avoid *Bt*-pollen, for representative inbreds and hybrids from your transformation event(s); in the field;

   (c) Determine and report whether monarch larvae avoid feeding on non-*Bt*-corn pollen under field conditions;

   (d) Determine and report whether monarchs are avoiding cornfields for preferred areas to feed.

(10) Determine and report whether there are practical ways of decreasing the potential of monarchs encountering or feeding upon *Bt*-pollen.

(11) Determine and report whether monarchs have a site preference for egg-laying;

   (a) Determine and report whether monarch adults oviposit on or avoid milkweeds near cornfields.

(12) Determine and report what is the effect of *Bt*-corn pollen presence, for representative inbreds and hybrids from your transformation event(s), on monarch oviposition behavior;

   (a) Determine and report whether monarchs deposit eggs on non-*Bt*-corn pollinated milkweed under field conditions,

   (b) Determine and report whether monarch adults deposit eggs on *Bt*-corn pollinated milkweed under field conditions,

   (c) Determine and report where monarch adults oviposit on milkweeds (under leaves, in inflorescents).

(13) Confirm and report where the various instars of monarch larvae feed on the milkweed plant:

   (a) upper vs. lower leaves,

   (b) also determine and report on feeding behavior regarding upper and undersides of leaves (changes potential exposure considerably),

   (c) shoot apex vs. tops of leaves, and

   (d) determine and report whether pollen will disseminate to and adhere to the undersides of leaves.

(14) Determine and report how long the lethal concentration of *Bt*-corn pollen, for representative inbreds and hybrids from your transformation event(s), stays on milkweed.

   (a) Determine and report how long *Bt* in corn pollen retains its toxicity,

   (b) Determine and report whether sunlight degrades *Bt* toxin in corn pollen on milkweed, and

   (c) Determine and report whether wind, rain or other environmental factors remove *Bt*-corn pollen from milkweed.

(15) Determine and report how soon after planting do representative inbreds and hybrids from your transformation event(s) pollinate.

(16) Determine and report whether the duration of pollination for each corn ear and the total field match the expected 3 and 13 days, respectively, for representative inbreds and hybrids from your transformation event(s).

(17) Determine and report whether the monarch larvae are feeding on milkweeds during pollen shed.

(a) If so, determine and report how long is the regional overlap of time when the monarch larvae are exposed to corn pollen.

(b) Determine and report what fraction of monarch larvae could be exposed to corn pollen, considering that in any specific region the corn is shedding pollen for only a week to 10 days each year.

(c) Determine and report the probability of monarch larvae encountering pollen from *Bt*-corn.

(18) Determine and report whether monarchs carry pollen on their exoskeleton and distribute it on milkweeds during egg deposition. If so, determine the quantity.

(19) Determine and report what is the risk of monarch exposure to *Bt*-corn pollen in the context of other significant risk factors impacting monarch survival and population size (e.g., conventional and microbial insecticides, herbicides, destruction of overwintering sites, predation, cars, etc.).

(20) Determine and report whether monarch populations travel linearly.

(21) Confirm and report whether 50% of monarchs pass through the corn belt.

(22) Develop and report a mathematical model to test the sensitivity of various environmental and biological risk factors, as well as to examine the risk to monarchs and other susceptible non-target insects at varying distances from *Bt*-corn fields.

(23) Define and report baseline monarch population levels and submit annual population level reports on a regional basis.

Appendix 6.2A: Letter to Bt-corn registrants regarding new regulations*

Letter to *Bt*-Corn Registrants

12/20/99

As you know, EPA has been actively considering resistance management and other issues associated with *Bt*-products developed through biotechnology. Our goal is to provide clear direction and timely information for the farming community on what the requirements will be for the 2000 growing season. As we agreed, the Agency provided additional time to consider further information with the stipulation that there is still time to make changes for the 2000 growing season.

This letter and the attachment provide the specific requirements for next season. EPA is committed to the necessary steps that fully protect public health and the environment, including additional measures to manage insect resistance and impacts to non-target organisms.

The additional measures for the next growing season are as follows:

Registrants must ensure that growers plant a minimum structured refuge of at least 20 percent non-*Bt*-corn.

* Available at http://www.epa.gov/pesticides/biopesticides/otherdocs/bt_corn_ltr.htm.

For *Bt*-corn grown in cotton areas, registrants must ensure that farmers plant at least 50 percent non-*Bt*-corn in these areas.

Registrants shall conduct expanded monitoring in the field as an early warning system to detect any potential resistance, and to implement voluntary measures, through the refugia requirements, that will protect non-target insects, particularly the Monarch butterfly. For certain products where *Bt* is not expressed at a high dose, there will be sales and planting restrictions on specific products in certain limited growing areas.

Attached to this letter you will find these and other actions that the Agency is requiring as part of the resistance management strategy for the 2000 growing season which are specific to your registrations. Please include the items in the attachment when you submit your revised amendment request.

In addition to these steps toward more effective insect resistance management, we believe that there is an opportunity here to use refugia areas to reduce exposure to the Monarch butterfly, given the concern raised in the past year about potential impacts of *Bt*-corn to this species. The Agency recognizes that the data on impacts to Monarch butterflies is preliminary. However, we suggest farmers locate refugia in such a manner as to serve to protect potentially vulnerable non-target insects. Refugia areas can serve as buffer zones between the corn field and the habitat of non-target insects. We strongly advise that registrants provide technical assistance to the growers to help place refugia where they can serve both goals of managing resistance and minimizing non-target exposure. EPA and USDA are also committed to providing the necessary education and outreach to encourage and assist growers to place refugia where they will also serve to protect non-target insects.

EPA is committed to making timely decisions on registrations for the 2001 growing season. We are soliciting public comments on the registrations through an open and public process. In order to initiate the review of the registrations for 2001, the Agency is requesting you submit any additional data you would like the Agency to consider by January 31, 2000. This will enable the Agency to have adequate time to evaluate any new scientific information you wish to submit in support of your registrations.

It is clear that as the scientific underpinnings of biotechnology continue to evolve, our decisions will need advice and input from all stakeholders, including the scientific community. The best available science is critical to our ability to reach registration decisions on new products as they are developed. The input of the agricultural community is equally critical, providing EPA with valuable information on actual growing conditions. As new data and information becomes available, EPA will work with the stakeholder community to modify registration requirements as necessary.

We appreciate the commitment shown by the registrants, the growers, the environmental community, and the scientific community, to continue to work together to resolve the issues associated with biotechnology. The Agency is committed to moving forward to protect public health and the environment, while providing the tools that the American farmer needs to produce a safe and healthy food supply.

If you have any questions regarding these new requirements, please don't hesitate to contact me.

Sincerely,

Janet L. Andersen, Director
Biopesticides and Pollution Prevention Division

Appendix 6.2B: *Excerpt from attachment to letter to Bt-corn registrants**

14. The Agency requests that [company name] instruct its customers who are planting a refugia beside *Bt*-corn, to place the refugia upwind and/or between the *Bt*-corn and sensitive habitats (e.g., roadsides and ditch banks) to provide increased protection for monarchs and other non-target Lepidoptera that might be in these habitats. This can be done either by instructions in the Grower Guides or by supplemental notice. The Agency recognizes that the data on impacts to monarch butterflies is preliminary. However, the Agency believes that consideration of non-target lepidopteran habitats when determining where to place a refugia is highly desirable. Cry1Ab protein has been shown to be toxic to monarchs and while the question of exposure remains under study, EPA strongly encourages a proactive approach.

* Available at http://www.epa.gov/pesticides/biopesticides/otherdocs/bt_corn_ltr_[company name].htm.

chapter seven

Transgenic host plant resistance and non-target effects

Angelika Hilbeck

Contents

7.1 Conventional breeding and genetic engineering for host plant resistance

One approach to improving plant resistance to insects is to alter the plant's chemical composition so that it will deter or intoxicate target herbivores. Traditional breeding to achieve this effect is restricted to the transfer/exchange of "desired" traits among sexually compatible, i.e., related, species. The crossing partners in this process already share genomes that are in large part homologous. Groups of functionally linked genes are moved; they include the relevant promoters, regulatory sequences, and associated genes involved in the regulated expression of the character of interest. The breeding process typically is slow, the change in traits often minimal, and the new levels of resistance to pests rarely high. During lengthy breeding programs, unintended effects can be identified and eliminated to the extent possible before a new cultivar is commercially released. In contrast, genetic engineering typically uses genes and gene constructs derived from completely unrelated organisms and adds them to the existing genome in an entirely novel genetic context (Regal 1994). This technology involves the random insertion of genes without the relevant promoter sequences and associated regulatory genes (Antoniou 1996). Currently, viral promoter genes from reproductively incompatible species — coding, for example, for antibiotic resistance — are inserted instead. Because genetic engineering is expected to yield commercial cultivars more readily than does conventional breeding, subtle unexpected effects may escape the attention of the seed producers and be discovered at the commercial stage, i.e., by farmers. Events like the "boll drop" of transgenic cotton or "stem splicing" of transgenic soybeans, both of which occurred in areas of the southeastern United States, may be the first such cases.

In altering the composition of plants, two types of changes can occur: intended and unintended. With genetic engineering, the added novel gene constructs, containing not only the "desired" trait but also marker genes, promoters, terminators, etc., and gene products interact with "native" genes and gene products and may exert so-called "position effects" or "pleiotropic effects" on the "native" genes. Such unintended effects can lead to altered plant chemical composition that, in turn, may affect the complex of non-target organisms associated with the plant, including, for example, insects other than the target herbivore. These changes can be truly novel and may not be anticipated beforehand. The other plant modifications — the intended, desired changes that an introduced gene construct is supposed to induce, e.g., the synthesis of novel insecticidal compounds which are the focus of this chapter — are expected to produce significant change in particular plant characteristics, i.e., high levels of resistance to target herbivores. As with the products of traditional breeding, much of the potential impact of these expected changes can be assessed in the laboratory prior to the release of the transgenic plants.

7.2   Non-target effects

"Non-target effects," for the purpose of this chapter, are any unintended effects of transgenic, insecticidal plants (i.e., the first trophic level) on organisms other than the target species. Unintended targets may include detrivorous organisms, pollinators, and other herbivores (i.e., the second trophic level) as well as higher-trophic-level organisms such as the insect natural enemies (predators or parasitoids) of both the non-target herbivores and the original target species. This chapter will focus on the tritrophic effects of transgenic *Bt*-plants on non-target natural enemies, especially those that are important for biological control of insect herbivores (i.e., pest species). Because lack of consideration of second-trophic-level effects will make it difficult, if not impossible, to explain effects observed at higher trophic levels (Price et al. 1980), some attention also will be given to what is known about effects on the herbivore level.

Non-target effects caused by transgenic crop plants can arise in managed and unmanaged ecosystems alike. However, effects in unmanaged ecosystems require that particular transgene(s) have escaped into unmanaged ecosystems via outcrossing and introgression into wild or weedy relatives. Much research has been done on gene flow and outcrossing — to the point where scientists agree that, where it can happen, it will happen. However, few data are available on the consequences of the outcrossing of transgenes into the wild (see Letourneau et al., Chapter 3 of this volume). Non-target effects can also occur through such events as mass dispersal or deposition of pollen containing one or more "new" protein(s) that are in some way toxic to fauna that come in contact with them. Essentially, this is a new form of "pesticide drift" and is covered elsewhere in this book (see Losey et al., Chapter 6 of this volume).

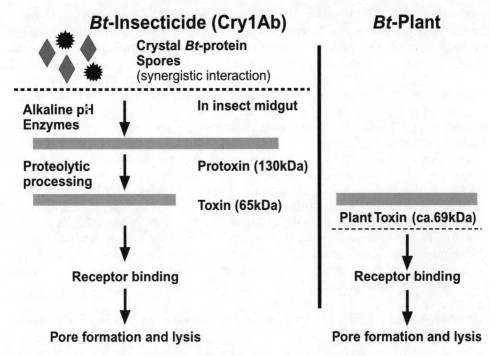

Figure 7.1   Differences between *Bt*-insecticides and transgenic *Bt*-plants.

7.2.1   Bt-*plants and* Bt-*insecticides*

Currently, gene constructs conferring resistance to pest insects express genes derived from *Bacillus thuringiensis* (*Bt*), a Gram-positive soil bacterium. These genes allow transgenic plants to produce highly bioactive toxins effective against a number of pest insects in corn, cotton, and potato (Perlak et al. 1990, Koziel et al. 1993). For decades, *Bt*-insecticides have been used for insect pest and mosquito control on agricultural land, forests, and swampland. There have been no reports of adverse effects related to these pesticides, and so they are generally considered safe for natural enemies. However, there are a number of differences between *Bt*-insecticides and transgenic *Bt*-plants that make it necessary to verify whether *Bt*-plants are also harmless to natural enemies. With *Bt*-insecticides, a mixture of spores and crystals is applied to plants. To be effective, the *Bt*-crystals must undergo a complex biochemical activation process in which spores and crystals interact synergistically (Figure 7.1) (Moar et al. 1990, Gill et al. 1992, Kumar et al. 1996). The mode of action in *Bt*-plants differs in a number of aspects, as illustrated in Figure 7.1. Ecologically perhaps even more interesting is the difference in temporal and spatial availability of *Bt*-toxins that originate in *Bt*-plants and those that originate in *Bt*-insecticides. With *Bt*-plants, there is significantly greater spatial and temporal availability of *Bt*-toxins in the agroecosystem than with *Bt*-insecticides.

In classic risk assessment, risk is the result of exposure × hazard, with "exposure" defined as a function of concentration, distribution, and time. With *Bt*-insecticides, the *Bt*-proteins are topically applied, whereas, in transgenic plants, the *Bt*-proteins are expressed constitutively, albeit at different concentrations, in essentially all plant parts, including pith, kernels, roots, and pollen (Koziel et al. 1993). Therefore, not only leaf-chewing, mandibulate insects are exposed to the toxin but also leaf-mining and cell-sucking insects such as fly larvae, thrips, leafhoppers, and spider mites. Further, unlike the *Bt*-toxin from *Bt*-insecticides, the *Bt*-toxin expressed in transgenic *Bt*-plants is present throughout the entire field season, from germination to plant senescence, with the level of expression dependent on general plant vigor. This again is due to constitutive expression of the *Bt*-toxin in plants. Topical *Bt*-insecticides, in contrast, degrade relatively quickly in the field as a result of ultraviolet light and typically lose substantial activity several days to 2 weeks after application (Ignoffo and Garcia 1978, Behle et al. 1997). Additionally, in compliance with the current pest resistance management strategy (high dose/refuge), transgenic technology strives to achieve the highest possible concentration of toxin expression in transgenic *Bt*-plants. The ultimate goal is to kill potential heterozygote resistant individuals (see Andow, Chapter 4 of this volume). The most promising technique in this regard is chloroplast transformation. It has increased plant expression levels of *Bt*-proteins multifold over the levels possible in current transgenic *Bt*-plants (McBride et al. 1995). This all leads to greatly extended temporal and spatial expression of *Bt*-protein in transgenic *Bt*-plants, which in turn results in considerably higher "exposure" to *Bt*-toxins. The ubiquitous and continuous availability of *Bt*-proteins in the agroecosystem, in addition to the modified form and mode of release of *Bt*-toxin in *Bt*-plants, necessitates new risk assessment techniques able to take these differences into account (Jepson et al. 1994). As suggested by Jepson et al. (1994), extended dietary exposure bioassays of natural enemies to plant-produced *Bt*-toxin would be one logical approach toward achieving this goal.

7.2.2   *A case study: Effects of transgenic* Bt-*plants and microbial* Bt-*preparations on* Chrysoperla carnea

In certain cases, the use of *Bt*-crop varieties will reduce the application rate of broad-spectrum insecticide, thus reducing the adverse effects of those insecticides on resident natural enemies. Many of these natural enemies play an important role in biological pest control, particularly in organic farming systems and in systems using integrated pest management (IPM). We have embarked on studies to determine if *Bt*-crops are as benign, in comparison, to natural enemies as was expected. Because of the ubiquitous presence of the *Bt*-toxin in a *Bt*-crop field, most, if not all, non-target herbivores colonizing transgenic *Bt*-plants in the field during the season will ingest plant tissue containing *Bt*-proteins and may pass these proteins on to their natural enemies in a more or less processed form. Concern for natural enemies was the motivation for a multiyear study of the effects of transgenic *Bt*-expressing corn and microbially produced *Bt*-proteins on an important natural enemy species *Chrysoperla carnea* (Order Neuroptera), the green lacewing. This predator feeds on soft-bodied insects, including aphids and caterpillars, as a larva. The study used a multitrophic and bitrophic model system approach (Figure 7.2). The results of the study and its potential ecological implications are summarized below (Hilbeck et al. 1998a, 1998b, 1999).

Three series of experiments on *C. carnea* were carried out using different *Bt*-delivery systems: (1) caterpillar prey fed transgenic *Bt*-corn and (2) artificial diet with incorporated *Bt*-protein (Figure 7.2). In the first series of experiments, which were repeated four times over time, *C. carnea* larvae were raised on two different lepidopteran herbivore species that had been fed either isogenic *Bt*-free corn or transgenic *Bt*-corn (Event 176) for 24 h prior to being provided as prey to *C. carnea* larvae. The two prey species used were the European corn borer, *Ostrinia nubilalis* (Hübner) (target pest), and the Egyptian cotton leaf-worm, *Spodoptera littoralis* (Boisduval) (for technical details, see Hilbeck et al. 1998a). *Spodoptera littoralis* is not commonly considered to be susceptible to the lethal effects of Cry1Ab, a *Bt*-protein expressed in Event 176 corn (Sneh et al. 1981, Höfte and Whiteley 1989, Müller-Cohn et al. 1996).

On average, 65% of the *C. carnea* larvae raised on *Bt*-corn-fed *O. nubilalis* and 59% of the larvae raised on *Bt*-corn-fed *S. littoralis* died. Only 37% of those raised on *Bt*-free prey

Tri-trophic feeding trials:

1st trophic level	2nd trophic level	3rd trophic level

Transgenic *Bt*-corn ⟹ Non-target herbivore ⟹ *Chrysoperla carnea*

Target herbivore

Bt-incorporated diet ⟹ Non-target herbivore ⟹ *Chrysoperla carnea*

Bi-trophic feeding trials:

1st trophic level	2nd trophic level

Bt-incorporated diet ⟹ *Chrysoperla carnea*

Figure 7.2 Overview of laboratory experiments. Non-target herbivore = *Spodoptera littoralis* (Egyptian cotton leaf worm); target herbivore = *Ostrinia nubilalis* (European corn borer). Larvae of *Chrysoperla carnea* are polyphagous predators.

Table 7.1 Summary Data of Bi- and Tritrophic Feeding Trials with *Chrysoperla carnea*; Mean Total Mortality and Developmental Time is Presented

	Bitrophic direct (toxin) (Hilbeck et al. 1998b)	Tritrophic *Bt*-corn (toxin) (Hilbeck et al. 1998a)	*Bt*-incorporated diet (Hilbeck et al. 1999)
Bt-concentration	100 μg/ml	~4–5 μg/g fresh weight[*]	25, 50, 100 μg toxin/g diet 50, 100, 200 μg protoxin/g diet
Total mean mortality in *Bt*-treatments	57% (only AD)[a] 29% (*E.k.*/AD)[b]	59% (*S. littoralis*)[a] 65% (*O. nubilalis*)[a]	55,[a,c] 68,[a,b] 78%[b] (Cry1Ab toxin) 60,[a,c] 45,[c] 55% [a,c] (Cry1Ab protoxin) 46%[c] (100 μg Cry2A protoxin/g diet) (*S. littoralis*)
Total mean mortality in control(s)	30% (only AD)[b] 17% (*E.k.*/AD)[c] 8% (*E.k.* only)[d]	37% (*S. littoralis*)[b] 37% (*O. nubilalis*)[b]	26%[d] (*S. littoralis*)
Total mean developmental time in *Bt*-treatment	37.5 days (only AD)[a] 28 days (*E.k.*/AD)[b]	31 days (*S. littoralis*)[a,b] 31.5 days (*O. nubilalis*)[b]	n.a.
Total mean developmental time in control	37.5 days (only AD)[a] 27.5 days (*E.k.*/AD)[b] 23 days (*E.k.*)[c]	31 days (*S. littoralis*)[a] 29 days (*O. nubilalis*)[c]	n.a.

Note: a,b,c,d = Different letters between rows within columns represent treatment means that are significantly different at $P = 0.05$ (LSMEANS); AD = artificial diet only; *E.k.*/AD = *Ephestia kuehniella* eggs during first instar and artificial diet (AD) during remaining larval stage; *E.k.* = *Ephestia kuehniella* eggs only; *S. littoralis*, *O. nubilalis* = type of prey used in trials; n.a.= not applicable (development was not determined in days but in percentage of population that had developed to the next stage at certain date).

[*] Fearing et al. 1997.

of both species died (Table 7.1). No significant differences in prey-mediated mortality of *C. carnea* between prey species were detected regardless of the plant variety used. Mean total developmental time of *Bt-O. nubilalis* fed *C. carnea* was 31.5 days and in the control (*Bt*-free *O. nubilalis*) 29 days. This difference was statistically significant and probably due to the intoxication of *Bt*-corn-fed *O. nubilalis* larvae. *Spodoptera littoralis*-fed *C. carnea* needed 31 days to complete their immature life stage regardless of whether their prey had fed on *Bt*-corn or *Bt*-free corn. No visible symptoms of disease in *S. littoralis* larvae were observed after feeding on *Bt*-corn foliage for 1 day (all prey larvae were exchanged daily).

To investigate a possible direct effect of *Bt*-toxins on *C. carnea*, a novel bioassay technique was used that allowed incorporation of activated Cry1Ab toxin into a liquid diet specifically developed for optimal nutrition of *C. carnea*. The liquid medium was encapsulated within small paraffin spheres and *C. carnea* larvae were allowed to feed on them. Three different treatments, including appropriate controls, were applied. In the first pair of treatments, all instars received a Cry1Ab toxin-containing artificial diet. Because first instars cannot penetrate the paraffin skins of the spheres, they were given small foam cubes soaked in non-encapsulated, liquid diet, with or without Cry1Ab toxin, until the second instar. In the second pair of treatments, *Ephestia kuehniella* (Hübner) eggs were provided as food during first instar. After reaching the second instar, all larvae received encapsulated, artificial diet with or without Cry1Ab. In the third treatment, chrysopid larvae were raised on *E. kuehniella* eggs only (for technical details, see Hilbeck et al. 1998b). The entire experiment was repeated five times over time.

The data revealed a significant, direct *Bt*-toxin effect and a significant diet-type effect. Because no statistically significant interaction effect was observed, we concluded that both factors acted independently. When all instars were reared only on the artificial diet containing Cry1Ab toxin, total mean immature mortality was significantly higher (57%) than in the respective untreated control (30%) (see Table 7.1). Also, significantly more of those larvae that received Cry1Ab beginning with their second instar (29%) died than did the respective control (17%). Only 8% of the larvae died when reared exclusively on *E. kuehniella* eggs. These results demonstrated that activated Cry1Ab was toxic to *C. carnea* at 100 µg/ml diet. Despite this fairly high concentration fed directly to *C. carnea* larvae, the mortality rate observed was remarkably similar to that achieved in the previous tritrophic trials (57 vs. 59 or 65% in the direct vs. the tritrophic feeding trials, respectively). This suggested that herbivore/*Bt*-protein interactions enhanced the toxicity of transgenic *Bt*-corn plants to *C. carnea* larvae. Mean total development time (hatch of larvae until adult eclosion) for *C. carnea* fed only on the artificial diet was 37 and 38 days and did not differ between the *Bt* and control treatments. Total developmental time for larvae raised on both *E. kuehniella* eggs and on the artificial diet lasted 27.5 to 28 days and, again, no differences between the *Bt* and the control treatments were observed. Finally, total developmental time for *C. carnea* raised on *E. kuehniella* eggs was only 23 days (see Table 7.1).

A third series of laboratory experiments was conducted in which *Bt*-protein-incorporated diets were the sole food source for the prey *S. littoralis*. Subsequent to the findings of the first series of experiments with transgenic *Bt*-corn, *S. littoralis* was the only prey studied. Activated Cry1Ab toxin and the inactive protoxins of Cry1Ab and Cry2A were incorporated into a standard meridic diet for *S. littoralis* larvae at various concentrations (Cry1Ab toxin: 100, 50, and 25 µg/g diet; Cry1Ab protoxin: 200, 100, and 50 µg/g diet, i.e., equivalent concentration of activated toxin; Cry2A protoxin: 100 µg/g diet). Similar to the feeding studies using transgenic *Bt*-corn (see above), individual *C. carnea* larvae were raised on *S. littoralis* larvae that had been fed one of the described treated meridic diets (for further technical details, see Hilbeck et al. 1999). The entire experiment was repeated four times over time.

Mean mortality of chrysopid larvae reared on *Bt*-toxin-fed prey was again significantly higher than in the control (26%) (see Table 7.1). Mean mortality of immature *C. carnea* reared on Cry1Ab toxin 100–fed prey was highest (78%) of all treatments. Mean mortality declined with decreasing toxin concentrations to 55% at the Cry1Ab toxin 25 concentration.

Prey-mediated mean mortality of Cry1Ab protoxin-exposed chrysopid larvae was 46 to 62% compared with 55 to 78% for Cry1Ab toxin–exposed larvae and 47% for Cry2A protoxin–exposed larvae (see Table 7.1). Mean total developmental times did not differ significantly among the treatments. After 31 days, 55 to 80% of *C. carnea* larvae had pupated in all treatments.

7.2.3   Discussion of results of case study

The results of all three series of experiments consistently demonstrated the susceptibility of immature *C. carnea* to *Bt*-proteins (Cry1Ab toxin, Cry1Ab protoxin, and Cry2A protoxin) whether ingested in prey or directly. The rate of mortality varied with the method of *Bt*-delivery. Prey-mediated mortality of immature *C. carnea* was highest when the prey's food source was transgenic *Bt*-corn (59 to 66%). This was despite the fact that the concentration of the *Bt*-toxin Cry1Ab was lower in *Bt*-plants (<4 to 5 µg/g fresh weight; Fearing et al. 1997) than in all other *Bt*-protein-incorporated diets. When the 100 µg Cry1Ab/ml artificial diet was fed directly to chrysopid larvae, the induced mean mortality of immature *C. carnea* was significantly higher than in the control but was lower than expected and reflected a mortality rate similar to the rate induced by prey fed transgenic *Bt*-corn. However, when the comparable *Bt*-toxin concentration was incorporated into a meridic diet (100 µg Cry1Ab toxin/g meridic diet) and provided via prey to *C. carnea*, mean mortality of immature *C. carnea* was 21% higher (78%) than when the larvae were fed this concentration directly (57%). At this high concentration (100 µg Cry1Ab toxin/g meridic diet), *C. carnea* mortality may also have been influenced by the increased intoxication of *S. littoralis* (42% mortality after 4 days) that was observed at this concentration only. However, similar effects were observed when lower concentrations of Cry1Ab toxin were incorporated into their meridic diet (50 and 25 µg/g). Although the lower concentrations were not lethal to *S. littoralis*, they did seem to cause such sublethal effects as stunting of growth (for more details, see Hilbeck et al. 1999).

In these experiments, *Bt*-protoxin-incorporated diets (Cry1Ab and Cry2A) caused significantly higher prey-mediated mortality in immature *C. carnea* than in the untreated control. However, the rate of mortality was not as high as in the experiments with Cry1Ab toxin–incorporated diet. The protoxins were not lethal to *S. littoralis*, regardless of the concentrations applied, although *S. littoralis* did exhibit sublethal effects similar to those in the toxin treatment.

Further, comparison of mortality rates among control subjects in all studies revealed that prey-mediated *C. carnea* mortality in trials using transgenic *Bt*-corn plants was approximately 11% higher (37% for both *S. littoralis*- and *O. nubilalis*-fed predator larvae) than mortality in trials using meridic diets (26%). This suggests that plant-fed prey larvae are less nutritious for immature *C. carnea* than *S. littoralis* raised on a meridic diet. This is commonly observed in laboratory rearing of herbivores, where optimized artificial diets are often found superior to a diet of their "preferred" host plant. In context, these results suggest complex three-way interactions, among *Bt*-proteins, herbivores, and plants, all contributing to the observed prey-mediated mortality of *C. carnea*.

7.3   Limits of existing knowledge

Based on current knowledge about the mode of action of *Bt*, the above-described results were unexpected. *Bt*-molecules are highly complex and exhibit quite different biochemical behavior in different gut environments, and we are still far from understanding all possible interactions with gut components.

7.3.1   *Regarding mode of action and metabolism of* Bt-*proteins*

Two explanations for the above-observed non-target effects of both microbially and plant-produced *Bt*-proteins seem reasonable, one addressing the herbivore level and the other addressing the natural enemy level.

At the herbivore level, *S. littoralis* may process the protoxin/toxin into a product lethal to *C. carnea*. Little information is available on the biochemical processing of *Bt*-proteins in so-called non-target herbivores, i.e., herbivores for whom the proteins are not lethal. In our studies, we found that *S. littoralis* exhibited sublethal symptoms at all concentrations. They exhibited lethal symptoms, i.e., increased mortality, at the highest Cry1Ab toxin concentration used (100 µg/g meridic diet) but not at the equivalent Cry1Ab protoxin concentration (200 µg/g meridic diet) (Hilbeck et al. 1999). This suggests that *S. littoralis* is affected to some degree by Cry1Ab proteins but may not efficiently process the Cry1Ab protoxin into toxin. Höfte and Whiteley (1989), Müller-Cohn et al. (1996), and Sneh et al. (1981) observed similar low toxicity and poor activity of Cry1Ab toxin in *S. littoralis* larvae. The reasons for this are unclear and could be a lack of receptors or inefficient enzyme composition and concentration or both. It is also unclear whether the *Bt*-toxin is processed/metabolized or accumulated in the larval intestinal tract. Unfortunately, this information is essential for understanding the *S. littoralis*-mediated effects of transgenic *Bt*-corn on *C. carnea* larvae.

Our lack of knowledge regarding the mode of action and metabolism of *Bt*-proteins in non-target insects results from focusing study of these effects primarily on target pest insects. The knowledge we have stems from research screening for susceptible target pest insects that are potentially controllable by these proteins (MacIntosh et al. 1990) or from research carried out to understand resistance mechanisms developing again in target pest insects (Moar et al. 1995). Both types of research are clearly important in an economic context. However, in an ecological context, information on fate and metabolism of *Bt*-proteins in non-target insects is crucial because the remaining insect community in an agroecosystem dominated by *Bt*-plants will consist primarily of non-target species that are not lethally affected by the *Bt*-proteins. These kinds of considerations have been of little interest and relevance prior to the large-scale deployment of transgenic plants capable of expressing high concentrations of *Bt*-toxins constitutively.

At the predator level, another explanation for the effects of *Bt*-proteins on immature *C. carnea* could be that the composition of digestive enzymes and the pH in the midguts of higher-trophic-level organisms is different and less alkaline than in the guts of herbivores (Waterhouse 1949, Srivastava and Srivastava 1961, Chapman 1982). Except for our own unpublished laboratory experiments, we are not aware of any studies of the modes of action of *Bt*-endotoxins in carnivorous insects. In our own trials, we followed a commonly used protocol for investigating binding of *Bt* Cry1 toxins in the midguts of susceptible herbivores. The resulting data provided evidence that Cry1Ab binds to *C. carnea* brush border membrane fragments, further confirming previous demonstrations that *Bt*-toxins directly affect larvae of the predator *C. carnea* (Hilbeck et al., unpublished).

Regardless of the mechanisms involved, our studies clearly demonstrate the importance of tritrophic level studies for the full assessment of the impact of transgenic insecticidal plants on important non-target species. While for herbivores, valuable information can be gained from direct feeding studies, such studies can only be part of the toxicity screening for higher-trophic-level organisms. This is because higher-trophic-level organisms may receive the compound in an altered form due to processing by the herbivores. If no tritrophic experiments are conducted, the effects of processing in the herbivore gut are ignored entirely and, thereby, important ecological interactions among plants, herbivores, and natural enemies may be missed. Our studies also demonstate that when conducting tritrophic trials, the actual transgenic plants must be used. Using the microbially produced equivalent does not permit the inclusion of important interaction effects. Our data showed that three-way interactions of the plant itself, the *Bt*-protein, and the sublethally affected herbivore all contributed to the observed toxicity in *C. carnea*. Without such ecologically important information in an overall risk assessment, effects seen in the laboratory may underestimate the effects experienced in the field.

7.3.2   Regarding ecotoxicological testing methodology

Previous research testing the toxicity of *Bt*-proteins to *C. carnea* did not reveal any adverse effects. This was likely due to the methodologies used. For example, in one study, *C. carnea* larvae were given meal moth eggs (*Sitotroga cerealella*) that were coated with a *Bt*-toxin solution (Sims 1995). This neglected to take into account the fact that the forceps-like mouthparts of *C. carnea* larvae only would allow them to suck out the liquefied contents of eggs or prey bodies. It has not yet been demonstrated that *Bt*-toxin can cross the corion of meal moth eggs and be present in significant amounts inside the *S. cerealella* egg. It seems reasonable to assume that in these tests, *C. carnea* larvae ingested very little or no *Bt*-toxin. Further, these experiments were terminated after 9 days, when control mortality exceeded 20% (Sims 1995). Our experiments lasted much longer, between 23 and 37.5 days. For transgenic *Bt*-plants expressing toxin during almost the entire field season, long-term trials better approximate realistic field conditions than do 1-week bioassays.

In other studies, *C. carnea* larvae were given *Bt*-corn pollen (Pilcher et al. 1997). However, *C. carnea* larvae are not known to feed on pollen because, as already mentioned, their mouthparts only permit them to ingest liquid food. In yet another experiment, the effects of *Bt*-corn were studied in a tritrophic design using aphids as prey (Lozzia et al. 1998). While this study investigated an ecologically important trophic system — *C. carnea* is an important predator of aphids in agroecosystems where aphids are present — the authors did not report testing the effects of *Bt*-fed prey on *C. carnea* larvae, as we did in our study. Most aphids of economic importance in agroecosystems, such as *Rhopalosiphum padi* used by Lozzia et al. (1998), are obligatory phloem-feeders. In recently conducted studies where phloem sap of *Bt*-corn plants was analyzed using a novel capillary technique, no *Bt*-proteins were detected (Raps et al. 2000). Hence, it is likely that no *Bt*-protein was ingested by the *C. carnea* larvae in experiments using aphids as prey. Clearly, protocols for ecotoxicity testing involving *C. carnea* must be constructed more carefully to ensure the uptake of the substance to be tested.

7.4   Long-term ecological implications: Lessons from the past

Previous research on effects of host plant resistance and synthetic pesticides on natural regulation of pest populations is useful for designing appropriate risk assessment. The effects on higher-trophic-level organisms of host plants with elevated levels of resistance to major pests have been studied for decades (for reviews, see Dilawari and Dhaliwal 1993, Hare 1992, Price et al. 1980). Similarly, a great deal of information has been published on the disruptive effects of large-scale use of insecticidal pesticides on naturally occurring biological control organisms (Croft 1990).

7.4.1   Host plant resistance and biological control

Historically, it was assumed that host plant resistance and biological control were compatible, independent pest management strategies that simultaneously introduced unrelated mortalities in pest populations (Duffey and Bloem 1986, Price 1986, Hare 1992). However, substantial evidence from research on multitrophic interactions suggests that this assumption does not always hold true (for review, see Price et al. 1980, Duffey and Bloem 1986, Kogan 1986, Price 1986, van Emden 1986, Hare 1992). Tritrophic interactions cannot be understood by simply adding together two bitrophic interactions, e.g., host plant–herbivore and herbivore–natural enemy interactions (Kareiva and Sahakian 1990). van Emden (1986) was among the first researchers to suggest that "with strongly resistant varieties, however, biological control may well suffer." Price (1986) further recommended that "during plant breeding programs such effects [tritrophic level interactions] should be evaluated if an integrated approach to crop yield improvement and pest management is to be used."

Host plants vary in the quality and quantity of chemical compounds they produce, i.e., their primary and secondary metabolites (e.g., allelochemicals). They also vary in their physical characteristics (e.g., hairiness, cuticular thickness) and other traits. This variation can occur within the same host plant species (i.e., intraspecific variation) or between different host plant species (i.e., interspecific variation), and can facilitate or impede a predator's search for its prey (Kareiva and Sahakian 1990, Eigenbrode et al. 1998, 1999). Searching behavior of natural enemies is also mediated by volatiles emitted by the host plant following herbivore injury, and it is conceivable that intraspecific host plant variability in volatile composition resulting from genetic transformation might also impact this type of complex interaction (Schuler et al. 1999). Further, inter- and intraspecific variation of secondary plant compounds are well known to influence the ability of herbivores to defend themselves, as well as to affect herbivore nutritional suitability and palatability to their natural enemies (for extensive review of these research areas, see Price et al. 1980, Vinson and Iwantsch 1980, Duffey and Bloem 1986, Nordlund et al. 1988, Hare 1992). With regard to transgenic plants, Hoy et al. (1998) provide an extensive overview and discussion of potential direct and indirect effects of insecticidal transgenic plants on biological control organisms.

Certain herbivores use sequestered secondary plant compounds in defense against natural enemies. A well-known case is the monarch butterfly whose larvae feed on milkweed (*Asclepias* sp.) and ingest cardenolides produced by their host plant. These compounds provide them with an effective defense against natural enemies (Brower 1969, 1984, Brower and Fink 1985). Inter- and intraspecific plant variation can affect the palatability of herbivores to natural enemies directly or indirectly via sequestration of plant allelochemicals. Since we do not know much about how *Bt*-proteins are being processed in herbivores that are not killed by the toxin, it is impossible to predict whether or not they are able to utilize *Bt*-toxins or derived compounds in, for example, defense against predation or parasitism.

In his review, Hare (1992) analyzed 28 research reports concerning interactions between conventional crop varieties exhibiting elevated levels of resistance to herbivores, and the natural enemies of those herbivores. Of those 28 reports, 16 investigated tritrophic interactions with parasitoids and 12 investigated the tritrophic interactions with predators. In 6 of the 16 cases involving parasitoids, antagonistic relationships were found (i.e., 37.5%), whereas a synergistic interaction was found in only 2 of the 16 cases (12.5%). No clear interactions were found in 5 cases (31.3%) and, in the remaining 3 cases (18.8%), the form of relationship varied with the resistance level of the cultivar. Of the 12 reports involving predators, 3 (25%) revealed synergistic interactions and 3 (25%) revealed antagonistic interactions. In the remaining 6 cases (50%), an additive effect of host plant resistance and predation was observed.

In two cases listed by Hare (1992), two hemipteran predators, *Geocoris punctipes* (Say) (Hemiptera: Lygaeidae) and *Podisus maculiventris* (Say) (Hemiptera: Pentatomidae), were negatively affected by resistance to caterpillars in soybean. *Geocoris punctipes* exhibited increased developmental time and increased mortality of nymphs when its diet included foliage from soybean cultivars and caterpillars that had been reared on the resistant cultivars (Rogers and Sullivan 1986). Similarly, *P. maculiventris* showed increased developmental time and reduced growth rates when raised on *Pseudoplusia includens* (Walker) (Lepidoptera: Noctuidae) fed with resistant soybean foliage (Orr and Boethel 1985). In a related study, the authors found that these effects extended to the fourth trophic level. The reproductive capacity of the egg parasitoid *Telenomus podisi* Ashmead (Hymenoptera: Sceleonidae) was lower when it was reared on eggs of *Podisus maculiventris* whose prey, *Pseudoplusia includens*, had been reared on resistant soybeans, than when it was reared on eggs of *Podisus maculiventris* whose prey had been fed susceptible soybeans (Orr and Boethel 1986).

In an older study, tomatine, an alkaloid in tomato plants, was found to be toxic to *Hyposoter exiguae* (Viereck), an ichneumonid parasitoid of a major lepidopterous pest, *Heliothis zea* (Boddie) (Campbell and Duffey 1979). In laboratory trials, the authors fed *H. zea* larvae an artificial diet containing either 0.3 or 0.5% tomatine. Parasitoid larvae reared on *H. zea* larvae that had fed on the 0.5% tomatine diet needed significantly more time to complete their larval stage than parasitoid larvae reared on *H. zea* larvae fed the lower concentration tomatine diet. Further, the 0.5% tomatine diet pupae weighed significantly less and significantly fewer of them developed to adults (Campbell and Duffey 1979). The authors speculated that this incompatibility could become greater if the pest population evolved tolerance to tomatine, while the parasitoid population remained sensitive.

Another study of herbivore–natural enemy relationships involved lines of cultivated tobacco (*Nicotiana tabacum* L.). Barbosa et al. (1986) found that parasitoid *Hyposoter annulipes* (Cresson) (Hymenoptera: Ichneumonidae) successfully parasitized a smaller proportion of its host, the fall armyworm, *Spodoptera frugiperda* (Smith) (Lepidoptera: Noctuidae), when its caterpillars were reared on an artificial diet containing nicotine than when its caterpillars were reared on a nicotine-free diet. The same study found the more specialized parasitoid *Cotesia congregata* (Say) (Hymenoptera: Braconidae) able to both tolerate higher quantities of nicotine in its tissues and more efficiently eliminate nicotine acquired from its host's hemolymph than could the more polyphagous *H. annulipes* (Barbosa et al. 1986).

In the above-listed studies, the effect of the plant-produced compounds acting through the food chain directly on the natural enemy of a pest herbivore was not differentiated from indirect effects of plant-produced compounds acting only on the herbivore, reducing its nutritional quality, and thereby indirectly affecting its natural enemy. In a more recent study, Reitz and Trumble (1996) assessed whether the linear furanocoumarins, psoralen, bergapten, and xanthotoxin (secondary plant compounds in *Apium* spp. and other taxa), exerted direct or indirect effects on a parastoid of *Trichoplusia ni* (Hübner), *Copidosoma floridanum* (Ashmead). Using diet incorporation bioassays, they found that increasing the concentrations of furanocoumarins in the diet prolonged the larval development of *T. ni* but affected *T. ni* mortality only marginally. However, increasing furanocoumarin concentrations in the diet of *T. ni* significantly increased mortality of its parasitoid. The authors concluded that the furanocoumarins directly affected the parasitoid larvae rather than indirectly as a consequence of furanocoumarin effects on the host.

All of the above-described studies dealt with traditionally bred host plant resistance. Similar effects have been observed in studies using transgenic plants containing novel *Bt*-toxins (Hilbeck et al. 1998a). When transgenic *Bt*-corn foliage was fed to two herbivore species, one susceptible to the *Bt*-toxin (*Ostrinia nubilalis*), and the other fairly unsusceptible (*Spodoptera littoralis*), both herbivore species induced similar mortalities in the predator *Chrysoperla carnea* (see above). Like Reitz and Trumble (1996), we (Hilbeck et al. 1998a) concluded that the *Bt*-toxin exerted a direct effect on *C. carnea* larvae. In further trials, we demonstrated that when increasing concentrations of the *Bt*-toxin were incorporated in an artificial diet for the "unsusceptible" herbivore (*S. littoralis*), the herbivore eventually suffered a slightly increased mortality (42%) at the highest toxin concentration but the herbivore's natural enemy suffered even greater mortality (78%). The toxic effect was amplified on the next trophic level.

Indirect evidence of effects similar to those observed in laboratory trials was collected in field studies with *Chrysomela aenicollis* (Shaeffer) (Coleoptera: Chrysomelidae). *Chrysomela aenicollis* utilizes salicin and perhaps other phenolglucosides of its host species, *Salix orestera* Schneider and *S. lasiolepis* Bentham, as precursors for the synthesis of salicylaldehyde, the active component of its defensive secretion. Willow clones can vary drastically in their phenolglucoside content. Although the authors did not measure predation directly, they did find that the larvae feeding on the genotypes producing low levels of phenolglucoside were themselves more extensively preyed upon when their host plants suffered less defoliation (Smiley et al. 1985). In another field study, Theodoratus and Bowers (1999) found that prairie wolf spiders (Lycosidae) ate caterpillars reared on *Plantago major* (containing low levels of iridoid glycosides, a secondary plant compound) significantly more often than they ate caterpillars reared on *Plantago lanceolata* (containing high levels of iridoid glycosides), although they attacked equal numbers of both types of prey. In laboratory experiments, Theodoratus and Bowers (1999) presented spiders with *P. lanceolata*-reared and *P. major*-reared caterpillars simultaneously for eight consecutive trials. Spiders consumed *P. major*-reared caterpillars significantly more often than *P. lanceolata*-reared caterpillars. Similarly, in another laboratory trial, Meier and Hilbeck (2001) also observed that *C. carnea* third instars preferred to eat *Bt*-free *S. littoralis* larvae rather than *Bt*-containing *S. littoralis* larvae.

In conclusion, the relationship between herbivores and their natural enemies is influenced by the relationship between herbivores and their host plant (Hufbauer and Via 1999). In nature, these effects will be further complicated by such phenomena as prey availability and composition, as well as abiotic factors (Weiser and Stamp 1998). Whereas in agricultural ecosystems the primary concern is that host plant resistance may offset the benefits from biological control through tritrophic effects, if natural enemies are adversely affected by that resistance in unmanaged, natural ecosystems, the ecosystem may suffer a decline in biodiversity due to the adversely affected insect species (regardless of the trophic level).

Although there have been many past reports of compatibility of host plant resistance and biological control organisms, including one involving a low-dose, experimental *Bt*-tobacco line (Johnson and Gould 1992), the evidence that both strategies can also yield antagonistic effects is too significant to be dismissed. Unfortunately, the concern over tritrophic effects associated with transgenic crops and the general lack of attention devoted to such effects involving conventionally bred crops, is viewed as a double standard by those who embrace genetic engineering. In answer to that concern, I would like to point out again that researchers did indeed raise this issue years ago but they went unheard by the regulatory authorities. But it must also be emphasized that it is precisely the achievements of genetic engineering that lead us to expect more tritrophic effects from *Bt*-crops than from conventionally bred crops. Through genetic engineering, season-long, high-level expression of toxins can be achieved throughout an entire plant. This was extremely difficult, if not impossible, to achieve in conventional breeding. With continued progress in chloroplast transformation, protein expression will most certainly reach unprecedented levels. Further, genetic engineering allows us to combine several such potent traits in the same plant. All this in addition to the earlier-discussed differences between genetic engineering and conventional breeding would seem to justify our further consideration of the effects of transgenic host plant resistance on natural enemies of pest herbivores.

7.4.2 The pesticide experience

In principle, whatever adversely affects the natural enemies of plant pests has the potential to disrupt natural plant pest control mechanisms. Despite the fact that pesticides and *Bt*-plants have different modes of action — *Bt*-proteins taking more time to bring death in target insects than do pesticides — and despite the lack of systematic monitoring in large-scale, transgenic plantings that might enable us to know whether comparisons are appropriate, in order to understand the range of possible impacts that *Bt*-plants may have on natural regulatory mechanisms, it is useful to review the lessons learned from intensive use of pesticides. In addition to the development of pest resistance (which is discussed elsewhere in this book), two major effects commonly occurred with intensive use of pesticides: target pest resurgence and secondary pest development. Further, many pesticides not only affect the target pest but, in varying degrees, also affect the target pest's natural enemies (Croft 1990). Both target pest resurgence and secondary pest development occur when pesticides alter the quantitative pest–natural enemy ratio in favor of the pest. In the case of pest resurgence, the target pest recovers from the pesticide impact faster than its natural enemies (because of its higher reproductive capacity). In the case of secondary pest development, the remaining herbivores, not affected by the pesticide to the same degree as the target pest, recover faster from the pesticide treatment than the target pest or its natural enemies. Many of the worst modern pest problems are clearly pesticide induced (Cohen et al. 1994, Cuong et al. 1997).

With transgenic *Bt*-plants, the *Bt*-toxin is present throughout the entire field season. Hence, target pest resurgence is a rather unlikely event until the target pest becomes *Bt*-resistant. Until such time, the target pest essentially is eliminated, except where large, nearby refuges are established for pest resistance management. However, every agricultural ecosystem contains many different herbivore species, most of which are of no economic importance. A number of those herbivores will remain rather unaffected or sublethally affected by the toxins in *Bt*-plants. When the target pest is a highly abundant and competitive species that is severely impacted by *Bt*-plants, a large niche consisting of abundant and nutritious food plants is opened for the remaining herbivores. Facing less competition from the target pest, the remaining herbivores will tend to utilize these resources (Figure 7.3). Which species will be successful in replacing the target pest and to what degree they will be successful will depend on a number of factors and will be difficult to predict. One factor will be the degree of susceptibility of particular non-target herbivore species to the expressed *Bt*-toxin. Another important factor will be the effectiveness of the remaining natural enemies. This in turn will also depend on the degree of their susceptibility to the *Bt*-toxins passed on to them by their prey and the degree to which they have been affected previously by the insecticides applied to control the target pest. Here is where laboratory data such as those presented above can provide insight into which species are at risk and therefore ought to be monitored.

Bt-toxins can impact natural enemy–prey/host relationships qualitatively and quantitatively. Qualitative impacts are those that result from disruption of the spatial and temporal relationships natural enemies have with their prey/host species. Quantitative impacts are those resulting from changes in the ratio of natural enemy density to prey/host density. It will be very difficult to verify directly such impacts in field trials (Hare 1992), especially if they last only one or two field seasons. The increased mortality of immature *C. carnea* that was observed in our experiments is likely to translate into multigenerational effects. Once *C. carnea* completes its immature life stage, the adults — fewer in number — leave the field and move to other habitats, first in search of food and later in search of mating partners and oviposition sites. All this occurs over several weeks, in varied habitats.

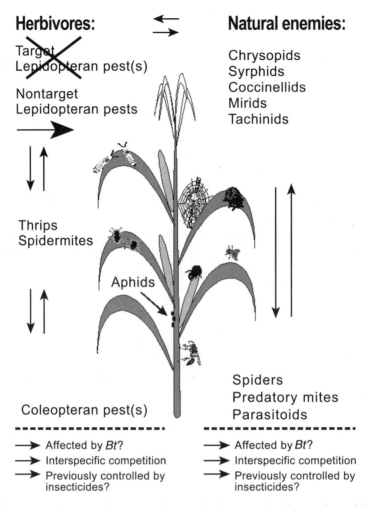

Herbivores:

Target Lepidopteran pest(s)

Nontarget Lepidopteran pests

Thrips
Spidermites

Aphids

Coleopteran pest(s)

→ Affected by *Bt*?
→ Interspecific competition
→ Previously controlled by insecticides?

Natural enemies:

Chrysopids
Syrphids
Coccinellids
Mirids
Tachinids

Spiders
Predatory mites
Parasitoids

→ Affected by *Bt*?
→ Interspecific competition
→ Previously controlled by insecticides?

Figure 7.3 Schematic drawing of a corn plant with its associated complex of herbivores and natural enemies.

Collecting indirect evidence may be more feasible and reliable. It is possible, for example, to survey non-target herbivore populations and to assess the potential for them to become secondary pests under current cultivation practices. There is already evidence that secondary pests may be emerging, at least locally. In summer 1999, the Economic Research Service of the U.S. Department of Agriculture (ERS-USDA, 1999) published a survey of pesticide use in transgenic *Bt*-cotton and conventional cotton fields in 1997. Insecticide use against the target pests in two of three regions surveyed ("Fruitful Rim," "Southern Seaboard," "Mississippi Portal") decreased significantly compared with use in conventional cotton production while it remained the same in one region ("Fruitful Rim"). This is the main benefit expected from *Bt*-cotton, a reduction in pesticide use with no associated reduction in yield. In fact, in some cotton-growing areas of the southeastern United States, due to pest resistance against essentially all used pesticides, *Bt*-cotton may provide the only avenue for growing that crop; pest damage would have prevented profitable returns if *Bt*-cotton varieties had not become commercially available (W. Moar, personal communication). However, in one region ("Mississippi Portal"), insecticide use against non-target pests in *Bt*-cotton almost doubled compared with insecticide use in conventional cotton (from 4.43 in conventional cotton to 8.19 insecticide acre-treatments in *Bt*-cotton). Use remained the same in the other two regions. These findings are consistent with the notion that, in the "Mississippi Portal," the pest complex differs from that in the other regions. It could represent an early sign that the secondary non-target pest species in the "Mississippi Portal" may replace the target pest of *Bt*-cotton. In *Bt*-corn, insecticide treatments against the target pest, *O. nubilalis*, decreased slightly because not much pesticide had been used against this pest to begin with (from 0.07 to 0.00 insecticide acre-treatments). Further, use of pesticides against non-target pests remained the same. Hence, with respect to pesticide savings, in corn, there seemed to be little benefit.

In another study, Riddick et al. (1998) compared fields grown in conventional, transgenic, and mixed conventional and transgenic potatoes. They reported a significant increase in leafhopper (*Empoasca fabae*) densities in the fields of both pure *Bt*-potato and mixed potatoes, compared with the control fields of conventional potatoes. This appears to be the result of (1) reducing some applications of pesticides that had been efficacious against both the target pest and other non-target herbivores and (2) the ineffectiveness of the remaining natural enemies in controlling secondary pests. Thus, regular disruption of natural biological controls through pesticide use may continue with use of transgenic crops. Whether there will be a net reduction in pesticide use will probably depend on the given herbivores, environmental conditions, and agricultural practice of individual farmers.

7.4.3 Implications for integrated pest management

The examples cited above remind us that intensive agricultural systems are often profoundly disturbed ecosystems that only produce acceptable yields with great external input. Simply taking away or reducing the application frequency of some of those inputs, e.g., pesticides, without a carefully designed, accompanying management strategy will not automatically lead to a more sustainable system. Nevertheless, even small reductions in pesticide inputs are beneficial, both economically and environmentally. As Lewis et al. (1997) stated, "As spectacular and exciting as biotechnology is, its breakthroughs have tended to delay our shift to long term, ecologically based pest management because the rapid array of new products provide a sense of security just as did synthetic pesticides at the time of their discovery in the 1940s."

One example of a pest management strategy that seeks to avoid high-input, chemical-intensive strategies is integrated pest management (IPM). According to Levins (1986), IPM represents a "softening of a stance of hostile confrontation with all of living nature except crop, and a groping toward a strategy of détente and coexistence with most species." This concept of IPM is based upon a variety of integrated methods designed to reduce the reliance on single-solution tactics. This variety of methods includes decision making based on economic threshold levels, cultural control methods (e.g., crop rotation, mulching), strategic use of host plant resistance (see above), mechanical control methods (e.g., plowing of infested plant residues), biological control methods (e.g., use of natural enemies), microbial control methods (e.g., *Bt*-based insecticides, etc.), and the judicious use of synthetic insecticides only as emergency, backup tools.

One of the great challenges in pest management of the coming decades is to evaluate how and whether transgenic plants can be incorporated as safe, effective components of sustainable IPM systems. Among the questions that will have to be answered is whether persistence of high concentrations of insecticidal toxin in almost all plant parts throughout an entire field season is compatible with the IPM philosophy of controlling insects at or below economic thresholds or whether such persistence represents just another eradication technique of the pesticide era, this time with the insecticide internally expressed rather than externally applied. If transgenic crops are to become an integral component of IPM systems, comprehensive plans for their incorporation have to be developed and communicated to farmers prior to large-scale, commercial deployment of those crops. Plans must be made to monitor non-target pests and natural enemies. A prerequisite for this, of course, is that we know early on what effects the transgenic, insecticidal plant has on the non-target insects. Economic thresholds developed for conventional varieties may not be valid in *Bt*-crop fields. Further, we will need to know whether and how natural enemies are affected by feeding on prey that contain the expressed toxin. The above-discussed experience in multitrophic interactions warns us to proceed cautiously and not to draw premature conclusions. The introduction of another new and powerful technology offers a unique opportunity to all involved, including farmers, scientists, regulators, industry, and the public to demonstrate that we have learned our lessons from the past.

Literature cited

Antoniou, M., Genetically engineered food — panacea or Pandora's box? *Nutr. Ther. Today,* 5, 8, 1996.

Barbosa, P., Saunders, J. A., Kemper, J., Trumbule, R., Olechno, J., and Martinat, P., Plant allelochemicals and insect parasitoids: effects of nicotine on *Cotesia congregata* (Say) (Hymenoptera:Braconidae) and *Hyposoter annulipes* (Cresson) (Hymenoptera:Ichneumonidae), *J. Chem. Ecol.*, 12, 1319, 1986.

Behle, R. W., McGuire, M. R., and Shasha, B. S., Effects of sunlight and simulated rain on residual activity of *Bacillus thuringiensis* formulations, *J. Econ. Entomol.*, 90, 1560, 1997.

Brower, L. P., Ecological chemistry, *Sci. Am.*, 220, 22, 1969.

Brower, L. P., Chemical defence in butterflies, in *The Biology of Butterflies*, Vane-Wright, R. I. and Ackery, P. R., Eds., Academic Press, London, 1984, 109.

Brower, L. P. and Fink, L. S., A natural defense system: cardenolides in butterflies versus birds. *Ann. N.Y. Acad. Sci.*, 443, 171, 1985.

Campbell, B. C. and Duffey, S. S., Tomatine and parasitic wasps: potential incompatibility of plant antibiosis with biological control, *Science*, 205, 700, 1979.

Chapman, R. F., *The Insects. Structure and Function*, Harvard University Press, Cambridge, MA, 1982, chap. 4.

Cohen, J. E., Schoenly, K., Heong, K. L., Justo, H., Arida, G., Barrion, A. T., and Litsinger, A., A food web approach to evaluating the effect of insecticide spraying on insect pest population dynamics in a Philippine irrigated rice ecosystem, *J. Appl. Ecol.*, 31, 747, 1994.

Croft, B. A., *Arthropod Biological Control Agents and Pesticides*, John Wiley & Sons, New York, 1990.

Cuong, N. L., Ben, P. T., Phuong, L. T., Chau, L. M., and Cohen, M. B., Effect of host plant resistance and insecticide on brown planthopper *Nilaparvata lugens* (Stal) and predator population development in the Mekong Delta, Vietnam, *Crop Prot.*, 16, 707, 1997.

Dilawari, V. K. and Dhaliwal, G. S., Host plant resistance to insects: novel concepts, in *Advances in Host Plant Resistance to Insects*, Dhaliwal, G. S. and Dilawari, V. K., Eds., Kalyani Publishers, New Delhi, 1993, chap. 12.

Duffey, S. S. and Bloem, K. A. Plant defense-herbivore-parasite interactions and biological control, in *Ecological Theory and Integrated Pest Management Practice*, Kogan, M., Ed., John Wiley & Sons, New York, 1986, chap. 6.

Economic Research Service (ERS)–U.S. Department of Agriculture (USDA), Impacts of Adopting Genetically Engineered Crops in the US — Preliminary Results, available at http://www.econ.ag.gov/whatsnew/issues/gmo/, 20 July 1999.

Eigenbrode, S. D., White, C., Rhode, M., and Simon, C. J., Behavior and effectiveness of adult *Hippodamia convergens* (Coleoptera: Coccinellidae) as a predator of *Acyrthosiphon pisum* (Homoptera: Aphididae) on a wax mutant of *Pisum sativum*, *Environ. Entomol.*, 27, 902, 1998.

Eigenbrode, S. D., Kabalo, N. N., and Stoner, K. A. Predation, behavior, and attachment by *Chrysoperla plorabunda* larvae on *Brassica oleracea* with different surface waxblooms, *Entomol. Exp. Appl.*, 90, 225, 1999.

Fearing, P. L., Brown, D., Vlachos, D., Meghji, M., and Privalle, L., Quantitative analysis of CryIA(b) expression in *Bt*-maize plants, tissues, and silage and stability of expression over successive generations, *Mol. Breeding*, 3, 169, 1997.

Gill, S. S., Cowles, E. A., and Pietrantonio, P. V., The mode of action of *Bacillus thuringiensis* endotoxins, *Annu. Rev. Entomol.*, 37, 615, 1992.

Hare, D. J., Effects of plant variation on herbivore-natural enemy interactions, in *Plant Resistance to Herbivores and Pathogens, Ecology, Evolution, and Genetics*, Fritz, R. S. and E. L. Simms, Eds., The University of Chicago Press, Chicago, 1992, chap. 12.

Hilbeck, A., Baumgartner, M., Fried, P. M., and Bigler, F., Effects of transgenic *Bacillus thuringiensis* corn-fed prey on mortality and development time of immature *Chrysoperla carnea* (Neuroptera: Chrysopidae), *Environ. Entomol.*, 27, 480, 1998a.

Hilbeck, A., Moar, W. J., Pusztai-Carey, M., Filippini, A., and Bigler, F., Toxicity of the *Bacillus thuringiensis* Cry1Ab toxin on the predator *Chrysoperla carnea* (Neuroptera: Chrysopidae) using diet incorporated bioassays, *Environ. Entomol.*, 27, 1255, 1998b.

Hilbeck, A., Moar, W. J., Pusztai-Carey, M., Filippini, A., and Bigler, F., Prey-mediated effects of Cry1Ab toxin and protoxin and Cry2A protoxin on the predator *Chrysoperla carnea, Entomol. Exp. Appl.*, 91, 305, 1999.

Hilbeck, A., Moar, W. J., Zwahlen, C., and Bigler, F., *Bt*-crops and their effects on non-target organisms — a story unfolds. I. Natural enemies, invited presentation at the annual meeting of the Society for Invertebrate Pathology, Irvine, CA, August 1999, unpublished.

Höfte, H. and Whiteley, H. R., Insecticidal crystal proteins of *Bacillus thuringiensis, Microbiol. Rev.*, 53, 242, 1989.

Hoy, C. W., Feldman, J., Gould, F., Kennedy, G. G., Reed, G., and Wyman, J. A., Naturally occurring biological controls in genetically engineered crops, in *Conservation Biological Control*, Barbosa, P., Ed., Academic Press, San Diego, 1998, chap. 10.

Hufbauer, R. A. and Via, S., Evolution of an aphid-parasitoid interaction: variation in resistance to parasitism among aphid populations specialized on different plants, *Evolution*, 53, 1435, 1999.

Ignoffo, C. M. and Garcia, M., UV-photoinactivation of cells and spores of *Bacillus thuringiensis* and effects of peroxidase on inactivation, *Environ. Entomol.*, 7, 270, 1978.

Jepson, P. C., Croft, B. A, and Pratt, G. E., Test systems to determine the ecological risks posed by toxin release from *Bacillus thuringiensis* genes in crop plants, *Mol. Ecol.*, 3, 81, 1994.

Johnson, M. T. and Gould, F., Interaction of genetically engineered host plant resistance and natural enemies of *Heliothis virescens* (Lepidoptera: Noctuidae) in tobacco, *Environ. Entomol.*, 21, 586, 1992.

Kareiva, P. and Sahakian, R., Tritrophic effects of a simple architectural mutation in pea plants, *Nature*, 345, 433, 1990.

Kogan, M., Plant defense strategies and host-plant resistance, in *Ecological Theory and Integrated Pest Management Practice*, Kogan, M., Ed., John Wiley & Sons, New York, 1986, chap. 5.

Koziel, M. G., Beland, G. L., Bowman, C., Carozzi, N. B., Crenshaw, R., Crossland, L., Dawson, J., Desai, N., Hill, M., Kadwell, S., Launis, K., Lewis, K., Maddox, D., McPherson, K., Meghji, M. R., Merlin, E., Rhodes, R., Warren, G. W., Wright, M., and Evola, S. V., Field performance of elite transgenic maize plants expressing an insecticidal protein derived from *Bacillus thuringiensis*, *Bio/Technology*, 11, 194, 1993.

Kumar, P. A., Sharma, R. P., and Malik, V. S., The insecticidal proteins of *Bacillus thuringiensis*, *Adv. Appl. Microbiol.*, 42, 1, 1996.

Levins, R., Perspectives in integrated pest management: from an industrial to ecological model of pest management, in *Ecological Theory and Integrated Pest Management Practice*, Kogan, M., Ed., John Wiley & Sons, New York, 1986, chap. 1.

Lewis, W. J., van Lenteren, J. C., Phatak, S. C., and Tumlinson III, J. H., A total system approach to sustainable pest management, *Proc. Natl. Acad. Sci. U.S.A.*, 94, 12243, 1997.

Lozzia, G. C., Furlanis, C., Manachini, B., and Rigamonti, I. E., Effects of *Bt*-corn on *Rhopalosiphum padi* L. (Rhynchota Aphididae) and on its predator *Chrysoperla carnea* (Neuroptera Chrysopidae), *Boll. Zool. Agrar. Bachic. Ser. II*, 30, 153, 1998.

MacIntosh, S. C., Stone, T. B., Sims, S. R., Hunst, P. L., Greenplate, J. T., Marrone, P. G., Perlak, F. J., Fischhoff, D. A., and Fuchs, R. L., Specificity and efficacy of purified *Bacillus thuringiensis* proteins against agronomically important insects, *J. Invertebr. Pathol.*, 56, 258, 1990.

McBride, K. E., Svab, Z., Schaaf, D. J., Hogan, P. S., Stalker, D. M., and Maliga, P., Amplification of a chimeric *Bacillus* gene in chloroplasts leads to an extraordinary level of an insecticidal protein in tobacco, *Bio/Technology*, 13, 362, 1995.

Meier, M. S. and Hilbeck, A., Influence of transgenic *Bacillus thuringiensis* corn-fed prey on prey preference of immature *Chrysoperla carnea* (Neuroptera: Chrysopidae), *Basic Appl. Ecol.*, 2, 35, 2001.

Moar, W. J., Masson, L., Trumble, J. T., and Brousseau, R., Toxicity to *Spodoptera exigua* and *Trichoplusia ni* of individual P1 protoxins and sporulated cultures of *Bacillus thuringiensis* subsp. *kurstaki* HD-1 and NRD-12, *Appl. Environ. Microbiol.*, 56, 2480, 1990.

Moar, W. J., Pusztai-Carey, M., van Faassen, H., Bosch, D., Frutos, R., Rang, C., Luo, K., and Adang, M. J., Development of *Bacillus thuringiensis* CryIC resistance by *Spodoptera exigua* (Hübner) (Lepidoptera: Noctuidae), *Appl. Environ. Microbiol.*, 61, 2086, 1995.

Müller-Cohn, J., Chaufaux, J., Buisson, C., Gilois, N., Sanchis, V., and Lereclus, D., *Spodoptera littoralis* (Lepidoptera: Noctuidae) resistance to CryIC and cross-resistance to other *Bacillus thuringiensis* crystal toxins, *J. Econ. Entomol.*, 89, 791, 1996.

Nordlund, D. A., Lewis, W. J., and Altieri, M. A., Influences of plant-produced allelochemicals on the host/prey selection behavior of entomophagous insects, in *Novel Aspects of Insect–Plant Interactions*, Barbosa, P. and Letourneau, D. K., Eds., John Wiley & Sons, New York, 1988, chap. 2.

Orr, D. B. and Boethel, D. J., Comparative development of *Copidosoma truncatellum* (Hymenoptera: Encyrtidae) and its host, *Pseudoplusia includens* (Lepidoptera: Noctuidae), on resistant and susceptible soybean genotypes, *Environ. Entomol.*, 14, 612, 1985.

Orr, D. B. and Boethel, D. J., Influence of plant antibiosis through four trophic levels, *Oecologia*, 70, 242, 1986.

Perlak, F. J., Deaton, R. W., Armstrong, E. A., Fuchs, R. L., Sims, S. R., Greenplate, J. T., and Fischhoff, D. A., Insect resistant cotton plants, *Bio/Technology*, 8, 939, 1990.

Pilcher, C. D., Obrycki, J. J., Rice, M. E., and Lewis, L. C., Preimaginal development, survival, and field abundance of insect predators on transgenic *Bacillus thuringiensis* corn, *Environ. Entomol.*, 26, 446, 1997.

Price, P. W., Ecological aspects of host plant resistance and biological control. Interactions among three trophic levels, in *Interactions of Plant Resistance and Parasitoids and Predators of Insects*, Boethel, D. J. and Eikenbary, R. D., Eds., Ellis Horwood, Chichester, U.K., 1986.

Price, P. W., Bouton, C. E., Gross, P., McPheron, B. A., Thompson, J. N., and Weis, A. E., Interactions among three trophic levels: influences of plants on interactions between insect herbivores and natural enemies, *Annu. Rev. Entomol.*, 11, 41, 1980.

Raps, A., Kehr, J. Gugerli, P., Moar, W. J., Bigler, F., and Hilbeck, A., Immunological analysis of phloem sap of *Bacillus thuringiensis* corn and of the non-target herbivore *Rhopalosiphum padi* (Homoptera: Aphidae) for presence of Cry1Ab, *Mol. Ecol.*, 10, 525, 2001.

Regal, P. J., Scientific principles for ecologically based risk assessment of transgenic organisms, *Mol. Ecol.*, 3, 5, 1994.

Reitz, S. R. and Trumble, J. T., Tritrophic interactions among linear furanocoumarins, the herbivore *Trichoplusia ni* (Lepidoptera: Noctuidae), and the polyembryonic parasitoid *Copidosoma floridanum* (Hymenoptera: Encyrtidae), *Environ. Entomol.*, 25, 1391, 1996.

Riddick, E. W. and Barbosa, P., Impact of Cry3A-intoxicated *Leptinotarsa decemlineata* (Coleoptera: Chrysomelidae) and pollen on consumption, development, and fecundity of *Coleomegilla maculata* (Coleoptera: Coccinellidae), *Ann. Entomol. Soc. Am.*, 91, 303, 1998.

Rogers, D. J. and Sullivan, M. J., Nymphal performance of *Geocoris punctipes* (Hemiptera: Lygaeidae) on pest-resistant soybeans, *Environ. Entomol.*, 15, 1032, 1986.

Schuler, T. H., Potting, R. P. J., Denholm, I., and Poppy, G. M., Parasitoid behavior and *Bt*-plants, *Nature*, 400, 825, 1999.

Sims, S. R., *Bacillus thuringiensis* var. *kurstaki* (CryIA(c)) protein expressed in transgenic cotton: effects on beneficial and other non-target insects, *Southwest. Entomol.*, 20, 493, 1995.

Smiley, J. T., Horn, J. M, and Rank, N. E., Ecological effects of salicin at three trophic levels: new problems from old adaptations, *Science*, 229, 649, 1985.

Sneh, B., Schuster, S., and Broza, M., Insecticidal activity of *Bacillus thuringiensis* strains against the Egyptian cotton leaf worm *Spodoptera littoralis* (Lep.: Noctuidae), *Entomophaga*, 26, 179, 1981.

Srivastava, U. S. and Srivastava, P. D., On the hydrogen-ion concentration in the alimentary canal of the Coleoptera, *Beitr. Entomol.*, 11, 15, 1961.

Theodoratus, D. H. and Bowers, M. D., Effects of sequestered iridoid glycosides on prey choice of the prairie wolf spider *Lycosa carolinensis*, *J. Chem. Ecol.*, 25, 283, 1999.

van Emden, H. F., The interaction of plant resistance and natural enemies: effects on populations of sucking insects, in *Interactions of Plant Resistance and Parasitoids and Predators of Insects*, Boethel, D. J. and Eikenbary, R. D., Eds., Ellis Horwood, Chichester, U.K., 1986.

Vinson, S. B. and Iwantsch, G. F., Host suitability for insect parasitoids, *Annu. Rev. Entomol.*, 25, 397, 1980.

Waterhouse, D. F., The hydrogen ion concentration in the alimentary canal of larval and adult Lepidoptera, *Aust. J. Sci. Res.*, 2, 428, 1949.

Weiser, L. A. and Stamp, N. E., Combined effects of allelochemicals, prey availability, and supplemental plant material on growth of a generalist insect predator, *Entomol. Exp. Appl.*, 87, 181, 1998.

chapter eight

Release, persistence, and biological activity in soil of insecticidal proteins *from* Bacillus thuringiensis

Guenther Stotzky

Contents

8.1 Introduction

Bacillus thuringiensis (*Bt*) is a Gram-positive, aerobic, spore-forming, rod-shaped bacterium that produces a parasporal, proteinaceous, crystalline inclusion during sporulation. This inclusion, which may contain more than one type of insecticidal crystal protein (ICP), is

ENVIRONMENTAL ASPECTS OF NORMAL AND GENETICALLY ENGINEERED INSECTICIDAL TOXINS FROM *BACILLUS THURINGIENSIS*

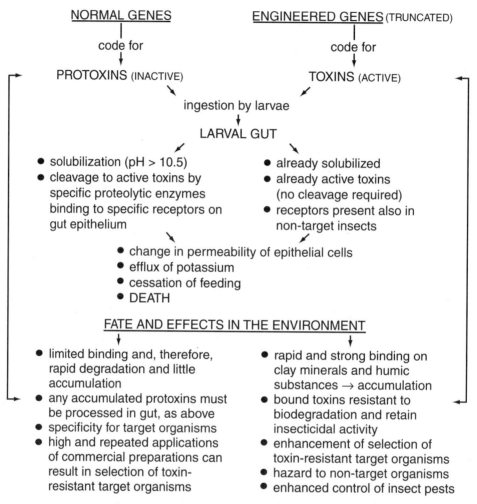

Figure 8.1 Environmental aspects of normal and genetically engineered insecticidal toxins from *Bacillus thuringiensis*. (Modified from Stotzky, G., *J. Environ. Qual.*, 29, 691, 2000. With permission.)

solubilized and hydrolyzed in the midgut of larvae of susceptible insects when ingested, releasing polypeptide toxins that eventually cause death of the larvae (Figure 8.1) (see Höfte and Whiteley 1989, Schnepf et al. 1998). More than 3000 isolates of *Bt* from 50 countries have been collected (see Feitelson et al. 1992, Crickmore et al. 1998, Schnepf et al. 1998). The ICPs have been classified on the bases of their structure, encoding genes, and host range and on the flagellar H-antigens of the bacteria that produce them (Höfte and Whiteley 1989, Crickmore et al. 1998). Numerous distinct crystal protein (*cry*) genes have been identified that code for insecticidal proteins (Cry proteins): CryI and CryIIB proteins are specifically toxic to Lepidoptera; CryIIA proteins to Lepidoptera and Diptera; CryIII proteins to Coleoptera; and four CryIV proteins to Diptera. In addition, two genes (*cytA, cytB*) that code for cytolytic proteins (CytA, CytB) are present with the CryIV proteins. This nomenclature has been revised (Crickmore et al. 1998) but will be retained here, as many of the published studies discussed in this chapter were done while the old nomenclature was used. Some ICPs also exhibit activity against other orders of insects (e.g., Homoptera, Hymenoptera, Orthoptera, Mallophaga), as well as against nematodes, mites, *Collembola*, protozoa, and other organisms (Feitelson et al. 1992, Addison 1993, Crickmore et al. 1998, Schnepf et al. 1998).

Preparations of *Bt*, usually as sprays that contain a mixture of cells, spores, and parasporal crystals, have been used as insecticides for more than 30 years. Until recently, no unexpected toxicities from such sprays have been recorded, probably because *Bt* does not survive or grow well in natural habitats such as soil (e.g., Saleh et al. 1970, West 1984, West et al. 1984a,b, 1985, West and Burges 1985, Petras and Casida 1985), and its spores are rapidly inactivated by ultraviolet (UV) radiation (Griego and Spence 1978, Ignoffo and Garcia 1978). Consequently, there is probably little production of toxins in soil, and the persistence of introduced toxins is a function primarily of the: (1) concentration added; (2) rate of consumption and inactivation by insect larvae; (3) rate of degradation by microorganisms; and (4) rate of abiotic inactivation. However, when the genes that code for these toxins are genetically engineered into plants, the toxins continue to be synthesized during growth of the plants. If production exceeds consumption, inactivation, and degradation, the toxins could accumulate to concentrations that may enhance the control of target pests or constitute a hazard to non-target organisms, such as the soil microbiota (see below), beneficial insects (e.g., pollinators, predators and parasites of insect pests) (e.g., Flexner et al. 1986, Goldburg and Tjaden 1990, Addison 1993, James et al. 1993, Johnson et al. 1995), and other animal classes. This hazard can be direct, e.g., larvae of the monarch butterfly (*Danaus plexippus*) killed by feeding on milkweed (*Asclepias curassavica*) contaminated with pollen from transgenic *Bt*-corn (*Z. mays*) (Losey et al. 1999), or indirect in tritrophic interactions, e.g., mortality and delayed development of the green lacewing (*Chrysoperla carnea*), a predator of insect pests, when fed larvae of the European corn borer (*Ostrinia nubilalis*) raised on transgenic *Bt*-corn expressing Cry1Ab toxin (Hilbeck et al. 1998a,b, 1999)]. The accumulation and persistence of the toxins could also result in the selection and enrichment of toxin-resistant target insects (e.g., Van Rie et al. 1990, McGaughey and Whalon 1992, Entwistle et al. 1993, Tabashnik 1994, Bauer 1995, Ferré et al. 1995, Tabashnik et al. 1997). Persistence is enhanced when the toxins are bound on surface-active particles in the environment (e.g., clays and humic substances) and, thereby, rendered less accessible for microbial degradation but still retentive of toxic activity (see below).

These potential hazards and benefits are affected by modifications (e.g., truncation and rearrangement of codons; see Schnepf et al. 1998) of the introduced toxin genes to code only for the synthesis of active toxins, or a portion of the toxins, rather than of nontoxic crystalline protoxins (Figure 8.1). Consequently, it will not be necessary for an organism that ingests the active toxins to have a high midgut pH (~10.5) for solubilization of the ICPs and specific proteolytic enzymes to cleave the protoxins into toxic subunits. Therefore, non-target insects and organisms in higher and lower trophic levels could be susceptible to the toxins, even if they do not have an alkaline gut pH and appropriate proteolytic enzymes. This leaves only the third of the three barriers that appear to be responsible for the host specificity of the ICPs: i.e., specific receptors for the toxins on the midgut epithelium that are often, but not always, present in larger numbers in susceptible larvae (e.g., Van Rie et al. 1990, Wolfersberger 1990, Garczynski et al. 1991).

8.2 Surface-active particles in soil

Surface-active particles in soil and other natural habitats (e.g., sediments in aquatic systems) are important for the persistence of organic molecules that, in the absence of such particles, would be rapidly degraded by the indigenous microbiota. These surface-active particles are primarily clay minerals and humic substances. Sand- and silt-size particles do not generally appear to be involved in the persistence of these molecules, as these particles are not surface active (i.e., they do not have a significant surface charge, because they are composed mainly of primary minerals), and they have a smaller specific surface area (Stotzky 1986). Many organic molecules are important in the ecology, activity, biodiversity, and evolution of microbes (e.g., as substrates, growth factors, siderophores), as well as in environmental protection and biological control of pests (e.g., Stotzky 1974, 1986, 1997).

The insecticidal proteins produced by various subspecies of *Bt* are discussed below as examples of the resistance to biodegradation of biomolecules when bound on surface-active particles (i.e., clays and humic substances), as well as of the retention of the biological activity of the bound molecules. Because of the large differences in the chemical composition and structure of the surface-active particles, these systems can serve as models for the study of the fate and effects of other biomolecules (e.g., products of transgenic microbes, plants, and animals, including antibodies, vaccines, and other bioactive compounds; toxins produced by fungi) that are also chemically and structurally diverse and that will eventually reach soil and other natural habitats in biomass, root exudates, feces, urine, and other forms. Surface-active particles are also involved in mediating the effects of anthropogenic pollution (e.g., acid precipitation, heavy metals, pesticides) (Babich and Stotzky 1983, 1986, Collins and Stotzky 1989, Stotzky 1986, 1997).

This chapter summarizes the effects of interactions of the insecticidal proteins from *Bt* with clays and humic acids on the persistence and biological activity of these biomolecules in soil. The methods used in these studies, as well as references to the studies of other investigators, are described in greater detail in the section on Literature Cited herein, especially in Stotzky (1986, 1989, 2000), Stotzky et al. (1993), and Doyle et al. (1995).

8.3   *Adsorption and binding of* Bt-*toxins on clays*

A series of our studies has described the equilibrium adsorption and binding of the purified toxins produced by *B. thuringiensis* subsp. *kurstaki* (*Btk*; 66 kDa; active against Lepidoptera) and subsp. *tenebrionis* (*Btt*; 68 kDa; active against Coleoptera) on the clay minerals montmorillonite and kaolinite and on the clay-, silt-, and sand-size fractions of soil (Venkateswerlu and Stotzky 1990, 1992, Tapp et al. 1994, Tapp and Stotzky 1995a,b, 1997, 1998, Koskella and Stotzky 1997), as well as the adsorption and binding of the toxin from *Btk* on humic acids from different soils (Crecchio and Stotzky 1998, 2001). Montmorillonite and kaolinite are the predominant clay minerals in many soils, and these clays differ in structure and numerous physicochemical characteristics (e.g., cation-exchange capacity, specific surface area) and in their effects on biological activity in soil (see Stotzky 1986). The toxins and protoxins were purified from pure cultures and commercial sources of the subspecies of *Bt*, and the clay minerals, various size fractions of soil, and humic acids were prepared as described in the references above. The availability to microbes of free and bound toxins as sources of carbon and/or nitrogen and the comparative insecticidal activity of free and bound toxins have also been studied (e.g., Koskella and Stotzky 1997, Crecchio and Stotzky 1998). The purpose of these *in vitro* studies with purified toxins and relatively defined clays and humic acids was to determine whether the toxins released *in situ* from transgenic plants or commercial spray preparations have the potential to be adsorbed and bound on such surface-active particles and to persist and retain insecticidal activity in soil. The results of these studies are summarized in Table 8.1.

The toxins and protoxins adsorbed rapidly (in <30 min, the shortest time studied) (Figure 8.2) on both "clean" and "dirty" montmorillonite and kaolinite and on the clay-size fraction of soil, indicating that toxins released in root exudates and upon disintegration

Table 8.1 Summary of Some of the Results of Studies in This Laboratory on the Persistence and Biological Activity in Soil of the Insecticidal Toxins from *Bacillus thuringiensis*

Insecticidal toxins from *Bacillus thuringiensis* subspp. *kurstaki* (*Btk*) and *tenebrionis* (*Btt*) bind rapidly and tightly on clays, humic acids, and clay–humic acid complexes; binding is pH dependent.

Bound toxins retain their structure, antigenicity, and insecticidal activity.

Biodegradability of the toxins is reduced when bound; microbial utilization as a sole source of carbon reduced more than use as a sole source of nitrogen.

Insecticidal activity of the toxin from *Btk* was detected 234 days after addition to nonsterile soils (the longest time studied).

Persistence of insecticidal activity was greater in acidic soils, probably in part because microbial activity was lower than in less acid soils; persistence was reduced when the pH of acidic soils was raised to ~7.0 with $CaCO_3$.

Persistence was similar under aerobic and anaerobic conditions and when soil was alternately wetted and dried or frozen and thawed.

Persistence in soil was demonstrated by dot-blot ELISA, flow cytometry, Western blots, and insect bioassays.

Transgenic corn containing Cry1Ab toxin reduced metabolic activity (CO_2 evolution) and affected the activity of some enzymes in soil.

The Cry1Ab toxin is released in root exudates and persists in rhizosphere soil *in vitro* and *in situ* but appears to have no effect on numbers of earthworms, nematodes, protozoa, bacteria, and fungi.

Toxins from *B. thuringiensis* introduced in transgenic plants and microbes could persist, accumulate, and remain insecticidal in soil as the result of binding on clays and humic substances and, therefore, pose a hazard to nontarget organisms, enhance selection of toxin-resistant target species, or enhance control of insect pests.

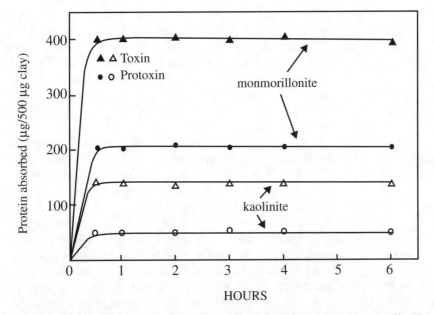

Figure 8.2 Rate of adsorption at equilibrium of 1 mg of the protoxin or toxin from *Bacillus thuringiensis* subsp. *kurstaki* on 0.5 mg of montmorillonite or kaolinite with a mixed cation complement. Data are expressed as the means ± the standard errors of the means, which are within the dimensions of the symbols. (From Venkateswerlu, G. and Stotzky, G., *Curr. Microbiol.*, 25, 225, 1992. With permission.)

of transgenic plant cells in soil would be only briefly in a free state susceptible to rapid biodegradation. The negative charges on the clean clays were compensated for by monomeric cations, either mono- or polyvalent and either homoionic or mixed, whereas the dirty clays were "coated" with two types of polymeric oxyhydroxides of iron and were

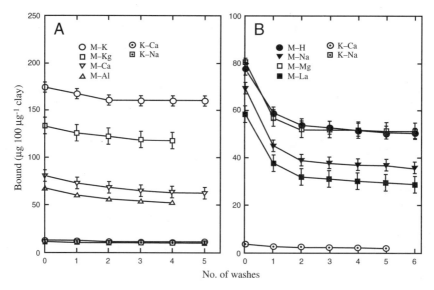

Figure 8.3   Desorption of the toxin from (A) *Bacillus thuringiensis* subsp. *kurstaki* and (B) *Bacillus thuringiensis* subsp. *tenebrionis* from homoionic montmorillonite (M)- and kaolinite (K)-toxin complexes after equilibrium adsorption. Data normalized to 100 μg of clay. Means ± 1 standard error of the mean. Values at 0 number of washes indicate amounts of toxin adsorbed at equilibrium. (From Tapp, H. et al., *Soil Biol. Biochem.*, 26, 225, 1992. With permission from Elsevier Science.)

probably more representative of clays in soil *in situ*. The greater adsorption of the toxins than of the protoxins was probably a result of differences in their molecular mass (M_r) (toxin = 66 kDa; protoxin = 132 kDa) and, possibly, in their conformation. Inasmuch as toxins, rather than protoxins, are expressed in most transgenic plants, this chapter will discuss only the results of studies obtained with toxins. Adsorption of the toxins increased with their concentration and then plateaued, indicating that the clays became saturated with the toxins, as observed with other proteins and organic compounds (Stotzky 1986, 2000); was maximal between pH 6 and 8 on clean clays and between pH 5 and 9 on dirty clays; was affected by the type of cations on the exchange complex of the clays (adsorption generally decreased as the valency of the cations increased, as has also been observed with other proteins; see Stotzky 1986); and was significantly lower on kaolinite than on montmorillonite. Larger amounts of the toxin from *Btk* than that from *Btt* were adsorbed. Only about 10 and 30% of the toxin from *Btk* and *Btt*, respectively, adsorbed at equilibrium, was desorbed by one or two washes with water (Figure 8.3). Additional washings desorbed no more toxins, indicating that the toxins were tightly bound on the clays and suggesting that the toxins would not be easily desorbed *in situ* and leached into ground water by rain, irrigation, snow melts, etc. Interaction of the toxins with the clays did not alter significantly their structure, as indicated by sodium dodecyl sulfate polyacrylamide gel electrophoresis (SDS-PAGE) (Tapp et al. 1994) and enzyme-linked immunosorbent assays (ELISA) (Tapp and Stotzky 1995a) of the equilibrium supernatants and desorption washes and by Fourier-transform infrared analyses (Table 8.2) and insect bioassays (Tables 8.3, and 8.4) of the bound toxins. The toxins partially intercalated montmorillonite, as determined by X-ray diffraction analyses; there was more intercalation by the toxin from *Btt* than by the toxin from *Btk*, but the entire protein from either *Btk* or *Btt* did not penetrate (Table 8.5). There was no intercalation of kaolinite, a nonexpanding clay mineral. Even though the M_r of the toxins from *Btk* and *Btt* was similar, the toxin from *Btt* intercalated montmorillonite more but adsorbed and bound less on montmorillonite and kaolinite than

Table 8.2   Frequencies (cm^{-1}) of the Infrared Bands (Amide I and II) of the Protoxin and Toxin from *Bacillus thuringiensis* subsp. *kurstaki* (*Btk*) and of the Toxin from *Bacillus thuringiensis* subsp. *tenebrionis* (*Btt*) Alone or Bound on Montmorillonite (M) or Kaolinite (K) Homoionic to Different Cations

Sample	Amide I	Amide II
Btk protoxin alone	1651	1538
M-Na	1658	1542
M-Ca	1651	1542
K-Na	1657	1546
K-Ca	1651	1543
Btk toxin alone	1651	1544
M-H	1641	1544
M-Ca	1650	1546
M-Al	1651	1540
Btt toxin alone	1651	1543
M-Na	1651	1536
M-H	1651	1536
M-K	1657	1536
M-Ca	1658	1541
M-Mg	1651	1537
M-La	1651	1537
M-Al	1646	1537
K-Na	1651	1543
K-Ca	1651	1543

Note: Difference spectra obtained by Fourier-transform infrared analyses.

Source: From Tapp, H. et al., *Soil Biol. Biochem.*, 26, 663, 1994. With permission from Elsevier Science.

Table 8.3   Lethal Concentration (LC$_{50}$) and 95% Confidence Interval (CI) of the Toxin from *Bacillus thuringiensis* subsp. *kurstaki*, Free, Adsorbed, or Bound on Clay Minerals or the Clay-Size Fraction from Soil, for Larvae of the Tobacco Hornworm (*Manduca sexta*) after 7 Days of Exposure

		Clay–toxin complex[a]			
		Adsorbed		Bound	
Clay[b]	Mortality with clay alone[c]	LC$_{50}$ (ng/100 µl)[d]	95% CI	LC$_{50}$ (ng/100 µl)[d]	95% CI
M-mix	0	22.0	16.42–27.80	30.7	20.10–50.43
K-mix	0	18.0	11.33–27.55	23.0	14.01–40.32
K-soil	0	21.9	18.67–25.77	21.4	15.82–29.12
K6M-soil	0	23.3	18.21–29.12	22.5	18.71–27.00

[a] The toxin was evaluated after both equilibrium adsorption and binding on the clays.

[b] The clays were montmorillonite (M) and kaolinite (K) with a mixed-cation complement (M-mix and K-mix) and the clay-size fraction separated from Kitchawan soil, unamended (K-soil), or amended to 6% (v/v) with montmorillonite (K6M-soil).

[c] Mortality at each clay concentration was not significantly different from the natural mortality of the controls.

[d] LC$_{50}$ of free toxin = 90.4 ng/100 µl with a 95% CI of 58.58 to 144.06 ng/100 µl.

Source: Tapp, H. and Stotzky, G., *Appl. Environ. Microbiol.*, 61, 1786, 1995. With permission.

did the toxin from *Btk*, indicating that the structure of these proteins differs sufficiently to result in different characteristics of adsorption on clays.

8.4   Detection of Bt-*toxins in soil*

A dot-blot ELISA method was developed to detect the toxins from *Btk* and *Btt* in soil, with a lower limit of detection of ~3 ng/g of soil (Tapp and Stotzky 1995a). Toxins added to nonsterile soil, unamended or amended with montmorillonite or kaolinite, were detected by this method on the clay-size fraction but not on the silt- and sand-size fractions on which the toxins do not appear to adsorb. The use of flow cytometry as a rapid and sensitive method with which to detect the toxins in soil was also developed (Tapp and Stotzky 1997). Currently, Lateral Flow Quickstix (EnviroLogix, Portland, ME), which are rapid (<10 min) Western blot detection systems with a lower detection limit of <10 parts per billion (10^9), are being used. The immunological determinations are confirmed by larvicidal assays (Saxena et al. 1999, Saxena and Stotzky 2000).

8.5   *Insecticidal activity of bound* Bt-*toxins*

The bound toxins retained insecticidal activity. The toxin from *Btk* or *Btt* bound on montmorillonite, kaolinite, and the clay-size fraction of soil was insecticidal to the larvae of the tobacco hornworm (*Manduca sexta*) (Table 8.3) and the Colorado potato beetle (*Leptinotarsa decemlineata*) (Table 8.4), respectively, in standard larvicidal assays wherein comparable amounts of purified adsorbed, bound, or free toxins were added to appropriate larval media (Tapp and Stotzky 1995b). Bound toxin from *Btk*, but not from *Btt*, had a higher toxicity (i.e., had lower LC_{50} values) than free toxin, possibly as the result of the toxin from *Btk* being concentrated on the clays, and the larvae ingested more clay-bound toxin than free toxin spread over the surface of the larval growth medium. The data are expressed as the apparent LC_{50} values (lethal concentration of toxin to kill 50% of a population of the assay larvae). Although the LC_{50} for a particular species of larva should theoretically not vary significantly, the modality of the actual assays reflects the amount of toxin that is present in a particular clay or soil system, which will vary with different clays and soils as well as with different batches of larvae. Consequently, the LC_{50} values in these studies indicate the amount of toxin present under specific conditions rather than the amount of toxin necessary to kill 50% of a larval population under standard assay conditions; e.g., with time, more toxin-containing soil was needed to kill 50% of the larvae, indicating a loss of toxicity, perhaps as the result of degradation of the toxin.

When free toxin from *Btk* was added to nonsterile soils in test tubes that had been maintained at minus 33-kPa water tension and 24 ± 2°C (i.e., under optimal conditions for microbial activity), insecticidal activity was detected after 234 days, the longest time evaluated, albeit with a reduction in activity during the incubation period (Table 8.6) (Tapp and Stotzky 1998). This persistence was considerably longer than persistences reported in the literature, which ranged in half-life from ~8 to 17 days for purified toxins and from ~2 to 41 days for biomass of transgenic corn, cotton, and potato (e.g., Palm et al. 1994, 1996, Sims and Holden 1996, Sims and Ream 1997). Insecticidal activity was greater and persisted longer in soil naturally containing or amended with kaolinite than in soils naturally containing or amended with montmorillonite (Table 8.6), possibly because montmorillonite-containing soils had a higher pH (5.8 to 7.3) and, therefore, more bacterial activity (Stotzky 1986), which may have resulted in greater biodegradation of the toxin. A decrease in the persistence of the toxin in kaolinite-containing soils adjusted to pH ~7 from pH 4.9 with $CaCO_3$ further indicated that pH is a factor in the persistence of this toxin in soil. The role of pH and other physicochemical and biological characteristics of

Table 8.4 Lethal Concentration (LC$_{50}$), 95% Confidence Interval (CI), and Relative Potency of the Toxin from *Bacillus thuringiensis* subsp. *tenebrionis*, Free, Adsorbed, or Bound on Clay Minerals or the Clay-Size Fraction from Soil, for Larvae of the Colorado Potato Beetle (*Leptinotarsa decemlineata*) after 3 Days of Exposure

| | | Clay-toxin complex[a] | | | | | |
| | | Adsorbed | | | Bound | | |
Clay[b]	Mortality with clay alone[c]	LC$_{50}$ (µg/50 µl)	95% CI	Relative potency[d]	LC$_{50}$ (µg/50 µl)	95% CI	Relative potency[d]
Batch 1[e]							
M-Na	0	1.5	1.07–2.05	0.86	17.8	12.93–24.57	0.07
M-K	0	1.7	1.13–2.52	0.77	13.4	8.95–20.18	0.10
M-Mg	0	1.2	0.81–1.71	1.05	9.5	6.51–13.43	0.13
M-La	0	2.6	1.80–3.56	0.52	96.1	40.69–264.32	0.01
M-mix	0	1.3	0.75–2.13	1.05	14.1	9.45–21.05	0.10
K-mix	0	4.5	3.05–6.61	0.28	1.4	0.95–1.91	0.91
K6M-soil	0	0.5	0.33–0.79	2.38	1.8	1.16–2.64	0.70
Batch 2[f]							
K-K	0	0.4	0.15–0.75	0.47	0.2	0.08–0.34	0.95
K-soil	0	0.3	0.16–0.38	0.88	2.3	1.30–4.40	0.10
K6K-soil	0	0.2	0.10–0.27	1.16	0.4	0.25–0.63	0.51

[a] The toxin was evaluated after both equilibrium adsorption and binding on the clays. Two different batches of *B. thuringiensis* subsp. *tenebrionis* toxin (purified from M-One) were used.

[b] The clays were montmorillonite (M) and kaolinite (K) with a mixed-cation complement (M-mix or K-mix) or homoionic to the indicated cations and the clay-size fraction separated from Kitchawan soil, unamended (K-soil), or amended to 6% (v/v) with montmorillonite (K6M-soil) or to 6% with kaolinite (K6K-soil).

[c] Mortality at each clay concentration was not significantly different from the natural mortality of the controls.

[d] Potency of adsorbed or bound toxin relative to the potency of free toxin at all concentrations.

[e] LC$_{50}$ of free toxin = 1.3 µg/50 µl with a 95% CI of 1.02 to 1.62 µg/50 µl.

[f] LC$_{50}$ of free toxin = 0.2 µg/50 µl with a 95% CI of 0.10 to 0.28 µg/50 µl.

Source: Tapp, H. and Stotzky, G., *Appl. Environ. Microbiol.*, 61, 1786, 1995. With permission.

Table 8.5   Interlayer Spacings (d_{001} value in nm) of Montmorillonite (M)
or Kaolinite (K) Homoionic to Different Cations and of Their Complexes
with the Protoxin or Toxin from *Bacillus thuringiensis* subsp. *kurstaki* (*Btk*)
and of the Toxin from *Bacillus thuringiensis* subsp. *tenebrionis* (*Btt*)

Clay	Clay alone	Clay-*Btt* complex	Clay-*Btk* complex Toxin	Protoxin
M-H	1.26 (1.19)[a]	2.63 (2.40)	ND[b]	ND
M-Na	1.25 (0.97)	2.57 (2.39)	1.28[c] (1.50)[d]	1.77 (1.69)
M-K	1.20 (1.01)	2.05 (1.94)	1.28 (1.22)	ND
M-Mg	1.50 (1.35)	2.05 (2.00)	RI[e] (RI)	ND
M-Ca	1.50 (0.99)	1.70 (1.63)	1.54 (1.44)	1.31 (1.26)
M-La	1.55 (1.24)	2.17 (2.08)	ND	ND
M-Al	1.51 (1.24)	1.51	1.78 (1.60)	ND
K-Na	0.72	0.71	0.72	0.72
K-Ca	0.72	0.71	0.71	0.72

[a] Values in brackets give d_{0001} spacings after heating at 110°C for monovalent cations or at 150°C for polyvalent cations.

[b] ND = not determined.

[c] Asymmetrical broad peak, suggesting the masking of peaks at higher spacings.

[d] Broad peak ranging from ~1.00 to 1.51 nm and indicating random interstratification.

[e] RI = random interstratification; tracings show a typical pattern of an interstratified clay, suggesting a random intercalation of the protein.

Source: Tapp, H. et al., *Soil Biol. Biochem.*, 26, 663, 1994. With permission from Elsevier Science.

soil in the persistence of toxins in soil must be further studied to explain the differences in persistence already reported.

8.6   Resistance of bound Bt-toxins to biodegradation

The binding of the toxins from *Btk* and *Btt* on clays reduced their availability to microbes, which is probably responsible for their persistence in soil (Koskella and Stotzky 1997). The free toxins were readily utilized as sources of carbon (Figure 8.4) and/or nitrogen by pure and mixed cultures of microbes, including soil suspensions, especially after the microbes had been "conditioned" (i.e., induction of proteases) on the toxins or other proteins. In contrast, the bound toxins were not utilized as sources of carbon, utilized only slightly as sources of nitrogen, and did not support growth in the absence of exogenous sources of both available carbon and nitrogen (Table 8.7). The utilization of the toxins was determined turbidimetrically (i.e., by measuring increases in protoplasm and cell numbers) and respirometrically (i.e., by measuring uptake of O_2, which reflects metabolic activity). After exposure of the bound toxins to microbes, *in vitro* and in soil, the toxins retained insecticidal activity, even after the soil was alternately frozen and thawed or wetted and dried for 40 days (wetting and drying did reduce the activity). Such treatments might have released toxins from surface-active particles and rendered them more susceptible to biodegradation (Table 8.8).

8.7   Insecticidal activity and resistance to biodegradation of Bt-toxins bound on humic acids

Results similar to those obtained with clays were observed when the toxin from *Btk* was reacted with humic acids prepared from four different soils and differing in total acidity (5.0

Table 8.6 Lethal Concentration (LC$_{50}$) and 95% Confidence Interval (CI) of the Toxin from *Bacillus thuringiensis* subsp. *kurstaki* after Incubation for 234 Days in Soil for Larvae of *Manduca sexta*

Days of incubation	Soil											
	MOP		K6M		K		K3K		K6K		K9K	
	LC$_{50}$[a]	95% CI	LC$_{50}$[a]	95% CI	LC$_{50}$[a]	95% CI	LC$_{50}$[a]	95% CI	LC$_{50}$[a]	95% CI	LC$_{50}$[a]	95% CI
0	53	42.6–65.4	71	NA[b]	34	25.7–43.3	26	19.9–34.3	16	11.6–21.5	17	11.6–24.0
0					55	31.4–97.0			10	4.0–19.1		
40	416	288.8–680.0	126	96.1–157.4	29	22.0–37.9	77	52.8–109.4	53	32.4–84.9	42	27.6–62.5
47					44	25.0–76.7			43	23.0–80.0		
71	677	517.9–940.0	494	NA	53	35.0–78.8	90	62.4–127.8	43	29.1–59.4	77	52.5–111.7
90					47	27.7–80.0			79	36.5–182.8		
112	ND[c]	ND	ND	ND	48	33.7–60.5	46	33.2–60.5	72	50.6–101.0	46	30.6–66.6
134			ND	ND	451	NA			101	55.0–201.5		
146	ND	ND	ND	ND	39	26.4–47.8	58	31.5–77.6	47	36.6–57.1	37	22.5–55.3
184					264	109.1–684.9			312	121.4–939.4		
195	ND	ND	ND	ND	63	NA	61	35.4–85.1	69	34.0–100.1	111	70.5–181.1
234					232	126.0–500.1			195	100.9–439.1		

Note: The soils were Mopala (MOP, a tropical soil in which montmorillonite is the dominant clay mineral) and Kitchawan (K, a soil from New York in which kaolinite is the dominant clay mineral), either unamended or amended to 6% (v/v) with montmorillonite (K6M) or to 3, 6, or 9% (v/v) with kaolinite (K3K, K6K, K9K). Values in brackets indicate the LC$_{50}$ and CI when the pH of the K and K6K soils was adjusted from 4.9 to approximately 7.0 with CaCO$_3$.

[a] LC$_{50}$ of free toxin = 91 ng 100/μl with a 95% CI of 62.0–132.4 ng 100/μl. There was no mortality with the soils alone. Two replicate tubes and 36 larvae per toxin concentration.

[b] NA = no data available; the 95% CI could not be calculated from the data.

[c] ND = not determined.

Source: Modified from Tapp, H. and Stotzky, G., *Soil Biol. Biochem.*, 30, 471, 1998. With permission from Elsevier Science.

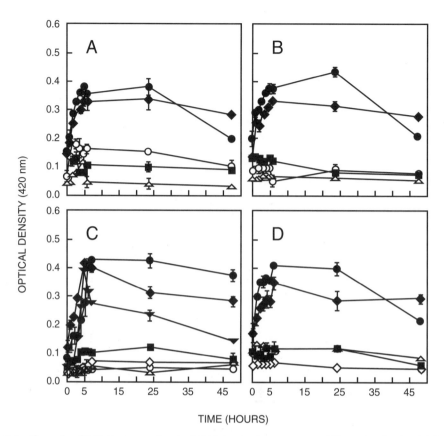

Figure 8.4 Growth of (A) *Proteus vulgaris*, (B) *Enterobacter aerogenes*, (C) a mixed microbial culture from a protein-enriched soil slurry, or (D) a mixed bacterial culture from a suspension of the protoxin from *Bacillus thuringiensis* subsp. *kurstaki* on pepsin (500 µg/ml) or the toxin from *B. thuringiensis* subsp. *kurstaki* (500 µg/ml), free or bound on montmorillonite homoionic to Ca. Means ± 1 standard error of the mean. ♦ = pepsin; ▼ = pepsin plus montmorillonite (1 mg/ml); ● = toxin; ■ = toxin bound on montmorillonite (1.4 mg/ml); ○ = control (pepsin, no microbes); ◇ = control (1 mg montmorillonite/ml, no microbes); △ = control (no toxin, no clay, microbes). (From Koskella, J. and Stotzky, G., *Appl. Environ. Microbiol.*, 63, 3561, 1997. With permission.)

to 7.4 cmol/kg), in the content of carboxyl and phenolic groups (3.2 to 4.2 and 1.0 to 3.4 cmol/kg, respectively), and in the degree of polymerization, as evaluated by the ratio of absorbance at 464 and 665 nm (E4:E6; 5.5 to 7.0): 75 to 80% of the toxin adsorbed at equilibrium was strongly bound, with the exception of the humic acids from soil on which sugar beets had been grown and which bound only ~43% of adsorbed toxin (Figure 8.5); the bound toxin was toxic to the larvae of *Manduca sexta*, with a LC_{50} lower than that of free toxin (Table 8.9); and the bound toxin did not support the growth of a mixed microbial culture from soil, although the free toxin was readily utilized as a source of carbon (Figure 8.6) (Crecchio and Stotzky 1998). When the toxin was bound on complexes of montmorillonite–humic acids–Al hydroxypolymers, which more closely approximate the presumed composition of surface-active particles in soil, similar results were obtained (Crecchio and Stotzky 2001).

8.8   Effect of free and bound Bt-toxins on microbes in vitro

The toxins from *Btk* and *Btt*, free or bound on clays, had no effect on the growth *in vitro* — as measured by standard disk-diffusion assays and dilution tests for bacteria (Harley

Table 8.7 Utilization of the Toxin from *Bacillus thuringiensis* subsp. *kurstaki*, Free or Bound on Homoionic Clay Minerals or the Clay-Size Fraction from Soil, as a Source of Carbon, Nitrogen, or Carbon plus Nitrogen by Various Microbial Cultures

Source of organisms	Clay or soil fraction bound[a]	Utilization of toxin as source of:[b]					
		Carbon		Nitrogen		Carbon + nitrogen	
		Free	Bound	Free	Bound[c]	Free	Bound
Mixed culture from a protein-enriched soil slurry[d]	M-Na	+	−	+	+	+	−
	M-Ca	+	−	ND[e]	ND	ND	ND
	M-Al	+	−	ND	ND	ND	ND
	K-Na	+	−	+	+	+	−
	K-Ca	+	−	+	+	+	−
	K-soil	+	−	+	+	+	−
	K6K-soil	+	−	+	+	+	−
	K6M-soil	+	−	+	+	+	−
Soil slurry[f]	M-Na	ND	ND	ND	ND	+	−
	K-Na	ND	ND	ND	ND	+	−
	K-soil	ND	ND	ND	ND	+	−
	K6K-soil	ND	ND	ND	ND	+	−
	K6M-soil	ND	ND	ND	ND	+	−

[a] K (kaolinite) or M (montmorillonite) homoionic to Na, Ca, or Al. See Table 8.4 for description of K, K6K, and K6M soils.

[b] + = utilization; − = no utilization (determined by measurement of O_2 uptake by the direct Warburg method).

[c] Utilization of bound toxin as a source of nitrogen was ~30% of the utilization of the free toxin.

[d] Mixed culture from a protein-enriched soil slurry incubated at 37°C in Davis citrate minimal medium for 50 h.

[e] ND = not determined.

[f] Mixed culture from a garden soil (1:4 [wt/vol] soil:tap water) amended with free pepsin or with free or bound toxin from *B. thuringiensis* subsp. *kurstaki* and incubated at 25°C for 7 h.

Source: Koskella, J. and Stotzky, G., *Appl. Environ. Microbiol.*, 63, 3561, 1997. With permission.

and Prescott 1999) and colony diameter and sporulation for fungi (Babich and Stotzky 1977, 1986) — of pure and mixed cultures of bacteria, both Gram-positive (*Bacillus subtilis, B. cereus, Btk, B. thuringiensis* subsp. *israelensis* (*Bti*), *Arthrobacter globiformis*) and Gram-negative (*Agrobacterium radiobacter, Pseudomonas aeruginosa, Proteus vulgaris, P. mirabilis, Escherichia coli, Enterobacter aerogenes, E. cloacae, Oscillatoria* sp.); fungi, both yeasts (*Saccharomyces cerevisiae, Candida albicans*) and filamentous forms (*Rhizopus nigricans, Cunninghamella elegans, Aspergillus niger, Fusarium solani, Penicillium* sp.); and algae, primarily green (*Chlamydomonas* sp., *Oedogonium* sp., *Euglena* sp.) and diatoms (Koskella and Stotzky, unpublished data). Although these results were not unexpected, they differed from those indicated by Yudina and co-workers (1990, 1996, 1997), who suggested that the toxins from *Bt* have antibiotic properties.

8.9   Binding of antidipteran Bt-toxins on clays

Preliminary studies with the toxins from *Bti* (25 to 130 kDa; active against Diptera) indicated that they also bind on montmorillonite and kaolinite and that the bound toxins retain activity against the larvae of mosquito (*Culex pipiens*) (Stotzky, unpublished data). Further studies with the toxins from *Bti* are in progress, as the larvae of mosquitoes, as well as of some other dipterans, are filter-feeders. Hence, toxins bound on clays and humic substances may be more effective for the control of mosquitoes and other undesirable dipterans and pose less risk to the environment than the use of transgenic *Bt*-cyanobacteria, which could transfer the toxin genes to other bacteria (see Stotzky 1989, Entwistle et al. 1993).

Table 8.8   Lethal Concentration (LC$_{50}$) and Relative Potency (RP) of the Toxin from *Bacillus thuringiensis* subsp. *kurstaki*, Free or Bound on Montmorillonite (M-Na) or Kaolinite (K-Na) Homoionic to Na, for Larvae of *Manduca sexta* after 40 days of Incubation in Soil in Test Tubes Maintained at Room Temperature and the –33-kPa Water Tension, Alternately Wetted and Dried, or Alternately Frozen and Thawed

| Treatment | LC$_{50}$ (µl) ± SEM[a] | | | RP of toxin bound on:[c] | |
| | | Toxin bound on:[b] | | | |
	Free toxin	M-Na	K-Na	M-Na	K-Na
Stock[d]	0.7 ± 0.01	0.7 ± 0.01	0.5 ± 0.00	1.0	1.4
Room temperature[e]	5.5 ± 1.55	3.0 ± 0.90	3.9 ± 1.00	1.8	1.4
Air-dried and rewetted[f]	11.2 ± 3.30	17.2 ± 5.36	17.0 ± 1.54	0.7	0.7
Frozen and thawed[g]	24.0 ± 2.25	13.4 ± 0.34	7.6 ± 0.70	1.8	3.2

Note: Data are expressed as the mean lethal concentration (LC$_{50}$) ± the standard errors of the means (SEM).

[a] LC$_{50}$ was measured after incubation of the toxin for 40 days in K-soil (100 µg toxin/g of soil) continuously at the –33-kPa water tension and room temperature, which was then alternately air-dried and rewetted or alternately frozen and thawed.

[b] M-Na, montmorillonite homoionic to sodium; K-Na, kaolinite homoionic to sodium.

[c] RP = relative potency: potency of bound toxin relative to potency of free toxin.

[d] Free toxin was stored as aliquots at –20°C, and bound toxin was stored as a clay-toxin pellet at 4°C; both were assayed for their LC$_{50}$ immediately after purification or binding on clay minerals.

[e] Maintained at room temperature at the –33-kPa water tension.

[f] Maintained at room temperature at the –33-kPa water tension for 7 days, air-dried for 7 days, and rewetted and maintained at the –33-kPa water tension for 7 days. This cycle was repeated twice.

[g] Maintained at room temperature at the –33-kPa water tension for 7 days and then at –20°C for 7 days. This cycle was repeated three times.

Source: Koskella, J. and Stotzky, G., *Appl. Environ. Microbiol.*, 63, 3561, 1997. With permission.

8.10   Biodegradation of Bt-corn in soil

The addition of biomass from transgenic corn expressing the Cry1Ab toxin from *Btk* resulted in a significantly lower gross metabolic activity (i.e., CO$_2$ evolution) of soil than did the addition of nontransgenic biomass (i.e., same variety but without the toxin gene) (Figure 8.7) (Flores, Saxena, and Stotzky, unpublished data). Soil from the Kitchawan Research Laboratory of the Brooklyn Botanic Garden augmented to 9% (vol/vol) with montmorillonite (see below) was amended to 0.5, 1, or 2% (wt/wt) with ground, air-dried leaves or stems of *Bt*-corn (NK6800Bt) or of the isogenic variety without the *cry1Ab* gene (the particle size distribution of the ground material was 70% <0.5 mm and 30% <1 mm). Subsamples of amended or unamended soil (25 or 50 g oven-dry equivalent, depending on the experiment) at minus 33-kPa water tension were placed in small jars (90-ml capacity), and 8 to 10 jars containing soil of each treatment were placed in individual 1-l "master" jars, attached to a respiratory train and incubated at 25 ± 2°C (Stotzky et al. 1993). The gross metabolic activity of the soil was determined by CO$_2$ evolution: CO$_2$ was trapped in NaOH, precipitated with BaCl$_2$, and the unneutralized NaOH titrated with HCl using an automatic titrator. Subsamples of soil, in the small jars, were removed periodically from the master jars, and the activities of proteases, dehydrogenases, alkaline and acid phosphatases, and arylsulfatases, as well as the numbers of total culturable bacteria and fungi, were measured (Stotzky et al. 1993). At the end of each incubation, the insecticidal activity of the soil was determined by bioassay using the larvae of *Manduca sexta* (Tapp and Stotzky 1998). There were three replicates of each treatment per experiment.

The amounts of C evolved as CO$_2$ increased in proportion to the amounts of biomass added and were greater than the amounts evolved from the unamended control soil.

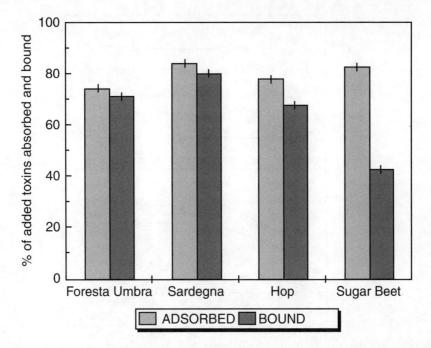

Figure 8.5 Percentage of added toxin from *Bacillus thuringiensis* subsp. *kurstaki* (140 µg) adsorbed or bound (after five washings with double-distilled water) on humic acids (200 µg) from four soils. Means ± SEMs. (Modified from Crecchio, C. and Stotzky, G., *Soil Biol. Biochem.*, 30, 463, 1998. With permission from Elsevier Science.)

Table 8.9 Lethal Concentration (LC_{50}) and 95% Confidence Interval (CI) of the Toxin from *Bacillus thuringiensis* subsp. *kurstaki*, Free or Bound on Humic Acids from Four Soils, for the Larvae of *Manduca sexta*

Source of humic acids (soil)	Mortality with humic acids alone[a]	Humic acid–toxin complex	
		LC_{50} (ng 100/µl)[b]	95% CI[c]
Sardegna	0	233	192.1–280.8
Foresta Umbra	0	254	209.4–307.5
Hop	0	272	225.6–326.5
Sugar beet	0	215	177.4–259.7

[a] Concentration of humic acids at 10× the concentration present in the humic acid-toxin complexes.

[b] LC_{50} of free toxin = 304 ng 100/µl with a 95% CI of 251.1 to 367.5 ng 100/µl.

[c] CI = confidence interval.

Source: Crecchio, C. and Stotzky, G., *Soil Biol. Biochem.*, 30, 463, 1998. With permission from Elsevier Science.

However, the amounts evolved were significantly lower throughout the incubations from soil amended with biomass of *Bt*-corn than from soil amended with biomass of non-*Bt*-corn (Figure 8.7). This difference occurred with stems and leaves from two separate batches of corn tissue, even when glucose was added with the tissue. Changes in the C:N ratio of the soil–biomass systems by the addition of glucose, NH_4NO_3, or glucose plus NH_4NO_3 did not alter the relative differences in CO_2 evolution between soil amended with biomass of *Bt*-corn and soil amended with biomass of non-*Bt*-corn.

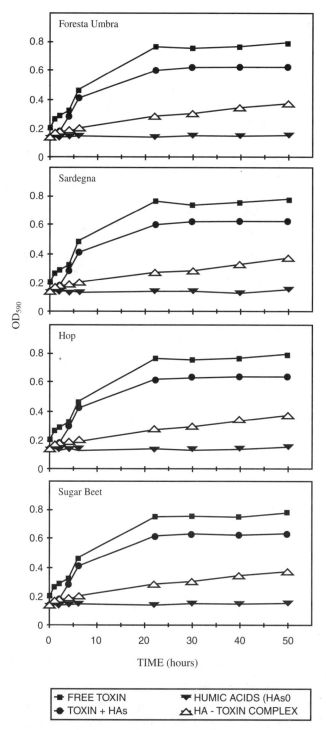

Figure 8.6   Utilization of the toxin from *Bacillus thuringiensis* subsp. *kurstaki*, free or bound on humic acids (HAs) from four soils, as a source of carbon and energy by a mixed microbial culture from soil. Means ± SEMs. (From Crecchio, C. and Stotzky, G., *Soil Biol. Biochem.*, 30, 463, 1998. With permission from Elsevier Science.)

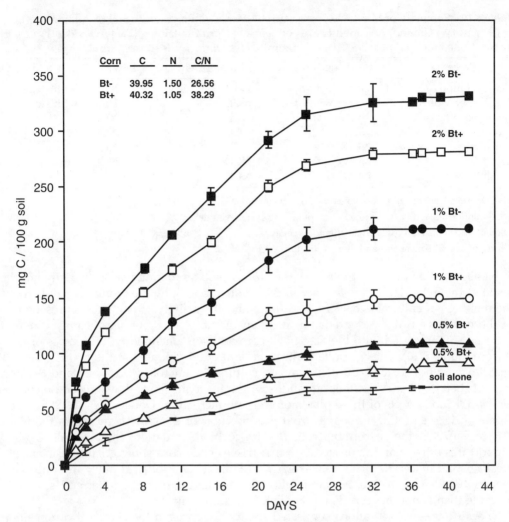

Figure 8.7   Gross metabolic activity (cumulative CO_2 evolution) of Kitchawan soil amended to 6% (v/v) with montmorillonite and with 0.5, 1, or 2% (w/w) ground, air-dried leaves of corn containing (*Bt*+) (NK6800Bt) or not containing (*Bt*–) the *cry1Ab* gene from *Bacillus thuringiensis* subsp. *kurstaki*. Soil was incubated at the –33-kPa water tension and 25 ± 1°C. Means ± SEMs. (From Flores, S., Saxena, D., and Stotzky, G., unpublished data.)

The activities of the enzymes and the numbers of culturable bacteria and fungi fluctuated throughout the incubations and differed with the various treatments, but there were no consistent statistically significant differences in activities and numbers between soil amended with biomass of *Bt*-corn and soil amended with biomass of non-*Bt*-corn (Table 8.10). All soil samples amended with biomass of *Bt*-corn were lethal to the larvae of *M. sexta* (LC_{50} values ranged from 0.27 to 0.59 mg of biomass, with confidence intervals of 0.144 to 0.495 and 0.351 to 1.070 mg of biomass, respectively). There was no mortality with soil amended with biomass of non-*Bt*-corn or with soil that was not amended.

The reasons for the lower biodegradation of the biomass of *Bt*-corn than of the biomass of non-*Bt*-corn are not known. It was not the result of differences in the C:N ratios of the biomass, as leaf and stem tissue of the second batch of both *Bt*-corn and non-*Bt*-corn had similar C:N ratios, and changes in the ratios, as well as the addition of an available carbon and energy source in the form of glucose, did not significantly alter the relative differences

Table 8.10 Summary of Effects on Some Microbe-Mediated Processes in Soil of Adding 0.5, 1.0, or 2.0% (w/w) Ground, Air-Dried Leaves or Stems of Corn Containing (*Bt+*) (NK6800*Bt*) or Not Containing (*Bt–*) the *cry1Ab* Gene from *Bacillus thuringiensis* subsp. *kurstaki*

Process	Among leaves	Among stems	Between leaves and stems
CO_2 evolution	Lower with *Bt+*	Lower with *Bt+*	Higher with stems
Alkaline phosphatase	Lower with *Bt+*	NCD[a]	Higher with stems
Acid phosphatase	Higher with *Bt+*	NCD	NCD
Protease	NCD	NCD	NCD
Dehydrogenase	NCD	NCD	NCD
Bacteria[b]	NCD	NCD	NCD
Fungi[c]	NCD	NCD	NCD

[a] NCD = no consistent differences over a 42-day incubation.
[b] Soil dilutions plated on soil extract agar containing cycloheximide.
[c] Soil dilutions plate on rose bengal-streptomycin agar.
Source: Flores, S., Saxena, D., and Stotzky, G., unpublished data.

in biodegradation between biomasses. The lower biodegradation was apparently not the result of the inhibition of the activity of the soil microbiota by the biomass of *Bt*-corn, as the numbers of culturable bacteria and fungi and the activity of enzymes representative of those involved in the degradation of plant biomass were not significantly different between soil amended with biomass of *Bt*-corn and soil amended with biomass of non-*Bt*-corn (Table 8.10). These results confirm *in vitro* observations that the toxins from *Btk* and *Btt* were not toxic to pure and mixed cultures of microbes (Koskella and Stotzky, unpublished data) and *in situ* observations with transgenic plants that showed no consistent and lasting effects of these plants on the soil microbiota (Donegan et al. 1995, 1996). Although it is tempting to suggest that the insertion of the *cry1Ab* gene into the plant genome affected the susceptibility of *Bt*-corn to biodegradation, there are no data to support this other than the observation that tissues of *Bt*-corn show greater resistance to breakage and maceration (Flores, Saxena, and Stotzky, unpublished data) and anecdotal reports that *Bt*-corn has greater standability (i.e., less lodging) and is preferred less as feed by cattle than is non-*Bt*-corn (P. Porter, personal communication).

Because these observations suggested some differences in the chemical composition between *Bt* and non-*Bt*-corn, the lignin content of ten different *Bt*-corn hybrids, representing three transformation events, and of their respective non-*Bt*-isolines, grown both in a plant-growth room and in the field, was evaluated. Uniform freehand sections of fresh stems of corn, harvested after tasseling and ear production and of the same age, from between the third and fourth node (thickness ~11 mm) were examined for lignin by fluorescence microscopy at 400 nm (Hu et al. 1999). A higher content of lignin was observed in the vascular bundle sheath and in the sclerenchyma cells surrounding the vascular bundle of all *Bt*-corn hybrids than of their respective non-*Bt*-isolines, which was confirmed by staining the sections with toluidine blue (Sylvester and Ruzin 1994). The average diameter of the vascular bundle and surrounding lignified cells in *Bt*-corn was 21.5 ± 0.84 µm, whereas that of non-*Bt*-corn was 12.4 ± 1.14 µm (Saxena and Stotzky, in press a).

The content of lignin of the same portion of the stems (oven-dried, ground, and passed through an 80-mesh sieve) was determined by the acetyl bromide method (Hatfield et al. 1999). The lignin content of all hybrids of *Bt*-corn, whether grown in the plant-growth room (Table 8.11) or in the field (Table 8.12), was significantly higher (33 to 97% higher) than that of their respective non-*Bt*-isolines. The lignin content of field-grown plants was higher than that of plants grown in the plant-growth room, which were smaller. There was a significantly higher lignin content ($P < 0.002$) in plants transformed by event Bt11

Table 8.11 Lignin Content of Different Hybrids of Corn with (*Bt*+) and without (*Bt*–) the *cry1Ab* Gene Grown in a Plant Growth Room

Company	Hybrid	Event	% Lignin	Hybrid	% Lignin	P
		Bt+		*Bt*–		
Novartis	N7590Bt	Bt11	7.2 ± 0.1	N7590	4.8 ± 0.14	0.00001
Novartis	N67-T4	Bt11	6.3 ± 0.25	N67-H6	3.9 ± 0.15	0.00175
Novartis	N3030Bt	Bt11	7.0 ± 0.22	N3030	4.4 ± 0.22	0.00003
Novartis	NC4990Bt	Bt11	6.6 ± 0.18	NC4880	3.4 ± 0.27	0.00020
Novartis	NK4640Bt	Bt11	6.3 ± 0.14	NK4640	3.2 ± 0.12	0.00001
Novartis	Maximizer	176	4.0 ± 0.15	—	—	—
Pioneer	P31B13	MON810	6.0 ± 0.24	P3223	3.2 ± 0.18	0.00032
DeKalb	DK647Bty	MON810	6.2 ± 0.25	DK647	4.4 ± 0.22	0.00174
DeKalb	DK679Bty	MON810	6.6 ± 0.11	DK679	3.8 ± 0.10	0.00005
DeKalb	DK626Bty	MON810	6.1 ± 0.20	DK626	3.2 ± 0.18	0.00006

Source: Saxena, D. and Stotzky, G., *Am. J. Bot.*, in press a. With permission.

Table 8.12 Lignin Content of Different Hybrids of Corn with (*Bt*+) and without (*Bt*–) the *cry1Ab* Gene Grown in the Field

Company	Hybrid	Event	% Lignin	Hybrid	% Lignin	P
		Bt+		*Bt*–		
Novartis	N7590Bt	Bt11	7.4 ± 0.15	N7590	5.2 ± 0.11	0.00007
Novartis	N67-T4	Bt11	7.1 ± 0.22	N67-H6	4.5 ± 0.14	0.00082
Novartis	NC4990Bt	Bt11	7.7 ± 0.06	NC4880	4.8 ± 0.22	0.00033
Novartis	NK4640Bt	Bt11	7.9 ± 0.10	NK4640	4.9 ± 0.09	0.00001
Novartis	Maximizer	176	5.9 ± 0.13	—	—	—
Pioneer	P32P76	MON810	6.8 ± 0.14	P32P75	4.8 ± 0.22	0.00016
Pioneer	P31B13	MON810	7.0 ± 0.04	P3223	4.9 ± 0.13	0.00007
DeKalb	DK626Bty	MON810	6.8 ± 0.15	DK626	5.1 ± 0.12	0.00085

Source: Saxena, D. and Stotzky, G., *Am. J. Bot.*, in press a. With permission.

(7.4 ± 0.10 and 6.7 ± 0.12% for field- and growth room-grown plants, respectively) than by event MON810 (6.9 ± 0.07 and 6.2 ± 0.10% for field- and growth room-grown plants, respectively). There were no significant differences in the lignin content of isogenic non-*Bt*-plants ($P > 0.67$ for field-grown plants and $P > 0.30$ for growth room–grown plants). The lignin content of the only available hybrid transformed by event 176 was lower than that of hybrids transformed by events Bt11 and MON810. These results differ from those reported by Faust (1999), which indicated no significant differences in lignin content between the dried biomass of whole plants of *Bt* (event MON810) and non-*Bt*-corn but which indicated that *Bt*-corn had a higher moisture content and a lower level of ammonia than non-*Bt*-corn ($P < 0.05$). However, Masoero et al. (1999) reported a 16% higher lignin content in *Bt*-corn than in non-*Bt*-corn.

Lignin is a major structural component of plant cells that confers strength, rigidity, and impermeability to water. Any modifications in lignin content could result in effects that may have ecological implications (Halpin et al. 1994). For example, the increase in lignin content in *Bt*-corn may be beneficial, as it can provide greater resistance to attack by second-generation European corn borer (Ostrander and Coors 1997), reduce suscepti- bility to molds (Masoero et al. 1999), and retard litter degradation and decomposition by microbes (Reddy 1984, Tovar-Gomez et al. 1997), as also indicated in our studies by the lower evolution of CO_2 from soils amended with biomass of *Bt*-corn than with biomass of non-*Bt*-corn. The lower degradation of the biomass of *Bt*-corn may be beneficial, as the

Table 8.13   Presence of Toxin in Root Exudates from Corn with (*Bt+*) (NK4640*Bt*) and without (*Bt–*) the *cry1Ab* Gene

	Day assayed after germination of seed[a]											
	7				15				25			
	Immuno. test[b]		Toxicity (LC$_{50}$)[c]		Immuno. test		Toxicity (LC$_{50}$)		Immuno. test		Toxicity (LC$_{50}$)	
Growth conditions	*Bt–*	*Bt+*	*Bt–*	*Bt+*	*Bt–*	*Bt+*	*Bt–*	*Bt+*	*Bt–*	*Bt+*	*Bt–*	*Bt+*
Hoagland's solution	–	+	NT[d]	5.6 µg (2.61–7.82)	–	+	NT	5.2 µg (2.81–8.21)	–	–	NT	NT
Soil	–	+	NT	2.3 µg (1.32–3.86)	–	+	NT	1.8 µg (0.96–3.40)	–	+	NT	1.6 µg (0.86–2.62)

[a] Assayed immunologically and by larvicidal assay (*Manduca sexta*) on indicated days after germination of seeds, which was usually 3 days after planting.

[b] Determined with Lateral Flow Quickstix: – = no toxin detected; + = toxin detected.

[c] Toxicity [concentration necessary to kill 50% of larvae (LC$_{50}$)] expressed in µg of total protein/bioassay vial and 95% confidence interval (in parentheses).

[d] NT = No toxicity.

Source: Modified from Saxena, D. and Stotzky, G., *FEMS Microbiol. Ecol.*, 33, 35, 2000. With permission from Elsevier Science.

organic matter derived from *Bt*-corn may persist and accumulate longer and at higher levels in soil, thereby improving soil structure and reducing erosion.

By contrast, the longer persistence of the biomass of *Bt*-corn may extend the time that the toxin is present in soil and, thereby, enhance the hazard to nontarget organisms and the selection of toxin-resistant target insects. Additional studies are needed to clarify the impacts of the lower degradation of the biomass of *Bt*-corn, especially as about 15 million acres of *Bt*-corn, approximately 20% of the total acreage of corn, were planted in the United States in 1998 (Wadman 1999).

8.11   *Release, binding, persistence, and insecticidal activity of* Bt-*toxin in root exudates of* Bt-*corn*

The Cry1Ab protein was found in root exudates from transgenic *Bt*-corn (NK4640Bt) grown in sterile hydroponic culture and in sterile and nonsterile soil in a plant-growth room (Saxena et al. 1999) (Table 8.13). The presence of the toxin was indicated by a major band migrating on SDS-PAGE to a position corresponding to a M_r of 66 kDa, the same position as that of the Cry1Ab protein; toxin presence was confirmed by immunological and larvicidal assays. After 25 days, when the hydroponic culture was no longer sterile, the band at 66 kDa was not detected (there were several new protein bands of smaller M_r) and the immunological and larvicidal assays were negative, indicating that microbial proteases had hydrolyzed the toxin. By contrast, the toxin was detected after 25 days in both sterile and nonsterile soil indicating that the released toxin was bound on surface-active particles in rhizosphere soil and was protected from hydrolysis. Similar results were observed with purified toxins (see above).

To verify these results and to estimate the importance of the clay mineralogy and other physicochemical characteristics, which are known to influence the activity and ecology of microbes in soil (Stotzky 1974, 1986, 1997), on the persistence of toxin released in root exudates from *Bt*-corn, studies were performed in a plant-growth room in soil amended with various concentrations of montmorillonite or kaolinite. Kitchawan soil, a sandy loam that naturally contains predominantly kaolinite, was collected at the Kitchawan

Research Laboratory of the Brooklyn Botanical Garden, Ossining, NY, and either not amended, control (C), or amended to 3, 6, 9, or 12% (vol/vol) with montmorillonite (3, 6, 9, and 12M soil) or kaolinite (3, 6, 9, and 12K soil). These stable soil–clay mixtures have been used extensively in studies on the effects of physicochemical and biological factors of soil on the activity, ecology, and population dynamics of microbes and viruses, on gene transfer among bacteria, on mediating the toxicity of heavy metals and other pollutants, and on the persistence of the toxins from *Btk* and *Btt* in soil. Therefore, there is a large database on these mixtures (e.g., Babich and Stotzky 1977, Stotzky et al. 1993, Tapp and Stotzky 1995b, 1998).

Seeds of *Bt*-corn (NK4640Bt) and of the isogenic variety without the *cry1Ab* gene were planted in test tubes containing 15 g of nonsterile soil amended or not amended with montmorillonite or kaolinite and in a sandy loam soil in the field. After 10, 20, 30, and 40 days of growth in a plant-growth room (26 ± 2°C, 12-h light–dark cycle), soil from randomly selected tubes (two tubes each of *Bt*-corn and non-*Bt*-corn for each soil–clay mixture) was analyzed by immunological and larvicidal assays. Rhizosphere soil from the field-grown plants (nonirrigated, nonfertilized) was similarly analyzed after the production of ears of corn (six plants each of *Bt*-corn and non-*Bt*-corn) and several months after the death of the plants and frost.

All samples of rhizosphere soil from plants of *Bt*-corn grown in the plant-growth room were positive 10, 20, 30, and 40 days after germination for the presence of the toxin when assayed immunologically with Lateral Flow Quickstix (Saxena and Stotzky 2000). No toxin was detected in any soil with plants of non-*Bt*-corn or without plants. All samples of soil in which *Bt*-corn was grown were toxic to the larvae of *Manduca sexta*, with mortality ranging from 25 to 100% on day 10 and increasing to 88 to 100% on day 40, whereas there was no mortality with any soil from non-*Bt*-plants and with soil without plants. In addition, the size and weight of surviving larvae exposed to soils from *Bt*-corn were significantly lower (~50 to 92% lower) than those exposed to soil from non-*Bt*-corn or to soil without plants, and these larvae usually died 2 to 3 days after weighing (Table 8.14). The larvicidal activity was generally higher in soil amended with montmorillonite than with kaolinite, probably because montmorillonite, a swelling 2:1, Si:Al, clay mineral with a significantly higher cation-exchange capacity and greater specific surface area than kaolinite, a nonswelling 1:1, Si:Al, clay, bound more toxin in the root exudates than did kaolinite. Similar results have been observed with pure toxin (e.g., Venkateswerlu and Stotzky 1992, Tapp et al. 1994, see above). Nevertheless, mortality in the montmorillonite and kaolinite soils was essentially the same after 40 days (see Table 8.14), indicating that over a longer time, the persistence of larvicidal activity may be independent of the clay mineralogy and other physicochemical characteristics of the soils. However, the increase in larvicidal activity between day 10 and day 40 indicated that the toxin in the root exudates was concentrated when adsorbed on surface-active components of the soils.

The immunological and larvicidal assays of soil from the rhizosphere of *Bt*-corn grown in the field were also positive, even in soil collected after the first frost from plants that had been dead for several months, whereas they were negative for non-*Bt*-corn (Table 8.15). Although larval mortality in rhizosphere soil from some plants of field-grown *Bt*-corn was only 38% and the coefficients of variation were large, the size and weight of the surviving *M. sexta* larvae were reduced by 40 to 50% when compared with those associated with soil from non-*Bt*-corn or from soil without plants. Moreover, most larvae from *Bt*-corn soils died a few days after being weighed.

To determine whether the release of the Cry1Ab protein is a common phenomenon with transgenic *Bt*-corn and is not restricted to the NK4640Bt hybrid, the release of the protein in the exudates of 12 additional *Bt*-hybrids, representing three different transformation events, and of their isogenic nontransgenic counterparts was studied with plants

Table 8.14 Larvicidal Activity, Expressed as % Mortality and Mean Weight (in g) of a Single Larva ± Standard Error of the Mean (in parentheses), of Rhizosphere Soil from Corn with (Bt+) (NK4640Bt) and without (Bt–) the *cry1Ab* Gene Grown in a Plant-Growth Room

Soil[b]	Day assayed after germination of seed[a]							
	10		20		30		40	
	Bt–	Bt+	Bt–	Bt+	Bt–	Bt+	Bt–	Bt+
C	0 (1.0 ± 0.05)	50 ± 10.2 (0.4 ± 0.08)	0 (1.1 ± 0.02)	50 ± 10.2 (0.2 ± 0.02)	0 (1.2 ± 0.02)	63 ± 7.2 (0.1 ± 0.01)	0 (0.9 ± 0.06)	100 ± 0.0 —
3K	0 (1.0 ± 0.04)	25 ± 10.2 (0.5 ± 0.07)	0 (1.1 ± 0.03)	38 ± 7.2 (0.2 ± 0.01)	0 (1.2 ± 0.08)	38 ± 7.2 (0.1 ± 0.01)	0 (1.0 ± 0.02)	88 ± 7.2 (0.1 ± 0.02)
6K	0 (1.0 ± 0.03)	25 ± 10.2 (0.4 ± 0.04)	0 (1.1 ± 0.08)	38 ± 7.2 (0.2 ± 0.02)	0 (1.1 ± 0.03)	50 ± 10.2 (0.1 ± 0.02)	0 (0.8 ± 0.06)	88 ± 7.2 (0.1 ± 0.01)
9K	0 (1.0 ± 0.04)	75 ± 10.2 (0.1 ± 0.04)	0 (1.0 ± 0.02)	75 ± 10.2 (0.2 ± 0.03)	0 (1.2 ± 0.04)	88 ± 7.2 (0.1 ± 0.01)	0 (0.9 ± 0.02)	100 ± 0.0 —
12K	0 (0.8 ± 0.06)	50 ± 10.2 (0.3 ± 0.01)	0 (1.1 ± 0.05)	63 ± 12.5 (0.2 ± 0.02)	0 (1.2 ± 0.03)	88 ± 7.2 (0.1 ± 0.03)	0 (0.9 ± 0.05)	100 ± 0.0 —
3M	0 (1.0 ± 0.04)	38 ± 7.2 (0.5 ± 0.02)	0 (1.0 ± 0.04)	50 ± 10.2 (0.2 ± 0.04)	0 (1.2 ± 0.02)	63 ± 12.5 (0.1 ± 0.01)	0 (0.9 ± 0.05)	100 ± 0.0 —
6M	0 (0.8 ± 0.01)	100 ± 0.0 —	0 (1.1 ± 0.04)	100 ± 0.0 —	0 (1.1 ± 0.05)	100 ± 0.0 —	0 (0.9 ± 0.03)	100 ± 0.0 —
9M	0 (1.0 ± 0.02)	100 ± 0.0 —	0 (1.0 ± 0.03)	100 ± 0.0 —	0 (1.2 ± 0.01)	100 ± 0.0 —	0 (1.0 ± 0.04)	100 ± 0.0 —
12M	0 (1.0 ± 0.01)	50 ± 10.2 (0.8 ± 0.04)	0 (1.0 ± 0.05)	100 ± 0.0 —	0 (1.0 ± 0.01)	100 ± 0.0 —	0 (1.0 ± 0.02)	100 ± 0.0 —

Note: Determined with the larvae of the tobacco hornworm (*Manduca sexta*); at least 16 larvae/assay; weight data normalized to one larva/treatment. No mortality with soils without plants (weight of a single larva: 0.8 to 1.3 ± 0.02 g).

[a] Usually 3 days after planting.

[b] Soil amended to 3, 6, 9, or 12 % (vol/vol) with kaolinite (K) or montmorillonite (M) or not amended with clay (C).

Source: Saxena, D. and Stotzky, G., *FEMS Microbiol. Ecol.*, 33, 35, 2000. With permission from Elsevier Science.

Table 8.15 Larvicidal Activity, Expressed as % Mortality and
Mean Weight (in g) of a Single Larva ± Standard Error of the Mean
(in parentheses), of Rhizosphere Soil from Corn with (*Bt*+) (NK4640*Bt*)
and without (*Bt*–) the *cry1Ab* Gene Grown *in situ* under Field Conditions

Plant	% Mortality[a]	
	Bt– corn	*Bt*+ corn
Before frost		
F1	0	38 ± 7.2
	(0.9 ± 0.01)	(0.3 ± 0.02)
F2	0	100 ± 0.0
	(0.9 ± 0.03)	—
F3	0	38 ± 12.5
	(1.1 ± 0.03)	(0.5 ± 0.09)
F4	0	100 ± 0.0
	(1.0 ± 0.02)	—
F5	0	38 ± 7.2
	(1.0 ± 0.03)	(0.6 ± 0.01)
F6	0	38 ± 7.2
	(1.0 ± 0.02)	(0.5 ± 0.02)
After frost		
F7	0	75 ± 10.2
	(1.0 ± 0.07)	(0.1 ± 0.01)
F8	ND[b]	88 ± 12.5
		(0.03 ± 0.00)

Note: Immunological tests with Lateral Flow Quickstix showed the presence
of toxin in the rhizosphere soil of all *Bt*+ plants but not of *Bt*– plants.

[a] Determined with the larvae of the tobacco hornworm (*Manduca sexta*); at least
16 larvae/assay; weight data normalized to one larva/treatment. No mortality
with soil without plants (weight of a single larva: 0.8 to 1.3 ± 0.02 g).

[b] Not determined.

Source: Saxena, D. and Stotzky, G., *FEMS Microbiol. Ecol.*, 33, 35, 2000. With
permission from Elsevier Science.

grown in the plant-growth room and in the field. In addition, the persistence of the protein
released from all *Bt*-hybrids in rhizosphere soil was evaluated.

All samples of rhizosphere soil from the 12 hybrids grown in the plant-growth room
were positive 40 days after germination for the presence of the toxin when assayed immunologically with Quickstix, and all samples were toxic to the larvae of *M. sexta*, with mortality
ranging from 38 to 100% (Table 8.16). No toxin was detected immunologically or by
larvicidal assay in any soil in which plants of non-*Bt*-corn or no plants had been grown
(Saxena et al., in press). In addition, the weight of surviving larvae exposed to soils from
Bt-corn was significantly lower (80 to 90%) than those exposed to soils from non-*Bt*-corn
or without plants, and these larvae usually died after an additional 2 to 3 days.

The immunological and larvicidal assays of soil from the rhizosphere of all *Bt*-hybrids
grown in the field and harvested after the production of ears of corn were also positive,
whereas they were negative for all non-*Bt*-corn isolines (Table 8.17). Although the larval
mortality in rhizosphere soil from some plants of field-grown *Bt*-corn was only 37%, the
weight of the surviving larvae was reduced by 85 to 98% when compared with soil from
non-*Bt*-corn or without plants, and most of these larvae died after a few more days. There
were no discernible or consistent differences in exudation of the toxin (as evaluated by
mortality, weight of surviving larvae, or immunology) between plants derived from

Table 8.16   Presence of the Cry1Ab Toxin in Rhizosphere Soil of Different Hybrids of Corn with (*Bt*+) and without (*Bt*-) the *cry1Ab* Gene Grown in a Plant-Growth Room

Company	Hybrid	Event	Immuno. test[a]	% Mortality[b]	Hybrid	Immuno. test	% Mortality
			Bt+			*Bt–*	
Novartis	N7590Bt	Bt11	+	93 ± 6.3 (0.03 ± 0.01)	N7590	–	0 (0.8 ± 0.03)
Novartis	N67-T4	Bt11	+	81 + 6.3 (0.05 ± 0.02)	N67-H6	–	0 (0.8 ± 0.02)
Novartis	N3030Bt	Bt11	+	38 ± 7.2 (0.15 ± 0.02)	N3030	–	6.3 ± 6.25 (0.8 ± 0.03)
Novartis	NC4990Bt	Bt11	+	81 ± 6.3 (0.11 ±0.09)	NC4880	–	0 (0.9 ± 0.04)
Novartis	NK4640Bt	Bt11	+	100 ± 0.0 (—)	NK4640	–	0 (1.1 ± 0.05)
Novartis	Maximizer	176	+	43 ± 11.9 (0.08 ± 0.02)	—	—	—
Pioneer	P32P76	MON810	+	81 ± 6.3 (0.07 ± 0.05)	P32P75	–	0 (0.8 ± 0.02)
Pioneer	P33B51	MON810	+	93 ± 6.3 (0.02 ± 0.01)	P33B50	–	0 (0.9 ± 0.06)
Pioneer	P31B13	MON810	+	50 ± 10.2 (0.07 ± 0.02)	P3223	–	0 (0.9 ± 0.04)
DeKalb	DK647Bty	MON810	+	56 ± 11.9 (0.10 ± 0.07)	DK647	–	6.3 ± 6.25 (0.9 ± 0.04)
DeKalb	DK679Bty	MON810	+	68 ± 18.8 (0.14 ± 0.02)	DK679	–	6.3 ± 6.25 (0.8 ± 0.02)
DeKalb	DK626Bty	MON810	+	68 ± 6.3 (0.08 ± 0.08)	DK626	–	0 (0.7 ± 0.09)

[a] Determined with Lateral Flow Quickstix; – = no toxin detected; + = toxin detected.

[b] Determined with the larvae of the tobacco hornworm (*Manduca sexta*); at least 16 larvae/assay; weight data normalized to one larva/treatment ± standard error of the mean (in parentheses). No mortality with soils without plants (weight of a single larva: 0.7 to 1.2 ± 0.02 g).

Source: Saxena, D. et al., *Soil Biol. Biochem.*, in press. With permission from Elsevier Science.

different transformation events, regardless of whether they were grown in the plant-growth room or in the field.

These results indicate that the toxin released in exudates from roots could accumulate in soil and retain insecticidal activity, especially when the toxin is bound on surface-active soil particles and, thereby, becomes resistant to degradation by microorganisms. Although some toxin was probably released from sloughed and damaged root cells, the major portion was derived from exudates, as was suggested by the fact that there was no discernible root debris after centrifugation of the Hoagland's solution from plants grown for 25 days in hydroponic culture.

In addition to the large amount of toxin that will be introduced to soil in plant biomass after harvest and some that will be introduced in pollen released during tasseling (Losey et al. 1999). The above results indicate that toxin will also be released to soil from roots during the entire growth of a *Bt*-corn crop. The presence of the toxin in soil could improve the control of insect pests, or the persistence of the toxin in soil could enhance the selection of toxin-resistant target insects and constitute a hazard to non-target organisms. Because *Bt*-corn contains truncated genes that encode toxins rather than the nontoxic crystalline protoxins produced by *Bt*, potential hazards are exacerbated, as it is not necessary for an

Table 8.17 Presence of the Cry1Ab Toxin in Rhizosphere Soil of Different Hybrids of Corn with (*Bt+*) and without (*Bt–*) the *cry1Ab* Gene Grown in the Field

Company	Hybrid	Event	Immuno. test[a]	% Mortality[b]	Hybrid	Immuno. test	% Mortality
			Bt+			Bt–	
Novartis	N7590Bt	Bt11	+	75 ± 10.2 (0.05 ± 0.02)	N7590	–	6.3 ± 6.25 (0.9 ± 0.06)
Novartis	N67-T4	Bt11	+	68± 11.9 (0.02 ± 0.01)	N67-H6	–	0 (1.2 ± 0.08)
Novartis	N3030Bt	Bt11	+	37 ± 7.2 (0.09 ± 0.04)	N3030	–	0 (0.9 ± 0.03)
Novartis	NC4990Bt	Bt11	+	62 ± 12.5 (0.07 ±0.02)	NC4880	–	6.3 ± 6.25 (1.1 ± 0.07)
Novartis	NK4640Bt	Bt11	+	81 ± 6.25 (0.09 ±0.01)	NK4640	–	0 (1.1 ± 0.09)
Novartis	0966 (Supersweet)	Bt11	+	63 ± 6.5 (0.09 ± 0.05)	PrimePlus (Supersweet)	–	6.3 ± 6.25 (0.8 ± 0.02)
Novartis	Maximizer	176	+	37 ± 7.2 (0.18 ± 0.03)	—	—	—
Pioneer	P32P76	MON810	+	68 ± 15.7 (0.08 ± 0.02)	P32P75	–	0 (0.9 ± 0.06)
Pioneer	P33B51	MON810	+	68 ± 15.7 (0.09 ± 0.01)	P33B50	–	0 (0.9 ± 0.04)
Pioneer	P31B13	MON810	+	43 ± 6.5 (0.06 ± 0.03)	P3223	–	6.3 ± 6.25 (0.9 ± 0.05)
DeKalb	DK647Bty	MON810	+	37 ± 12.5 (0.08 ± 0.04)	DK647	–	0 (1.2 ± 0.08)
DeKalb	DK679Bty	MON810	+	56 ± 11.9 (0.04 ± 0.01)	DK679	–	12.5 ± 6.25 (0.8 ± 0.03)
DeKalb	DK626Bty	MON810	+	50 ± 10.2 (0.09 ± 0.03)	DK626	–	6.3 ± 6.25 (0.9 ± 0.01)

[a] Determined with Lateral Flow Quickstix; – = no toxin detected; + = toxin detected.

[b] Determined with the larvae of the tobacco hornworm (*Manduca sexta*); at least 16 larvae/assay; weight data normalized to one larva/treatment ± standard error of the mean (in parentheses). No mortality with soils without plants (weight of a single larva: 0.8 to 1.2 ± 0.04 g).

Source: Saxena, D. et al., *Soil Biol. Biochem.,* in press. With permission from Elsevier Science.

organism ingesting the toxins to have a high gut pH (~10.5) for solubilization of the protoxins and for specific proteases to cleave the protoxins into toxins (see Figure 8.1). Moreover, receptors for the toxins are present in both target and non-target insects (Höfte and Whiteley 1989). Consequently, nontarget insects and organisms in higher and, perhaps also, in lower trophic levels could be susceptible to the toxins.

8.12 Effect of Bt-toxin from root exudates and biomass of Bt-corn on earthworms, nematodes, protozoa, bacteria, and fungi in soil

To determine the effects of the Cry1Ab toxin released in root exudates from *Bt*-corn on various organisms in soil, ~4.5 kg of a loam soil from a farm in East Marion, Long Island, New York, was placed in plastic pots, and 20 medium-size (3.5 ± 0.32 g) earthworms (*Lumbricus terrestris*), purchased from Carolina Biological Supply Company, were introduced to each pot. Seeds (three per pot) of *Bt*-corn (NK4640Bt) and isogenic non-*Bt*-corn were planted, and some pots remained unplanted. After 40 days of growth in a plant-growth

Table 8.18   Effect on Earthworms (*Lumbricus terrestris*) of Cry1Ab Toxin Released in Root Exudates and from Biomass of *Bt*-Corn (NK4640*Bt*) in Soil (expressed as % mortality and mean weight of a single earthworm ± standard error of the mean)

Soil	% Mortality	Weight (g)
	Root exudates[a]	
No plants	8 ± 2.5	3.6 ± 0.10
Bt– plants	14 ± 1.0	3.6 ± 0.03
Bt+ plants	10 ± 1.5	3.9 ± 0.27
	Biomass[b]	
No biomass	30 ± 10.0	3.9 ± 0.06
Bt– biomass	28 ± 4.8	4.3 ± 0.15
Bt+ biomass	24 ± 7.5	4.0 ± 0.19

[a] Mortality and weight of earthworms were determined in soil with no plants or with *Bt*+ or *Bt*– corn after 40 days of plant growth ($n = 20$ worms × 5 replications/treatment).

[b] Mortality and weight of earthworms were determined after 45 days of incubation with 1% (wt/wt) ground biomass of *Bt*+ or *Bt*– corn or with no biomass ($n = 5$ worms × 5 replications/treatment).

Source: Saxena, D. and Stotzky, G., *Soil Biol. Biochem.*, 33, 1225, 2001b. With permission from Elsevier Science.

room (26 ± 2°C, 12-h light–dark cycle), the plants were gently removed and rhizosphere soil was collected. To determine the effects of biomass from *Bt*-corn and non-*Bt*-corn on the organisms, 500 g of soil amended with 1% (wt/wt) of ground, air-dried biomass of *Bt* (NK4640Bt) or non-*Bt*-corn (leaves, stems, and roots) was placed in glass containers, and five medium-size earthworms were added to each container. Control containers received no biomass. The containers were kept in the plant-growth room for 45 days. Before the introduction of the worms, the numbers of nematodes (Van Gundy 1982) and of culturable protozoa, fungi, and bacteria, including actinomycetes (Stotzky et al. 1993), in the soil in the pots and containers were determined.

In both experiments, casts produced by the worms were collected. After incubation, the numbers of earthworms were counted and their weight determined. Three representative worms from each pot with plants and two worms from each container with biomass were dissected, and soil from the guts, as well as from the casts, was analyzed for the presence of the Cry1Ab protein by immunological and larvicidal assays.

There were no significant differences in the percent mortality and weight of earthworms after 40 days in soil planted with *Bt*-corn or non-*Bt*-corn or not planted or after 45 days in soil amended with biomass of *Bt*-corn or non-*Bt*-corn or not amended (Table 8.18) (Saxena and Stotzky, in 2001b). However, the toxin was present in both casts and guts of worms in soil planted with *Bt*-corn or amended with biomass of *Bt*-corn (Table 8.19), whereas it was absent in casts and guts of worms in soil planted with non-*Bt*-corn or not planted and in soil amended with biomass of non-*Bt*-corn or not amended. When worms from pots of *Bt*-corn or from containers of soil amended with biomass of *Bt*-corn were transferred to pots containing fresh soil, the toxin was cleared from the guts in 2 to 3 days. All samples of soil amended with biomass of *Bt*-corn and from the rhizosphere of *Bt*-corn were positive for the presence of the toxin and were lethal to the larvae of *Manduca sexta*, even after 45 and 40 days, respectively. There was no mortality in soil amended with biomass of non-*Bt*-corn, in rhizosphere soil of non-*Bt*-corn, or in soil with no plants or not amended; these soils were also negative in the immunological assays (Table 8.19).

There were no statistically significant differences ($P > 0.5$) in the total numbers of nematodes and culturable protozoa, bacteria (including actinomycetes), and fungi between rhizosphere soil of *Bt*-corn and non-*Bt*-corn or between soil amended with *Bt* or non-*Bt*-biomass (Table 8.20).

Table 8.19 Cry1Ab Toxin in Soil, Casts, and Guts of Earthworms (*Lumbricus terrestris*) in the Presence of Corn Plants with (*Bt*+) (NK4640*Bt*) and without (*Bt*−) the *cry1Ab* Gene, in Soil Amended with Ground Biomass of *Bt*+ or *Bt*− Corn, and in the Absence of Plants or Biomass

| Sample | No plants | | Root exudates[a] | | | | No biomass | | Biomass[b] | | | |
	Immuno. test[c]	% Mortality[d]	*Bt*− Immuno. test	% Mortality	*Bt*+ Immuno. test	% Mortality	Immuno. test	% Mortality	*Bt*− Immuno. test	% Mortality	*Bt*+ Immunotest	% Mortality
Soil	−	0 (1.0 ± 0.10)	−	5 ± 5.0 (1.0 ± 0.09)	+	92.5 ± 3.06 (0.02 ± 0.00)	−	6.3 ± 6.25 (0.9 ± 0.07)	−	2.5 ± 2.5 (0.9 ± 0.07)	+	97.5 ± 2.5 (0.01 ± 0.00)
Cast	−	4.2 ± 4.16 (0.9 ± 0.07)	−	2.5 ± 2.50 (0.8 ± 0.04)	+	100	−	6.3 ± 6.25 (0.9 ± 0.05)	−	5 ± 5.0 (0.8 ± 0.04)	+	90 ± 6.12 (0.02 ± 0.00)
Gut	−	0 (0.8 ± 0.10)	−	0 (0.8 ± 0.08)	+	97.5 ± 2.50 (0.05 ± 0.00)	−	6.3 ± 6.25 (0.8 ± 0.10)	−	2.5 ± 2.50 (0.9 ± 0.08)	±	92.5 ± 3.06 (0.01 ± 0.00)

[a] Samples from soil with no plants or with *Bt*+ or *Bt*− corn were evaluated after 40 days of plant growth.

[b] Samples from soil with ground residues of *Bt*+ or *Bt*− corn or with no residues were evaluated after 45 days of incubation.

[c] Determined with Lateral Flow Quickstix: − = no toxin detected; + = toxin detected.

[d] Determined with the larvae of the tobacco hornworm (*Manduca sexta*); 8 to 16 larvae/assay; expressed as % mortality and mean weight, in g, of a single larva ± standard error of the mean (in parentheses).

Source: Saxena, D. and Stotzky, G., *Soil Biol. Biochem.*, 33, 1225, 2001b. With permission from Elsevier Science.

Table 8.20 Effect on Various Groups of Organisms of Cry1Ab Toxin Released in Root Exudates[a] and from Biomass[b] of Bt-Corn (NK4640Bt) in Soil (mean ± standard error of the mean)[c]

Organism	No plants	Root exudates		No biomass	Biomass	
		Bt-	Bt+		Bt-	Bt+
Bacteria[d]	$1.4 \pm 0.16 \times 10^8$	$7.6 \pm 0.14 \times 10^7$	$5.8 \pm 0.23 \times 10^7$	$1.1 \pm 0.01 \times 10^8$	$8.1 \pm 0.20 \times 10^7$	$6.3 \pm 0.10 \times 10^7$
Actinomycetes[d]	$4.2 \pm 0.45 \times 10^5$	$2.3 \pm 0.20 \times 10^5$	$2.3 \pm 0.36 \times 10^5$	$2.4 \pm 0.22 \times 10^5$	$2.8 \pm 0.53 \times 10^5$	$2.5 \pm 0.34 \times 10^5$
Fungi[d]	$3.3 \pm 0.41 \times 10^6$	$2.7 \pm 0.56 \times 10^6$	$1.8 \pm 0.21 \times 10^6$	$3.8 \pm 0.56 \times 10^6$	$3.6 \pm 0.53 \times 10^6$	$3.4 \pm 0.72 \times 10^6$
Protozoa[e]	$1.3 \pm 0.05 \times 10^4$	$1.8 \pm 0.02 \times 10^4$	$1.4 \pm 0.02 \times 10^4$	$1.1 \pm 0.07 \times 10^4$	$1.1 \pm 0.02 \times 10^4$	$1.0 \pm 0.08 \times 10^4$
Nematodes[f]	$1.0 \pm 0.03 \times 10^3$	$1.3 \pm 0.04 \times 10^3$	$1.3 \pm 0.08 \times 10^3$	$1.4 \pm 0.01 \times 10^3$	$1.5 \pm 0.05 \times 10^3$	$1.4 \pm 0.06 \times 10^3$

[a] Soil with no plants or with Bt+ or Bt- corn was evaluated after 40 days of plant growth.

[b] Soil was evaluated after 45 days of incubation with ground residues of Bt+ or Bt- corn or with no residues.

[c] $P > 0.5$ for all comparisons between Bt+ and Bt- corn.

[d] Numbers/g soil, oven-dry equivalent; serial dilution plate method.

[e] Numbers/g soil, oven-dry equivalent; most probable number method.

[f] Numbers/50 g soil, oven-dry equivalent; Baermann method.

Source: Saxena, D. and Stotzky, G., *Soil Biol. Biochem.,* 33, 1225, 2001b. With permission from Elsevier Science.

These results suggested that the toxin released in root exudates of *Bt*-corn or from the degradation of the biomass of *Bt*-corn is not toxic to a variety of organisms in soil. However, the toxin was detected in the guts and casts of earthworms grown with *Bt*-corn and in soil amended with biomass of *Bt*-corn (Table 8.19), indicating again that the released toxin bound on surface-active particles in soil, which protected the toxin from biodegradation, similar to what had been observed with purified toxins. Because only one species of earthworms and only total culturable microorganisms and nematodes were evaluated, more detailed studies on the composition and diversity of these groups of organisms are necessary, including studies using techniques of molecular biology (e.g., single-strand conformation polymorphism, denaturing gradient and temperature gradient gel electrophoresis), the BIOLOG or equivalent system for bacteria, speciation of fungi, and nutritional groups of protozoa and nematodes, to confirm the absence of effects of the Cry1Ab toxin on biodiversity in soil.

8.13   Uptake of Bt-*toxin from soil by plants*

Because of public concern that plants will take up *Bt*-toxins from soil (e.g., a supermarket chain in the United Kingdom would not sell produce from plants that have been grown on soils previously planted with *Bt*-crops (Nuttall 2000), non-*Bt*-corn, carrot, radish, and turnip were grown in soil in which *Bt*-corn had previously been planted or which was amended with ground biomass of *Bt*-corn. Non-*Bt*-corn and turnip were harvested after 20, 30, 50, and 120 days and carrot and radish after 30, 50, 120, and 180 days, and the leaves, stems, and roots were analyzed immunologically and by bioassay with the larvae of *Manduca sexta*. No Cry1Ab protein was detected in the tissues of any of the plants, whereas the protein was present in all samples of soil (Tables 8.21 and 8.22) (Saxena and Stotzky 2001a, in press b).

To determine whether the Cry1Ab protein is taken up in hydroponic culture, *Bt*-corn was grown aseptically in Hoagland's solution for 15 days in the plant-growth room, and 2-day-old seedlings derived from surface-sterilized seeds of non-*Bt*-corn that had been germinated on agar were transferred aseptically and grown in the same solution for 7 and 15 days. The tissues of three representative plants and the solution were analyzed after 7 and 15 days for the toxin by immunological and larvicidal assays. No Cry1Ab protein was detected in the tissues of non-*Bt*-corn, whereas the toxin was easily detected in the hydroponic solution (Table 8.23). These results indicated that the lack of uptake of the toxin from soil was not the result of the binding of the protein on surface-active particles, as no such surfaces were present in the hydroponic solution.

These results were not unexpected, as it is doubtful that plants can take up molecules as large as 66 kDa. Moreover, these results demonstrated that the toxin released in root exudates and from biomass of *Bt*-corn persists in soil for at least 120 to 180 days, probably because it is bound on surface-active particles.

8.14   Conclusions

Insecticidal proteins produced by various subspecies of *B. thuringiensis* bind rapidly and tightly on both pure mined clay minerals and soil clays and on humic acids extracted from soil. This binding reduces the susceptibility of these proteins to microbial degradation, and the bound toxins retain their biological activity. Both purified toxins and toxins released from the biomass of transgenic *Bt*-corn and in root exudates of growing *Bt*-corn exhibit binding and persistence in soil. Moreover, biomass of transgenic *Bt*-corn decomposes less in soil than does biomass of isogenic non-*Bt*-corn, possibly because biomass of *Bt*-corn has a significantly higher content of lignin than does biomass of non-*Bt*-corn. The

Table 8.21   Uptake of Cry1Ab Toxin by non-*Bt* (*Bt*–) Corn Grown in Soil Containing the Cry1Ab Toxin Released as Root Exudates by Previously Grown *Bt* (*Bt+*) Corn and from Ground Biomass of *Bt*-Corn

Days of growth	Soil Immuno. test[a]	Soil % Mortality[b]	Leaves Immuno. test	Leaves % Mortality	Stems Immuno. test	Stems % Mortality	Roots Immuno. test	Roots % Mortality
				Root exudate[c]				
20	+	75 ± 10.2 (0.1 ± 0.01)	–	0 (1.0 ± 0.06)	–	0 (0.6 ± 0.05)	–	0 (0.6 ± 0.06)
30	+	63 ± 7.2 (0.1 ± 0.02)	–	0 (0.9 ± 0.03)	–	0 (0.7 ± 0.04)	–	0 (0.6 ± 0.05)
50	+	75 ± 10.2 (0.04 ± 0.02)	–	0 (0.9 ± 0.08)	–	6.3 ± 6.25 (0.6 ± 0.01)	–	6.3 ± 6.25 (0.7 ± 0.03)
120	+	63 ± 7.2 (0.08 ± 0.01)	–	6.3 ± 6.25 (0.9 ± 0.05)	–	0 (0.6 ± 0.02)	–	6.3 ± 6.25 (0.8 ± 0.05)
				Biomass[c]				
20	+	88 ± 7.2 (0.05 ± 0.01)	–	0 (0.9 ± 0.02)	–	0 (0.5 ± 0.05)	–	0 (0.5 ± 0.03)
30	+	88 ± 7.2 (0.06 ± 0.01)	–	6.3 ± 6.25 (0.8 ± 0.03)	–	0 (0.6 ± 0.04)	–	0 (0.6 ± 0.04)
50	+	88 ± 7.2 (0.09 ± 0.01)	–	6.3 ± 6.25 (0.8 ± 0.08)	–	0 (0.7 ± 0.06)	–	6.3 ± 6.25 (0.7 ± 0.04)
120	+	75 ± 10.2 (0.07 ± 0.01)	–	6.3 ± 6.25 (0.9 ± 0.08)	–	75 ± 14.4 (0.6 ± 0.08)	–	6.3 ± 6.25 (0.6 ± 0.08)

[a] Determined with Lateral Flow Quickstix: – = no toxin detected; + = toxin detected.

[b] Determined with the larvae of the tobacco hornworm (*Manduca sexta*); at least 16 larvae/assay; expressed as % mortality and mean weight, in g, of a single larva ± standard error of the mean (in parentheses).

[c] *Bt*– corn was grown in soil in which *Bt+* corn had been grown or in which ground biomass of *Bt+* corn had been incorporated.

Source: Saxena, D. and Stotzky, G., *Plant Soil*, in press b. With permission.

toxins do not appear to have any consistent effects on organisms (earthworms, nematodes, protozoa, bacteria, fungi) in soil or *in vitro*. The toxins are not taken up from soil by non-*Bt*-corn, carrot, radish, or turnip grown in soil in which *Bt*-corn has been grown or into which biomass of *Bt*-corn has been incorporated.

These studies on the interaction of insecticidal proteins with two types of surface-active particles that differ greatly in composition and structure further demonstrate the importance of surface-active particles to the biology of natural habitats. These studies also confirm and extend previous observations on the influence of clays and other surface-active particles on the activity, ecology, and population dynamics of microbes (including viruses) in soil and other natural habitats, as well as on the transfer of genetic information among bacteria by conjugation, transduction, and transformation (Table 8.24) (e.g., Stotzky 1986, 1989, 2000, Yin and Stotzky 1997, Vettori et al. 1999).

Moreover, the results obtained with these proteins indicate their potential environmental importance when bound on surfaces in soil. For example, the persistence of the bound toxins from *Bt* could pose a potential hazard to non-target organisms and result in the selection of toxin-resistant target insects, thereby negating the benefits of using a biological insecticide rather than a synthetic chemical insecticide. However, the persistence of the bound toxins could also enhance the control of target pests. These aspects require more extensive study.

In addition to suggesting potential hazards and benefits of bound toxins from *Bt*, these studies indicate that caution must be exercised before transgenic plants and animals

Table 8.22   Uptake of Cry1Ab Toxin by Carrot (*Daucus carota*) Grown in Soil Containing the Cry1Ab Toxin Released as Root Exudates by Previously Grown *Bt*-Corn and from Ground Biomass of *Bt*-Corn

Days of growth	Soil Immuno. test[a]	% Mortality[b]	Leaves Immuno. test	% Mortality	Stems Immuno. test	% Mortality	Roots Immuno. test	% Mortality
			Root exudate[c]					
30	+	68 ± 11.9 (0.1 ± 0.02)	–	0 (0.8 ± 0.02)	–	0 (0.8 ± 0.09)	–	0 (0.8 ± 0.09)
50	+	68 ± 6.3 (0.08 ± 0.03)	–	0 (0.9 ± 0.06)	–	6.3 ± 6.25 (0.8 ± 0.06)	–	0 (0.7 ± 0.02)
120	+	62 ± 12.3 (0.06 ± 0.01)	–	0 (0.8 ± 0.08)	–	0 (0.7 ± 0.05)	–	6.3 ± 6.25 (0.6 ± 0.09)
180	+	56 ± 11.9 (0.1 ± 0.03)	–	0 (1.1 ± 0.09)	–	6.3 ± 6.25 (0.8 ± 0.02)	–	6.3 ± 6.25 (0.6 ± 0.08)
			Biomass[c]					
50	+	88 ± 12.5 (0.08 ± 0.06)	–	0 (0.8 ± 0.05)	–	0 (0.7 ± 0.02)	–	0 (0.6 ± 0.05)
120	+	75 ± 14.4 (0.04 ± 0.02)	–	0 (0.8 ± 0.01)	–	0 (0.7 ± 0.06)	–	0 (0.7 ± 0.03)
180	+	68 ± 11.9 (0.05 ± 0.01)	–	0 (0.9 ± 0.02)	–	0 (0.9 ± 0.06)	–	12.5 ± 12.5 (0.8 ± 0.02)

[a] Determined with Lateral Flow Quickstix: - = no toxin detected; + = toxin detected.

[b] Determined with the larvae of the tobacco hornworm (*Manduca sexta*); at least 16 larvae/assay; expressed as % mortality and mean weight, in g, of a single larva ± standard error of the mean (in parentheses).

[c] Carrots were grown in soil in which *Bt*+ corn had been grown or in which ground biomass of *Bt*+ corn had been incorporated.

Source: Saxena, D. and Stotzky, G., *Plant Soil*, in press b. With permission.

Table 8.23   Uptake of Cry1Ab Toxin by non-*Bt*-Corn Grown in Hydroponic Solution Containing the Cry1Ab Toxin Released as Root Exudates by Previously Grown *Bt*-Corn

Days of growth	Solution Immuno. test[a]	% Mortality[b]	Leaves Immuno. test	% Mortality	Stems Immuno. test	% Mortality	Roots Immuno. test	% Mortality
7	+	87.5 ± 7.21 (0.07 ± 0.01)	–	0 (0.8 ± 0.08)	–	6.3 ± 6.25 (0.6 ± 0.02)	–	0 (0.6 ± 0.02)
15	+	93.7 ± 6.25 (0.9 ± 0.00)	–	6.3 ± 6.25 (0.8 ± 0.02)	–	6.3 ± 6.25 (0.6 ± 0.2)	–	12.5 ± 7.21 (0.6 ± 0.02)

[a] Determined with Lateral Flow Quickstix: – = no toxin detected; + = toxin detected.

[b] Determined with the larvae of the tobacco hornworm (*Manduca sexta*); at least 16 larvae/assay; expressed as % mortality and mean weight, in g, of a single larva ± standard error of the mean (in parentheses).

Source: Saxena, D. and Stotzky, G., *Plant Soil*, in press b. With permission.

genetically modified to function as "factories" for the production of vaccines, hormones, antibodies, toxins, pharmaceuticals, and other bioactive compounds are released to the environment (Table 8.25). Because of the large differences in the chemical composition and structure of proteins and other molecules studied, as well as between clays and humic acids, these studies can serve as models for the potential fate and effects of other biomolecules, which are also chemically and structurally diverse, that will be introduced to soil from such factories. As with *Bt*-plants, where only a portion of the plants is harvested (e.g., ears of corn, bolls of cotton, kernels of rice, potatoes) and the remainder of the biomass

Table 8.24   Summary of Effects of Interactions of Some Biomolecules with Surface-Active Particles in Soil

---

Biomolecules (e.g., DNA, viruses, proteins) bind rapidly and tightly on clay minerals and humic substances

Bound biomolecules are protected against degradation and inactivation

Bound biomolecules retain biological activity

- Bound DNA transforms
- Bound bacteriophages transduce
- Bound viruses lyse host cells
- Bound insecticidal toxins from *B. thuringiensis* kill insect larvae
- Bound enzymes convert substrates

Source: Stotzky, G., *J. Environ. Qual.*, 29, 691, 2000. With permission.

Table 8.25   Potential Hazards to the Environment from Biological "Factories" (transgenic plants, animals, and microbes)

Bioactive products

   For example, vaccines, hormones, antibodies, toxins, enzymes, various other pharmaceuticals, genetically modified foods

Plants (primarily soil)

- Biomass (eventually incorporated into soil)
- Root exudates, e.g., green fluorescent protein, human placental alkaline phosphatase, and xylanase (from *Clostridium thermocellum*) from genetically engineered tobacco (signal peptide coding region from the endoplasmic reticulum protein, calreticulin) in hydroponic culture;[a] Cry1Ab insecticidal protein from *Bt*-corn
- Pollen, e.g., monarch butterflies killed by feeding on milkweed contaminated with pollen from *Bt* corn
- Tritrophic interactions (predators killed by ingesting prey feeding on genetically engineered plants; e.g., green lacewings on European corn borers fed *Bt*-corn; ladybird beetles on aphids fed potatoes expressing the lectin from snowdrop anemone)[b]

Animals (soil, surface water, and groundwater)

- Feces
- Urine
- Carcasses

Microbes (soil, surface water, and groundwater)

- Waste media
- Waste cells
- Fermentor breakdowns

---

[a] Borisjuk et al. 1999.

[b] Birch et al. 1999.

Source: Modified from Stotzky, G., *J. Environ. Qual.*, 29, 691, 2000. With permission.

is incorporated into soil wherein the toxins released from disintegrating biomass are rapidly bound on surface-active particles, some of the biomass of these plant factories will also be incorporated into soil. With transgenic animal factories, feces, urine, and subsequently even carcasses containing bioactive compounds will eventually reach soil and other natural habitats (e.g., surface water and groundwater). If these bioactive compounds bind on clays and humic substances — and since many of these compounds are proteinaceous, they most likely will — they may also persist in natural habitats. If they retain their bioactivity, they could affect the biology of those habitats. Consequently, before the use of such plant and animal factories (and, probably, also microbial factories), the persistence of their products and the potential effects of the products on the inhabitants of soil and on other habitats must be thoroughly evaluated.

Acknowledgments

Most of the studies from this laboratory discussed in this chapter were supported, in part, by grants from the U.S. Environmental Protection Agency (most recently R826107-01), the U.S. National Science Foundation, and the NYU Research Challenge Fund. The opinions expressed herein are not necessarily those of the Agency, the Foundation, or the Fund. Sincere appreciation is expressed to colleagues who contributed to these studies.

Literature cited

Addison, J. A., Persistence and nontarget effects of *Bacillus thuringiensis* in soil: a review, *Can. J. For. Res.*, 23, 2329, 1993.

Babich, H. and Stotzky, G., Effects of cadmium on fungi and on interactions between fungi and bacteria in soil: influence of clay minerals and pH, *Appl. Environ. Microbiol.*, 33, 1059, 1977.

Babich, H. and Stotzky, G., Developing standards for environmental toxicants: the need to consider abiotic environmental factors and microbe-mediated ecologic processes, *Environ. Health Perspect.*, 49, 247, 1983.

Babich, H. and Stotzky, G., Environmental factors affecting the utility of microbial assays for the toxicity and mutagenicity of chemical pollutants, in *Toxicity Testing Using Microorganisms*, Dutka, B. J. and Bitton, G., Eds., CRC Press, Boca Raton, FL, 1986, 19.

Bauer, L. S., Resistance: a threat to the insecticidal crystal proteins of *Bacillus thuringiensis*, *Fla. Entomol.*, 78, 414, 1995.

Birch, A. N. E., Geoghegan, I. E., Majerus, M. E. N., McNicol, J. W., Hackett, C. A., Gatehouse, A. M. R., and Gatehouse, J. A., Tritrophic interactions involving pest aphids, predatory 2-spot ladybirds and transgenic potatoes expressing snowdrop lectin for aphid resistance, *Mol. Breed.*, 5, 75, 1999.

Borisjuk, N. V., Borisjuk, L. G., Logendra, S., Peterson, F., Gleba, Y., and Raskin, I., Production of recombinant proteins in plant root exudates, *Nat. Biotechnol.*, 17, 466, 1999.

Collins, Y. E. and Stotzky, G., Factors affecting toxicity of heavy metals to microbes, in *Metal Ions and Bacteria*, Beveridge, T. J. and Doyle, R. J., Eds., John Wiley & Sons, New York, 1989, 31.

Crecchio, C. and Stotzky G., Insecticidal activity and biodegradation of the toxins from *Bacillus thuringiensis* subsp. *kurstaki* bound to humic acids from soil, *Soil Biol. Biochem.*, 30, 463, 1998.

Crecchio, C. and Stotzky, G., Biodegradation and insecticidal activity of the toxin from *Bacillus thuringiensis* subsp. *kurstaki* bound on complexes of montmorillonite-humic acids-Al hydroxy-polymers, *Soil Biol. Biochem.*, 33, 573, 2001.

Crickmore, N., Zeigler, D. R., Feitelson, J., Schnepf, E., Van Rie, J., Lereclus, D., Baum, J., and Dean, D. H., Revision of the nomenclature for the *Bacillus thuringiensis* pesticidal crystal proteins, *Microbiol. Mol. Biol. Rev.*, 62, 807, 1998.

Donegan, K. K., Palm, C. J., Fieland, V. J., Porteous, L. A., Ganio, L. M., Schaller, D. L., Bucao, L. Q., and Seidler, R. J., Changes in levels, species and DNA-fingerprints of soil microorganisms associated with cotton expressing the *Bacillus thuringiensis* var. *kurstaki* δ-endotoxin, *Appl. Soil Ecol.*, 2, 111, 1995.

Donegan, K. K., Shaller, D. L., Stone, J. K., Ganio, L. M., Reed, G., Hamm, P. B., and Seidler, R. J., Microbial populations, fungal species diversity and plant pathogen levels in field plots of potato plants expressing the *Bacillus thuringiensis* var. *tenebrionis* endotoxin, *Transgenics Res.*, 5, 25, 1996.

Doyle, J. D., Stotzky, G., McClung, G., and Hendricks, C. W., Effects of genetically engineered microorganisms on microbial populations and processes in natural habitats, *Adv. Appl. Microbiol.*, 40, 237, 1995.

Egorov, N. S., Yudina, T. C., and Baranov, A. Y., Correlation between the insecticidal and antibiotic activities of the parasporal crystals of *Bacillus thuringiensis*, *Mikrobiologiya*, 59, 448, 1990.

Entwistle, P. F., Cory, J. S., Bailey, M. J., and Higgs, S., Eds., *Bacillus thuringiensis, an Environmental Biopesticide: Theory and Practice*, John Wiley & Sons, Chichester, U.K., 1993.

Faust, M. A., Research update on *Bt* corn silage, in *Proceedings of the Four-State Applied Nutrition and Management Conference*, Iowa State University, Ames, 1999.

Feitelson, J. S., Payne, J., and Kim, L., *Bacillus thuringiensis*: insects and beyond, *Bio/Technology*, 10, 271, 1992.

Ferré, J., Escriche, B., Bel, Y., and Van Rie, J., Biochemistry and genetics of insect resistance to *Bacillus thuringiensis* insecticidal crystal proteins, *FEMS Microbiol. Lett.*, 132, 1, 1995.

Flexner, J. L., Lighthart, B., and Croft, B. A., The effects of microbial pesticides on non-target, beneficial arthropods, *Agric. Ecosyst. Environ.*, 16, 203, 1986.

Garczynski, S. F., Crim, J. W., and Adang, M. J., Identification of putative brush border membrane-binding protein specific to *Bacillus thuringiensis* δ-endotoxins by protein blot analysis, *Appl. Environ. Microbiol.*, 57, 2816, 1991.

Goldburg, R. J. and Tjaden, G., Are B.t.k. plants really safe to eat? *Bio/Technology*, 8, 1011, 1990.

Griego, V. M. and Spence, K. D., Inactivation of *Bacillus thuringiensis* spores by ultraviolet and visible light, *Appl. Environ. Microbiol.*, 35, 906, 1978.

Halpin, C., Knight, M. E., Foxon, G. A., Campbell, M. M., Boudet, A. M., Boon, J. J., Chabbert, B., Tollier, M., and Schuch, W., Manipulation of lignin quality by down regulation of cinnamyl alcohol dehydrogenase, *Plant J.*, 6, 339, 1994.

Harley, J. P. and Prescott, L. M., *Laboratory Exercises in Microbiology*, 4th ed., W.C. Brown, Dubuque, IA, 1999.

Hatfield, R. D., Grabber, J., Ralph, J., and Brei, K., Using the acetyl bromide assay to determine lignin concentration in herbaceous plants: Some cautionary notes, *J. Agric. Food Chem.*, 47, 628, 1999.

Hilbeck, A., Baumgartner, M., Fried, P. M., and Bigler, F., Effects of transgenic *Bacillus thuringiensis* corn-fed prey on mortality and development time of immature *Chrysoperla carnea* (Neuroptera: Chrysopidae), *Environ. Entomol.*, 27, 480, 1998a.

Hilbeck, A., Moar, W. J., Pusztai-Carey, M., Filippini, A., and Bigler, F., Toxicity of *Bacillus thuringiensis* Cry1Ab toxin to the predator *Chrysoperla carnea* (Neuroptera: Chrysopidae), *Environ. Entomol.*, 27, 1255, 1998b.

Hilbeck, A., Moar, W. J., Pusztai-Carey, M., Filippini A., and Bigler, F., Prey-mediated effects of Cry1Ab toxin and protoxin and Cry2A protoxin on the predator *Chrysoperla carnea*, *Entomol. Exp. Appl.*, 91, 305, 1999.

Höfte, H. and Whiteley, H. R., Insecticidal crystal proteins of *Bacillus thuringiensis*, *Microbiol. Rev.*, 53, 242, 1989.

Hu, W., Harding, S. A., Lung, J., Popko, J. L., Ralph, J., Stokke, D. D., Tsai, C., and Chiang, V. L., Repression of lignin biosynthesis promotes cellulose accumulation and growth in transgenic trees, *Nature Biotech.*, 17, 808, 1999.

Ignoffo, C. M. and Garcia, C., UV-photoinactivation of cells and spores of *Bacillus thuringiensis* and effects of peroxide on inactivation, *Environ. Entomol.*, 7, 270, 1978.

James, R. R., Miller, J. C., and Lighthart, B., *Bacillus thuringiensis* var. *kurstaki* affects a beneficial insect, the Cinnabar moth (Lepidoptera: Arctiidae), *J. Econ. Entomol.*, 86, 334, 1993.

Johnson, K. S., Scriber, J. M., Nitas, J. K., and Smitley, D. R., Toxicity of *Bacillus thuringiensis* var. *kurstaki* to three nontarget lepidoptera in field studies, *Environ. Entomol.*, 24, 288, 1995.

Koskella, J. and Stotzky, G., Microbial utilization of free and clay-bound insecticidal toxins from *Bacillus thuringiensis* and their retention of insecticidal activity after incubation with microbes, *Appl. Environ. Microbiol.*, 63, 3561, 1997.

Losey, J. E., Rayor, L. S., and Carter, M. E., Transgenic pollen harms monarch larvae, *Nature*, 399, 214, 1999.

Masoero, F., Moschini, M., Rossi, F., Prandini, A., and Pietri, A. Nutritive value, mycotoxin contamination and *in vitro* rumen fermentation of normal and genetically modified corn (Cry1A(B)) grown in northern Italy, *Maydica*, 44, 205, 1999.

McGaughey, W. H. and Whalon, M. E., Managing insect resistance to *Bacillus thuringiensis* toxins, *Science*, 258, 1451, 1992.

Nuttall, N., Tesco to ban produce from GM trials sites, *London Times*, 7 January 2000, 9.

Ostrander, B. N. and Coors, J. G., Relationship between plant composition and European corn borer resistance in three maize populations, *Crop Sci.*, 37, 1741, 1997.

Palm, C. J., Donegan, K. K., Harris, D., and Seidler, R. J., Quantification in soil of *Bacillus thuringiensis* var. *kurstaki* δ-endotoxin from transgenic plants, *Mol. Ecol.*, 3, 145, 1994.

Palm, C. J., Schaller, D. L., Donegan, K. K., and Seidler, R. J., Persistence in soil of transgenic plant produced *Bacillus thuringiensis* var *kurstaki* δ-endotoxin, *Can. J. Microbiol.*, 42, 1258, 1996.

Petras, S. F. and Casida, L. E., Survival of *Bacillus thuringiensis* spores in soil, *Appl. Environ. Microbiol.*, 50, 1496, 1985.

Reddy, C. A., Physiology and biochemistry of lignin degradation, in *Current Perspectives in Microbial Ecology*, Klug, M. J. and Reddy, C. A., Eds., American Society of Microbiology, 558, 1984.

Saleh, S. M, Harris, R. F., and Allen, O. N., Recovery of *Bacillus thuringiensis* from field soils, *J. Invertebr. Pathol.*, 15, 55, 1970.

Saxena, D. and Stotzky, G., Insecticidal toxin from *Bacillus thuringiensis* is released from roots of transgenic *Bt* corn *in vitro* and *in situ*, *FEMS Microbiol. Ecol.*, 33, 35, 2000.

Saxena, D. and Stotzky, G., *Bt* toxin uptake from soil by plants, *Nature Biotech.*, 19, 199, 2001a.

Saxena, D. and Stotzky, G., *Bacillus thuringiensis* (*Bt*) toxin released from root exudates and biomass of *Bt* corn has no apparent effect on earthworms, nematodes, protozoa, bacteria, and fungi in soil, *Soil Biol. Biochem.*, 33, 1225, 2001b.

Saxena, D. and Stotzky, G., *Bt* corn has a higher lignin content than non-*Bt* corn, *Am. J. Bot.*, in press a.

Saxena, D. and Stotzky, G., *Bt* toxin is not taken up from soil or hydroponic culture by corn, carrot, radish, and turnip, *Plant Soil*, in press b.

Saxena, D., Flores, S., and Stotzky, G., Insecticidal toxin from *Bacillus thuringiensis* in root exudates of transgenic corn, *Nature*, 402, 480, 1999.

Saxena, D., Flores, S., and Stotzky, G., *Bt* toxin is released in root exudates from 12 transgenic corn hybrids representing three transformation events, *Soil Biol. Biochem.*, in press.

Schnepf, E., Crickmore, N., Van Rie, J., Lereclus, D., Baum, J., Feitelson, J., Zeigler, D. R., and Dean, D. H., *Bacillus thuringiensis* and its pesticidal crystal proteins, *Microbiol. Mol. Biol. Rev.*, 62, 775, 1998.

Sims, S. R. and Holden, L. R., Insect bioassay for determining soil degradation of *Bacillus thuringiensis* subsp. *kurstaki* CryIA(b) protein in corn tissues, *Environ. Entomol.*, 25, 659, 1996.

Sims, S. R. and Ream, J. E., Soil inactivation of the *Bacillus thuringiensis* subsp. *kurstaki* CryIIA insecticidal protein within transgenic cotton tissue: laboratory microcosms and field studies, *J. Agric. Food Chem.*, 45, 1502, 1997.

Stotzky, G., Activity, ecology, and population dynamics of microorganisms in soil, in *Microbial Ecology*, Laskin, A. I. and Lechevalier, H., Eds., CRC Press, Cleveland, OH, 1974, 57.

Stotzky, G., Influence of soil mineral colloids on metabolic processes, growth, adhesion, and ecology of microbes and viruses, in *Interactions of Soil Minerals with Natural Organics and Microbes*, Huang, P. M. and Schnitzer, M., Eds., Soil Science Society of America, Madison, WI, 1986, 305.

Stotzky, G., Gene transfer among bacteria in soil, in *Gene Transfer in the Environment*, Levy, S. B. and Miller, R. V., Eds., McGraw-Hill, New York, 1989, 165.

Stotzky, G., Soil as an environment for microbial life, in *Modern Soil Microbiology*, van Elsas, J. D., Wellington, E. M. H., and Trevors, J. T., Eds., Marcel Dekker, New York, 1997, 1.

Stotzky, G., Persistence and biological activity in soil of insecticidal proteins from *Bacillus thuringiensis* and of bacterial DNA bound on clays and humic acids, *J. Environ. Qual.*, 29, 691, 2000.

Stotzky, G., Broder, M. W., Doyle, J. D., and Jones, R. A., Selected methods for the detection and assessment of ecological effects resulting from the release of genetically engineered microorganisms to the terrestrial environment, *Adv. Appl. Microbiol.*, 38, 1, 1993.

Sylvester, A. W. and Ruzin, S. E., Light microscopy I: dissection and microtechnique, in *The Maize Handbook*, Freeling, M. and Walbot, V., Eds., Springer-Verlag, New York, 1994, 83.

Tabashnik, B. E., Evolution of resistance to *Bacillus thuringiensis*, *Annu. Rev. Entomol.*, 39, 47, 1994.

Tabashnik, B. E., Liu, Y.-B., Malvar, T., Heckel, D. G., Masson, L., Ballester, V., Granero, F., Ménsua, J. L., and Ferré, J., Global variation in the genetic and biochemical basis of diamondback moth resistance to *Bacillus thuringiensis*, *Proc. Natl. Acad. Sci. U.S.A.*, 94, 12780, 1997.

Tapp, H. and Stotzky, G., Dot blot enzyme-linked immunosorbent assay for monitoring the fate of insecticidal toxins from *Bacillus thuringiensis* in soil, *Appl. Environ. Microbiol.*, 61, 602, 1995a.

Tapp, H. and Stotzky, G., Insecticidal activity of the toxins from *Bacillus thuringiensis* subspecies *kurstaki* and *tenebrionis* adsorbed and bound on pure and soil clays, *Appl. Environ. Microbiol.*, 61, 1786, 1995b.

Tapp, H. and Stotzky, G., Monitoring the fate of insecticidal toxins from *Bacillus thuringiensis* in soil with flow cytometry, *Can. J. Microbiol.*, 43, 1074, 1997.

Tapp, H. and Stotzky, G., Persistence of the insecticidal toxins from *Bacillus thuringiensis* subsp. *kurstaki* in soil, *Soil Biol. Biochem.*, 30, 471, 1998.

Tapp, H., Calamai, L., and Stotzky, G., Adsorption and binding of the insecticidal proteins from *Bacillus thuringiensis* subsp. *kurstaki* and subsp. *tenebrionis* on clay minerals, *Soil Biol. Biochem.*, 26, 663, 1994.

Tovar-Gomez, M. R., Emile, J. C., Michalet-Doreau, B., and Barriere, Y., *In situ* degradation kinetics of maize hybrid stalks, *Anim. Food Sci. Technol.*, 68, 77, 1997.

Van Rie, J., McGaughey, W. H., Johnson, D. E., Barnett, B. D., and Van Mellaert, H., Mechanism of insect resistance to the microbial insecticide of *Bacillus thuringiensis*, *Science*, 247, 72, 1990.

Venkateswerlu, G. and Stotzky, G., A simple method for the isolation of the antilepidopteran toxin from *Bacillus thuringiensis* subsp. *kurstaki*, *Biotech. Appl. Biochem.*, 12, 245, 1990.

Venkateswerlu, G. and Stotzky, G., Binding of the protoxin and toxin proteins of *Bacillus thuringiensis* subsp. *kurstaki* on clay minerals, *Curr. Microbiol.*, 25, 225, 1992.

Vettori, C., Stotzky, G., Yoder, M., and Gallori, E., Interaction between bacteriophage PBS1 and clay minerals and transduction of *Bacillus subtilis* by clay-phage complexes, *Environ. Microbiol.*, 1, 347, 1999.

Wadman, M., US groups sue over approval of *Bt* crops, *Nature*, 397, 636, 1999.

West, A. W., Fate of the insecticidal, proteinaceous, parasporal crystal of *Bacillus thuringiensis* in soil, *Soil Biol. Biochem.*, 16, 351, 1984.

West, A. W. and Burges, H. D., Persistence of *Bacillus thuringiensis* and *Bacillus cereus* in soil supplemented with grass or manure, *Plant Soil*, 83, 389, 1985.

West, A. W., Burges, H. D., and Wyborn, C. H., Effect of incubation in natural and autoclaved soil upon potency and viability of *Bacillus thuringiensis*, *J. Invertebr. Pathol.*, 44, 121, 1984a.

West, A. W., Burges, H. D., White, R. J., and Wyborn, C. H., Persistence of *Bacillus thuringiensis* parasporal crystal insecticidal activity in soil, *J. Invertebr. Pathol.*, 44, 128, 1984b.

West, A. H., Burges, H. D., Dixon, T. J., and Wyborn, C. H., Survival of *Bacillus thuringiensis* and *Bacillus cereus* spore inocula in soil: effects of pH, moisture, nutrient availability and indigenous microorganisms, *Soil Biol. Biochem.*, 17, 657, 1985.

Wolfersberger, M. G., The toxicity of two *Bacillus thuringiensis* δ-endotoxins to gypsy moth larvae is inversely related to the affinity of binding sites on midgut brush border membranes for the toxins, *Experientia*, 46, 475, 1990.

Yin, X. and Stotzky, G., Gene transfer among bacteria in natural environments, *Adv. Appl. Microbiol.*, 45, 153, 1997.

Yudina, T. G., Comparison of antibacterial activity of the parasporal bodies from various bacilli, *Izv. Akad. Nauk Ser. Biol.*, 5, 451, 1996.

Yudina, T. G. and Burtseva, L. I., Activity of δ-endotoxins of four *Bacillus thuringiensis* subspecies against prokaryotes, *Mikrobiologiya*, 66, 25, 1997.

Yudina, T. G., Mil'ko, E. S., and Egorov, N. S., Sensitivity of *Micrococcus luteus* dissociation variants to δ-endotoxins of *Bacillus thuringiensis*, *Mikrobiologiya*, 63, 365, 1996.

chapter nine

Survival, persistence, transfer: The fate of genetically modified microorganisms and recombinant DNA in different environments

Beatrix Tappeser, Manuela Jäger, and Claudia Eckelkamp

Contents

9.1   Introduction

Genetic engineering provides a very powerful and sophisticated means to create new combinations of genes and traits which, as far as we have been able to assess, mainly have developed separately and remained separate throughout their evolution. Separate development and the development and maintenance of barriers are prerequisites of the development of diversity — from species to ecosystem diversity. There are exceptions, especially in the microbial world. But even there, barriers have developed that normally restrict genetic exchange to a limited community of bacterial species. Unfortunately, we only have superficial understandings of those selective and regulating forces guiding horizontal transfer and, more importantly, of those factors helping new genes and traits to become established in new hosts. Further, engineered traits are not part of the existing cellular and organismal regulating mechanisms and feedback interactions. Recombinant genes are essentially inputs developed outside the "usual" organismal and environmental regulating interactions.

We concentrate on genetically engineered microorganisms (GEMs) rather than non-manipulated bacteria and focus on the overall ability of GEMs to survive, persist, and spread regardless of the type of trait that has been introduced. We analyze current knowledge of the fate of GEMs in a variety of environments including soil, aquatic, wastewater, and the human digestive tract to understand if some environments may be safer or more predictable than others. Clearly, each of these environments is relevant with the variety of GEMs used currently and considered for release. Does survival in one environment help predict the situation for GEMs in another environment? Are any generalizations or overall assumptions possible? To analyze these questions, we use the following tenets and tools:

1. GEMs and recombinant genes/traits taken up by autochthonous organisms may have the potential to interfere with the existing organismal interactions and balances because they may "provide" properties and newly combined traits never before present in a given environment. That does not mean that each GEM or each recombinant gene will necessarily pose a risk but, as of yet, we have no means and criteria by which to assess the possibility.
2. Our knowledge of microbial ecology and of biotic and abiotic factors influencing survival, gene transfer, population dynamics, etc. is very limited and scattered.
3. There is no real chance to assess all possible parameters or different combinations of parameters in different environments to discern all possible impacts of a GEM. If the goal is to minimize the probability of adverse effects, we have to minimize exposure/access to the environment, especially for those GEMs designed for contained use and never tested for environmental effects.

As a consequence of the above and based on the Precautionary Principle, survival and spread of GEMs and their recombinant sequences are perceived as risks.

At present, GEMs are discussed as tools for solving diverse problems. Already, large quantities of enzymes and drugs are produced with the aid of GEMs. Although these production processes take place in fermenters, they inevitably lead from time to time to inadvertent release of genetically modified production organisms into the environment (De Vos 1998). Nevertheless, containment of GEMs has been deregulated and safety measures have been relaxed throughout the industrial world, as genetic engineering has not yet resulted in any obvious accident or visible negative impact during two decades of rapid development and constantly increasing use.

Further, GEMs are released quite intentionally for such purposes as soil pollution remediation. GEMs are also used as biological pesticides in agriculture (for example, *Pseudomonas* sp. carrying a *Bacillus thuringiensis* toxin) and active vaccines in medicine.

At the same time as use of GEMs has increased, another development has taken place, according to the 1996 WHO Report (World Health Organization 1996), notably an increase in frequency of outbreaks of new and re-emerging infectious diseases. Current pathogen strains are often resistant to known treatments, some even resistant to all commonly used antibiotics. Horizontal gene transfer is now recognized as the main avenue of exchange of genetic material in the microbial world, and hence also of the exchange and spread of antibiotic resistance genes.

These developments give rise to two questions: (1) Does the extensive use of antibiotic-resistant genes in genetic engineering contribute to the increase in frequency of antibiotic resistance in bacterial pathogens? (2) What will be the outcome of a spread of recombinant genes analogous to the spread of antibiotic-resistant genes?

Although the past few years have seen a considerable increase in knowledge of microbial ecology, it is still not possible to predict the behavior of GEMs and/or their recombinant genes in the environment. On the one hand, estimating the consequences of a GEM release requires a definition of characteristic ecosystem structures and functions. On the other hand, a core prerequisite for predicting ecological effects of microorganisms is knowledge of their ecological functions in their natural habitat (Förster 1998). As yet, there are no generally accepted ecosystem criteria for assessing the ecological risk of releasing GEMs. While it is true that many methods and parameters have been proposed and applied for determining the ecological effects of GEM releases, many of these methods are so crude as to be able to detect only "disastrous" changes in an ecosystem at best (Förster 1998).

As yet, it is not possible to study the overall effects of GEMs and their new recombinant genes on the ecosystem. However, it is possible to study specific properties of microorganisms, such as short generation times, adaptability to unfavorable environmental conditions, and the capacity to exchange genetic material. This makes it possible to develop indicators for determining the potential impact of a GEM release. Kluepfel (1992) distinguishes five ecological parameters that he considers relevant to questions of biosafety related to GEMs and recombinant DNA:

1. Viability or survival ability;
2. Spread;
3. Population dynamics;
4. Competition between different microorganisms;
5. Effect on biocoenoses in the environment.

In this following chapter we first give an overview of the survival ability of GEMs and the persistence of isolated DNA in different environmental media. We then use this overview as the basis for summarizing present knowledge of the possibilities and limits of natural gene transfer and the spread of transgenes.

9.2 Survival and spread of genetically engineered microorganisms — Overview

On the whole, the literature suggests that the ability of GEMs to survive often has been greatly underestimated both in the case of GEMs engineered explicitly for use in the field as well as in the case of those designed for use in closed systems. Characteristics of GEMs in both categories are discussed in detail in Jäger and Tappeser (1996) and in Eckelkamp et al. (1998).

The much-cited argument that GEMs are generally at a disadvantage because of the "extra burden" constituted by their additional gene(s) has been definitively refuted by many studies (Bouma and Lenski 1988, Regal 1988, 1994, Awong et al. 1990, Chao and Feng 1990, Barcina et al. 1992, Sobecky et al. 1992, Van Elsas 1992, Lenski et al. 1994, Goldschmidt et al. 1994, Kozdroj and Piotrowska-Seget 1995, Kozdroj 1996, among others). In addition, there is still no evidence of plasmid-bearing microorganisms being less fit than their native parents. For example, Fujimura et al. (1994) tested genetically modified *Saccharomyces cerevisiae* and their parent strains under different environmental conditions but were unable to prove any negative impact on survival probability. Under selective pressure from, for example, antibiotic substances, GEMs have even been shown to profit from possessing the corresponding resistance genes (Van Elsas 1992). This was also demonstrated by Ramos et al. (1994) with recombinant *Pseudomonas putida* (pWW0-EB62) in soil systems. In unsterile soil and under conditions where selective pressure was assumed, cells numbers even increased after inoculation (Ramos et al. 1994).

This is not to say that selective pressure is a necessary prerequisite for survival of GEMs. Their survival and persistence depend on various interacting factors, including biotic factors such as nutrient supply, predators, bacteriophage density, competing microorganisms, and population density, as well as abiotic factors such as temperature, pH, oxygen demand, water and salt content, and soil characteristics (Smalla et al. 1989, Van Elsas 1992, Doyle et al. 1995). Other important factors may include the special species-specific, genetically determined capabilities of the GEM (e.g., development of permanent stages, spores) and the new properties conferred on it by inserted transgenes. Some microorganisms can survive for very long periods under unfavorable environmental conditions by developing spores or cysts. Other microorganisms, rather than developing permanent stages, take on a state in which they are viable but not culturable (VNC) (Eckelkamp et al. 1998, Förster 1998, Tappeser et al. 1999). New properties conferred by transgenes can either promote persistence (fitness-enhancing traits) or limit their existence (suicide genes).

This long list of factors influencing survival should indicate the difficulty of monitoring and making statements on the ability of the GEM to survive in the environment. The validity of statements about the survival and dispersal of an organism depends on the detection method used and on the possibility of studying different habitats and developmental stages of the organism.

9.2.1   Survival

The difficulty of making reliable statements about GEM survival and dispersal is aptly demonstrated by examples of the release of genetically modified *Pseudomonas fluorescens*. In 1993 and 1994, Thompson and colleagues (Thompson et al. 1995a, 1995b) released sugar beet seeds inoculated with genetically modified *P. fluorescens*. The GEMs successfully colonized the roots and leaves of the sugar beets and remained detectable over the entire season (270 days). However, the GEM population dynamics on the sugar beet leaves varied considerably between study years. In 1993, the proportion of GEMs in the entire pseudomonad population was low, accounting for less than 6%. Despite the fact that genetically modified *P. fluorescens* had not been detectable either in the soil or on the overwintering plants, their proportion rose to 81% in the course of the following year. De Leij et al. (1995) found similar effects after the release of wheat grains inoculated with transgenic *P. fluorescens*. After 319 days, GEM abundance decreased to below the detection threshold. In spite of this, wheat that had germinated after the harvest and had then been resown was successfully colonized by the GEM. Lilley and Bailey (1997) similarly found that fluctuations in transgenic *P. fluorescens* populations in the hothouse and in the fields depend on the developmental stage of the host plant. Following inoculation of leaves and root surfaces of young sugar beet plants, the GEMs initially showed reduced fitness compared with the wild type. As the inoculated plants grew older, this effect was reversed in the field; GEMs on leaves and roots of older plants outgrew the wild type, whereas in the hothouse this was only the case on the roots. Apparently, microorganisms can reappear and successfully colonize habitats in which they were previously undetected.

Attempts to assess the survival ability of a GEM under natural environmental conditions are complicated by changes in their physiological properties, their adaptability, and differences in the initial physiological situation and environmental parameters. Although it is frequently assumed that GEMs that are released into the environment for certain purposes, e.g., pollutant degradation, will suffer heavy mortality and disappear from the environment when conditions change, this view disregards the fact that GEMs are just as capable of adapting to certain environmental conditions as are other microorganisms. For example, Velicer (1999) showed that the probability of survival of GEMs bearing a transgene for 2,4-D degradation did not depend on the presence or absence of 2,4-D. The author concluded from his experiments that GEMs can adapt to altered environmental conditions and in this way persist. In addition, adaptation often becomes apparent only after an extended period of time. Thus, safety studies must take account of long-term processes to permit valid statements of any kind. To date, most studies (1) are discontinued after only a few weeks, (2) find that the number of released GEMs has decreased rapidly and that only a small proportion of the released GEMs has been able to survive, and (3) conclude that the GEMs were unable ultimately to establish themselves in the environment under study. Such studies warrant more careful scrutiny. There are numerous examples in which changes in environmental parameters after an initial reduction in cell numbers down to very low values, possibly even below detection threshold, were followed by population growth and an eventual increase in cell numbers (Sobecky et al. 1992, Van Elsas 1992, Kluepfel et al. 1994, Ramos et al. 1994, Clegg et al. 1995, Gillespie et al. 1995, Sjorgen 1995, Thompson et al. 1995a, 1995b).

Efforts to control the survival rate of GEMs in the field have led to the development of various biological containment strategies. One type of safety modification is aimed at decimating previously released bacteria by making them eliminate themselves after a certain period of time. As yet, these "suicide" systems, as they are called, do not work 100% effectively, and there remains a chance that some mutants will survive (Ramos et al. 1995). For example, in a field trial in Germany, *Sinorhizobium meliloti* was genetically modified to shorten its life span. The aim was to achieve "biological containment" by integrating a luciferase gene in the chromosomal *recA* gene, which codes for recombination, among other functions (Selbitschka and Pühler 1994, Selbitschka et al. 1994). However, in these bacteria reduced survival ability was only observed in model ecosystems. Under conditions of a real release, this trait did not distinguish the modified strain from the original, unmodified one (Dresing et al. 1995).

9.2.2 Spread of GEMs

An important consideration in assessing risk from GEMs is the potential for GEMs to access habitats other than those anticipated by their point of release. Microorganisms may be transported from unfavorable environments to ones that offer better chances of survival. Microorganisms can be transported in a number of ways, including transport by wind, by running waters or rainwater, or as entities attached to objects or agricultural produce. They can also be carried by such vectors as protozoa, springtails, bees, ants, and earthworms (Schmidt 1991, Clegg et al. 1995, Heijnen and Marinissen 1995, Weiskel et al. 1996). These vectors are capable of movement over variable distances and have intestinal flora that can interact with ingested microorganisms (Byzov et al. 1993). Large animals such as mammals and birds also can contribute to the spread of microorganisms by carrying them over long distances in their intestinal tract or on the surface of their body (e.g., in fur or on feathers).

9.2.2.1   Dispersal by water

The possibility of vertical transport of microorganisms in the soil poses different hazards including health risks from contamination of groundwater by GEMs and their products and the potential for transport of GEMs to habitats offering better chances of survival (Natsch et al. 1996). Previously, researchers had concluded from laboratory and microcosm studies that GEMs are retained by the soil matrix at the point of release, essentially remaining stationary. These conclusions are debatable, however. Smith et al. (1985), for example, have shown that bacteria under certain conditions are capable of rapid vertical transport via macroporous flows. Analogous studies on GEMs have yielded varied results on this point. In microcosm trials involving inoculation of wheat grains with transgenic *P. aureofaciens*, De Leij et al. (1994) found that whereas most GEMs remained in the upper 15 cm of the rhizosphere of the wheat plants, they could be discerned in small numbers down to a depth of 60 cm. Similarly, in field tests, it has been found that GEMs tend to remain in the uppermost soil layers. However, recent studies indicate that small numbers of GEMs can be transported to depths of 15 to 75 cm (Heidenreich 1998). GEMs are also capable of reaching even deeper soil horizons. For example, after inoculating 2.5-m-long field lysimeter columns with *P. fluorescens*, Troxler et al. (1997) found that vertical transport of bacteria into deeper soil layers depended on precipitation intensity. With normal rainfall, transport proceeded very slowly. A heavy rainstorm that occurred after 200 days transported the *Pseudomonas* strain through the lysimeter. When heavy rainfall was simulated directly after inoculation, bacteria were transported quickly into deeper soil layers (Natsch et al. 1996). The authors concluded that the likelihood of vertical transport of *Pseudomonas* into the groundwater cannot be dismissed under natural conditions, especially not after heavy rainfall (Natsch et al. 1998).

Gillespie et al. (1995) simulated rainfall on wheat plants or naked soil in greenhouse experiments after inoculation with genetically modified *P. aureofaciens*. The number of cells detectable in the soil was found to correlate with the number of cells in water running from the plots. One exception to this occurred 245 days after inoculation: 10^8 cfu (colony-forming units) were measured in the water probes, whereas cell counts in the soil were below the detection limit. The authors concluded that standard techniques for detection of GEMs in soil may not be suitable for registering bacteria in all soil layers and that short and heavy rainfalls can cause quite high numbers of GEMs (about 10^{14} cfu/ha) to be washed away from the place of release (Gillespie et al. 1995). Similar testing was done by Hekman et al. (1995) with *P. fluorescens* and *Burkholderia cepacia*. In the case of planted microcosms, these authors demonstrated enhanced distribution of inoculated bacteria in deeper soil layers after moderate rainfall (Hekman et al. 1995).

GEMs can be dispersed not only vertically but also laterally. In studies on lateral dispersal it was found that whereas some GEMs remained stationary at the point of release, others spread by a distance of up to 40 cm (e.g., Amarger and Delgutte 1990, Cook et al. 1990, Kluepfel et al. 1991, Thompson et al. 1995a, 1995b). De Leij et al. (1995) detected *P. fluorescens* released with wheat grains at a distance of up to 2 m from the point of release. Dane and Shaw (1996) even found transgenic bacteria *Xanthomonas campestris* at distances of 2 to 8 m from the inoculated *Brassica* plants on which the GEMs had been released.

9.3.2.2 Dispersal by animal vectors

Lilley et al. (1997) have found that caterpillars of cabbage moths (*Mamestra brassicae*) can transmit leaf-colonizing *P. fluorescens* from infected to non-infected sugar beet plants. Insects not only can transmit leaf-colonizing GEMs but also root-inhabiting ones. Root-inhabiting GEMs can enter a plant via root lesions and from there can be transported into aboveground plant parts (Kluepfel et al. 1994) where they eventually can be ingested by plant-feeding insects. In insects (southern corn root worm) fed on maize colonized by transgenic *P. aureofaciens*, the GEMs remained detectable for 12.5 days. The insects were also capable of transmitting the GEMs to non-infected plants (Kluepfel et al. 1994). Similarly, red-legged grasshoppers (*Melanoplus femurrubrum*) were found capable, both in field and microcosm trials, of taking up rhizospheric GEMs from aboveground plant parts, retaining them in their gastrointestinal tract for approximately a week, and then passing them with their feces on to noninfected maize plants. In the field trials, the probability of successful transmission was greater the more time had elapsed since GEM uptake, despite a decrease in absolute GEM abundance in the insects (Snyder et al. 1999). Kluepfel et al. (1994) and Snyder et al. (1999) concluded that leaf-eating insects are capable of taking up and transmitting bacteria originating from the rhizosphere. Considering that classical plant pathogens are capable of long-range dispersal through insects, it is reasonable to assume that it is also possible for released rhizospheric GEMs to be transported to new host plants and new habitats by insects.

All these data indicate that it will be impossible to restrict GEMs to the point of their release. As long as we are unable to assess the potential impact of a GEM on different environments, one approach to operationalizing the Precautionary Principle might be to assess the possibility of long-term GEM survival and persistence. Survival and persistence are preconditions for developing an impact.

9.3 Survival and persistence in different environments

The following section presents in detail the data on survival of GEMs in different environments, persistence of recombinant DNA, and what is currently known of the different gene transfer mechanisms involved.

9.3.1 Survival in the digestive tract

The digestive systems of both vertebrates and invertebrates offer oportunities for bacteria ingested with food to come in close contact with each other and with the microflora of the animal. Digestive environments may prolong bacterial survival compared with other environments and, because of the high density of microorganisms present, may increase the probability of gene transfer (Adamo and Gealt 1996). Further, bacteria can be disseminated by their host animals, e.g., earthworms (Clegg et al. 1995). Some bacteria may even change their metabolic activities and capacity for survival during their passage through the digestive tract (Clegg et al. 1995).

Brockmann et al. (1996) investigated the possibility of different strains of *Lactococcus lactis* colonizing the rat intestine and transferring their plasmids among one another. Although the digestive tract of germ-free animals was found to be rapidly colonizable by all kinds of bacteria, colonization only succeeded temporarily (for 2 to 3 days) in non-germ-free controls. Studies with *L. lactis* in the human intestine yielded similar results (Klijn et al. 1995a). In spite of the short persistence time, the conjugative plasmid pAMß1, which has a broad host range, proved transferable both to a *Lactococcus* recipient strain in the intestine of germ-free rats and an endogenous strain of *Enterococcus faecalis* in that of non-germ-free animals (Brockmann et al. 1996). The authors concluded that even those microorganisms only temporarily detectable in the digestive tract are able to transfer their conjugative plasmids to endogenous microflora. The rate of such events primarily depends on the structure of the plasmids themselves.

9.3.2   Survival in wastewater and sludge

The most important factor influencing bacterial survival in wastewater is the presence of competitors. Because competitive predator populations fluctuate seasonally, time of year is an important factor in evaluating the possibility of GEM survival (Inamori et al. 1992). If they are not eliminated by their competitors, *Escherichia coli* K12 and *P. putida* species have the best chances of survival when associated with snow particles containing sludge material (McClure et al. 1989, Overbeck 1991). In addition, whereas viable cell counts in the aqueous phase were found to decrease quickly after introduction of recombinant *E. coli* K12 into the aeration basin of a model wastewater treatment plant, such a reduction in viable *E. coli* K12 was not usually the case for those remaining attached to sludge particles (Heitkamp et al. 1993).

Most studies on survival and gene transfer in wastewater and sludge use *in vitro* systems or models of settlement tanks or activated sludge for investigation. Ashelford et al. (1995) concentrated on percolating filter beds, which are layered with living cells (a so-called biofilm) similar to the river epilithon. Two inoculated *Pseudomonas* strains survived the whole investigation time of 145 days. The authors demonstrated that a conjugative plasmid (pQKH6) harbored by one of these strains was transferred to another strain that initially lacked the plasmid (Ashelford et al. 1995). Similar results were published by Feldmann and Sahm (1994) who investigated the survival of four different recombinant microorganisms in laboratory wastewater systems. Their experiments, sponsored by Bayer AG, Germany, were part of the "Verbundprojekt Sicherheitsforschung Gentechnik," which aimed at investigating the safety of GEMs released from production facilities. Although none of the GEMs (*Saccharomyces cerevisiae, Zymomonas mobilis, Corynebacterium glutamicum, Hansenula polymorpha*) persisted in the settlement tank, the two yeasts *S. cerevisiae* and *H. polymorpha* tested in the aeration basin remained detectable for more than 25 days.

9.3.3   Survival in aquatic ecosystems

Many of the above phenomena involving GEM persistence occur in aquatic ecosystems. The chance of survival of, e.g., *E. coli* and *Campylobacter jejuni* increases in filtered water taken from lakes, since filtering eliminates predators and competitors (Korhonen and Martikainen 1991). Sterile tap water also allowed the persistence of four transgenic microorganisms tested within the scope of the Verbundprojekt Sicherheitsforschung Gentechnik. In this medium, the microorganisms survived the whole testing period of 440 days (Tebbe et al. 1994a). The longer a GEM can survive, the better are the possibilities of its adaptation to the new habitat, its spread to other habitats, and the transfer of its recombinant DNA to other microorganisms — all with unknown consequences.

Survival of bacteria in aquatic ecosystems can also depend on abiotic factors such as temperature, nutrient availability, and osmotic conditions. Ahl et al. (1995) studied the survival of *P. fluorescens* Ag1 in experiments with seawater microcosms (0.5 l water) and seawater mesocosms (5300 l water). They demonstrated less pronounced reduction of bacterial cells inoculated in mesocosms than had been found in earlier microcosmic experiments. In the study, 25% of inoculated pseudomonads survived the whole testing period of 14 days.

Intensity and seasonal changes of light are also important abiotic factors influencing the survival of bacteria in natural environments (Canteras et al. 1995). Five different strains of *E. coli* kept in complete darkness persisted for 96 h in sterile river water with only a minor reduction of cfu. Differences in persistence were only apparent upon illumination. However, sterile conditions cannot be taken to represent natural aquatic ecosystems. In the comparative study of Alvarez et al. (1996b) on the survival of *E. coli* strains with or without plasmids (DH1 and JM103) in tropical river water systems, differences in survival only showed in unsterile test chambers.

Bacteria taken from their natural environment, transformed with a plasmid, and then released again as GEMs have a very high probability of establishing themselves in their original ecosystem. For example, Sobecky et al. (1996) demonstrated that a marine bacterium (*Achromobacter* sp.) is able to form stable populations in seawater microcosms within 2 to 3 weeks after being transformed with a plasmid bearing a gene encoding an alkaline phosphatase. Even the enzyme was proved to have retained its activity. Bacteria not adapted to the environmental conditions of different water systems, like typical enteropathogenic microorganisms (e.g., *E. coli* ETEC, *Yersinia enterolitica*, and *Campylobacter jejuni*) are not only able to survive within their hosts at body temperature but may also persist in a cultivable stage for weeks in aquatic environments at temperatures between 6 and 16°C (Terzieva and McFeters 1991). Davies et al. (1995) demonstrated that the time of survival of enteropathogenic bacteria (fecal *Streptococci* and Coliforme) in sediments exceeds that in open water. Although Gram-negative bacteria inoculated in open seawater soon entered a stage of non-culturability, *E. coli* cells residing in sediments remained cultivable during the entire 68-day testing period (Davies et al. 1995). The same seems to be true of groundwater aquifers and freshwater environments; sediments provide a better chance of survival than does open water. This was demonstrated by Winkler et al. (1995) with *Burkholderia cepacia* in a groundwater microcosm experiment and by Fish and Pettibone (1995) with *E. coli*. The latter species remained cultivable in sediments for 56 days, i.e., for the entire duration of the test.

In conclusion, quite independently of the engineered trait, GEMs apparently have a good chance of survival in the sediments of different water systems. As yet, there are no reliable criteria to predict the probability of their survival.

9.3.4 Survival in soil

Since soil environments are especially complex and heterogeneous systems, it is difficult to predict the chance of survival they offer to newly introduced GEMs. The density of soil bacteria is estimated at about 10^8 to 10^9 cfu/g soil. However, less than 10% of these are cultivable using current techniques, and only around 1% have been characterized to date (Hugenholz and Pace 1996, Pace 1997). In clay sediments, bacteria are detectable at depths down to 224 m (Boivin-Jahns et al. 1996). Microbial populations from the laboratory decrease drastically during the first days after their introduction into natural soil and afterward often remain at a constant level (Schmidt 1991).

In assessing and comparing published data on the survival of GEMs in soil systems, it is important to analyze thoroughly the design of the microcosm or mesocosm and the detection method used. The variety of microcosms studied makes it is rather difficult to compare the results collected (Angle et al. 1995). An undisturbed soil texture is an important prerequisite for optimal imitation of natural environmental conditions, as Angle et al. (1995) demonstrated. Those authors compared survival over a 2-year period of *P. aureofaciens* in intact soil cores, disturbed soil microcosms (50 g of fresh, sieved soil), and in experimental field release. The disturbed soil microcosm showed a much shorter survival time (0 to 34 days, after which the cell number reached below the detection threshold) than the intact soil cores or the field release experiments. In bulk soil, the pseudomonads could be detected for 63 days; in the rhizospere, they could be detected even after 93 days.

Validation of the microcosms used is also important with aquatic systems. Leser (1995) found 63 to 76% agreement between data obtained from a lake water microcosm and a natural lake. After inoculation of *Alcaligenes eutrophus* in both systems, the degree of coincidence was reduced to 2 to 27%, but increased again with decreasing cell counts (Leser 1995).

As with aquatic and wastewater systems, predation in soils by protozoa is one of most important factors influencing the persistence of inoculated GEMs. In soils, protozoal activity requires a certain degree of soil moisture and free water in soil pores of suitable size (Van Elsas 1992, Wright et al. 1995). Experiments in soil microcosms with *P. fluorescens* and the ciliate *Colopoda steinii* demonstrated that small pores (<6 μm) may offer effective protection zones against predators (Wright et al. 1995). Although bacterial populations in larger pores (<30 μm) diminish more rapidly, those bacteria that do survive show a higher level of metabolic activity in response to the richer supply of nutrients there (Wright et al. 1995). Van Der Hoeven et al. (1996) used a computer simulation model to evaluate the chance of survival of GEMs in soil pores of different sizes. Bacteria introduced into the soil were eliminated very slowly, in some cases taking years for total elimination (Van Der Hoeven et al. 1996).

The number of bacteria inoculated into the soil is a very important determinant of duration of survival, as differences as small as 100 *E. coli* K12 cells/g soil suffice to change the results of an assay (Recorbet et al. 1992). This strong correlation between inoculation number and survival is also found in aquatic ecosystems (Karapinar and Sahika 1991). Another criterion influencing survival is the depth of inoculation. In depths from 10 to 50 cm, the chance of survival of the phytopathogenic bacterium *Erwinia carotovora* increases with growing inoculation depth. This correlation largely reflects differences in competitive pressure (Armon et al. 1995). Recorbet et al. (1995) demonstrated uneven distribution of genetically modified *Escherichia coli* K12 which had been inoculated in soil systems. Furthermore, GEMs were concentrated in different soil layers than autochthonous bacteria.

Spore-forming bacteria like *Bacillus subtilis* are particularly likely to persist for prolonged time periods. Spores of *B. subtilis* of both native and genetically modified parent strains remained culturable in sterile soil for at least 50 days (Tokuda et al. 1995). The ability of bacteria, including GEMs, to adapt to natural environments makes it more difficult than with higher organisms to estimate their chance of survival after release. Cells of *P. fluorescens* showed improved stress resistance against temperature or osmotic pressure 1 day after inoculation into two different types of soil (Van Overbeek et al. 1995). When cultivated under starving conditions in the laboratory, these bacteria may withstand stress factors such as hunger for 70 days by reducing their cell size. These physiological changes soon reverse when conditions improve (Clegg et al. 1996).

In a very profound and very long-term study, Sjogren (1995) examined survival of *E. coli* using a multiresistant native strain isolated from a lake. The bacteria were inoculated into different field plots seeded with rye. After 41 days, there were no more bacteria detectable on the rye grass. After 2 months, bacteria were re-isolated from soil at depths of 20 to 50 cm. At 60 cm depth, they reached groundwater and persisted there at the water–soil interface for 2 more years. Six years later 1.5 cfu/ml were still detectable, and after 8 years, cell counts had increased to 8 cfu/ml. After 13 years, at the end of the experiment, population density was still at 0.25 cfu/ml (Sjogren 1995). The author calculated the speed of distribution as about 2 cm/day. In analogous experiments performed under laboratory conditions with microcosms of the same soil, the maximum survival time had been 2.5 years (Sjogren 1995).

9.3.5 Survival on plants

It should be evident that any analysis of survival of microorganisms, including GEMs, in soil has to include analysis of those microorganisms persisting on the plants growing in that soil. For example, a recombinant strain of *Bacillus amyloliquefaciens*, used to produce alpha-amylase, was no longer detectable in the soil of planted microcosms 6 days after inocul[...] the grass for at least 70 days (Wendt-Potthoff et al[...] f the phytopathogenic bacteria *Xanthomonas cam*[...] host plants survived for 6 months and were [...] nrelated plants (Dane and Shaw 1996).

[...]s may depend on the plant species used. [...]*orescens* remained at a constant level on [...]g a 30-day glasshouse experiment, but [...] Cirvilleri et al. 1996). Genetically modified [...]lated from sugar beets, persisted on leaf [...] least 531 days. Fluctuations in the number [...]uations in the endemic pseudomonads. In [...]minent population on the leaves, GEMs [...]es in the glasshouse and being distributed [...]mpson et al. 1995a). Analogous deliberate [...] these GEMs for 270 days (the entire veg-[...]) spread further (Thompson et al. 1995b). [...]ason were colonized, although cell counts [...] during the winter season in both soil and

[...]d dispersal of *P. fluorescens* on wheat plants [...]ger in field experiments than in microcosm [...]r plants and to the soil (De Leij et al. 1995). [...]ents and microcosm experiments were also [...] microcosm experiments on the persistence [...]ot correlate with data from release experi-[...]ed smaller or larger populations than those [...] which plant species was used.

[...]*ncy and adaptation ability*

[...]tivable and dormant cells, regardless of the [...]1991b, Turpin et al. 1993). The problem of [...]n, as results are found to differ tremendously [...]ble cells are counted (Pickup 1991, Stotzky [...]ulturable or cultivable cells are viable and [...]r metabolism is restricted so that they are

[...]NC cells and cultivable cells. Counts of VNC [...]y rather variable (Byrd and Colwell 1990, [...]liano 1996, Jensen et al. 1996). VNC bacteria [...]h high salt concentration (Byrd and Colwell 1990, Garcia-Lara et al. 1993). The problem of poor culturability remains to be addressed in the case of bacteria in activated sludge or soil probes (Binnerup et al. 1993, Wagner et al. 1993). VNC cells can be distinguished from dead cells by metabolism and protein synthesis parameters. VNC cells, rather than losing their plasmids, may even enrich their intracellular plasmid content (Nwoguh et al. 1995, 1996, Arturo-Schaan et al. 1996). The VNC cells can be activated again when environmental conditions change. Improved nutritional state is not the only factor responsible for reactivation; temperature fluctuations in natural environments may also play a role (Ferguson et al. 1995, Oliver et al. 1995a, Colwell et al. 1996, Jiang and Chai 1996, Rahman et al. 1996). The presence of plasmids can further contribute to induction of the VNC stage (Oliver et al. 1995b). Bacteria detection methods, which rely on the expression of marker genes and are thus blind to VNC cells, will underestimate cell concentrations unless they are combined with techniques that record VNC cells (Oliver et al. 1995b, Lee et al. 1996, Leff and Leff 1996).

Obviously, the choice of a detection method greatly influences the level of detection of bacteria released in natural environments. The ability of microorganisms to adsorb on particles and sediments renders detection more difficult, although application of polymerase chain reaction (PCR) has certainly lowered detection limits (Bej et al. 1990, Vahjen and Tebbe 1994, Hodson et al. 1995, Lindahl and Bakken 1995, Tsushima et al. 1995, Lindahl 1996). One disadvantage of the PCR technique is the remaining difficulty in distinguishing between isolated DNA and DNA associated with living or VNC organisms (England et al. 1995, Brockmann et al. 1996, Tamanai-Shacoori et al. 1996).

A further point to consider in the context of survival of GEMs in natural environments is their ability to adapt to unfavorable environmental conditions. When given a chance to adapt slowly, bacterial populations in aquatic microcosms that imitate natural surroundings show prolonged persistence compared with non-adapted populations of the same strain (Awong et al. 1990, Sobecky et al. 1992, Mezrioui et al. 1995). Bacteria cultivated in the laboratory, when exposed to various primary stress factors, such as unfavorable pH or nutrient deficiency, are able to adapt, and also to demonstrate higher resistance against different secondary stress factors (Ferianc et al. 1995, Lou and Yousef 1996). Consequently, any design of microcosm or mesocosm experiments should include at least some environmental stress factors to check responsiveness of GEMs to them. In determining the timescale of such experiments, one should consider that adaptation may be a slow process and that populations of GEMs, undetectable after a short time, may multiply and become apparent again over longer time periods (Ramos et al. 1994, Clegg et al. 1995, Gillespie et al. 1995, Sjorgen 1995, Thompson et al. 1995b). Stress factors may even serve as positive selection pressures for those strains that become adapted.

9.4   *Spread of recombinant genes*

Risk assessment should not rest with the assumption or proof that a given GEM has or has not survived. Risk assessment should also take into account the fate of the DNA of the GEM, which may be stably integrated in, eventually expressed by, and capable of providing advantages to indigenous microorganisms (Henschke and Schmidt 1990, Istock 1991, Schmidt et al. 1994). GEMs may only require a short time of survival to be transported to other ecological niches or to transfer some of their nucleic acids to other constituents of a given ecosystem (Byzov et al. 1993, Heijnen and Marinissen 1995, Weiskel et al. 1996). Although conjugation and transduction are restricted to living cells, transformation may also take place after the death of donor cells. The probability of transformation events rises with the stability and quantity of isolated DNA (Fulthorpe and Wyndham 1991), although selective forces like nutrient composition and stress factors such as toxic components should be taken into account. They may play a crucial role.

Pukall et al. (1996) isolated plasmids with mobilizing capability from soil probes and pig manure in numbers substantially higher than previously estimated. Such plasmids can transform competent cells living in soil or manure environments under suboptimal conditions. These mobilizing plasmids may even be able to mobilize recombinant plasmids without mobilizing genes in these cells (Lorenz and Wackernagel 1987, Pukall et al. 1996).

9.4.1   Persistence of "naked" DNA

Isolated DNA not only originates from dead and lysed cells but can also be actively secreted from living cells (Paget and Simonet 1994). Quantifying extracellular DNA in environmental media seems to pose difficulties, judging by the great variability among published data (Lorenz and Wackernagel 1994). Some authors assume that bacteria in natural environments release large quantities of high-molecular-weight DNA and plasmids into the extracellular gene pool (Romanowski et al. 1990, Wackernagel et al. 1992, Lorenz and Wackernagel 1994). Secretion of DNA and competence for nucleic acid uptake can be induced, especially under unfavorable environmental conditions like those causing starvation (Wackernagel et al. 1992, Lorenz and Wackernagel 1994).

9.4.1.1   Persistence in wastewater treatment plants (water/sludge)

Wastewater treatment plants guarantee rapid and almost complete inactivation and degradation of isolated DNA suspended in open water. Nucleic acids adsorbed to sludge particles, however, find some degree of protection against degradation and also against detection (Aardema et al. 1983, Gross et al. 1994, Bauda et al. 1995). Because efficiency of inactivation of naked DNA depends on different factors, such as temperature, efficiency varies with the season and with the type of treatment plant under study (Bergemann et al. 1994, Lorenz and Wackernagel 1994).

9.4.1.2   Persistence in aquatic systems

Some recent studies have dealt with the origin and distribution of naked DNA in aquatic systems (Beebee 1993, Hermansson and Lindberg 1994, Jiang and Paul 1995, Weinbauer et al. 1995). As in wastewater environments, isolated and dissolved nucleic acids are rapidly degraded in aquatic systems by enzymatic activity (Lorenz and Wackernagel 1994, Alvarez et al. 1996a). Adsorption to sand or clay minerals protects DNA from nucleases that can degrade the molecules, even if the nucleases are bound to the same particles (Aardema et al. 1983, Romanowski et al. 1993a). Prolonged protection of nucleic acids against inactivation is also provided by VNC cells (Byrd et al. 1992, Lorenz and Wackernagel 1994). If naked DNA remains biologically active, it can give rise to the transformation of competent cells. Therefore, transformation experiments are suitable for detecting intact molecules, regardless of whether they were adsorbed to particles or not (Chamier et al. 1993, Romanowski et al. 1993b, Mieschendahl and Danneberg 1994, Alvarez et al. 1996a).

9.4.1.3   Persistence in soils

Isolated DNA in non-sterile soils can persist for some time, depending on such factors as temperature, pH, salt concentration, type of soil, and nucleic acid material, all of which may influence the binding of DNA on mineral or quartz particles (Lorenz and Wackernagel 1992). As in aquatic systems, when DNA is introduced into non-sterile soils, particle-adsorbed nucleic acids are protected against degradation by nucleases, even if both are bound to the same particle samples (Lorenz and Wackernagel 1987, Khanna and Stotzky 1992). Nucleic acids can bind to minerals quite rapidly and relatively independently of pH (Lorenz and Wackernagel 1994). One factor determining the speed of association in aquatic systems is the structure of the DNA (Lorenz et al. 1988, Khanna and Stotzky 1992, Lorenz and Wackernagel 1992, Paget et al. 1993). Factors more relevant to soil environments include the humidity and granular structure of the soil (Hobom 1995).

Data on the persistence of isolated DNA in soils show great variation. Smalla (1995) detected recombinant DNA from sugar beets in soil 18 months after inoculation of plant material. Romanowski et al. (1993b) found particle-associated plasmids persisting in different soils for at least 60 days. Plasmid conformation remained intact during the first 2 days of the study. According to Wackernagel et al. (1992) and Romanowski et al. (1992), the time limit for detecting plasmids in unsterile soil probes without PCR application is 5 to 10 days. Intact genes or plasmids can frequently be detected for as long as the originating bacteria persist.

9.4.1.4   Persistence in the digestive system

If isolated DNA is able to persist in different environmental media, then it is quite possible for the digestive systems of humans or other animals to take up these nucleic acids with food or drinking water. Moreover, as Schubbert and Doerfler (1996) demonstrated in feeding experiments with mice, nucleic acids reaching the gastrointestinal tract may not necessarily be completely fragmented, and thereby inactivated, but instead may reach the bloodstream and temporarily be detectable in leukocytes and liver cells. The results found in mice experiments may be extrapolated to other mammalian digestive systems, including that of humans. For example, while investigating the persistence of bacterial cells (*Lactococcus lactis*) and their DNA in the human intestine, Klijn et al. (1995b) were able to isolate the bacterial DNA from feces 4 days after administering inoculated milk drinks.

9.4.2   Avenues and barriers of genetic transmission

To date, gene transfer has been found mostly by means of specially designed plasmids, as in a number of studies carried out in microcosms (Mazodier and Davies 1991, Paul et al. 1991, Miller et al. 1992, Paget et al. 1992, Paul 1992, Collard et al. 1993, Del Solar et al. 1993). However, the spread of antibiotic resistance markers throughout bacterial communities would indicate that gene transfer is likely to happen not only in more or less artificial settings but also under natural conditions (Saye and Miller 1989, Kruse and Sørum 1994, Götz et al. 1996). Moreover, some plasmids originating from Gram-positive bacteria have also been isolated in Gram-negative bacteria, stressing the possibility of a wide distribution of genetic information. It was not possible to ascertain in this case whether gene transfer took place via conjugation, transformation, or transduction (Del Solar et al. 1993). The avenues for genetic transmission and barriers that prevent it are not very well understood (Istock 1991).

9.4.2.1   Conjugation

Conjugation, i.e., the transfer of nucleic acids by direct contact between a living bacterial donor and recipient cells, is assumed to be the most effective mechanism of genetic transmission under natural conditions. Wastewater treatment plants offer ideal conditions for this kind of transmission because they contain an abundance of different cell populations (Schneider et al. 1994). Analogously, conjugation in the gastrointestinal tract is promoted by high population densities of potential bacterial donor and receptor cells (Morrison 1996). Several studies have dealt with conjugation in digestive systems (Klijn et al. 1995a, 1995b, Adamo and Gealt 1996, Brockmann et al. 1996, Rang et al. 1996, Rybachenko et al. 1996). Microcosm experiments or *in situ* testing with soil or aquatic systems have also demonstrated the transfer of plasmids by conjugation in other environments (Bale et al. 1988, Lukin and Prozorov 1992, Lebaron et al. 1994, Sandaa and Enger 1994, Pukall et al. 1996).

Barkay et al. (1995) have pointed out the importance of testing with autochthonous bacteria as donors or recipients for efficient detection of transfer through conjugation in natural environments. Other authors have discussed the influence of different soil systems and supplements of plant material, temperature, and soil humidity (Van Elsas et al. 1990, Temann et al. 1992, Klingmüller 1993, Hermansson and Linberg 1994, Ashelford et al. 1995). The main factors inhibiting or promoting conjugation in natural environments are thought to be selective forces from, e.g., heavy metals, herbicides, or antibiotics (Gadkari 1991, De Rore et al. 1994a, 1994b, Klingmüller and Rieder 1994, Kozdroj and Pietrowska-Seget 1995). Lafuente et al. (1996) have pointed out that conjugation in sterile soil microcosms is most efficient under "natural" abiotic conditions. Since conjugation is an energy-consuming process, it can be enhanced by any of a variety of nutrients (Götz and Smalla 1997). An environment fulfilling the nutrient requirement, and additionally characterized by high numbers of bacterial cells, will provide optimal conditions for conjugation (Hermansson and Linberg 1994, Barkay et al. 1995, Amabile-Cuevas and Chicurel 1996).

9.4.2.2 Transduction

Transduction, a gene transmission process mediated by bacteriophages, requires living cells. It is more likely to occur under conditions of high metabolic activity (Ripp and Miller 1995). Transduction may contribute to the spread of plasmids that are not transferable by conjugation because they lack essential sequences (Replicon et al. 1995). Transduction is not restricted to plasmids, however; chromosomal DNA can be transferred by transduction as well (Lorenz and Wackernagel 1993). Where that happens, the host range of the phages determines the possible radius of distribution of genetic material. Empirical evidence of gene transfer by transduction occurring in natural environments (Ogunseitan et al. 1992, Ripp et al. 1994, Stotzky and Babich 1994, Schicklmayer and Schmieger 1995) has been further substantiated by experiments carried out under natural conditions (Miller et al. 1992, Kidambi et al. 1994) and by the detection of high numbers of phage particles in some environmental media (Kokjohn and Miller 1992, Lorenz and Wackernagel 1993, Schäfer 1996). Because transduction requires a certain minimum concentration of phage particles and corresponding bacterial host cells to be efficient, it is more likely to occur in environmental compartments characterized by high cell density and plentiful nutrients (Ogunseitan et al. 1992, Replicon et al. 1995). These conditions, however, should not be taken as prerequisite since oligothrophic conditions and natural cell densities also allow transduction (Goodman et al. 1994, Ripp et al. 1994). Adsorption of phages and bacteria to particles further increases the chances of survival, interaction, as well as transduction (Pickup et al. 1993, Ripp and Miller 1995).

9.4.2.3 Transformation

Transformation, the third mechanism of gene transfer, does not depend on living cells as donors of nucleic acids. It requires the active uptake of isolated DNA by recipients in a physiological stage of competence (Lorenz and Wackernagel 1994). According to present knowledge, induction of competence is determined by growth stage, stress factors, and/or presence of diffusible competence factors secreted by bacteria and, not least, by characteristics of the species in question (Goodman 1994, Lorenz and Wackernagel 1994, Baur et al. 1996, Schlüter and Potrykus 1996, Cheng et al. 1997, Prozorov 1997). Establishment of foreign nucleic acids — i.e., uptake by and integration into other microorganisms — involves recombination to integrate the new sequences into the recipient genome. Homologous sequences have better chances of transformation, although uptake does not usually depend on the presence of similar DNA (Lorenz and Wackernagel 1994, Prozorov 1997). Establishment of plasmids that will not be integrated into the host genome does not depend on homology, although the transferred plasmid must be characterized by suitable origins of replication (Lorenz and Wackernagel 1994). However, there are restriction systems that act to degrade or "silence" foreign DNA, preventing integration of foreign DNA (Heinemann 1991). Experimental and empirical data both suggest that transformation is common in natural environments (Hermansson and Linberg 1994, Lorenz and Wackernagel 1994, Boyle-Vavra and Seifert 1996, Schäfer 1996). Microorganisms in aquatic and soil systems can also be transformed by isolated DNA, as has been demonstrated by Frischer et al. (1994), Lorenz and Wackernagel (1994), Paget and Simonet (1993), and others. Transformation of endogenous bacteria in the gastrointestinal tract occurred in the intestine of springtails (*Folsomia candida*) after feeding on GEMs (Tebbe et al. 1994b, Hobom 1995). Baur et al. (1996) postulate that natural environments, not wholly imitable in the laboratory, can provide optimal conditions for transformation. Environmental factors influencing transformation include either lack of or an oversupply of nutrients and certain minerals, ionic strength of water, and temperature (Frischer et al. 1993, Chamier et al. 1993, Hermansson and Linberg 1994, Baur et al. 1996, Lundsford and London 1996, Williams et al. 1996). As in the case of conjugation and transduction, enhanced rates of transformation are found on surfaces and biofilms (Hermansson and Linberg 1994, Baur et al. 1996).

Reports of intense natural gene transfer may appear to challenge the ability of bacterial strains and varieties to maintain their genetic makeup. However, there are restriction systems that act to degrade or "silence" foreign DNA, preventing integration rather than uptake of foreign DNA (Heinemann 1991). Plasmid incompatibility also acts to hinder genetic transfer (Naik et al. 1994). Stressful conditions seem to reduce the activity of restriction systems in bacteria and to stimulate genes mediating integration via recombination (Saunders and Saunders 1993, Schäfer et al. 1994).

Special plasmids are being designed to reinforce the constraints of biological containment (safety vectors without transfer genes), at the same time cloning increasingly relies on shuttle vectors capable of transgressing existing borders between bacterial classes and even kingdoms (Trieu-Cuot et al. 1987, Henschke and Schmidt 1990, Doucet-Populaire 1992, Schäfer et al. 1994). Recombinant plasmids are being constructed to undermine the different levels of restriction they encounter under normal conditions in natural environments. Thus, their persistence and integration is ensured (Heinemann 1991, Dunn et al. 1993).

9.5   Relevance of recent advances in microbiology to risk assessment of gems

Current risk assessment of GEMs and their recombinant genes may be based on outdated data and perceptions. GEMs, even those only constructed for contained use, can survive or transfer their transgenes to indigenous organisms. DNA is more stable than has been hitherto imagined, and DNA taken up with the food is not completely degraded in the gastrointestinal tract, but may enter white blood cells as well as spleen and liver cells. DNA may even be transferred to the cells of fetuses, as has been shown in newborn mice, where transfer probably took place via the placenta (Doerfler and Schubbert 1997). Further, there are now nucleic acid constructs containing sequences that contribute not only to effective replication in different cellular backgrounds but also to stability and integration via recombination, transfer, and extraordinary expression. They are — this is the essence of the art — especially designed to fulfill these functions. These shuttle vectors open new, unprecedented avenues for transfer. They substantially increase the probability for genetic integration across species and even kingdoms never before able to exchange genetic material. Nearly all these vectors, including those used in plant and animal cell transformation, contain bacterial sequences, thereby providing the recombinant genes better access to the bacterial community. The consequences of such access are unknown.

Neither laboratory nor *in situ* studies of survival, DNA persistence, and gene transfer provide a realistic measure of what can happen in complex terrestrial, aquatic, and gut environments (Stotzky et al. 1991, Spielman et al. 1996). Crude as they may be, validated microcosm and mesocosm studies, imitating natural conditions as accurately as possible, are indispensable for testing and prognosticating the influence of some of the biotic and abiotic factors operating on GEMs and their recombinant genes in natural environments (Pickup et al. 1993, Spielman et al. 1996). But such studies are far from presenting an overall picture. What has been neglected until now in designing such studies has been the inclusion of selective forces that may act on the spread of recombinant genes. For example, contaminants like heavy metals or high salt concentrations may facilitate gene transfer, particularly in disturbed or polluted environments.

In a seminar organized in 1997 by the Norwegian Biotechnology Advisory Board on Antibiotic Resistance, Marker Genes and Transgenic Plants, one of the invited speakers summarized:

> The extent and consequences of horizontal gene transfer are apparent in the evolution of antibiotic resistant microorganisms and evidence suggests that horizontal gene transfer may be equally frequent among multicellular eukaryotic organisms. But the actual and potential frequencies of gene transfer are poor indicators of risk; very common genes are not maintained in nature if unselected and rare genes become common extremely quickly if they are the subject of selection. What remains essential to assessing risk is identifying all potential selective pressures a recombinant gene might be suited to neutralize. New evidence suggests that current knowledge of evolutionary theory is inadequate to predict the fate of recombinant organisms or recombinant genes (Heinemann 1997).

To this we would add that, until there is sufficient scientific evidence that a given GEM or its recombinant genes will not pose any environmental stress or health impact, we should use the precautionary approach in regulating its contained use and deliberate release. This implies that authorities put a stop to deregulation of contained use and to releases of production plants where environmental impacts are not routinely assessed.

Further, we should bear in mind that soil ecology, for example, has much to do with mineralization and nutrient flow, which again are dependent on the enzymatic properties of the organismal network. Genes currently perceived as harmless, like many enzyme-coding genes (or the organisms containing them), may alter soil chemistry and subsequently the soil food web and the organismal network essential for soil fertility. Further, such changes may create selective pressure or conditions favorable to the survival of GEMs or may facilitate transfer of (recombinant) genes/plasmids with unwanted ecological impacts (see Holmes 1995, and a review by Doyle et al. 1995). Fermenter sludge quite often is used as fertilizer on agricultural fields. It is shortsighted to restrict safety assessments to toxic impacts of a given organism as is often done in classification schemes for contained use. There is no way to extrapolate the results of deliberate release of GEMs from one region or environment to another. This is especially true for GEMs that are transferred to ecosystems and climates that differ greatly from those in which they were first developed, tested, and used. This inability to extrapolate is evident, although not always acknowledged, for deliberate release; it also holds true for contained use in production plants where there are multiple pathways for escape.

Literature cited

Aardema, B. W., Lorenz, M. G., and Krumbein, W. E., Protection of sediment-adsorbed transforming DNA against enzymatic inactivation, *Appl. Environ. Microbiol.*, 46, 417, 1983.

Adamo, J. A. and Gealt, M. A., A demonstration of bacterial conjugation within the alimentary canal of *Rhabditis* nematodes, *FEMS Microbiol. Ecol.*, 20, 15, 1996.

Ahl, T., Christoffersen, K., Riemann, B., and Nybroe, O., A combined microcosm and mesocosm approach to examine factors affecting survival and mortality of *Pseudomonas fluorescens* Ag1 in seawater, *FEMS Microbiol. Ecol.*, 17, 107, 1995.

Alvarez, A. J., Yumet, G. M., Santiago, C. L., and Toranzos, G. A., Stability of manipulated plasmid DNA in aquatic environments, *Environ. Toxicol. Water Qual.*, 11, 129, 1996a.

Alvarez, A. J., Yumet, G. M., Santiago, C. L., Hazen, T. C., Chaudhry, R., and Toranzos, G. A., *In situ* survival of genetically engineered microorganisms in a tropical aquatic environment, *Environ. Toxicol. Water Qual.*, 11, 21, 1996b.

Amábile-Cuevas, C. F. and Chicurel, M. E., A possible role for plasmids in mediating the cell-cell proximity required for gene flux, *J. Theor. Biol.*, 181, 237, 1996.

Amarger, N. and Delgutte, D., Monitoring genetically manipulated *Rhizobium leguminosarum* bv. *viciae* released in the field, in *Proceedings of the 1st International Symposium on the Biosafety Results of Field Tests of Genetically Modified Plants and Microorganisms*, Mackenzie, D. R. and Henry, S. C., Eds., Kiawah Island, Bethesda, MD, 1990, 221.

Angle, J. S., Levin, M. A., Gagliardi, J. V., and Mcintosh, M. S., Validation of microcosms for examining the survival of *Pseudomonas aureofaciens* (lacZY) in soil, *Appl. Environ. Microbiol.*, 61, 2835, 1995.

Armon, R., Dosoretz, C., Yoirish, A., Shelef, G., and Neeman, I., Survival of the phytopathogen *Erwinia carotovora* subsp. *carotovora* in sterile and non-sterile soil, sand and their admixture, *J. Appl. Bacteriol.*, 79, 513, 1995.

Arturo-Schaan, M., Tamanai-Shacoori, Z., Thomas, D., and Cormier, M., Variations in R-plasmid DNA concentrations of *Escherichia coli* during starvation in sewage and brackish waters, *J. Appl. Bacteriol.*, 80, 117, 1996.

Ashelford, K. E., Fry, J. C., and Learner, M. A., Plasmid transfer between strains of *Pseudomonas putida*, and their survival, within a pilot scale percolating-filter sewage treatment system, *FEMS Microbiol. Ecol.*, 18, 15, 1995.

Awong, J., Bitton, G., and Chaudhry, G. R., Microcosm for assessing survival of genetically engineered microorganisms in aquatic environments, *Appl. Environ. Microbiol.*, 56, 977, 1990.

Bale, M. J., Day, M. J., and Fry, M. C., Novel method for studying plasmid transfer in undisturbed river epilithon, *Appl. Environ. Microbiol.*, 54, 2756, 1988.

Barcina, I., Arana, I., Fernandez-Astorga, A., Iriberri, J., and Egea, L., Survival strategies of plasmid-carrier and plasmidless *Escherichia coli* strains under illuminated and non-illuminated conditions, in a fresh water ecosystem, *J. Appl. Bacteriol.*, 73, 229, 1992.

Barkay, T., Kroer, N., Rasmussen, L. D., and Sorensen, S. J., Conjugal transfer at natural population densities in a microcosm simulating an estuarine environment, *FEMS Microbiol. Ecol.*, 16, 43, 1995.

Bauda, P., Lallement, C., and Manem, J., Plasmid content evaluation of activated sludge, *Water Res.*, 29, 371, 1995.

Baur, B., Hanselmann, K., Schlimme, W., and Jenni, B., Genetic transformation in freshwater: *Escherichia coli* is able to develop natural competence, *Appl. Environ. Microbiol.*, 62, 3673, 1996.

Beebee, T. J. C., Identification and analysis of nucleic acids in natural freshwaters, *Sci. Total Environ.*, 135, 123, 1993.

Bej, A. K., Steffan, R. J., DiCesare, J., Haff, L., and Atlas, R. M., Detection of coliform bacteria in water by polymerase chain reaction and gene probes, *Appl. Environ. Microbiol.*, 56, 307, 1990.

Bergemann, K., Noe, W., and Walz, F., Identifizierung und Messung von Parametern zur Beurteilung der potentiellen Risiken bei der Entsorgung von rekombinanten tierischen Zellkulturen im Produktionsmaßstab, in *Biologische Sicherheitsforschung Biotechnologie*, Vol. 3, Bundesministerium für Forschung und Technologie (BMFT), Bonn, Germany, 1994, 579.

Bianchi, A. and Giuliano, L., Enumeration of viable bacteria in the marine pelagic environment, *Appl. Environ. Microbiol.*, 62, 174, 1996.

Binnerup, S. J., Jensen, D. F., Thordal-Christensen, H., and Sorensen, J., Detection of viable, but non-culturable *Pseudomonas flourescens* DF57 in soil using a microcolony epiflourescence technique, *FEMS Microbiol. Ecol.*, 12, 97, 1993.

Boivin-Jahns, V., Ruimy, R., Bianchi, A., Daumas, S., and Christen, R., Bacterial diversity in a deep-subsurface clay environment, *Appl. Environ. Microbiol.*, 62, 3405, 1996.

Bouma, J. E. and Lenski, R. E., Evolution of a bacteria/plasmid association, *Nature*, 335, 351, 1988.

Boyle-Vavra, S. and Seifert, H. S., Uptake-sequence-independent DNA transformation exists in *Neisseria gonorrhoeae*, *Microbiology*, 142, 2839, 1996.

Brockmann, E., Jacobsen, B. L., Hertel, C., Ludwig, W., and Schleifer, K. H., Monitoring of genetically modified *Lactococcus lactis* in gnotobiotic and conventional rats by using antibiotic resistance markers and specific probe or primer based methods. *Syst. Appl. Microbiol.*, 19, 203, 1996.

Byrd, J. J. and Colwell, R. R., Maintenance of plasmids pBR322 and pUC8 in nonculturable *Escherichia coli* in the marine environment, *Appl. Environ. Microbiol.*, 56, 2104, 1990.

Byrd, J. J., Leahy, J. G., and Colwell, R. R., Determination of plasmid DNA concentration maintained by nonculturable *Escherichia coli* in marine microcosms, *Appl. Environ. Microbiol.*, 58, 2266, 1992.

Byzov, B. A., Ngugen, T., and Babjeva, I. P., Yeasts associated with soil invertebrates, *Biol. Fertil. Soils*, 16, 183, 1993.

Canteras, J. C., Juanes, J. A., Perez, L., and Koev, K. N., Modelling the coliforms inactivation rates in the Cantabrian Sea (Bay of Biscay) from *in situ* and laboratory determinations of T-90, *Water Sci. Technol.*, 32, 37, 1995.

Chamier, B., Lorenz, M. G., and Wackernagel, W., Natural transformation of *Acinetobacter calcoaceticus* by plasmid DNA adsorbed on sand and groundwater aquifer material, *Appl. Environ. Microbiol.*, 59, 1662, 1993.

Chao, W. L. and Feng, R. L., Survival of genetically engineered *Escherichia coli* in natural soil and river water, *J. Appl. Bacteriol.*, 68, 319, 1990.

Cheng, Q., Campbell, E. A., Naughton, A. M., Johnson, S., and Masure, H. R., The com locus controls genetic transformation in *Streptococcus pneumoniae*, *Mol. Microbiol.*, 23, 683, 1997.

Cirvilleri, G., Marino, R., Catara, V., and Di Silvestro, S., Monitoring of *Pseudomonas flourescens* released in the phylloplane of different species in microcosm, *Z. Pflanzenkr. Pflanzenschutz*, 103, 507, 1996.

Clegg, C. D., Anderson, J. M., Lappinscott, H. M., VanElsas, J. D., and Jolly, J. M., Interaction of a genetically modified *Pseudomonas fluorescens* with the soil-feeding earthworm *Octolasion cyaneum* (Lumbricidae), *Soil Biol. Biochem.*, 27, 1423, 1995.

Clegg, C. D., Van Elsas, J. D., Anderson, J. M., and Lappinscott, H. M., Survival of parental and genetically modified derivatives of a soil isolated *Pseudomonas flourescens* under nutrient-limiting conditions, *J. Appl. Bacteriol.*, 81, 19, 1996.

Collard, J. M., Provoost, A., Dijkmans, R., and Mergeay, M., Heavy metal-resistance genes as markers for measuring gene transfer in soils, presented at Bridge/Biotech, Final Sectorial Meeting on Biosafety and First Sectorial Meeting on Microbial Ecology, Granada, 1993, 43.

Colwell, R. R., Brayton, P., Herrington, D., Tall, B., Huq, A., and Levine, M. M., Viable but non-culturable *Vibrio cholerae* 01 revert to a cultivable state in the human intestine, *World J. Microbiol. Biotechnol.*, 12, 28, 1996.

Cook, R. J., Weller, D. M., Kovacevich, P., Drahos, D., Hemming, B., Barnes, G., and Pierson, E. L., Establishment, monitoring, and termination of field tests with genetically altered bacteria applied to wheat for biological control of take-all, in *Proceedings of the 1st International Symposium on the Biosafety Results of Field Tests of Genetically Modified Plants and Microorganisms*, Mackenzie, D. R. and Henry, S. C., Eds, Kiawah Island, Bethesda, MD, 1990, 221.

Dane, F. and Shaw, J. J., Survival and persistence of bioluminescent *Xanthomonas campestris* pv. *campestris* on host and non-host plants in the field environment, *J. Appl. Bacteriol.*, 80, 73, 1996.

Davies, C. M., Long, J. A. H., Donald, M., and Ashbolt, N. J., Survival of fecal microorganisms in marine and freshwater sediments, *Appl. Environ. Microbiol.*, 61, 1888, 1995.

De Leij, F. A. A. M., Sutton, E. J., Whipps, J. M., and Lynch, J. M., Effect of a genetically modified *Pseudomonas aureofaciens* on indigenous microbial populations of wheat, *FEMS Microbiol. Ecol.*, 13, 249, 1994.

De Leij, F. A. A. M., Sutton, E. J., Whipps, J. M., Fenlon, J. S., and Lynch, J. M., Impact of field release of genetically modified *Pseudomonas fluorescens* on indigenous microbial populations of wheat, *Appl. Environ. Microbiol.*, 61, 3443, 1995.

Del Solar, G., Moscoso, M., and Espinosa, M., Promiscuous plasmids from Gram-positive bacteria as vehicles for horizontal gene transfer, presented at Bridge/Biotech; Final Sectorial Meeting on Biosafety and First Sectorial Meeting on Microbial Ecology, Granada, 1993, 23.

De Rore, H., Demolder, K., De Wilde, K., Top, E., Houwen, F., and Verstraete, W., Transfer of the catabolic plasmids RP4: Tn4271 to indigenous soil bacteria and its effect on respiration and biphenyl breakdown, *FEMS Microbiol. Ecol.*, 15, 71, 1994a.

De Rore, H., Top, E., Houwen, F., Mergeay, M., and Verstraete, W., Evolution of heavy metal resistant transconjugants in a soil environment with a concomitant selective pressure, *FEMS Microbiol. Ecol.*, 14, 263, 1994b.

De Vos, W. M., Introduction to risk assessment when employing GMM's: coping with uncertainties, in *Past, Present and Future Considerations in Risk Assessment When Using GEM's*, De Vries, G., Ed., Commission on Genetic Modification, Bilthoven, the Netherlands, 1998.

Doerfler, W. and Schubbert, R., Fremde DNA im Säugersystem. *Dtsch. Arztebl.*, 51/52A, 3465, 1997.

Dott, W., Khoury, N., Ankel-Fuchs, D., Henninger, W., and Kämpfer, P., Überlebensfähigkeit von genetisch veränderten *Escherichia coli*-Stämmen. 1. Mitteilung: Physiologische Charakterisierung und Einfluß verschiedener physikochemischer Bedingungen, *Zentralbl. Hyg.*, 191, 539, 1991a.

Dott, W., Khoury, N., Ankel-Fuchs, D., Henninger, W., and Kämpfer, P., Überlebensfähigkeit von genetisch veränderten *Escherichia coli*-Stämmen. 2. Mitteilung: Überleben von Reinkulturen in verschiedenen Wasser- und Bodenmatrices. *Zentralbl. Hyg.*, 192, 1, 1991b.

Doucet-Populaire, F., Conjugal transfer of genetic information in gnotobiotic mice, in *Microbial Releases*, Gauthier, M. J., Ed., Springer-Verlag, Berlin, 1992, 161.

Doyle, J. D., Stotzky, G., McClung, G., and Hendricks, C. W., Effects of genetically engineered microorganisms on microbial populations and processes in natural habitats, *Adv. Appl. Microbiol.*, 40, 237, 1995.

Dresing, U., Damman-Kalinowski, T., Keller, M., Selbitschka, W., Pühler, A., Tebbe, C., Schwieger, F., and Munch, J. C., Persistence of two bioluminescent *Rhizobium meliloti* strains in model ecosystems and in field release experiments, in *Abstracts of the Annual Meeting of the Genetic Society 1995 in Bielefeld*, Kalinowski, J., Pühler, A., and Schäfer, A., Eds., Gesellschaft für Genetik, Cologne, Germany, 1995, 18.

Dunn, A., Day, M., and Randerson, P., Plasmid replication and maintenance in binary fissile micro-organisms, *Endeavour*, 17, 21, 1993.

Eckelkamp, C., Jäger, M., and Tappeser, B. Verbreitung und Etablierung rekombinanter Desoxy-ribonukleinsäure (DNS) in der Umwelt, in *UBA-Texte*, 51/98, Umweltbundesamt, Berlin, 1998.

England, L. S., Lee, H., and Trevors, J. T., Recombinant and wild-type *Pseudomonas aureofaciens* strains introduced into soil microcosms — effect on decomposition of cellulose and straw, *Mol. Ecol.*, 4, 221, 1995.

Feldmann, S. D. and Sahm, H., Untersuchungen zum Verhalten gentechnisch veränderter Mikro-organismen in Laborkläranlagen, *BIOforum*, 6, 220, 1994.

Ferguson, Y., Glover, L. A., McGillivray, D. M., and Prosser, J. I., Survival and activity of lux-marked *Aeromonas salmonicida* in seawater, *Appl. Environ. Microbiol.*, 61, 3494, 1995.

Ferianc, P., Polek, B., Godocikova, J., and Toth, D., Metabolic activity of cadmium-stressed and/or starved *Vibrio* sp., *Fol. Microbiol.*, 40, 443, 1995.

Fish, J. T. and Pettibone, G. W., Influence of freshwater sediment on the survival of *Escherichia coli* and *Salmonella* sp. as measured by three methods of enumeration, *Lett. Appl. Microbiol.*, 20, 277, 1995.

Förster, B., Studie zur Ökologie ausgewählter Mikroorganismen, in *UBA-Texte*, 64/98, Umwelt-bundesamt, Berlin, 1998.

Frischer, M. E., Thurmond, J. M., and Paul, J. H., Factors affecting competence in a high frequency of transformation marine *Vibrio*, *J. Gen. Microbiol.*, 139, 753, 1993.

Frischer, M. E., Stewart, G. J., and Paul, J. H., Plasmid transfer to indigenous marine bacterial populations by natural transformation, *FEMS Microbiol. Ecol.*, 15, 127, 1994.

Fujimura, H., Sakuma, Y., and Amann, E., Survival of genetically-engineered and wild-type strains of the yeast *Saccharomyces cerevisiae* under simulated environmental conditions — a contribution on risk assessment, *J. Appl. Bacteriol.*, 77, 689, 1994.

Fulthorpe, R. A. and Wyndham, R. C., Transfer and expression of the catabolic plasmid pBRC60 in wild bacterial recipients in a freshwater ecosystem, *Appl. Environ. Microbiol.*, 57, 1546, 1991.

Gadkari, D., Conjugation between two *Escherichia coli* strains and between *Escherichia coli* (donor) and *Azospirillum brasilence* (recipient) in the presence of various herbicides, *Toxicol. Environ. Chem.*, 30, 211, 1991.

Garcia-Lara, J., Martinez, J., Vilamu, M., and Vives-Rego, J., Effect of previous growth condition on the starvation-survival of *Escherichia coli* in seawater, *J. Gen. Microbiol.*, 139, 1425, 1993.

Gillespie, K. M., Angle, J. S., and Hill, R. L., Runoff losses of *Pseudomonas aureofaciens* (lacZY) from soil, *FEMS Microbiol. Ecol.*, 17, 239, 1995.

Goldschmidt, J., Kassel, K., Taleghani, K. M., Mehling, A., Rössler, C., Schmidt, E., Wehmeier, U., and Piepersberg, W., Ökologische und analytische Studien an gentechnisch veränderten industriellen Produktionsorganismen und Abbaustämmen in Klärfloren, in *Biologische Sicherheitsforschung Biotechnologie*, Vol. 3, Bundesministerium für Forschung und Technologie (BMFT), Bonn, Germany, 1994, 691.

Goodman, A. E., Marshall, K. C., and Hermansson, M., Gene transfer among bacteria under conditions of nutrient depletion in simulated and natural aquatic environments, *FEMS Microbiol. Ecol.*, 15, 55, 1994.

Götz, A. and Smalla, K., Manure enhances plasmid mobilization and survival of *Pseudomonas putida* introduced into field soil, *Appl. Environ. Microbiol.*, 63, 1980, 1997.

Götz, A., Pukall, R., Smit, E., Tietze, E., Prager, R., Tschäpe, H., VanElsas, J. D., and Smalla, K., Detection and characterization of broad-host-range plasmids in environmental bacteria by PCR, *Appl. Environ. Microbiol.*, 62, 2621, 1996.

Gross, A., Wurz, A., and Willmund, R., Untersuchungen zum Verbleib rekombinanter Plasmid-DNA in einer Modellkläranlage, *Korrespondenz Abwasser*, 4, 15, 1994.

Heidenreich, B., Invasivität transgener Mikroorganismen, in *Nutzung der Gentechnik im Agrarsektor der USA — Die Diskussion von Versuchsergebnissen und Szenarien zur Biosicherheit* — Vol. 2, Schütte, G., Heidenreich, B., and Beusmann, V., Eds., Auftrag des Umweltbundesamtes, Bonn, Germany, 1998.

Heijnen, C. E. and Marinissen, J. C. Y., Survival of bacteria introduced into soil by means of transport by *Lumbricus rubellus*, *Biol. Fertil. Soils*, 20, 63, 1995.

Heinemann, J. A., Genetics of gene transfer between species, *Trends Genet.*, 7, 181, 1991.

Heinemann, J. A., Assessing the risk of interkingdom DNA transfer, in *Nordic Seminar on Antibiotic Resistance Marker Genes and Transgenic Plants*, Norwegian Biotechnology Advisory Board, Ed., Norwegian Biotechnology Advisory Board, Oslo, 1997, 17.

Heitkamp, M. A., Kane, J. F., Morris, P. J. L., Bianchini, M., Hale, M. D., and Bogosian, G., Fate in sewage of a recombinant *Escherichia coli* K-12 strain used in the commercial production of bovine somatotropin, *J. Ind. Microbiol.*, 11, 243, 1993.

Hekman, W. E., Heijnen, C. E., Burgers, S. L. G. E., VanVeen, J. A., and VanElsas, J. D., Transport of bacterial inoculants through intact cores of two different soils as affected by water percolation and the presence of wheat plants, *FEMS Microbiol. Ecol.*, 16, 143, 1995.

Henschke, R. B. and Schmidt, F. R. J., Plasmid mobilization from genetically engineered bacteria to members of the indigenous soil microflora *in situ*, *Curr. Microbiol.*, 20, 105, 1990.

Hermansson, M. and Linberg, C., Gene transfer in the marine environment, *FEMS Microbiol. Ecol.*, 15, 47, 1994.

Hobom, B., Kein Platz für verirrte Mikroben, Ergebnisse des Projektbeirates des Verbundprojektes Sicherheitsforschung Gentechnik, Bayer AG, Leverkusen, 1995.

Hodson, R. E., Dustman, W. A., Garg, R. P., and Moran, M. A., *In situ* PCR for visualization of microscale distribution of specific genes and gene products in prokaryotic communities, *Appl. Environ. Microbiol.*, 61, 4074, 1995.

Holmes, M. T., Ecological Assessment after the Addition of Genetically Engineered *Klebsiella planticola* SDF20 into Soil, Ph.D. thesis, Oregon State University, Corvallis, 1995.

Hugenholz, P. and Pace, N. R., Identifying microbial diversity in the natural environment: a molecular phylogenetic approach, *Trends Biotechnol.*, 14, 190, 1996.

Inamori, Y., Murakami, K., Sudo, R., Kurihara, Y., and Tanaka, N., Environmental assessment method for field release of genetically engineered microorganisms using microcosm systems, *Water Sci. Technol.*, 26, 2161, 1992.

Istock, C. A., Genetic exchange and genetic stability in bacterial populations, in *Assessing Ecological Risks of Biotechnology*, Ginzburg, R., Ed., Butterworth-Heinemann, Stoneham, MA, 1991, 123.

Jäger, M. and Tappeser, B., Politics and science in risk assessment, in *Coping with Deliberate Release — The Limits of Risk Assessment*, Van Dommelen, A., Ed., International Centre for Human and Public Affairs, Tilburg/Buenos Aires, 1996, 63.

Jensen, G. B., Andrup, L., Wilcks, A., Smidt, L., and Poulsen, O. M., The aggregation-mediated conjugation system of *Bacillus thuringiensis* subsp. *israelensis*: host range and kinetics of transfer, *Curr. Microbiol.*, 33, 228, 1996.

Jiang, S. C. and Paul, J. H., Viral contribution to dissolved DNA in the marine environment as determined by differential centrifugation and kingdom probing, *Appl. Environ. Microbiol.*, 61, 317, 1995.

Jiang, X. P. and Chai, T. J., Survival of *Vibrio parahaemolyticus* at low temperatures under starvation conditions and subsequent resuscitation of viable, nonculturable cells, *Appl. Environ. Microbiol.*, 62, 1300, 1996.

Karapinar, M. and Sahika, A. G., Survival of *Yersinia enterocolitica* and *Escherichia coli* in spring water, *Int. J. Food Microbiol.*, 13, 315, 1991.

Khanna, M. and Stotzky, G., Transformation of *Bacillus subtilis* by DNA bound on montmorillonite and effect of DNase on the transforming ability of bound DNA, *Appl. Environ. Microbiol.*, 58, 1930, 1992.

Kidambi, S. P., Ripp, S., and Miller, R. V., Evidence of phage-mediated gene transfer among *Pseudomonas aeruginosa* strains on the phylloplane, *Appl. Environ. Microbiol.*, 60, 496, 1994.

Kinkel, L. L., Wilson, M., and Lindow, S. E., Utility of microcosm studies for predicting phylloplane bacterium population sizes in the field, *Appl. Environ. Microbiol.*, 62, 3413, 1996.

Klijn, N., Weerkamp, A. H., and De Vos, W. M., Detection and characterization of lactose-utilizing *Lactococcus* spp. in natural ecosystems, *Appl. Environ. Microbiol.*, 61, 788, 1995a.

Klijn, N., Weerkamp, A. H., and De Vos, W. M., Genetic marking of *Lactococcus lactis* shows its survival in the human gastrointestinal tract, *Appl. Environ. Microbiol.*, 61, 2771, 1995b.

Klingmüller, W., Plasmid transfer in natural soil, as stimulated by sucrose, wheat rhizosphere, and ground sugar beets: a study with nitrogen-fixing enterobacter, *Microb. Releases*, 1, 229, 1993.

Klingmüller, W. and Rieder, G., A) Konjugativer Transfer gentechnisch veränderter Plasmide. B) Konjugativer Transfer gentechnisch veränderter Plasmide bei Freisetzungssimulierung, in *Biologische Sicherheitsforschung Biotechnologie*, Vol. 3, Bundesministerium für Forschung und Technologie (BMFT), Bonn, Germany, 1994, 83.

Kluepfel, D. A., The behaviour of nonengineered bacteria in the environment: what can we learn from them? in *Proceedings of the 2nd International Symposium on the Biosafety Results of Field Tests of Genetically Modified Plants and Microorganisms*, Caspar, R. and Landsmann, J., Eds., Braunschweig, 1992, 37.

Kluepfel, D. A., Kline, E. L., Skipper, H. D., Hughes, T. A., Gooden, D. T., Drahos, D. J., Barry, G. F., Hemming, B. C., and Brandt, E. J., The release and tracking of genetically engineered bacteria in the environment, *Phytopathology*, 81, 348, 1991.

Kluepfel, D. A., Lamb, T. G., Snyder, W. E., and Tonkyn, D. W., Six years of field testing a lacZY modified fluorescent pseudomonad, in *Proceedings of the 3rd International Symposium on the Biosafety Results of Field Tests of Genetically Modified Plants and Microorganisms*, Jones, D. D., Ed., Monterey, CA, 1994, 169.

Kokjohn, T. A. and Miller, R. V., Gene transfer in the environment: transduction, in *Release of Genetically Engineered and Other Micro-Organisms*, Fry, J. C. and Day, M., Eds., Cambridge University Press, Cambridge, U.K., 1992, 54.

Korhonen, L. K. and Martikainen, P. J., Survival of *Escherichia coli* and *Campylobacter jejuni* in untreated and filtered lake water, *J. Appl. Bacteriol.*, 71, 379, 1991.

Kozdroj, J., Survival of lux-marked bacteria introduced into soil and the rhizosphere of bean (*Phaseolus vulgaris* L.), *World J. Microbiol. Biotechnol.*, 12, 261, 1996.

Kozdroj, J. and Piotrowska-Seget, Z., Indigenous microflora and bean responses to introduction of genetically modified *Pseudomonas fluorescens* strains into soil contaminated with copper, *J. Environ. Sci. Health A*, 30, 2133, 1995.

Kragelund, L. and Nybroe, O., Competition between *Pseudomonas flourescens* Ag1 and *Alcaligenes eutrophus* JMP134 (pHP4) during colonization of barley roots, *FEMS Microbiol. Ecol.*, 20, 41, 1996.

Kruse, H. and Sørum, H., Transfer of multiple drug resistance plasmids between bacteria of diverse origins in natural microenvironments, *Appl. Environ. Microbiol.*, 60, 4015, 1994.

Lafuente, R., Maymó-Gatell, X., Mas-Castellà, J., and Guerrero, R., Influence of environmental factors on plasmid transfer in soil microcosms, *Curr. Microbiol.*, 32, 213, 1996.

Lebaron, P., Batailler, N., and Baleux, B., Mobilization of a recombinant nonconjugative plasmid at the interface between wastewater and the marine coastal environment, *FEMS Microbiol. Ecol.*, 15, 61, 1994.

Lee, S. Y., Bollinger, J., Bezdicek, D., and Ogram, A., Estimation of the abundance of an uncultured soil bacterial strain by a competitive quantitative PCR method, *Appl. Environ. Microbiol.*, 62, 3787, 1996.

Leff, L. G. and Leff, A. A., Use of green fluorescent protein to monitor survival of genetically engineered bacteria in aquatic environments, *Appl. Environ. Microbiol.*, 62, 3486, 1996.

Lenski, R. E., Souza, V., Duong, L. P., Phan, Q. G., Nguyen, T. N. M., and Bertrand, K. P., Epistatic effects of promoter and repressor functions of the Tn10 tetracycline-resistance operon on the fitness of *Escherichia coli*, *Mol. Ecol.*, 3, 127, 1994.

Leser, T. D., Validation of microbial community structure and ecological functional parameters in an aquatic microcosm designed for testing genetically engineered microorganisms, *Microb. Ecol.*, 29, 183, 1995.

Lilley, A. K. and Bailey, M. J., Impact of plasmid pQBR103 acquisition and carriage on the phytosphere fitness of *Pseudomonas fluorescens* SBW25: burden and benefit, *Appl. Environ. Microbiol.*, 63, 1584, 1997.

Lilley, A. K., Hails, R. S., Cory, J. S., and Bailey, M. J., The dispersal and establishment of pseudomonad populations in the phyllosphere of sugar beet by phytophagous caterpillars, *FEMS Microbiol. Ecol.*, 24(2), 151, 1997.

Lindahl, V., Improved soil dispersion procedures for total bacterial counts, extraction of indigenous bacteria and cell survival, *J. Microbiol. Methods*, 25, 279, 1996.

Lindahl, V. and Bakken, L. R., Evaluation of methods for extraction of bacteria from soil, *FEMS Microbiol. Ecol.*, 16, 135, 1995.

Lorenz, M. G. and Wackernagel, W., Adsorption of DNA to sand and variable degradation rates of adsorbed DNA, *Appl. Environ. Microbiol.*, 53, 2948, 1987.

Lorenz, M. G. and Wackernagel, W., DNA binding to various clay minerals and retarded enzymatic degradation of DNA in a sand/clay microcosm, in *Microbial Releases*, Gauthier, M. J., Ed., Springer-Verlag, Berlin, 1992, 103.

Lorenz, M. G. and Wackernagel, W., Bacterial gene transfer in the environment, in *Transgenic Organisms*, Wöhrmann, K. and Tomiuk, J., Eds., Birkhäuser, Basel, Switzerland, 1993, 43.

Lorenz, M. G. and Wackernagel, W., Bacterial gene transfer by natural genetic transformation in the environment, *Microbiol. Rev.*, 58, 563, 1994.

Lorenz, M. G., Aardema, B. W., and Wackernagel, W., Highly efficient genetic transformation of *Bacillus subtilis* attached to sand grains, *J. Gen. Microbiol.*, 134, 107, 1988.

Lou, W. Q. and Yousef, A. E., Resistance of *Listeria monocytogenes* to heat after adaptation to environmental stresses, *J. Food Prot.*, 59, 465, 1996.

Lukin, S. A. and Prozorov, A. A., Conjugate transfer of plasmid DNA between bacteria in soil, *Microbiology*, 61, 493, 1992.

Lunsford, R. D. and London, J., Natural genetic transformation in *Streptococcus gordonii*: comX imparts spontaneous competence on strain wicky, *J. Bacteriol.*, 178, 5831, 1996.

Mazodier, P. and Davies, J., Gene transfer between distantly related bacteria, *Annu. Rev. Genet.*, 25, 147, 1991.

McClure, N. C., Weightman, A. J., and Fry, J. C., Survival of *Pseudomonas putida* UWC1 containing cloned catabolic genes in a model activated-sludge unit, *Appl. Environ. Microbiol.*, 55, 2627, 1989.

Mezrioui, N., Baleux, B., and Troussellier, M., A microcosm study of the survival of *Escherichia coli* and *Salmonella typhimurium* in brackish water, *Water Res.*, 29, 459, 1995.

Mieschendahl, M. and Danneberg, G., Untersuchungen zur in vivo — Transformation von Mikro-organismen in Oberflächengewässern, in *Biologische Sicherheitsforschung Biotechnologie*, Vol. 3, Bundesministerium für Forschung und Technologie (BMFT), Bonn, Germany, 1994, 901.

Miller, R. V., Ripp, S., Replicon, J., Ogunseitan, O. A., and Kokjohn, T. A., Virus-mediated gene transfer in freshwater environments, in *Microbial Releases*, Gauthier, M. J., Ed., Springer-Verlag, Berlin, 1992, 51.

Morrison, M., Do ruminal bacteria exchange genetic material? *J. Dairy Sci.*, 79, 1476, 1996.

Naik, G. A., Bhat, L. N., Chopade, B. A., and Lynch, J. M., Transfer of broad-host range antibiotic resistance plasmids in soil microcosms, *Curr. Microbiol.*, 28, 209, 1994.

Natsch, A., Keel, C., Troxler, J., Zala, M., von Albertini, N., and Défago, G., Importance of preferential flow and soil management in vertical transport of a biocontrol strain of *Pseudomonas fluorescens* in structured field soil, *Appl. Environ. Microbiol.*, 62(1), 33, 1996.

Natsch, A., Troxler, J., and Défago, G., Baselines for regulation which may be derived from a detailed case study on the biosafety of genetically improved bacteria for biological control, in *Past, Present and Future Considerations in Risk Assessment When Using GMO's*, De Vries, G., Ed., Commission on Genetic Modification, Bilthoven, the Netherlands, 1998.

Nwoguh, C. E., Harwood, C. R., and Barer, M. R. Detection of induced beta-galactosidase activity in individual non-culturable cells of pathogenic bacteria by quantitative cytological assay, *Mol. Microbiol.*, 17, 545, 1995.

Nybroe, O., Einarson, K., and Ahl, T., Growth and viability of *Alcaligenes eutrophus* JMP134 in seawater as affected by substrate and nutrient amendment, *Lett. Appl. Microbiol.*, 22, 366, 1996.

Ogunseitan, O. A., Salyer, G. S., and Miller, R. V., Application of DNA probes to the analysis of bacteriophage distribution in the environment, *Appl. Environ. Microbiol.*, 58, 2046, 1992.

Oliver, J. D., Hite, F., McDougald, D., Andon, N. L., and Simpson, L. M., Entry into, and resuscitation from, the viable but nonculturable state by *Vibrio vulnificus* in an estuarine environment, *Appl. Environ. Microbiol.*, 61, 2624, 1995a.

Oliver, J. D., McDougald, D., Barrett, T., Glover, L. A., and Prosser, J. I., Effect of temperature and plasmid carriage on nonculturability in organisms targeted for release, *FEMS Microbiol. Ecol.*, 17, 229, 1995b.

Overbeck, J., Ermittlung und Bewertung des ökologischen Risikos im Umgang mit gentechnisch veränderten Organismen, in *Vortrag und Diskussion zum Vortrag, Forschungsbericht:* 10802 086/01, Umweltbundesamt, Eds., Bonn, Germany, 1991.

Pace, N. R., A molecular view of microbial diversity and the biosphere, *Science*, 276, 734, 1997.

Paget, E. and Simonet, P., On the track of natural transformation in soil, *FEMS Microbiol. Ecol.*, 15, 109, 1994.

Paget, E. and Simonet, P., Evidence of gene transfer in soil via transformation, presented at Bridge/Biotech; Final Sectorial Meeting on Biosafety and First Sectorial Meeting on Microbial Ecology, Granada, 1993, 32.

Paget, E., Monrozier, L. J., and Simonet, P., Adsorption of DNA on clay minerals: protection against DNase l and influence on gene transfer, *FEMS Microbiol. Lett.*, 97, 31, 1992.

Paget, E., Jouan, D., Lebrun, M., and Simonet, P., Persistence of plant DNA in soil, presented at Bridge/Biotech; Final Sectorial Meeting on Biosafety and First Sectorial Meeting on Microbial Ecology, Granada, 1993, 33.

Paul, J. H., Intergeneric natural plasmid transformation between *Escherichia coli* and a marine *Vibrio* species, in *Microbial Releases*, Gauthier, M. J., Ed., Springer-Verlag, Berlin, 1992, 63.

Paul, J. H., Frischer, M. E., and Thurmond, J. M., Gene transfer in marine water column and sediment microcosmos by natural plasmid transformation, *Appl. Environ. Microbiol.*, 57, 1509, 1991.

Pickup, R. W., Development of molecular methods for the detection of specific bacteria in the environment, *J. Gen. Microbiol.*, 137, 1009, 1991.

Pickup, R. W., Morgan, J. A. W., and Winstanley, C., *In situ* detection of plasmid transfer in the acquatic environment, in *Monitoring Genetically Manipulated Microorganisms in the Environment*, Edwards, C., Ed., John Wiley & Sons, Chichester, U.K., 1993, 61.

Prozorov, A. A., Cell differentiation related to genetic transformation in bacteria and its regulation, *Microbiology*, 66, 1, 1997.

Pukall, R., Tschäpe, H., and Smalla, K., Monitoring the spread of broad host and narrow host range plasmids in soil microcosms, *FEMS Microbiol. Ecol.*, 20, 53, 1996.

Rahman, I., Shahamat, M., Chowdhury, M. A. R., and Colwell, R. R., Potential virulence of viable but nonculturable *Shigella dysenteriae* type 1, *Appl. Environ. Microbiol.*, 62, 115, 1996.

Ramos, J. L., Diaz, E., Dowling, D., DeLorenzo, V., Molin, S., O'Gara, F., Ramos, C., and Timmis, K. N., The behavior of bacteria designed for biodegradation (review), *Bio/Technology*, 12, 1349, 1994.

Rang, C. U., Kennan, R. M., Midtvedt, T., Chao, L., and Conway, P. L., Transfer of the plasmid RP1 *in vivo* in germ free mice and *in vitro* in gut extracts and laboratory media, *FEMS Microbiol. Ecol.*, 19, 133, 1996.

Recorbet, G., Steinberg, C., and Faurie, G., Survival in soil of genetically engineered *Escherichia coli* as related to inoculum density, predation and competition, *FEMS Microbiol. Ecol.*, 101, 251, 1992.

Recorbet, G., Richaume, A., and Jocteur-Monrozier, L., Distribution of a genetically-engineered *Escherichia coli* population introduced into soil, *Lett. Appl. Microbiol.*, 21, 38, 1995.

Regal, P. J., The adaptive potential of genetically engineered organisms in nature, in *Planned Release of Genetically Engineered Organisms, Trends in Biotechnology/Trends in Ecology and Evolution*, Hodgson, J. and Sugden, A. M., Eds., Special Publication, Elsevier, Cambridge, U.K., 1988, 36.

Regal, P. J., Scientific principles for ecologically based risk assessment of transgenic organisms, *Mol. Ecol.*, 3, 5, 1994.

Replicon, J., Frankfater, A., and Miller, R.V., A continuous culture model to examine factors that affect transduction among *Pseudomonas aeruginosa* strains in freshwater environments, *Appl. Environ. Microbiol.*, 61, 3359, 1995.

Ripp, S. and Miller, R. V., Effect of suspended particulates on the frequency of transduction among *Pseudomonas aeruginosa* in a freshwater environment, *Appl. Environ. Microbiol.*, 61, 1214, 1995.

Ripp, S., Ogunseitan, O. A., and Miller, R. V., Transduction of a freshwater microbial community by a new *Pseudomonas aeruginosa* generalized transducing phage UT1, *Mol. Ecol.*, 3, 121, 1994.

Romanowski, G., Lorenz, M. G., and Wackernagel, W., Bakterieller Gentransfer durch freie DNA im natürlichen Lebensraum: Studien an einem Bodenmodell, in *Biologische Sicherheitsforschung Biotechnologie*, Vol. 2, Bundesministerium für Forschung und Technologie (BMFT), Bonn, Germany, 1990, 11.

Romanowski, G., Lorenz, M. G., Sayler, G., and Wackernagel, W., Persistence of free plasmid DNA in soil monitored by various methods, including a transformation assay, *Appl. Environ. Microbiol.*, 58, 3012, 1992.

Romanowski, G., Lorenz, M. G., and Wackernagel, W., Plasmid DNA in a groundwater aquifer microcosm — adsorption, DNase resistance and natural genetic transformation of *Bacillus subtilis*, *Mol. Ecol.*, 2, 171, 1993a.

Romanowski, G., Lorenz, M. G., and Wackernagel, W., Use of polymerase chain reaction and electroporation of *Escherichia coli* to monitor the persistence of extracellular plasmid DNA introduced into natural soils, *Appl. Environ. Microbiol.*, 59, 3438, 1993b.

Rybachenko, N. F., Stepanova, T. V., Rumer, L. M., Mozgovaya, I. N., Byzov, B. A., and Zvyagintsev, D. G., Conjugal transfer of plasmid pAM beta 1 from *Streptococci* to *Bacilli* in vermicompost, *Microbiology*, 65, 663, 1996.

Sandaa, R. A. and Enger, O., Transfer in marine sediments of the naturally occurring plasmid pRAS1 encoding multiple antibiotic resistance, *Appl. Environ. Microbiol.*, 60, 4234, 1994.

Saunders, J. R. and Saunders, V. A., Genotypic and phenotypic methods for the detection of specific released microorganisms, in *Monitoring Genetically Manipulated Microorganisms in the Environment*, Edwards, C., Ed., John Wiley & Sons, New York, 1993, 27.

Saye, D. J. and Miller, R. V., The aquatic environment: consideration of horizontal gene transmission in a diversified habitat, in *Gene Transfer in the Environment*, Levy, S. B. and Miller, R. V., Eds., McGraw-Hill, New York, 1989, 223.

Schäfer, A., Horizontaler Gentransfer — Mechanismen und biologische Sicherheit, *Biospektrum*, 6, 23, 1996.

Schäfer, A., Kalinowski, J., and Pühler, A., Increased fertility of *Corynebacterium glutamicum* recipients in intergeneric matings with *Escherichia coli* after stress exposure, *Appl. Environ. Microbiol.*, 60, 756, 1994.

Schicklmaier, P. and Schmieger, H., Frequency of generalized transducing phages in natural isolates of the *Salmonella typhimurium* complex, *Appl. Environ. Microbiol.*, 61, 1637, 1995.

Schlüter, K. and Potrykus, I., Horizontaler Gentransfer von transgenen Mikroorganismen (Bakterien und Pilzen) und seine ökologische Relevanz, in *Gentechnisch veränderte krankheits- und schädlings- resistente Nutzpflanzen. Eine Option für die Landwirtschaft?* Vol. 1, *Materialien*, Schulte, E. and Käppeli, O., Eds., Schwerpunktprogramm Biotechnologie des Schweizerischen Nationalfonds, Bern, Switzerland, 1996.

Schmidt, F. R. J., Die Bewertung von Daten zu den Risiken bei der Freisetzung von gentechnisch veränderten Bodenmikroorganismen, *BIOforum*, 312, 1991.

Schmidt, F., Brokamp, A., Henschke, R., Henschke, E., and Happe, B., Möglichkeiten und Risiken der landwirtschaftlichen Nutzung gentechnologisch veränderter Mikroorganismen, in *Biologische Sicherheitsforschung Biotechnologie*, Vol. 3, Bundesministerium für Forschung und Technologie (BMFT), Bonn, Germany, 1994, 985.

Schneider, E., Schaefer, B., and Feuerpfeil, I., Konjugativer Gentransfer und Überleben gentechnisch veränderter Bakterien in Abwasser und Oberflächenwasser unter simulierten Freisetzungs- bedingungen, in *Biologische Sicherheitsforschung Biotechnologie*, Vol. 3, Bundesministerium für Forschung und Technologie (BMFT), Bonn, Germany, 1994, 107.

Schubbert, R. and Doerfler, W., Food ingested foreign DNA is taken up via gastrointestinal wall epithelia and Peyer's patches into peripheral leucocytes, spleen and liver, in *Transgenic Organisms and Biosafety*, Schmidt, E. R. and Hankeln, T. H., Eds., Springer-Verlag Berlin, 1996, 132.

Selbitschka, W. and Pühler, A., Nachweis von Gentransfer in Umweltmedien, in *Überwachungs- methoden Gentechnik — Nachweisverfahren für Mikroorganismen, Viren und Gene in der Umwelt*, Forschungsbericht 108 02 089/01, Umweltbundesamt, Berlin, 1994.

Selbitschka, W., Hagen, M., Niemann, S., and Pühler, A., Risikoanalysen zur Freisetzung gentechnol- ogisch veränderter Rhizobien: Analyse der Wechselwirkung zwischen gentechnisch verändertem Organismus und Modell-Ökosystemen, in *Biologische Sicherheitsforschung Biotechnologie*, Vol. 3, Bundesministerium für Forschung und Technologie (BMFT), Bonn, Germany, 1994, 137.

Sjogren, R. E., Thirteen-year survival study of an environmental *Escherichia coli* in field mini-plots, *Water Air Soil Pollut.*, 81, 315, 1995.

Smalla, K., Horizontal gene transfer from transgenic plants into plant associated bacteria and soil bacteria, in *Safety of Transgenic Crops, Environmental and Agricultural Considerations, Proceedings of the Basel Forum on Biosafety*, Basel, Switzerland, 1995.

Smalla, K., Isemann, M., Levy, R., and Thriene, B., Risikoabschätzung der industriellen Nutzung gentechnisch veränderter Mikroorganismen, *Z. Gesamte Hyg.*, 35, 475, 1989.

Smith, M. S., Thomas, G. W., Withe, R. E., and Ritonga, D., Transport of *Escherichia coli* through intact and disturbed soil columns, *J. Environ. Qual.*, 14, 87, 1985.

Snyder, W. E., Tonkyn, D. W., and Klupfel, D. A., Transmission of a genetically engineered rhizo- bacterium by grasshoppers in the laboratory and field, *Ecol. Appl.*, 9(1), 245, 1999.

Sobecky, P. A., Schell, M. A., Moran, M. A., and Hodson, R. E., Adaption of model genetically engineered microorganisms to lake water: growth rate enhancements and plasmid loss, *Appl. Environ. Microbiol.*, 58, 3630, 1992.

Sobecky, P. A., Schell, M. A., Moran, M. A., and Hodson, R. E., Impact of a genetically engineered bacterium with enhanced alkaline phosphatase activity on marine pytoplankton communities, *Appl. Environ. Microbiol.*, 62, 6, 1996.

Spielman, A., Regal, P., Mundt, C., Klinger, T., Kapuscinski, A. R., Istock, C., Ingham, E. R., Holmes, M., and Fagan, J. B., Draft — Assessment of Genetically Engineered Organisms in the Environment: The Puget Sound Workshop Biosafety Handbook, Edmonds Institute, Edmonds, WA, 1996.

Stotzky, G. and Babich, H., Fate of genetically-engineered microbes in natural environments, *Recombinant DNA Tech. Bull.* 7, 163, 1994.

Stotzky, G., Zeph, L. R., and Devanas, M. A., Factors affecting the transfer of genetic information among microorganisms in soil, in *Assessing Ecological Risks of Biotechnology*, Ginzburg, R., Ed., Butterworth-Heinemann, Stoneham, MA, 1991.

Tamanai-Shacoori, Z., Jolivet-Gougeon, A., and Cormier, M., Comparison of direct PCR and PCR amplification after DNA extraction for the detection of viable enterotoxigenic *Escherichia coli* in laboratory microcosms, *J. Microbiol. Methods*, 26, 21, 1996.

Tappeser, B., Jäger, M., and Eckelkamp, C., Survival, persistence, transfer — an update on current knowledge on GMOs and the fate of their recombinant DNA, in *TWN Biotechnology and Biosafety,* Series 3, Third World Network, Penang, Malaysia, 1999.

Tebbe, C. C., Vahjen, W., Munch, J. C., Feldmann, S. D., Ney, U., Sahm, H., Gelissen, G., Amore, R., and Hollenberg, C. P., Überleben der Untersuchungsstämme und Persistenz ihrer rekombinanten DNA, *BioEngineering,* 6, 14, 1994a.

Tebbe, C. C., Vahjen, W., Munch, J., Meier, B., Gellissen, G., Feldmann, S., Sahm, H., Amore, R., Hollenberg, C. P., Blum, S., and Wackernagel, W., Mesokosmenuntersuchungen und Einfluß der Habitatbedingungen auf die Expression, Überdauerung und Übertragung des Aprotinin-Gens, *BioEngineering,* 6, 22, 1994b.

Temann, U. A., Hösl, R., and Klingmüller, W., Plasmid transfer between *Pseudomonas aeruginosa* strains in natural soil, in *The Release of Genetically Modified Microorganisms,* Stewart-Tull, D. E. S. and Sussmann, M., Eds., Plenum Press, New York, 1992.

Terzieva, S. I. and McFeters, G. A., Survival and injury of *Escherichia coli, Campylobacter jejuni,* and *Yersinia enterocolitica* in stream water, *Can. J. Microbiol.,* 37, 785, 1991.

Thompson, I. P., Ellis, R. J., and Bailey, M.J., Autoecology of a genetically modified fluorescent pseudomonad on sugar beet, *FEMS Microbiol. Ecol.,* 17, 1, 1995a.

Thompson, I. P., Lilley, A. K., Ellis, R. J., Bramwell, P. A., and Bailey, M. J., Survival, colonization and dispersal of genetically modified *Pseudomonas fluorescens* SBW25 in the phytosphere of field grown sugar beet, *Bio/Technology,* 13, 1493, 1995b.

Tokuda, Y., Ano, T., and Shoda, M., Survival of *Bacillus subtilis* NB22 and its transformant in soil, *Appl. Soil Ecol.,* 2, 85, 1995.

Trieu-Cuot, P., Carlier, C., Martin, P., and Courvalin, P., Plasmid transfer by conjugation from *Escherichia coli* to Gram-positive bacteria, *FEMS Microbiol. Lett.,* 48, 289, 1987.

Troxler, J., Zala, M., Natsch, A., Moenne-Loccoz, Y., and Défago, G., Autoecology of the biological strain *Pseudomonas fluorescens* CHA0 in the rhizosphere and inside roots at later stages of plant development, *FEMS Microbiol. Ecol.,* 23, 119, 1997.

Tsushima, S., Hasebe, A., Komoto, Y., Carter, J. P., Miyashita, K., Yokoyama, K., and Pickup, R. W., Detection of genetically engineered microorganisms in paddy soil using a simple and rapid nested polymerase chain reaction method, *Soil Biol. Biochem.,* 27, 219, 1995.

Turpin, P. E., Maycroft, K. A., Rowlands, C. L., and Wellington, E. M. H., Viable but non-culturable salmonellas in soil, *J. Appl. Bacteriol.,* 74, 421, 1993.

Vahjen, W. and Tebbe, C. C., Enhanced detection of genetically engineered *Corynebacterium glutamicum* pUN1 in directly extracted DNA from soil, using the T4 gene 32 protein in the polymerase chain reaction, *Eur. J. Soil Biol.,* 30, 93, 1994.

Van Der Hoeven, N., Van Elsas, J. D., and Heijnen, C. E., A model based on soil structural aspects describing the fate of genetically modified bacteria in soil, *Ecol. Modelling,* 89, 161, 1996.

Van Elsas, J. D., Environmental pressure imposed on GEMMOS in soil, in *The Release of Genetically Modified Microorganisms,* Stewart-Tull, D. E. S. and Sussman, M., Eds., Plenum Press, New York, 1992, 1.

Van Elsas, J. D., Trevors, J. T., Starodub, M. E., and van Overbeek, L. S., Transfer of plasmid RP4 between *Pseudomonas* after introduction into soil; influence of spatial and temporal aspects of inoculation, *FEMS Microbiol. Ecol.,* 73, 1, 1990.

Van Overbeek, L. S., Eberl, L., Givskov, M., Molin, S., and van Elsas, J. D., Survival of, and induced stress resistance in, carbon-starved *Pseudomonas fluorescens* cells residing in soil, *Appl. Environ. Microbiol.,* 61, 4202, 1995.

Velicer, G. J., Pleiotropic effects of adaptation to a single carbon source for growth on alternative substrates, *Appl. Environ. Microbiol.,* 65, 264, 1999.

Wackernagel, W., Romanowski, G., and Lorenz, M. G., Studies on gene flux by free bacterial DNA in soil, sediment and groundwater aquifer, in *The Release of Genetically Modified Microorganisms,* Stewart-Tull, D. E. S. and Sussman, M., Eds., Plenum Press, New York, 1992, 171.

Wagner, M., Amann, R., Lemmer, H., and Schleifer, K. H., Probing activated sludge with oligo-nucleotides specific for proteobacteria: inadequacy of culture-dependent methods for describing microbial community structure, *Appl. Environ. Microbiol.,* 59, 1520, 1993.

Weinbauer, M. G., Fuks, D., Puskaric, S., and Peduzzi, P., Diel, seasonal, and depth-related variability of viruses and dissolved DNA in the northern Adriatic Sea, *Microb. Ecol.,* 30, 25, 1995.

Weiskel, P. K., Howes, B. L., and Heufelder, G. R., Coliform contamination of a coastal embayment: sources and transport pathways, *Environ. Sci. Technol.*, 30, 1872, 1996.

Wendt-Potthoff, K., Backhaus, H., and Smalla, K., Monitoring the fate of genetically engineered bacteria sprayed on the phylloplane of bush beans and grass, *FEMS Microbiol. Ecol.*, 15, 279, 1994.

Williams, H. G., Day, M. J., Fry, J. C., and Stewart, G. J., Natural transformation in river epilithon, *Appl. Environ. Microbiol.*, 62, 2994, 1996.

Winkler, J., Timmis, K. N., and Snyder, R. A., Tracking the response of *Burkholderia cepacia* G4 5223-PR1 in aquifer microcosms, *Appl. Environ. Microbiol.*, 61, 448, 1995.

World Health Organization, *WHO Report*, Geneva, 1996.

Wright, D. A., Killham, K., Glover, L. A., and Prosser, J. I., Role of pore size location in determining bacterial activity during predation by protozoa in soil, *Appl. Environ. Microbiol.*, 61, 3537, 1995.

Zweifel, U. L. and Hagstrom, A., Total counts of marine bacteria include a large fraction of non-nucleoid-containing bacteria (ghosts), *Appl. Environ. Microbiol.*, 61, 2180, 1995.

chapter ten

The spread of genetic constructs in natural insect populations

Henk R. Braig and Guiyun Yan

Contents

10.1   Introduction

This chapter is concerned with insects of medical, veterinary, and economic importance. Special emphasis is given to the mosquitoes that are the most important vectors of human disease such as malaria, dengue fever, yellow fever, and filariasis, all of which constitute major public health problems in tropical and subtropical countries (Nchinda 1998, Patz et al. 1998). To interrupt disease transmission by insect vectors, it may be desirable to permanently establish a different genotype or a transgene in a natural population. This approach requires:

1. Identification and cloning of parasite-inhibiting genes in the mosquito vectors,
2. Development of stable and efficient mosquito transformation tools, and
3. Development of strategies to spread parasite-inhibiting genes into natural populations (Collins and Besansky 1994).

10.2   Genetic driving mechanisms for natural and genetically engineered insects in natural populations

This chapter focuses on the deliberate spread of genetic constructs into natural mosquito populations with the aim of driving these constructs to fixation. In most instances, this will not be possible without driving mechanisms. To be able to assess the impact of genetic manipulation and intervention on the environment and public health requires a thorough understanding of the genetic mechanisms by which these drives act. A comprehensive overview is given of potential driving mechanisms. The drives are analyzed for their suitability to be linked to a payload of transgenes and for their stability over time. The factors that critically affect the spread rate of the genetic constructs are discussed, and the viability of the genetic approach for vector-borne disease control is evaluated. Throughout the chapter, the discussion takes cognizance of one of the most important future applications, the campaign against malaria transmission by *Anopheles* mosquitoes.

10.2.1   Potential environmental applications of transgenic insects

Although only a few transgenic insects have been engineered thus far, their numbers will undoubtedly rise in coming years. Table 10.1 lists examples of insect germ-line transformations. With the notable exception of *Drosophila* into which myriad transgenes have been cloned since 1982, almost all the transgenes cloned into non-drosophilid insects thus far have been marker or reporter genes, inserted either (1) to reintroduce an eye pigment in a mosquito or fruit fly mutant strain that had no eye pigment or (2) to express the green fluorescence protein (GFP) from the jellyfish, *Aequorea victoria*, or the firefly luciferase gene or the bacterial gene, *lacZ*, that codes for the enzyme β–galactosidase, the activity of which can easily be made visible in a staining reaction. The proposed uses of many of these engineered insects in vector and pest control imply a dimension of environmental impact that sets them fundamentally apart from all genetically engineered organisms (GEO) discussed thus far. Instead of restricting or controlling the release of these GEOs, the goal, in the case of these insects, at least in their medical application for controlling disease vector insects, is to spread the GEOs as widely as possible, perhaps even worldwide.

Table 10.1 Examples of Transgenic Insects

Species		Transposable element (family)	Ref.
Model insects			
Drosophila melanogaster	Vinegar fly	*P* element	Rubin and Spradling 1982, Lombardo et al. 2000, Heinrich and Scott 2001
		Hobo (*hAT*)	Berg and Howe 1989
		*Mos*1 (*mariner*)	Lidholm et al. 1993
		Minos (*Tc1*)	Loukeris et al. 1995a
		Hermes (*hAT*)	O'Brochta et al. 1996
		piggyBac (TTAA)	Berghammer et al. 1999, Handler and Harrell 1999
Drosophila virilis	Vinegar fly	*Hobo*	Lozovskaya et al. 1996
		*Mos*1	Lohe and Hartl 1996b
Agricultural pest insects			
Anastrepha suspensa	Caribbean fruit fly	*piggyBac*	Handler and Harrell 2000
Apis mellifera	honey bee	Sperm-mediated transformation, inherited over three generations, but no integration	Robinson et al. 2000
Bactrocera dorsalis	Oriental fruit fly	*piggyBac*	Handler and McCombs 2000
Bombyx mori	Silk worm	*piggyBac*	Tamura et al. 2000
		*Mos*1, in progress	Wang et al. 2000
Ceratitis capitata	Mediterranean fruit fly (medfly)	*Minos*	Loukeris et al. 1995b, Christophides et al. 2001
		piggyBac	Handler et al. 1998
		Hermes	Michael et al. 2001
Helicoverpa zea	Corn earworm	*Hobo*	DeVault et al. 1996
Musca domestica	house fly	*Mos*1	Yoshiyama et al. 2000
		piggyBac	Hediger et al. 2001
Pectinophora gossypiella	Pink bollworm	*piggyBac*	Peloquin et al. 2000, Thibault et al. 2000
Triboleum castaneum	Red flour beetle	*Hermes*, *piggyBac*	Berghammer et al. 1999
Stomoxys calcitrans	Stable fly	*Hermes*	O'Brochta et al. 2000
Vector insects			
Aedes aegypti	Yellow fever mosquito	*Mos*1	Coates et al. 1998
		Hermes	Jasinskiene et al. 1998, Kokoza et al. 2000, Moreira et al. 2000, Pinkerton et al. 2000
Anopheles stephensi	Malaria mosquito	*Minos*	Catteruccia et al. 2000a
A. gambiae	Malaria mosquito	*Minos*, in progress	Catteruccia et al. 2000b

These insects are meant to exhibit maximum environmental impact by eliminating and replacing the natural, transgene-free population as completely as possible. The capability for active spread and population replacement does not derive from the transgenes of insects but from driving mechanisms that are added to those transgenes. The most promising of these driving mechanisms will be genetically engineered transgenes themselves.

The following sections provide the background information necessary for assessing environmental impact of transgenic arthropods. Based on proposed applications of genetically engineered insects and the insect species involved, the importance of the release of transgenic insects for public health is compared with more traditional methods. An overview is given of potential driving mechanisms and some of their properties. One of the most widespread and presumably most versatile mechanisms, cytoplasmic incompatibility induced by the endosymbiotic bacterium *Wolbachia pipientis*, is described in more detail.

10.2.1.1 *Potential applications in vector control*

Some of the most devastating tropical diseases are transmitted by insects. Incidents of insect-transmitted diseases like malaria, dengue fever, or leishmaniasis are increasing in numbers and the diseases are threatening to invade new countries. It is very likely only a matter of time before large-scale international epidemics of vector-borne diseases such as yellow fever will occur (Gubler 1998). Table 10.2 lists global estimates of present human

Table 10.2 Major Human Diseases Transmitted by Insects

Disease	Vector insects		People infected	Annual mortality	Morbidity
Malaria	*Anopheles* spp.	Malaria mosquitoes	>500,000,000	1,500,000–2,700,000	300,000,000–500,000,000 clinical cases
Lymphatic filariasis	*Culex* spp., *Aedes* spp.	Mosquitoes	120,000,000	Low	44,000,000 chronic cases, elephantiasis
Loaiasis (African eye worm, *Loa loa*)	*Chrysops* spp.	Deer flies	20,000,000	Low	—
Onchocerciasis (river blindness)	*Simulium* spp.	Black flies	18,000,000	Low	270,000 cases of blindness
Leishmaniasis Kala-azar, Visceral Cutaneous	*Phlebotomus* spp. *Lutzomyia* spp.	Sand flies	12,000,000	80,000 Low	Very high —
Sleeping sickness (African trypanosomiasis)	*Glossina* spp.	Tsetse flies	>300,000	20,000	>300,000
Chagas' disease (American trypanosomiasis)	*Triatoma* spp., *Rodnius* spp.	Assassin bugs	18,000,000	>45,000	3,000,000
Dengue, dengue haemorrhagic fever (DHF)	*Aedes aegypti, A. albopictus*	Mosquitoes	>10,000,000	30,000	2,000,000 clinical cases, 500,000 cases of DHF
Yellow fever	*Aedes aegypti*	Yellow fever mosquito	Few	Low	High risk of international epidemics
Bubonic plague	*Xenopsylla cheopsis*	Rat flea	3,000	Low	High risk of international epidemics
Epidemic typhus	*Pediculus humanus*	Body louse	Few	Low	High risk of local epidemics

death tolls due to the major insect-borne diseases. These vector insects are or should become priorities in the application of transgenic technology. Over the last couple of decades, enormous resources have been dedicated to combating these diseases by killing the transmitting insects; yet the diseases are still spreading. In the short term, the development of new vaccines is more elusive than ever, especially in the case of malaria. Even the availability of a very good and reliable yellow fever vaccine for over 60 years has not had much impact on the transmission of yellow fever in endemic areas and certainly has not contributed to the eradication of yellow fever. There is a strong need for novel approaches to vector control. Rendering these insects refractory to disease transmission through genetic engineering may circumvent such past problems as insect resistance induced by vector control measures like insecticides. Because the loss of vector competence will have as little effect as possible on the fitness of the vector insect, no density-dependent, population-regulating mechanisms are likely to be affected. Since there would be no change in the ecological niche occupied by the vector species, the danger of immigrant vectors of the same or different, perhaps worse, species would not arise. Manipulation of vector insects is a particularly attractive strategy for vector-borne diseases that have an animal reservoir, as is the case for yellow fever and Japanese encephalitis, among others. Such a strategy could cut the transmission cycle in a relatively stable way ecologically on sites in which eradication of the parasite is not a viable option, as is true for most zoonoses. For example, rabies has many reservoir animals. An eradication of the rabies virus is practically impossible. The immunization of foxes against rabies would render the vector incompetent without affecting its ecology. However, the immunization of insects is not a good idea. Insects are too short-lived for a direct immunization. Insects cannot synthesize normal antibodies, only single-chain antibodies. Parasites like malaria and dengue fever are so heterogeneous and are constituted by so many different strains that they would likely overcome the hurdle of a single antibody almost instantly. For reasons of transmission and population dynamics discussed later, a sympatric coexistence of significant numbers of natural vectors and genetically engineered, incompetent vectors is neither stable nor desirable. Only the replacement of a natural vector population will improve public health.

More and more tools for the transformation of mosquitoes have become available (see, e.g., Matsubara et al. 1996, Ashburner et al. 1998, Coates et al. 1998, Jasinskiene et al. 1998, Zhao and Eggleston 1998, Atkinson and O'Brochta 1999, Johnson et al. 1999, Pinkerton et al. 2000, Handler and James 2000). The control of disease transmission through the genetic engineering of mosquitoes and bacterial symbionts of vector insects has become an active field of research (James 2000, Beard et al. 2000, Sinkins and O'Neill 2000). Very recently, the gut-specific gene expression of the firefly *luciferase* reporter gene in a transgenic yellow fever mosquito was accomplished (Moreira et al. 2000). Since all parasites must be taken up through the mosquito midgut before they can be transmitted, a gut-specific expression of a transgene represents a major breakthrough. Raikhel and colleagues (Kokoza et al. 2000) succeeded in engineering an increased, blood meal–activated systemic immune response in the same mosquito. This is the first case in which the transgene is not a reporter gene but a factor of the insect's innate immune system, the *defensin* gene. In order for a genetic strategy to be successful, these genetically modified insects not only have to be at least as fit, and preferentially even more fit, than the natural population, but they must also actually replace the natural population. This level of overall fitness and the consequent population replacement can only be accomplished with the help of a driving mechanism.

10.2.1.2   *Potential applications in pest management*

Some agricultural pest insects like the Colorado potato beetle, *Leptinotarsa decemlineata*, have developed resistance against almost all available insecticides. Cross-resistance is a rapidly increasing problem for agricultural pest management as well as for vector control. It will not be long before genetically engineered insects become part of an integrated pest management strategy, although adoption of such a strategy has been much slower than anticipated (Hoy 2000). As herbicide resistance was one of the first goals for transgenic crop plants, agriculturally beneficial insects like predators and parasitoids might be the first to be genetically engineered to gain insecticide resistance (Beckendorf and Hoy 1985, Hoy 1990a, 1990b, 1991, Caprio et al. 1991, Spollen and Hoy 1992, Edwards and Hoy 1995, Pfeifer and Grigliatti 1996). In light of the widespread resistance of pest and vector insects, one would expect that there might be easier ways to accomplish increased insecticide resistance in beneficial predators of pest insects, yet we are not aware of any conventional attempt in this direction, with the exception of three predatory mite species to be used in apple orchards (Kostiainen and Hoy 1994, 1995, Edwards and Hoy 1995, Thistlewood et al. 1995, Solomon et al. 2000). The commercial exploitation of insecticide-resistant, beneficial insects and other arthropods has not yet occurred. Thus, we have yet to discern which risks are greater: the risks associated with the consumption of, for example, genetically engineered potatoes, the transgenes of which are more or less locally confined, or the risks associated with natural potatoes, which are "protected" by genetically engineered, beneficial insects that cannot be confined in any way.

One transgenic strategy actively pursued for insect pest control is the application of inducible fatality genes (IFG). Examples of IFGs are dominant conditional lethal mutations (Klassen et al. 1970, Smith 1971, Fitz-Earle and Suzuki 1975). This strategy is also known as autocidal biological control. It is actively followed for the pink bollworm, *Pectinophora gossypiella*, a major pest of cotton (Fryxell and Miller 1995) and for the Mediterranean fruit fly, *Ceratitis capitata*, which is a worldwide pest (Kerremans and Franz 1995a, 1995b). The applicability of the sterile insect technique also depends on how often the female of the species in question mates during her lifetime. For species that will mate only once, like the screwworm fly, the tobacco hornworm, and the tsetse fly, the sterile insect technique is an ideal strategy.

The future introgression of inducible fatality genes and inducible sterility genes into wild populations is not limited to insects and mites; models of introgression to the common carp, *Cyprinus carpio*, and the mosquitofish, *Gambusia affinis*, which in Australia are considered major freshwater pests, have been developed (Davis et al. 1999, 2000).

The sterile insect technique (SIT), also known as SIRM (sterile insect release method) or SMR (sterile male release), is the oldest strategy employing genetic engineering and constitutes the release of vast numbers of insects, mainly carrying genetically engineered insecticide resistance, into the natural environment. It is also one of the most successful techniques in eliminating natural insect populations (Krafsur 1998). It has been successfully used in the eradication of screwworms *Cochliomyia hominivorax* from the southern United States, Central America, and Libya. Recently, the tsetse fly, *Glossina austeni*, a vector of African trypanosomiasis or sleeping sickness, has been eradicated from Zanzibar. This technique is increasingly used for agricultural pests. Natural populations of the boll weevil, *Anthonomus grandis grandis*, and the pink bollworm, *Pectinophora gossypiella*, both important cotton pests, the Mediterranean fruit fly, *Ceratitis capitata*, the Mexican fruit fly, *Anastrepha ludens*, and the tsetse fly, *Glossina morsitans morsitans*, are currently suppressed with the help of sterile insect releases. Technically, the undesired insect species is mass-reared; then, where possible, males are separated from females, the males sterilized through ionizing radiation, transported within hours to the target area, and then released. There are two major obstacles to this technique. First, most agricultural pest insects, especially Lepidoptera (butterflies and moths), cannot be sterilized by radiation or chemicals without losing so much fitness that they no longer are able to compete with their natural counterparts. Genetically engineered male sterility likely will overcome this problem. Second, in insect species where the female either contributes to disease transmission or agricultural damage through feeding or probing on fruits, the release of sterile females is highly undesirable. Rearing of females also doubles the costs of insect factories. Further, sexing of insects, the mechanical separation of very large numbers of females and males, is often difficult or impossible. The application of genetic sexing might overcome this problem. A transgene will kill the female during development under certain inducible conditions. Strains with these properties are called genetic sexing strains (GSS). Thus, the future expansion of the sterile insect technique will very much depend on the application of transgenic insect technology (Robinson and Franz 2000).

Genetically enhanced sterile insect techniques might allow genetically engineered insects carrying transgenes for male sterility and inducible female-killing to be released in great numbers into the natural environment, with the goal of eliminating natural populations, without the need for an additional driving mechanism. In that case, the released, transgenic population would not be expected to propagate and establish itself.

10.2.1.3   Release of genetically altered insects

There is a very fine and, at the end, a totally artificial and unjustifiable line between transgenic organisms and genetically altered or genetically modified organisms. Transgenic organisms carry a gene, a transgene, which cannot be found in natural populations of that species. In genetically altered organisms, a native gene has been altered or parts of a chromosome have been relocated, rearranged, or translocated with the help of chemicals or ionizing radiation; the end product is again an organism carrying a gene or a chromosome that cannot be found in natural populations of that species. If one moves from the term *transgenic insects* to *genetically engineered insects*, it becomes impossible to pinpoint where genetic engineering starts. One could argue that silencing a gene through chemical mutagenesis or radiation is not to be considered genetic engineering whereas removing a gene by cutting it out with the help of restriction enzymes or silencing it with a natural transposable element might be considered genetic engineering. Is the engineering of sterile insects, especially of sterile male insects and female insects with lethal mutations, genetic engineering? Artificially causing mutations and chromosome relocations are genetic manipulations, and techniques like sterile insect release are commonly known as genetic control strategies. In genetics, the radiation or chemicals causing male sterility are considered mutagens leading to mutagen-induced dominant lethality. First proposed by the Russian geneticist Serebrovskii in 1940 for the control of insect populations (Serebrovskii 1940), genetic control has become an integral part of vector and pest management (e.g., Smith and von Borstel 1972, Davidson 1974, Pal and Whitten 1974, Sharma 1974, Whitten and Foster 1975, Curtis 1985, Rai 1996, Seawright and Cockburn 1996). Further, discussions of transgenic organisms tend to focus on the genetic manipulation of the nucleus, practically ignoring any changes to the inheritable genetic information of the cytoplasm.

It is important to remember when discussing the release of transgenic insects that there is a long and active history of releases of genetically altered insects into the environment with both nuclear and cytoplasmic alterations or modifications. Table 10.3 provides an overview of some releases of genetically altered insects in field trials for genetic control of vector species and large-scale population suppression programs of agricultural pest insects. The majority of the cited literature is rather old. Most field releases were done in the 1960s and 1970s. Some of the literature is not yet accessible in all computerized databases. This table, while by no means comprehensive, emphasizes the extent of releases in nature, both in terms of number of different species and in terms of widespread geographical distribution. Some field trials were limited to 100 insects, whereas others involved releases of more than 7 million insects a day for half a year. Some of the strains bear highly complex and sophisticated alterations, which required several years of genetic

Table 10.3 Examples of Releases of Genetically Altered Insects

Species	Outcome/problem	Genetic manipulation	Location	Organization[a]	Ref.
Agricultural pest insects					
Blatella germanica, German cockroach	Too few released	Double translocation	Ship, Norfolk, Virginia, U.S.A.	USNONR	Ross et al. 1981
Chrysomya bezziana, Old World screwworm fly	Suppression	Radiation sterilization	Musa Valley, Papua New Guinea		Spradbery et al. 1989
Cochliomyia hominivorax, New World screwworm fly	Eradication	Radiation sterilization	U.S.A., Curaçao, Mexico, Belize, Guatemala, El Salvador, Honduras		Meyer 1994
	Ongoing release		Nicaragua		Galvin and Wyss 1986; Wyss 1998
	Ongoing release		Panama		Grant et al 1998
	Planned release		Cuba, Hispaniola, Jamaica Libya		Anonymous 1992, Krafsur and Lindquist 1996; Anonymous 1998a
Cydia pomonella, codling moth	Ongoing release	Radiation sterilization	British Columbia, Canada		Anonymous 1998a
Dacus cucurbitae, melon fly	Eradication	Radiation sterilization	Rota, Mariana Islands	USDA, U.S. Navy	Steiner et al. 1965
Bactrocera cucurbitae, melon fly	Eradication	Radiation sterilization	Okinawa, Japan		Kakinohana et al. 1997
Ceratitis capitata, Mediterranean fruit fly	Ongoing release	Radiation sterilization	California, Florida, Hawaii, U.S.A., South and Central America, Europe, Japan, Africa, Australia		Economopoulos 1996, Anonymous 1998b, Cayol and Zaraï 1999
Delia antiqua, onion fly	Ongoing release	Radiation sterilization	The Netherlands		Anonymous 1998a
Drosophila melanogaster, vinegar fly	Unsuccessful	Compound chromosome	Maryland, U.S.A.		Cantelo and Childress 1974
	Unsuccessful	Compound chromosome	Brindabella Valley, New South Wales, Australia		Mckenzie 1976

Species	Result	Method	Location	Agency	Reference
Lucilia cuprina, sheep blowfly	Unsuccessful	Compound chromosome	Victoria, Australia		Foster et al. 1985b, Smith and Morton 1985
	Immigration	Translocation, mutation, insecticide resistance (dieldrin)			Vogt et al. 1985, Foster et al. 1985a, Foster and Smith 1991
Musca domestica, house fly	Pilot study	Chemical sterilization	Lipari Islands, Italy		Sacca et al. 1967
	Pilot study	Chemical sterilization	Vulcano Island, Italy		Magaudda et al. 1969
	Unsuccessful, dispersal instead of competition	Double translocation	Florida, North Dakota, U.S.A.		Wagoner et al. 1973
Pectinophora gossypiella, pink bollworm	Unsuccessful, as above	Double translocation	Florida, U.S.A.		Morgan et al. 1973
	Ongoing release	Radiation sterilization	California, U.S.A.		Anonymous 1998a
Vector insects					
Aedes aegypti, yellow fever mosquito	Too high dose of radiation	Radiation sterilization	Pensacola, Florida, U.S.A.	USPHS	Morlan et al. 1962
	Pilot study	Mutation (*Si*, silver mesonotum)	Meridian, Mississippi, U.S.A.	USAEC	Bond et al. 1970
	Success, persistence, pilot study	Translocation, mutation (*Si*)	Delhi, India	WHO, ICMR	Rai et al. 1973
	Pilot study, no indigenous mosquitoes	Translocation	Florida, U.S.A.	USDA	Seawright et al. 1975
	Pilot study, no indigenous mosquitoes, recombination, construct breakdown	Double translocation, no inversion	Florida, U.S.A.	USDA	Seawright et al. 1976
	Promising	Chemical sterilization	Delhi, India	WHO, ICMR	Grover et al. 1976
	Not competitive	Double translocation			
	Promising	Double translocation, sex ratio distorter			
	Unsuccessful, inbreeding	Translocation, homozygous	Rabai, Kenya coast	ICIPE, USAID	Lorimer et al. 1976

Table 10.3 (continued) Examples of Releases of Genetically Altered Insects

Species	Outcome/problem	Genetic manipulation	Location	Organization[a]	Ref.
	Inconclusive, part. suppres., rainy seas.	Double translocation	Mombasa, Kenya	ICIPE, USAID	McDonald et al. 1977
	Successful for type strain, dry season, but unaffected subspecies	Double translocation	Rabai, Kenya coast	ICIPE, USAID	Petersen et al. 1977
	Unsuccessful, no hybrid vigor	Chemical sterilization, hybrid males	Florida, U.S.A.	USDA	Seawright et al. 1977
Anopheles albimanus, malaria mosquito	Successful (isolated population)	Chemical sterilization	Lake Apastepeque, El Salvador	USDA	Lofgren et al. 1974
	Reduction, but immigration of up to 20% of fertile females	Chemical sterilization, chromosome inversion, male insecticide resistance (propoxur)	Coastal plains, El Salvador	USDA	Dame et al. 1981
A. culicifacies, malaria mosquito	Unsuccessful, wrong photoperiod	Y-linked multiple chromosomal translocations, X-linked inversion	Kot Baghicha Singh Walla, Punjab, Pakistan		Baker et al. 1980
	Unsuccessful, males not competitive	Chemical sterilization, male insecticide resistance gene (dieldrin)	Same		Reisen et al. 1981
A. gambiae, malaria mosquito	Unsuccessful, males not recognized as mates	Interspecies hybrid sterility of an artificial cross	Bobo Dioulasso, Burkina Faso	WHO, ORSTOM	Davidson et al. 1970
A. quadrimaculatus, malaria mosquito	Laboratory inbreeding, no longer competitive	Radiation sterilization	Florida, U.S.A.	USDA	Weidhaas et al. 1962
	Unsuccessful, too toxic	Chemical sterilization			Dame et al. 1964

Species	Method	Outcome	Location	Organization	Reference
Culex pipiens, house mosquito	Translocation	Inconclusive, almost eradication, persistence, contributing factors, pilot study	Notre Dame, Montpellier, France	EIDLM	Laven et al. 1971, 1972, Cousserans and Guille 1974
C. quinquefasciatus, house mosquito	Radiation sterilization	Unsuccessful	New Delhi, India	Indian Government	Krishnamurthy et al. 1962
	Wolbachia	Eradication	Okpo, Rangoon, Burma	WHO	Laven 1967, 1971
	Chemical sterilization	Almost eradication (small island), pilot study	Seahorse Key, Florida, U.S.A.	USDA	Patterson et al. 1970, Lowe et al. 1974
	Radiation sterilization	Unsuccessful, males not competitive	Florida, U.S.A.	USDA	Patterson et al. 1975
	Radiation sterilization	Successful	Florida, U.S.A.	USDA	Patterson et al. 1977
	Chemical sterilization	Promising, immigration	New Delhi, India	WHO, ICMR	Sharma et al. 1976
	Chemical sterilization	Unsuccessful, immigration of inseminated females	New Delhi, India	WHO, ICMR	Yasuno et al. 1978
	Double translocation, *Wolbachia*	Unsuccessful, immigration of inseminated females	New Delhi, India	WHO, ICMR	Curtis et al. 1982
C. tarsalis, mosquito	Chemical sterilization	Not competitive	California, U.S.A.	CA DPH	Lewallen et al. 1965
	Radiation sterilization	Too few released, inbreeding	California, U.S.A.	U.S. Army, USPHS	Asman et al. 1980, Reisen et al. 1981a, 1982
	Translocation	Too few released	California, U.S.A.	U.S. Army, USPHS	Asman et al. 1979
	Eye pigment mutation	Reduced larval fitness, immigration	California, U.S.A.	U.S. Army, USPHS	Reisen et al. 1985
C. tritaeniorhynchus, mosquito	Dominant temperature-sensitive female lethal, translocation	Unsuccessful, laboratory inbreeding	Punjab, Pakistan	USPHS	Baker et al. 1979
	Similar, but outcross	Unsuccessful, laboratory inbreeding	Punjab, Pakistan	USPHS	Reisen et al. 1980

Table 10.3 (*continued*) Examples of Releases of Genetically Altered Insects

Species	Outcome/problem	Genetic manipulation	Location	Organization[a]	Ref.
Glossina austeni, tsetse fly	Eradication	Radiation sterilization	Zanzibar	IAEA	Dyck et al. 1997
G. morsitans morsitans, tsetse fly	Reduction, but less competitive	Radiation sterilization	Tanzania, Zimbabwe	IAEA	Dame et al. 1981
G. palpalis gambiensis, tsetse fly	Reduction, but less competitive	Radiation sterilization	Burkina Faso	IAEA	Cuisance et al. 1978, Politzar et al. 1979
	Reduction, but less competitive	Radiation sterilization	Nigeria	IAEA	Lindquist 1984
G. palpalis palpalis, tsetse fly	Eradication	Radiation sterilization	Central Nigeria	IAEA	Takken et al. 1986
G. tachinoides, tsetse fly	Reduction	Radiation sterilization	Burkina Faso	IAEA	Politzar and Cuisance 1984

[a] Organization abbreviations: CA DPH = California Department of Public Health; EIDLM = Entente Interdépartementale pour la Démoustication du Littoral Méditerranéen, Montpellier, France; IAEA = International Atomic Energy Agency, Vienna, Austria; ICIPE = International Centre of Insect Physiology and Ecology, Nairobi, Kenya; ICMR = Indian Council for Medical Research, New Delhi, India; IRD = Institut de Recherche pour le Development, Paris, France; ORSTOM = Institut Français de Recherche Scientifique pour le Developpement en Cooperation, now IRD; USAEC = U.S. Atomic Energy Commission; USAID = U.S. Agency for International Development; USDA = U.S. Department of Agriculture; USNONR = U.S. Navy Office of Naval Research; USPHS = U.S. Public Health Service; WHO = World Health Organization, Geneva, Switzerland.

engineering to produce. Most released insects carry more than one alteration; in more modern, large-scale releases, it is often the combination of radiation-induced sterility with either an insecticide resistance gene or a genetic means of killing females during rearing.

Marjorie A. Hoy pioneered the official release of the first transgene in an arthropod in the United States. In spring of 1996, roughly a thousand females of the Western orchard predatory mite, *Metaseiulus occidentalis* (Acari: Phytoseiidae), were released onto pinto bean plants in an orchard in Alachua County, Florida. *Metaseiulus occidentalis* is a beneficial arthropod. It preys on phytophagous spider mites like the two-spotted spider mite, *Tetranchus urticae*, an important pest in deciduous orchards of apples, almonds, walnuts, pears, and peaches, as well as in vineyards and strawberry and hop plantations. The predatory mites released by Hoy were carrying the *Escherichia coli* reporter gene *lacZ* and a *Drosophila melanogaster* heatshock promoter hsp70 as transgenes. Within 3 weeks, many mites were killed as heavy rains washed them off the plants. A freeze followed the rain, reducing the quality of the bean plants for the mites and thereby reducing mite numbers sufficiently to cause the crash of the transgenic predator population. Six additional colonies of transgenic mites were released in October of the same year. Although under laboratory conditions, the transgenic lines had been stable for more than 120 generations (Li and Hoy 1996, McDermott and Hoy 1997, Presnail et al. 1997, Jeyaprakash et al. 1998), under field conditions the transgenes did not seem to be stable. The transgene was rapidly lost from all released colonies, suggesting it might have been episomal rather than chromosomal. It is assumed that a transformation by illegitimate recombination might have occurred since no transposase was injected and there was no evidence of native, active P-elements. However, the nature of the transformation has never been established.

The first release of a transgenic, agricultural pest insect will presumably involve the pink bollworm, *Pectinophora gossypiella* (Lepidoptera: Gelechiidae), an important pest of cotton. An application is pending to release these moths into a cotton field of the Phoenix Plant Protection Center of the U.S. Department of Agriculture (USDA). The moth has been transformed with the help of the transposon *piggyBac* to carry the marker gene *green fluorescent protein* (GFP) (Peloquin et al. 2000). If the release is successful, the transgenic pink bollworm will be used in sterile insect releases of the bollworm in the United States. This might become the first official mass release of transgenic insects. At the moment, it is impossible in field catches to distinguish between released sterile bollworms and native bollworms. The transgenic moths are seen as an ideal way to mark the released sterile bollworms. The GFP will glow under ultraviolet light even in dead moths caught in traps. The long-term goal is to clone an autocidal biological control gene based on a truncated *Notch* gene into these bollworms (Fryxell and Miller 1995).

Detailed information about all releases of transgenic arthropods and other invertebrates in the United States can be found on the Web pages of the USDA Animal and Plant Health Inspection Service (APHIS), http://www.aphis.usda.gov/biotech/arthropod/. USDA–APHIS is responsible for permits and the regulation of the release of transgenic organisms (Young et al. 2000). Comprehensive information about both applications for permits to release as well as evaluation reports of past releases can be found on the site.

General safety considerations and the risks and challenges for the safe use of transgenic insects in future pest control and management programs were addressed by Hoy (e.g., Hoy 1995, 1998, 2000). However, genetic driving mechanisms have not yet been considered.

10.2.2   The need for driving mechanisms

A driving mechanism actively introduces a genetic trait into a population by directly or indirectly eliminating individuals without that trait. Dilution of a natural population by the simple mass release of insects carrying the desired genes seems to be a proven method and the only proven method in the numerous pest management campaigns using sterile insect releases. Although for the sterile insect technique the release of males only is obvious, the question of whether the release of males only or both sexes should be attempted for population replacement is still debated by some workers. The mere fact that female mosquitoes have a much longer life expectancy than male mosquitoes is regarded as sufficient reason to strongly favor the release of both sexes. However, sterile insect releases and dilution or mass releases of non-sterile insects constitute two fundamentally different techniques. The sterile insect technique uses a very strong driving force by reducing the fitness of the resident population with every mating between a fertile, resident female and a sterile, released male. Sterile techniques aim at eradication and not at replacement of populations. Eradication is ecologically the most unstable condition possible. In contrast, the simple mass release of a refractory mosquito strain, with no support by any driving force and equal fitness of released and wild-type individuals, would not affect the fitness of the feral population at all, and therefore would not be expected to spread. Further, the genetic load of the introduction of any transgene not related to a driving mechanism is likely to reduce the overall fitness of the transgenic insect, making the survival or maintenance of any released mosquitoes in a natural population highly unlikely. The release of transgenic insects would only make sense if, at the very least, these insects were able to maintain themselves in a population; this requires that they be at least as fit as the natural population. Any loss of fitness due to the genetic manipulation or the particular genetic load has to be compensated for in some way. Ideally, it will be overcompensated for to facilitate population replacement.

It is important to consider whether partial exchange or replacement of a vector population by a refractory strain might be as good as no replacement or no intervention at all. Based on the basic reproductive rate (R_0) of parasites like the malaria protozoa, the dengue fever virus, or the filarial worms in a certain environment or country, one can make predictions about the impact of control or management interventions (Anderson and May 1991). For example, to reduce the transmission of an infectious agent in a population by vaccination in the case of malaria in hyperendemic regions of Nigeria, where one can assume a R_0 value of 100, one would need to have successfully immunized 99% of the total population before the age of 6 months to make progress toward eradication of the infectious agent. For rubella (German measles) in Wales, where one may assume an R_0 value of 5, a successful immunization level of 70% of the population has to be reached by roughly age 17 to make progress toward an eradication of the infectious agent. Similar calculations hold true for the vector insects. Unless the intervention successfully affects far more than 99% of the vector species in areas with stable malaria transmission, no reduction of transmission will be expected. It may be difficult to imagine that removing, for example, 90% of a vector species, one might not have any lasting positive effect on the transmission of a certain disease, but such is the case. The facts strongly argue the need for a very strong spreading force to ensure that any control strategy will have any chance of success at all. On a population level, success means that the driving mechanism has propelled the transgenes to fixation. If fixation in a population cannot be attained, the mechanism is of no practical use for vector insects.

If the replacement of natural vector species of mosquitoes with strains naturally refractory to human malaria is the goal, we are facing an almost insurmountable challenge. It does not matter whether the refractory strain has been collected in the field or selected for in the laboratory; all refractory strains known so far are recessive or partially dominant for the resistance phenotype (Curtis 1994, Severson et al. 1995, Feldmann et al. 1998, Somboon et al. 1999). Any mating with a wild-type mosquito will result in the loss of the refractoriness. Thus, simple release strategies without any manipulations are out of the question.

The single most important vector-borne disease, malaria, is perhaps the most challenging disease to combat with a transgene strategy. Snow and Marsh (1995, 1998, Snow 2000) predict that even a 100-fold reduction in transmission of *Plasmodium falciparum* may not significantly decrease mortality in African children younger than 5 years. The number of infectious mosquito bites per person per year, the entomological inoculation rate (EIR), will not reduce the long-term impact on mortality and morbidity from malaria unless this rate falls below one infectious bite per person per 10 years (Trape and Rogier 1996). In Tanzania, for example, the average EIR ranges from 10 to 300, with a minimum of 0 and a maximum of 3000 infectious bites per person per year. The interpretation of these data is obviously highly controversial. It will take several more years before enough data become available (e.g., Coleman et al. 1999). Any intervention strategy in an area of stable or endemic malaria that is falling short of eradication might only make the situation worse in the long run. Young children who have been protected by, for example, insecticide-impregnated bednets will not have built up any protective immunity. Malaria mortality just might be shifted to an older age, an age when the disease might hit even harder and lead to greater overall mortality. Large-scale eradication programs bear a high probability of ending in disappointment (Spielman et al. 1993). For these reasons, some recommend not to focus research on transgenic approaches in the fight against malaria (Spielman 1994a, 1994b). Reemphasizing the recommendations of Spielman et al. (1993), Pettigrew and O'Neill (1997) propose to initially evaluate genetic manipulation methods of insect populations on diseases such as African trypanosomiasis transmitted by tsetse flies or unstable malaria in India. The complex epidemiology of malaria together with the intense transmission rates of malaria in Africa underline the importance of driving mechanisms for any transgenic strategy. If fixation of a transgene in a population cannot be obtained by a drive, the transgene may not be of much use in vector control.

10.2.3 Genetic driving mechanisms for population replacement

There are many natural strains of vector insects that are refractory to disease transmission. However, strains that are refractory to disease transmission are rare in nature compared with strains capable of disease transmission. One hypothesis for explaining the comparative rarity is that the refractory strains are less fit than their counterparts in the environments where disease transmission occurs. Thus, even with use of naturally refractory strains of vector mosquitoes, there may be no way to avoid artificial manipulation of the strains, i.e., genetic engineering to make the strains useful for vector and disease control!

Most, if not all, forms of meaningful natural resistance in vector insects against human parasites are coded for by more than one gene (Curtis 1994, Feldmann et al. 1998, Somboon et al. 1999). Several animal parasites may represent exceptions to that rule. To minimize the possibility of the parasite developing resistance to a transgene and thereby defeating the control strategy, it is highly desirable that an engineered vector insect carry more than one transgene and that the different transgenes act quite differently. However, linking more than one transgene to any genetic driving mechanism so that the linkage is not likely to be broken during the first 200 generations after the release is a genetic engineering problem that has yet to be solved. The problem effectively overshadows any discussion of conceivable driving mechanisms. All driving mechanisms are germ-line-based, selfish genetic elements that are inherited in a non-Mendelian way.

10.2.3.1 Competitive displacement and positive heterosis

Conceptually, the easiest form of drive would be a replacement strain that is fitter than the natural vector population. However, the reason some diseases are such huge problems is because vector competent strains are fitter than refractory strains. The term *heterosis* was coined to capture the fact that heterozygotes, crossings between two populations, are fitter than homozygotes, crossings within populations.

One still could argue that, even if there are no refractory strains of equal or superior fitness, there might be ecologically competitive, non-vector species that can outcompete the vector species. Or, even more elegantly, there might be a very closely related species that is still capable of mating with the native vector species and producing offspring that either are not viable or are sterile. Unfortunately, such ecologically competitive species have not yet been found for the major disease vectors. The strategy, known as competitive displacement or competitive exclusion, would constitute an almost ideal vector control strategy. Rosen et al. (1976) reported a field trial conducted on a Pacific atoll to replace the native population of *Aedes polynesiensis*, the mosquito that transmits the nematode *Wuchereria bancrofti* that causes filariasis, with *A. albopictus*, a species that does not transmit filarial worms. The results of this attempt and many other were at best inconclusive. Earlier experiments with large walk-in cages simulating the natural habitat of the two mosquito species were very encouraging (Gubler 1970, Rozeboom 1971). Equally good results were obtained in greenhouses by flooding the confined space with 10 to 40 times the number of the local population with a different race of the Hessian fly, *Mayetiola destructor*, a major pest of wheat (Foster and Gallun 1972). They were never introduced into the field. Ellis et al. (1997) discuss the possibility of competitive exclusion of the rodent species *Calomys musculinus*, the principal reservoir of Junin virus, which is the etiological agent of Argentine hemorrhagic fever, by other rodent species aided by invasive grasses capable of changing the microhabitat. The only documented displacement successes have been several cases of freshwater pulmonate gastropods, which sustain the life cycle of blood and liver flukes, that were replaced by ampulariid species. Competitive displacement of the freshwater snail *Biomphalaria glabrata*, an intermediate host of the blood flukes that cause schistosomiasis in the New World, is the best example of applied biological control of vectors (e.g., Pointier and Augustin 1999). However, suspicion is rising that some of these beneficial snail species might be or become intermediate hosts of several less prominent animal and human parasites such as the trematode *Paragonimus westermani*, a lung fluke of humans (Pointier 1999). In any case, population replacement of insect vector species using ecological strategies is not a viable option at the moment.

Because insecticide resistance has developed and spread in many important vector species worldwide, one might be tempted to suggest use of insecticide resistance as a strong and proven force to drive transgenes around the world. However, to approach complete replacement of a natural population, any degree of resistance or cross-resistance in the wild population toward the insecticide to be used for the driving would have to be ruled out. That would require that a totally new and powerful insecticide be developed and effectively banned from any use other than use in association with release of the transgene. Whitten (1970) proposed to add resistance to already existing insecticides over and above other driving mechanisms to provide an advantage to released mosquitoes. The increased fitness should speed the spread and population replacement. However, the development of new insecticides without any cross-resistance is far from trivial and is, in essence, unpredictable. Curtis emphasizes that one would require great confidence in the effectiveness of the replacement strategy to persuade malaria control authorities to deliberately release resistance genes and prevent the use of the insecticide for traditional chemical control (Curtis and Graves 1988).

10.2.3.2   Meiotic drives

If a given chromosome is transmitted to more than 50% of the offspring, then that chromosome is inherited in a non-Mendelian way; it has experienced preference during meiosis. That is, the chromosome is occurring in more gametes than would be predicted if both sister chromosomes of a diploid organism had been treated equally. The segregation of sister chromatids or heterozygous alleles or heteromorphic chromosomes into gametes would have been distorted during meiosis. Genes responsible for this are called *segregation distorters*. Any genes associated with or transgenes linked to such a chromosome are preferentially inherited or segregated as well; they are subject to meiotic drive. If the population dynamics of a phenotype or its genetic locus is considered, the term used is *genic meiotic drive*. If for structural reasons a whole chromosome is achieving an advantage during segregation, for example, expediency in movement on the spindle, one speaks of a *chromosomal meiotic drive* (Lyttle 1991). If such a chromosome is a sex chromosome, it constitutes a sex drive. In the absence of any interference, such genes or chromosomes would eventually approach fixation. Theoretically, an exponential increase in the frequency of the driven chromosome and transgenes should be possible. However, fixation of transgenes would be expected to occur faster in systems that employ reduced heterozygote fitness.

Most insects, like vertebrates, have heterogametic males (XY or X0) and homogametic females (XX). This makes the heterogametic sex chromosome, the Y chromosome, the limiting factor in a population. Conversely, some insect orders such as Lepidoptera (butterflies and moths) and Trichoptera (caddisflies) as well as birds, reptiles, and some fish display the opposite of sex determination, heterogametic females and homogametic males. A strong X chromosome drive would result in a higher frequency of females and that, in turn, could increase the population size by almost tenfold before it abruptly crashed because of lack of males. If even a minimal influx of immigrating males cannot be ruled out, X chromosome drives not only would replace a local population but also would lead to an explosion of the local population size after some 30 generations, depending on parameters. If the released females have any residual vectorial capacity left or are a nuisance as blood-sucking insects, X drives can become a big problem. Strong Y chromosome drives, on the other hand, will almost immediately cause a decrease in population size. Unless there is considerable immigration, strong Y drives may push populations to extinction in fewer than 15 generations. Presumably for this very reason, examples of strong Y drives have rarely been observed (Hamilton 1967). X and Y drives denote only which chromosome is preferentially segregated and which sex is benefiting; it does not indicate in which sex the distorted segregation is actually occurring. Unlike meiotic sex drives, strong autosomal drives should lead to fixation without a major impact on the population size.

There are very few examples of active meiotic drives in insects. However, the importance of segregation distortions as an evolutionary force have been studied in *Drosophila* (Sandler and Novitski 1957, Hamilton 1967, Sandler and Golic 1985, Wilkinson et al. 1998, Carvalho et al. 1998). Five naturally occurring phenomena resembling sex drives have been studied in detail. A weak X drive occurs in the wood lemming, *Myopus schisticolor*; it is remarkable because it causes chromosomal males to develop as females (Eskelinen 1997, Liu et al. 1998). At least 14 species of butterflies have populations that produce only female offspring (Ishihara 1994). For a long time, this was explained with a W drive in the butterfly genera *Danaus* and *Acraea* capable of producing almost female-only populations. However, doubts have arisen about whether sex drives do actually exist outside the order Diptera (Jiggins et al. 1999). For example, in the species *A. encedon* and *A. encedana*, a male-killing endosymbiont *Wolbachia pipientis* has been detected (Jiggins et al. 2000a, 2000b). The sex ratio can be cured with antibiotics. The sex ratio in the butterfly *D. chrysippus* is also of secondary origin and not the result of a sex drive after all. It is infected by a male-killing *Spiroplasma* bacterium and a normal sex ratio can also be restored with antibiotics (Jiggins et al. 2000c). An X drive is widespread and well studied in many species of the genus *Drosophila*. It is known as the SR (sex ratio) drive (Carvalho and Vaz 1999). Recently, a meiotic sex drive has been discovered in stalk-eyed flies (Lande and Wilkinson 1999). And a male drive has been found in mosquitoes and is described in detail below.

Autosomal drives are even rarer but are more widely distributed. The *t* haplotypes in mice carry a *tailless* (*t*) allele that is responsible for the transmission-ratio distortion (TRD). *Drosophila melanogaster* carries the SD (segregation distortion) system. And the *Sk* (*Spore killer*) drive is found in the ascomycete fungus *Neurospora*. No autosomal drives are known in mosquitoes.

10.2.3.2.1   Biased gene conversion.   Hastings has suggested that biased gene conversions might be developed into powerful techniques for population control (Hastings 1994). A single-copy gene, a cluster of genes, or multigene families can show biased conversion. The allele carrying these genes becomes selfish in biasing conversion in its favor. The selfish gene does not need to be beneficial to its carrier in order to spread; the only requirement is that the loss of fitness be less than the conversion bias. At first, this phenomenon seems to constitute only a minor form of meiotic drive, affecting just one gene, but it in fact might control large segments of one or more chromosomes. Multigene families are likely to be the most effective for invading a population. The use of multiple copies would allow a drive to be successful even if the fitness falls below 0.5 (Hastings 1994). Walsh and Slatkin (Walsh 1985, 1986, 1992, Slatkin 1986) were the first to investigate the population dynamics of biased conversion in multigene families. Although it has been studied primarily in animals and plants, biased gene conversion has been found in bacteria as well (Yamamoto et al. 1992). Research into biased gene conversion can help us understand molecular evolution and explain the unexpected molecular homogeneity of many genes in all members of a given species, i.e., the cohesive mode of species evolution (Dover 1982, Hillis et al. 1991). Increasing access to whole genome sequences will help in analyzing conversion bias (Akashi et al. 1998). Hybrid individuals of two subspecies of the Australian grasshopper, *Caledia captiva*, show a biased exchange of ribosomal DNA (Arnold et al. 1988). This form of asymmetric introgression or population replacement suggests a potential means of introducing a transgene. However, the molecular mechanisms underlying biased gene conversion are still little understood. In its simplest form, a DNA repair enzyme is believed to change the sequence on one allele in part or in whole according to the sequence on its counterpart on a homologous chromosome. It is not known what initiates and regulates this mechanism (Lamb and Helmi 1989, Zwolinski and Lamb 1995). Unfortunately, any molecular manipulation or application of gene conversion is far from possible at the moment. The term *biased gene conversion* does not suggest fixation. Yet, gene conversion may be responsible for the fact that several hundred copies of ribosomal DNA are identical in each species of all animals. It has been shown that all 20,000 members of a large gene family, the mouse interspersed family (MIF-1), were identical within each of two species of mice but different between the two species (Brown and Dover 1981, Dover 1982, Tremblay et al. 2000). In gene conversion, the genes do not have to be physically linked or close to each other but can be widely separated in the genome, and can even reside on different chromosomes (Brown and Dover 1981). This phenomenon — important for genetic engineering — makes gene conversion unique among the driving systems. Use of gene conversion becomes even more desirable in cases where resistance to disease transmission is caused by mutations in household genes. Thus, an extremely powerful mechanism, a "molecular drive" as Dover coined it, looms on the horizon but we have yet to learn how to use it.

10.2.3.2.2   Distorter genes and drives linked to sex chromosomes.   Distorted sex ratios were recorded early on in laboratory-reared mosquitoes. A male-biased sex ratio was discovered in the yellow fever mosquito, *Aedes aegypti*, in 1922 (Gordon 1922, Young 1922).

The male bias is inherited through the nucleus of the male (Craig et al. 1960, Wood 1961). The sex distortion was attributed to a gene called *Distorter* (*D*), which is closely linked to the sex-determining locus of the sex chromosomes (Hickey and Craig 1966). Meiotic drive was only apparent if *D* were linked to the dominant sex determinant *M*, meaning when *D* was on the Y chromosome. In 1969, Ralph Barr noticed that a mutation called divided-eye in the tropical house mosquito, *Culex quinquefasciatus/pipiens*, was restricted to males (Barr 1969). This mutation was driven by a distorter gene similar to the distorter gene described in *A. aegypti* (Sweeny and Barr 1978). Barr suggests that *D* in *Culex* might be a mutation of *m* (the female-determining factor) (Sweeny et al. 1987).

In both mosquitoes *A. aegypti* and *C. quinquefasciatus/pipiens* it is assumed that the mechanism of meiotic drive is associated with a sex chromosome breakage during male meiosis. More than 90% of the breakage occurs in the X chromosome rather than in the Y chromosome. Isochromatic breakages occur before diplotene, in chiasmic rather than achiasmic arms of bivalents, with the result that most acentric fragments remain attached to the unbroken homologue by a chiasma. Most breaks occur at four discrete sites (Wood and Newton 1991). The almost exclusive breakage of the X chromosome results in the transmission of many more Y chromosomes than X chromosomes to the progeny. *Distorter* males produce fewer spermatozoa than normal males. There is also a slightly greater post-zygotic mortality associated with distortion. However, the overall influence of distortion in reducing fertility is far less than what one might expect and is not significant compared with normal crosses, leaving a very strong driving force with almost pure sex distortion. Such a drive and the natural population in which it occurs would be driven rapidly to extinction through sheer lack of females. Therefore, it is not surprising that several mod-ifying genes have been discovered in natural populations. Most of these modifying genes are resistance genes against sex-linked distorter genes, although enhancers of meiotic drives have been identified as well. Extensive information on distorter genes is only available for *A. aegypti*. Some *A. aegypti* populations carry distorter genes on all their Y chromosomes and resistance genes on all their X chromosomes, eliminating meiotic drive totally from the population; other populations, with varying degrees of resistance on their X chromosomes, exhibit some degree of active drive; and still other populations exhibit no distorter genes on the Y chromosome with all their X chromosomes highly sensitive to the action of distorter genes. The latter populations are suitable targets for population replace-ment by sex-linked meiotic drives. It is important to remember, however, that in the wake of a comprehensive genetic analysis of all the genes influencing meiotic drive, there were cases of increased recombination across translocation junctions and greater variation in fertility between families, thus establishing any tight linkage between transgenes and meiotic drive is difficult (Pearson and Wood 1980; Wood and Newton 1991).

Crosses between distorter males and sensitive females can result in as few as 3% females in the offspring. Introduction of such males will almost certainly lead to replace-ment of natural populations; however, it will not lead to the introduction of potential transgenes in any of the females. Consequently, the desired transgenes have to be linked not only to the distorter gene on the Y chromosome but also to the sensitivity gene(s) on the X chromosome. One breeding scheme to isolate X chromosomes sensitive to meiotic drive revealed X chromosomes with the capability to drive against Y chromosome–based distorter genes (Owusu-Daaku et al. 1997). As much as this may create new possibilities, it also highlights how little understood and how complex natural mosquito populations are. Meiotic drives linked to the X chromosome instead of the Y chromosome and pro-ducing almost totally female populations have been described in more than ten *Drosophila* species but in no mosquito populations (Capillon and Atlan 1999, Cazemajor et al. 2000).

In a laboratory cage experiment, a red eye mutation (*re*) in *A. aegypti* that is linked to the *Distorter* gene *D* was driven into a sensitive population by Wood of the University of

Manchester (Wood et al. 1977). However, the experiment also revealed an unexpectedly high fitness of the red eye mutation under these laboratory conditions, thus making an evaluation of the driver difficult.

Curtis from the London School of Tropical Medicine and Hygiene linked a double translocation to the *Distorter* gene in *A. aegypti* (Curtis et al. 1976). In a large field cage experiment in India, the meiotic drive proved highly effective in suppressing the cage population. The importance of the meiotic drive in suppressing naive populations was proved in a control experiment with the double translocation alone. It was shown that the combination of meiotic drive and double translocation in homozygotes could overcome a resistance to distortion on the X chromosome in at least 20% of the field cage population (Grover et al. 1976, Suguna et al. 1977a, 1977b). These suppression experiments suggest that a replacement strategy might have been possible as well.

For the time being, one is limited to the use of natural, active distorter genes in species in which they have been discovered. As far as vector insects are concerned, distorter genes have been found only in *A. aegypti* and *C. quinquefasciatus*. No sex-linked meiotic drive has been described for anopheline mosquitoes. Further applications are limited to pockets of natural populations highly sensitive to meiotic drive. Uniformly sensitive populations might be exceptional. Any introduction through a meiotic drive would be a definitive experiment without a second chance. Theoretically, it should be possible to modify existing genes or design new distorter genes or add enhancer genes like the A factor to overcome natural resistance genes and species barriers, but at the moment there appears to be no active research in this area.

10.2.3.2.3 Hybrid dysgenesis and mobile elements.   Green (1969a, 1969b) was the first to point out clearly, against all prevailing notions in genetics at the time, that genetic elements in *Drosophila* do sometimes spontaneously change their position in the genome and that the phenomena observed by McClintock on corn and by himself on *Drosophila* might both be caused by movable DNA segments originally called controlling elements but now known as transposable elements.

Crosses between strains of *D. melanogaster* collected in south Texas and long established laboratory colonies produced hybrids with reduced fitness (Hiraizumi 1971). The hybrids displayed male recombination, which was very unusual because normally male *Drosophila* do not undergo meiotic recombination. The reduced fitness of the hybrids, called hybrid dysgenesis (Sved 1976), stems from several factors: a 100-fold higher mutation rate, recombination in males, chromosome breakage, and distorted segregation during meiosis (Kidwell et al. 1977, Grigliatti 1998). The most obvious phenotypic consequence of hybrid dysgenesis is sterility, which is often preceded by gonadal atrophy or gonadal dysgenesis. Hybrid dysgenesis shows remarkable similarities to bacteriophage-induced zygotic induction in bacteria and *Wolbachia*-induced unidirectional cytoplasmic incompatibility in insects and mites. The latter will be discussed in more detail in Section 10.2.3.3.3.

Theoretically, if an active or mobile element is introduced into a new, naive, and susceptible host, it can transpose in the germ line from one haplotype of a genome to the other. If the transposable element is introduced by one of the parents, the heterozygote will become homozygous for that element. This constitutes the meiotic part of the driving force of a mobile element. This meiotic part of the drive does not depend on an increase in fitness of the bearer or a decrease in the fitness of individuals not bearing the element. However, in natural populations, a factor evolves rapidly in the host that is expressed in the cytoplasm of eggs and will repress or silence the activity of the transposon. If in a cross the element is introduced by the female, the mobile element will not transpose and the hybrid will stay heterozygous. The element may cause hybrid dysgenesis, a form of reduced heterozygote fitness. On the other hand, if the element originates from the father, the sperm cell will not contribute any cytoplasm to the zygote and therefore no repressor; the transposon will become mobile. It will jump into the other haplotype. If the transposon integrates into a portion of the genome that is not active, into a heterochromatic part, or into a position where it does not disrupt any genes or regulatory elements, then the transposition presumably will have no detrimental effects on its host; the transposon will exhibit meiotic drive, and hybrid dysgenesis will be absent. If, on the other hand, the transposon integrates at a position where it disrupts or changes genes, it can strongly reduce the fitness of its host, leading to hybrid dysgenesis. A female carrying a mobile element can mate with both males carrying these elements and males without them; the progeny of either cross will be normal. A female without such an element, however, can avoid hybrid dysgenesis only by mating with a male also without a transposable element. This effectively halves the mating possibilities for the female without a transposable element. Although this reproductive advantage will favor the female with mobile elements, only half the gametes of that female will transfer those elements to the next generation.

One of the first-discovered and best-studied hybrid dysgenesis systems is the P-element system in *D. melanogaster* (Ashburner 1989a). Active P-elements are also known as P-factors. Because active transposition and, with it, hybrid dysgenesis are only transmitted by the paternal side and the maternal side does not contribute to an obvious phenotype, strains with the transposable element are called P strains whereas strains without the element are called M strains. In a cross between a P male and an M female, hybrid dysgenesis will affect both male and female offspring. Sterility increases with temperature.

The P-element is extensively used to transform *Drosophila* (Rubin and Spradling 1982, Ashburner 1989b). For transformation, the transgene is cloned into a P-element from which the transposase has been removed. The transposase is injected either as an active enzyme or coded on a helper plasmid. This assures that there will be only a single integration event of the transgene and that no further transpositions or any spread will occur.

The P-element was not detected in natural populations until the 1950s. Flies captured before 1950 and which still exist in laboratory colonies do not carry any P-elements. Since the 1950s, P-elements have spread worldwide, becoming predominant by 1980 (Kidwell 1983). Under laboratory conditions, most strains carrying P-elements went to fixation for P when P was introduced at a low frequency into naive populations, but some populations did not maintain the element at fixation (Kidwell 1987). In one case, 8 years after an active P element from *D. melanogaster* was introduced into a sister species, *D. simulans*, hybrid dysgenesis levels varying from 0 to 96% were observed. There was no relationship found between any known parameters and the observed level of hybrid dysgenesis (Higuet et al. 1996). There is evidence for horizontal transmission between *Drosophila* species (Daniels et al. 1990a, Biemont and Cizeron 1999) and to and among leaf miner flies (*Scaptomyza*), blowflies (*Lucilla*), and houseflies (*Musca*) (Anxolabéhère and Periquet 1987, Clark et al. 1994, Hagemann et al. 1996, Lee et al. 1999). Kidwell found evidence that the mite *Proctolaelaps regalis*, when cultured together with *Drosophila*, could acquire P-element DNA from the flies (Houck et al. 1991).

A second hybrid dysgenesis system, I-R hybrid dysgenesis, was described in the early 1970s (Picard and L'Heritier 1971). Here, *D. melanogaster* strains fall again into two categories denoted inducer (I) and reactive (R). The dysgenesis is observed only in F_1 females, resulting from crosses between inducer males and naive, reactive females. The transposition burst in the female offspring, called SF females, renders them partly sterile, and the surviving progeny exhibit strong dysgenesis. Sterility decreases with temperature. In contrast to the P-M system, all crosses give rise to normal male progeny. The I-factor, a non-LTR retrotransposon or retroposon similar to mammalian *LINEs*, was shown to be responsible for the I-R hybrid dysgenesis phenomenon (Fawcett et al. 1986). This system is not suitable for transformations (Hartl et al. 1997). The transposition is regulated by several host factors and an additional factor, ORF1, of the *I*-element, the function of which is not yet fully understood (de La Roche Saint André and Bregliano 1998, Seleme et al. 1999). Flies collected in the 1920s lack an *I*-element. By the 1930s, the element appears in fly stocks and by 1980 almost all *D. melanogaster* strains tested carried an *I*-element (Kidwell 1983). The *I*-element shows the most-restricted range of horizontal transfer (Stacey et al. 1986, Biemont and Cizeron 1999).

A third hybrid dysgenesis system is based on the *hobo* element, again in *D. melanogaster*, and is composed of H strains (*hobo*) and E strains (Ashburner 1989a, Bazin and Higuet 1996). It seems to have spread in natural populations around the 1940s and 1950s, spreading in time just between the *I*-element and the *P*-element (Pascual and Periquet 1991). Horizontal transmission only occurred in a limited range (Daniels et al. 1990b; Bonnivard et al. 2000).

Recently, a fourth system of hybrid dysgenesis was discovered in *D. virilis*. In this system, at least four unrelated transposable elements are all mobilized following a dysgenic cross. One of the elements that is mobilized, is a *mariner/Tc1*-like element designated *Paris* (Petrov et al. 1995). The data suggest a mechanism in which the mobilization of *Paris* triggers the mobilization of other elements, perhaps through chromosome breakage. For the *mariner* transposable system, mutant transposases have been described that antagonize the activity of wild-type transposases. These may help explain why the vast majority of *mariner* elements in natural populations undergo "vertical inactivation" by multiple mutations and, eventually, undergo stochastic loss (Hartl et al. 1997). *Mariner/Tc1*-like elements are the most widespread elements and, because of this, may become *the* universal transformation tool (Robertson 1993, Plasterk et al. 1999). However, extensive horizontal transfer has to be assumed for *mariner* (Kidwell 1993, Lohe et al. 1995). For risk assessment, it is important to know that similar *mariner/Tc1*-like transposable elements have been detected in Lepidoptera and their insect viruses. There is increasing evidence that baculoviruses may play a role as an interspecies vector for the horizontal transmission of insect transposons (Jehle et al. 1998). Examples date to 1982 of insect viruses that have picked up from their host other transposable elements, such as the *copia* element, that do not induce hybrid dysgenesis (Miller and Miller 1982, Bauser et al. 1996).

A transposable element, perhaps a new element, at this very moment is sweeping through *D. willistoni* populations. Crosses between newly collected flies and an old laboratory stock exhibit hybrid dysgenesis (Regner et al. 1999).

From the discussion above, it seems that at least three, and presumably more than five, worldwide sweeps of transposable elements have occurred in *Drosophila* populations during the last 50 years alone. The question remains whether it is a stroke of remarkable luck to have witnessed these sweeps during an extremely short period of evolutionary time or whether global sweeps of mobile elements such as these occur almost constantly (Powell 1997).

Transposons or their remains are extremely widespread and are found in both prokaryotes and eukaryotes, including humans (Berg and Howe 1989, Smit 1999). It might be difficult to find a species totally without transposons. Yet, so far, true, intraspecies, intrastrain hybrid dysgenesis has only been described for *Drosophila*. This may reflect how well *Drosophila* has been studied and/or how little we know about everything else. Many species have numerous classes, families, and copies of transposable elements. Altogether there are more than 40 different families of transposable elements in *Drosophila* alone (Biemont and Cizeron 1999). It is important to consider whether or not all species might harbor large numbers of defective elements, elements that no longer can transpose on their own. After all, transposases from other mobile elements can sometimes activate, i.e., transmobilize, defective transposable elements with unpredictable consequences.

Transposable elements are, at the moment, the only rational tools, and *P*-elements, for example, very easy tools, to use for germ-line transformation. At the same time, these elements have been found to have a genuine global distribution. These facts taken together have sparked much interest in transposable elements as genetic drives for population replacement. Computer simulations have suggested that transposable elements with a wide range of properties might be used successfully to drive genes into populations (Uyenoyama and Nei 1985, Kidwell and Ribeiro 1992, Ribeiro and Kidwell 1994; for a discussion of the aforementioned article, see also Eichner and Kiszewski 1995, Ribeiro and Kidwell 1995). The rapid spread of *P*-elements introduced at very low frequencies into *Drosophila* laboratory cage populations encouraged this assumption (Kidwell et al. 1981, Kiyasu and Kidwell 1984, Good et al. 1989). However, early on there were indications that the size of the transgenic load would strongly affect the mobilization of the element. Robertson et al. (1988) found that a *P*-element carrying the *white* eye color gene was far less mobile than an element carrying the smaller *rosy* transgene insert; similar results have been reported for other transposable elements like *mariner* (Robertson et al. 1988, Lohe and Hartl 1996a). And even with the smaller *rosy* gene as an insert, deletions and, as a consequence, instability of the transgene were proportional to the *P*-element-induced dysgenesis (Daniels et al. 1985). The first cage experiment to spread an actual transgene through a *D. melanogaster* population with the help of a *P*-element was reported by Meister and Grigliatti (1993). As transgene, the enzyme alcohol dehydrogenase (Adh) was chosen. Instead of analyzing the genetic integrity of the transgenic construct, the presence or absence of enzyme activity was scored with a detection limit of 20% of natural enzyme activity. The enzyme activity was not quantified. Two populations were seeded with 1% transgenic flies, and two additional populations were seeded with 10% transgenic flies. The experiment covered only the period of increasing enzyme activity associated with detectable amounts of *P*-element sequence, exceeding 80% of the population after 13 generations. However, for one population seeded with 10% transformed flies, the association of *P*-element and alcohol dehydrogenase activity peaked as early as generation 6 in just 70% of the test population. Only two generations later, the phenotype was exhibited in less than 60% of the population.

The fate of the eye color gene *rosy* in *D. melanogaster* populations, where either autonomous or non-autonomous *P*-element constructs were used as driving forces, was followed over many more generations by Kidwell (Carareto et al. 1997). Introductions of 1% as a starting frequency often failed to establish themselves, whereas 5 and 10% introductions exhibited an initial rapid increase in their numbers within the first five generations. In these initial experiments, the *rosy* construct spread equally fast and to fixation with and without a transposable element, presumably because of a positive selection for the marker gene itself, underlining how important it is to control these experiments carefully. By using a different genetic background better suited to measure the performance of the driving system, all eight different constructs used in the experiment failed to replace the naive population or to fix the transgene in the population. All lines showed a breakdown of the transgenic construct. The *P*-element without the transgene continued to rise in frequency in the population, whereas the percentage of the transgene itself in the population started to decline. This experiment documented a strong negative correlation between transposition rate, which equals driving force, and construct stability, meaning useful transgenes. On average, each transposition event of a *P*-element in *Drosophila* reduces the viability of the fly by 12% in the homozygous state and by 5.5% in the heterozygous state (Mackay et al. 1992). Deletions in transgenic transposable elements as well as in natural constructs like mitochondria had been predicted to result in shorter elements that in turn may replicate faster and therefore be favored by selection (Rand and Harrison 1986).

One is tempted to look for ways to outrun the degeneration of the transgene. If that were possible, it might not be necessary to maintain the transgene in a population forever, but just long enough to interrupt the transmission cycle. In that case, transposable elements might still have a chance. Regardless of how risky and unrealistic that scenario might seem from a vector ecology point of view, even a transposable element close to ideal might not be able to live up to the outrun degeneration strategy. Most model systems have only tried to predict the behavior of a transposable element in time. Kiszewski and Spielman (1998) argue that a model should account also for space in addition to time and for discontinuities in both. Temporal discontinuities between dry and wet seasons and uncertainties of transgenerational perpetuation are characteristic for malaria mosquitoes. The spatial separation of villages and peculiarities in the mating behavior and the pattern of dispersal of anopheline mosquitoes underpin the importance of these parameters. Using a spatially explicit cellular automata system to simulate the transposon-based genetic drive of a transgene into fluctuating vector populations, the authors conclude that even under the ideal situation of the transgenic insect being as fit as the wild-type vector, fixation of the transgene may require 150 generations or more (Kiszewski and Spielman 1998). Transposable elements may be ideal tools for the transformation of insects; however, they cannot yet be used as a genetic driving force for transgenes where the goal is population replacement.

10.2.3.2.4   B chromosomes.   B chromosomes, as opposed to A chromosomes, which are the autosomes and sex chromosomes, are selfish, extra chromosomes, additional to the standard complement. They are also referred to as supernumerary, dispensable, conditionally dispensable (CD), or accessory chromosomes. B chromosomes may originate from autosomes or sex chromosomes in intra- and interspecies crosses. As early as 1907, B chromosomes were described in Hemiptera by Wilson (1907). They are much more widely spread than commonly anticipated. B chromosomes are as easily detected in plants (1300 species) as in fungi and invertebrate and vertebrate animals (500 species) (Jones and Rees 1982, Camacho et al. 2000). For example, the vector of African sleeping sickness or trypanosomiasis, the tsetse flies *Glossina* spp., and the vector of African river blindness or onchocerciasis, the blackflies *Simulium* spp., carry B chromosomes. In anopheline mosquitoes, however, no supernumerary chromosomes have been detected. The various species complexes in the genus *Anopheles* might form an ideal basis for attempts to generate supernumerary chromosomes in interspecific crosses.

The driving mechanisms of supernumerary chromosomes can be complex. They depend for their success initially on the origin of the chromosomes and the effects on host's genome fitness and later on the evolutionary state of the host's defense against supernumerary chromosomes. To overtake a Mendelian transmission rate of 0.5, supernumerary chromosomes have to increase their presence in gametes. Three different modes for accomplishing this have been described (Jones 1991). A premeiotic accumulation or a premeiotic drive of supernumerary chromosomes occurs during spermatogonial mitoses in male Orthoptera. Mitotic non-disjunction changes a stable somatic complement of one into two B chromosomes, whereas cells with no B chromosomes seem to disappear at a ratio of 15 to 1 in the grasshopper *Calliptamus palaestinensis*. Here, B chromosomes are male driven. In a meiotic accumulation or drive, the B chromosomes take advantage of the fact that only one of the two potential egg cells formed during meroistic oogenesis will progress to become the oocyte. A structure called the fusome is found only in the cell destined to become an ovum; it makes that cell asymmetrical. Univalent B chromosomes lie outside the metaphase plate and orient themselves to the spindle pole that remains with the fusome. Nevertheless, all cells will develop into gametes during spermatogenesis and there are no known provisions for increasing the copy number of B chromosomes in spermatogonia or spermatocytes. Thus, meiotic drive of supernumerary chromosomes is female driven. In the mottled grasshopper, *Myrmeleotettix maculatus*, almost all egg cells carry B chromosomes instead of the expected 50% without accumulation, whereas only 30% of the sperm cells transfer B chromosomes. In this case, there is meiotic drag in males. Post-meiotic accumulation or drive is a mirror image of pre-meiotic accumulation. It is by far the most common mode of B drive in flowering plants. During the first mitosis, directed non-disjunction takes place, making use of an asymmetrical spindle apparatus. Post-meiotic drive can be male or female driven.

There are several variations on these three driving modes, for example, in mealybugs (Hemiptera), which have holokinetic chromosomes and in which the meiotic sequence is inverted, with the first division equational and the second division reductional. Not all supernumerary chromosomes show signs of accumulation. Most notably, the grasshopper *Eyprepocnemis plorans*, in which the three different types of supernumerary chromosomes lack accumulation mechanisms, appears to have chromosomes that either were once parasitic or selfish and have lost their drive and accumulation mechanism, or were once mutual and have lost their beneficial properties.

The parasitoid wasp *Nasonia vitripennis*, the jewel wasp, harbors the most extreme example of a B chromosome: a paternal sex-ratio chromosome (PSR). In Hymenoptera, females are diploid and males haploid. The PSR chromosome is transmitted in almost all sperm cells. In the fertilized egg cell, it eliminates the paternal chromosome set with which it was transferred. Diploid females become haploid males. The PSR chromosome has a nearly 100% transmission rate and reduces the fitness of the paternal chromosomes to zero. As such, it is considered by its discoverers to be the most selfish genetic element so far described (Skinner 1982, Werren et al. 1987, Nur et al. 1988, Shaw and Hewitt 1990). Interestingly, this same wasp harbors a second drive mechanism, the maternally inherited endosymbiont *Wolbachia*, which causes cytoplasmic incompatibility. Because of the haplodiploid nature of Hymenoptera, cytoplasmic incompatibility has the same net effect as the PSR chromosome, notably, loss of the paternal chromosomes (Reed and Werren 1995, Dobson and Tanouye 1996, 1998a). The two drive mechanisms do not interact with each other. However, the PSR chromosome shares the fate of the paternal chromosomes in *Wolbachia*-induced cytoplasmic incompatible crosses; it is eliminated. Dobson and Tanouye (1998b) have shown that the PSR chromosome can easily be moved from *N. vitripennis* into closely related species *N. giraulti* and *N. longicornis* through hybrid crosses. The transfers occur at high rates and the meiotic drive system of the PSR chromosome continues to function in both new hosts, thus making the PSR chromosome ideal for a transgene drive.

Supernumerary chromosomes are largely composed of repetitive sequences; for some chromosomes, they may be the only sequences. Tandem repeats of ribosomal DNA sequences have frequently been described, most of them as genetically inactive. They are often compared with the silenced parts of sex chromosomes. Presumably, many factors contribute to the heterochromatic state of these supernumerary chromosomes. The heterochromatin acts as a sink for the accumulation of transposable elements. Nevertheless, there are many examples of active gene products; the more prominent examples are genes that control resistance to rust in plants, genes that increase germination vigor, genes that regulate the expression of A chromosomes, and genes that confer antibiotic resistance in fungi. Transcription also has been noted in animals, but the products of the gene activity are often not known. In many cases, the chromosomes are expected to exert slightly detrimental effects on their hosts. The detrimental effects will be more than compensated for by the accumulation mechanisms that allow the chromosomes to spread in spite of a heavy genetic load. The PSR chromosome is an extreme example of this.

Thus far, little is known about the population dynamics of supernumerary chromosomes. Variation in transmission rates between individuals and in intra- and interpopulation crosses reflect an interaction between A and B chromosomes. Lines can be selected for high and low transmission rates. An extreme form of fixation has been detected in the zebra finch, *Taeniopygia guttata*. There, all females and males carry precisely one copy of a B chromosome, which is restricted to the germ-line tissue (Pigozzi and Solari 1998). The antibiotic resistance conferred by the B chromosome in the fungus *Nectria haematococca* presents another extreme for obtaining fixation in a population (Miao et al. 1991, Enkerli et al. 2000). In the Brazilian wasp, *Trypoxylon albitarse,* most females support two copies of a B chromosome, whereas most males carry only one copy (Camacho et al. 2000, Araujo et al. 2000). Since females are diploid and males haploid, this particular B chromosome is presumably about to become an autosome.

Being supernumerical and heterochromatic, crossing-over and recombination are strongly suppressed in supernumerary chromosomes. This makes supernumerary chromosomes almost ideal driving mechanisms for transgenes and population replacement. They might have the largest imaginable capacity for carrying transgenes, practically the better part of a chromosome. The linkage of transgenes and drive seems to be perfect. The threat that the transgene might be inactivated through the accumulation of mutations is certainly possible on an evolutionary timescale. The suppression of crossing-over and recombination would lead one to expect that B chromosomes would behave as chromosomes in parthenogenic organisms and be doomed to Müller's ratchet. The little information that is available, however, suggests that B chromosomes escape this fate by means as yet unknown, much as parthenogenic organisms have done for several million years.

Wild populations often carry two different supernumerary chromosomes. In corn plants, *Zea mays,* as many as 34 different supernumerary chromosomes have been counted. However, corn has only ten different autosomes. The New Zealand frog, *Leiopelma hochstetteri*, has 15 different, mitotically stable extra chromosomes. The upper limit for extra chromosomes is not known. The extra load does not seem to have much effect on the fitness of the carrier. This is of immense importance for a genetic driving mechanism. It would allow supernumerary chromosomes to be used over and over again as vehicles for transgenes. With the exception of *Wolbachia*-induced cytoplasmic incompatibility, which will be discussed in more detail in Section 10.2.3.3.3, the impossibility of performing repeated population sweeps or replacements is a major shortcoming of all other driving mechanisms.

Supernumerary chromosomes are for the most part heterochromatic. Heterochromatin does not allow gene expression. Engineering euchromatic regions into a supernumerary chromosome is an interesting endeavor that may teach a great deal about chromatin regulation. Insecticide resistance genes, as they are now used in many sterile insect release projects, might offer an easy means for keeping the chromosomal segment transcriptional active while adding a beneficial factor to the chromosome.

10.2.3.2.5 Medea factors. Medea is an acronym for Maternal Effect Dominant Embryonic Arrest. Medea was also a revengeful and murderous figure in Greek mythology, famous for killing her own children. Four different Medea factors are found in the red flour beetle, *Tribolium castaneum*. They are forms of hybrid dysgenesis or semi-sterility. Lethality factors are inherited from the mother through the egg cytoplasm and will kill all offspring that do not inherit a rescuing M allele from either parent. Medea factors are widespread in wild populations of the beetle in Europe, North and South America, Africa, and Southeast Asia. They are rare or absent in Australia and the Indian subcontinent. Several strains have more than one Medea factor (Beeman et al. 1992). Beeman and Friesen (1999) were the first to succeed in knocking out the gene for the cytoplasmic lethality factor while retaining the zygotic rescue activity of the M allele. The molecular basis of Medea is unknown. In Indian populations of the red flour beetle, a hybrid incompatibility factor has been discovered when mated with American strains (Thomson et al. 1995). The hybrid incompatibility (HI) factor causes the death of hybrids with a paternally derived H gene and a maternally derived Medea factor. The incompatibility between strains with Medea and strains with hybrid incompatibility is caused by the HI factor, which suppresses the rescuing factor of Medea. Thomson and Beeman (1999) explain the conspicuous absence of Medea from India as the outcome of the interaction of Medea and HI. With hybrid incompatibility, Medea changes from a selfish element into a self-destructive or suicidal element. This factor has inspired several models to explain its population behavior (Wade and Beeman 1994, Hastings 1994, Smith 1998, Hatcher 2000).

In mice, a similar factor has been identified. Named *scat*, it causes lethality through a severe combination of anemia and thrombocytopenia (Peters and Barker 1993, Hurst 1993a). *scat* displays the same pattern of a modification-and-rescue system with a cytoplasmic lethality factor and a gene-based zygotic rescue factor as Medea. The *scat* locus has been mapped on chromosome 8. It is believed that *scat* originated from a spontaneous mutation in a BALB/cBY laboratory colony. It is not known outside the laboratory. The gene of the *scat* locus has not been identified, and the molecular mechanism for rescue is not known.

A prime example of how plant and animal sciences can live parallel lives without any overlap or any notice of each other is the mutant of the plant *Arabidopsis thaliana*, which is also called *medea* (MEA) (Grossniklaus et al. 1998). *Arabidopsis thaliana* is a model organism and a member of the mustard family. The mutant *medea* has maternal control over embryogenesis. Self-fertilization of heterozygotes with *medea* produces 50% abortion. The *A. thaliana medea* encodes a SET domain protein similar to *Enhancer of zeste* of *Drosophila*, a member of the *Polycomb* group of proteins, which in animals ensure the stable inheritance of expression pattern through cell division. Paternally inherited MEA alleles are transcriptionally silent in the young embryo. Mutations at the *decrease in DNA methylation 1* (*ddm1*) locus are able to rescue MEA by functionally reactivating paternally inherited MEA alleles during development. The maintenance of the genomic imprint at the *medea* locus requires zygotic *ddm1* activity. Because *ddm1* encodes a putative chromatin remodeling factor, chromatin structure is likely to be interrelated with genomic imprinting (Vielle-Calzada et al. 1999, Mora-Garcia and Goodrich 2000).

Although Medea in flour beetles has shown that it can conquer most continents, the lack of any molecular insight into its workings precludes use of Medea in any applications in the foreseeable future. It is difficult to predict how easy it might be to link a transgene stably to a Medea factor or gene. However, it seems unlikely that Medea-like factors are limited to a single insect species. We should expect to discover many more nuclear-based modification-and-rescue systems.

10.2.3.3   Reduced heterozygote fitness
Inherent in some definitions of a species is the fact that conspecifics recognize each other as potential mates. Evolving species increasingly develop hybrid sterility and, presumably, at a much faster rate, stop recognizing each other as mating partners. Eventually, this results in reproductive isolation.

10.2.3.3.1   Subspecies hybrid sterility.   However, sometimes reduced fitness can also result in "hybrids" from crosses between subspecies and races. Davidson et al. (1970) attempted to apply subspecies hybrid sterility with the *Anopheles gambiae* complex in Burkina Faso. Laboratory crosses of what are now known as *A. melas* from Liberia and *A. arabiensis* from Nigeria led to sterile males. These sterile males were released. Unfortunately, feral females of *A. gambiae* did not recognize the released males as mates. The release was a failure. Today, the six constituent strains of the *A. gambiae* complex are recognized as six different species. In another attempt to take advantage of hybrid sterility as a means of facilitating population replacement, the result was successful. In 1942, Vanderplank, way ahead of his time, attempted to eradicate *Glossina swynnertoni* in an arid trypanosomiasis hot spot of Tanzania. *Glossina swynnertoni* are allopatric with the very closely related *G. morsitans centralis* that prefer a more humid habitat. Field-collected pupae of *G. m. centralis* were transported into the *G. swynnertoni* region. Crosses produce semi-sterile females and completely sterile males. The two species recognized each other as mates, and *G. m. centralis* replaced the natural *G. swynnertoni*. *Glossina m. centralis* could not withstand the arid climate and the population quickly collapsed, leaving the region practically free of tsetse flies and African sleeping sickness (Vanderplank 1944, 1947).

10.2.3.3.2   Inter-racial hybrid sterility and negative heterosis.   If the hybrids of two otherwise identical strains or populations are less fit or fertile than within-strain or within-population matings, then the strain or population that is in a numerical majority will increase in frequency until it completely replaces the other strain. In other words, if cross- or outbred heterozygote individuals are less fit than inbred, homozygote individuals, the majority breed will eliminate the minority breed. Since this is the opposite of hybrid vigor or overdominance, it is known as underdominance or negative heterosis. Inasmuch as negative heterosis in a natural environment will lead to reproductive isolation and eventually to speciation, it will be very difficult to find natural populations expressing negative heterosis that do not differ in any other way. Therefore, the challenge lies in finding an elegant method to induce inter-racial hybrid sterility. The other requirement for the success of such a strategy is to mass-rear and mass-release the species at several times the numbers of the wild population. This has been accomplished for several insects with sterile insect methods, even for relatively demanding vector species like tsetse flies (e.g., *G. morsitans, G. austeni*). However, only a few *Anopheles* species can be kept in laboratory colonies and keeping them alive is not always easy. After Serebrovskii's (1940) original discussion of possible strategies for inducing inter-racial hybrid sterility lay dormant for more than 25 years, Curtis (1968), Laven (1968) and Rai and Asman (1968), began to promote negative heterosis as an option for vector control. Soon after, chromosome translocations were applied to cause hybrid sterility.

Chromosomal translocations.   Translocations occur when a segment from one chromosome is detached or broken off and reattached to a different, non-homologous chromosome and a corresponding segment of the second chromosome replaces the broken-off segment of the first chromosome. The chromosomal translocations are reciprocal. Genes on the relocated segments are transferred from one chromosome to another. The linkage relationships of the genes are altered. Translocations can be induced by radiation, transposable elements, or mechanical sheer.

Most progeny with homozygous chromosome translocations, if viable, are completely fertile. Progeny with heterozygous chromosome translocations, on the other hand, are semi-sterile. Inherited semi-sterility in which 50% of the zygotes abort was first described for *Stizolobium deeringianu*, the Georgia velvet bean (Belling 1914). During meiosis in translocation heterozygotes, translocated chromosomes pair with their untranslocated homologues in a structure resembling the four arms of a cross, with translocated and non-translocated chromosomes alternating with each other. This cruciform pairing involves four centromeres pointing in four different directions. When these four centromeres are pulled in only two opposite directions of the spindle poles, two of the four centromeres point at a 90° angle to the poles and their final location cannot be predicted. There is a 50% chance that the resulting gametes will not receive a full complement of chromosomes. They will be unbalanced or aneuploid, and therefore unable to produce viable offspring.

The other half of the gametes will be balanced in their set of chromosomes, euploid, and fertile. Accordingly, a translocation heterozygote is 50% sterile. A transgene could be introduced onto a translocation chromosome near the breakpoint of the translocation such that variable and fertile homozygotes would be formed and heterozygotes would display reduced fertility and viability. Crossing-over near translocation breakpoints is often suppressed. This would link the transgene to the driving mechanism. This strategy, however, is limited to a single transgene and the linkage would not be absolute. Field applications of simple chromosomal translocations to secure a higher degree of linkage, so far, have been limited to only one sex of a very few species. Unlike most animals, several insect orders exhibit achiasmatic meiosis in the heterogametic sex. Examples include males in Diptera, especially *Drosophila* flies and *Aedes* mosquitoes, and females in Lepidoptera. Achiasmatic meiosis means that crossing-over is nearly absent in the sex carrying two different sex chromosomes. For use in releases of non-inverted chromosomal translocations, male mosquitoes and female moths have been used. The translocations often involve the heterogametic sex chromosome. Some mosquitoes, similarly to many other insects, do not have reduced or specialized sex chromosomes but carry sex-determining loci on larger chromosomes. Most field trials have been performed with translocation heterozygotes, although the optimal choice would be translocation homozygotes. However, homozygotes are difficult to isolate and often suffer severely from fitness problems once obtained. Increasing the effectiveness of heterozygotes by combining two different translocations into double translocations has proved successful. In mosquitoes with only three chromosomes, double translocations are necessarily linked to each other since one of the chromosomes must be common to the two different translocations. The sterility is higher than 50% in double translocations, usually about 70%, because there are more ways to segregate lethal gametes. In mosquitoes, double translocations will always involve the sex or heteromorphic chromosomes. Multiple translocations with concomitant high levels of sterility are superior for population-suppression strategies; single translocations with lower levels of sterility are preferable for population replacement and driving strategies. Whereas density-dependent factors may compensate for the imposed genetic load of translocations on a population and may make eradication impossible, they are not relevant to the success of translocations in driving transgenes into a population. Once translocations approach fixation in a population, they are supposed to protect themselves against a reasonable degree of immigration of wild type without further tending.

Less than 1 year after the reappreciation of chromosome translocations as genetic vector control methods, Laven (1969) demonstrated in a cage experiment that, by the release of a few translocation heterozygotes, a mosquito population can be diminished and finally exterminated after only a few generations. In the summer of 1970, Laven released up to 20,000 translocation heterozygotes a day of the mosquito *Culex pipiens* near a well in the relatively isolated village of Notre Dame near Montpellier in the south of France (Laven et al. 1971, 1972). With the help of high predator and parasite pressure, the natural population was brought close to eradication after only one release campaign. However, the interpretation of results and the contribution of changes in experimental design became controversial after 4 years of monitoring (Cousserans and Guille 1974).

In a large field cage experiment with the yellow fever mosquito, *Aedes aegypti*, the population-suppressing efficiency of chemosterilized males, of a double translocation, and of the same double translocation linked to a sex distorter gene was compared. All three genotypes succeeded in competing for mates with the wild males and eventually brought the wild population to eradication (Curtis et al. 1976). After only 11 weekly releases, the distorter double translocation heterozygote males had introduced so much genetic load into the population that it could not recover; after 12 weekly releases, this was accomplished by the chemosterilized males. However, it took the double translocation alone

23 weeks of releases to do the same job. This suggests that, in this case, the combination of sex distorter gene and double translocation was twice as strong a driving force as the chromosomal translocation alone.

Field trials with the Australian sheep blowfly, *Lucilia cuprina*, carrying deleterious mutations and Y-autosome translocations revealed genetic deterioration as the result of an unexpected high frequency of male crossing-over. It revealed a breakdown of the linkage between autosomal recessive eye color mutations and the multibreak translocation (Foster et al. 1980, 1991b). The field releases of *A. aegypti* in India (Rai et al. 1973) and Kenya (Lorimer 1981) and of *C. pipiens* in France (Laven et al. 1972) showed that these translocations persisted for one to several years in natural populations. In the case of *C. pipiens*, however, the translocation frequency dropped from 89% to less than 1% within 3 years (Cousserans and Guille 1974). Notwithstanding the extent of the drop, a decline is expected through natural selection against the male-linked translocation (Curtis 1975).

Inversions and pericentric inversions.   When a segment is cut out of a chromosome or cut off from a chromosome, turned around 180°, and reattached to the chromosome, the order of the genes in that segment is reversed in comparison with the rest of the chromosome. The segment becomes an inversion. If the segment includes the centromere of the chromosome, the inversion is called a pericentric inversion. Pericentric inversions are mostly generated through radiation and screening (Steffens 1983a, Foster et al. 1991a). Centromeres hold sister chromatids together until they serve as attachment points for spindle fibers during mitosis and meiosis. Insects in several orders such as Lepidoptera, Hemiptera (bugs, hoppers, aphids, whiteflies), and Phthiraptera (biting and sucking lice), do not have chromosomes with localized centromeres; instead, they have holokinetic or holocentric chromosomes and cannot produce pericentric inversions.

Simple inversions and pericentric inversions have an advantage over non-inverted chromosomal translocations in that they provide a much tighter linkage of transgene and drive because of stronger suppression of crossing-over at the break points of the translocation. Inversions will also ensure semi-sterility in animals in whom crossing-over has been expressed in both sexes.

The exploration of chromosomal translocations, and especially exploration of their use in malaria mosquitoes, has been hampered by insurmountable fitness problems (Robinson 1976, Boussy 1988). In contrast, the majority of pericentric inversions in *Drosophila* do not cause significant reduction in fertility. Expression of underdominance has been associated more with the position of the inversion than with the length of the inversion (Coyne et al. 1993). Although more, and even homozygous, translocations for *Anopheles* mosquitoes were reported during the 1980s, none of them made it to a field trial (Kaiser et al. 1982, 1983, Pleshkova 1984, Sakai and Mahmood 1985, Krafsur and Davidson 1987, Benedict et al. 1999). Insecticide resistance genes had been anchored with translocations on the Y chromosome of several anopheline species to generate genetic sexing lines in anticipation of mass rearing and releases, e.g., propoxur resistance in male *A. albimanus* (Seawright et al. 1981), malathion resistance in *A. quadrimaculatus* (Kim et al. 1987), and dieldrin resistance in *A. stephensi* (Robinson and Van Lap 1987). In some lines, the genetic leakage through recombination was kept below 0.02%.

Several researchers proposed combining chromosomal translocations with a second genetic manipulation. Substerilizing radiation combined with inheritable, translocation-induced partial sterility was put forward as a means to achieve in mosquitoes an effect similar to the F_1 sterility known for Lepidoptera and other holokinetic insects. In those insects, sterility after exposure to radiation does not become evident in the irradiated specimen but in its F_1 progeny (Steffens 1983b). Based on experience with the sheep blowfly, a system was simulated, incorporating a homozygous viable pericentric inversion on the same chromosome as deleterious mutations but repelled by a Y-linked translocation. The idea was that released males would transmit the pericentric inversion and mutations to their daughters, thereby rendering the daughters semi-sterile in matings with wild-type and engineered males due to crossing-over within the inversion-carrying, inviable duplications and deficiencies. Sons would stay unaffected because of the repulsion by the Y-linked translocation (Foster 1991). Both systems described above, unfortunately, are meant for greater persistence of genetic load in a population and would not lend themselves easily as driving mechanisms for population replacements.

Compound chromosomes.   Compound chromosomes are formed through the fusion of homologous chromosome segments or sister chromatids. The compound chromosome carries two copies of a certain segment of the chromosome unless the compound chromosome is comprised of two sex chromosomes, which are homologous as well. Compound chromosomes differ from translocations in that translocations always involve fusions between non-homologous chromosome segments. This distinction is important because it explains the sterility in the heterozygous hybrids. Compound chromosomes are stable as long as they have only a single functional centromere. If two homologous segments are united, for example, two right arms of a chromosome, and fused at one centromere, a compound half-chromosome is formed, which is also known as an isochromosome.

Compound chromosomes are not easily obtained by random radiation treatment but require a good understanding of the species genetics. The first compound chromosome, the fusion of two X chromosomes in *D. melanogaster*, was discovered in 1922 by Morgan. A compound chromosome strain exhibits between 25 and 50% fertility, but there is 100% sterility in its outcrosses. Introduced "compound" populations should be able to tolerate a significant rate of migration. With 25% fertility, compound strains should withstand immigration rates of 5% (Asman et al. 1981). The limited number of chromosomes in many insects becomes a major problem in engineering compound chromosome strains. It will be difficult to assure competitiveness with wild populations. Because of the limited number of chromosomes, outcrosses can be severely restricted and elaborate systems with bridging strains might have to be employed to ensure their competitiveness (Foster 1982).

Compound chromosomes have been constructed for only a very few insect species, e.g., the vinegar flies *D. melanogaster* (Fitz-Earle and Holm 1978) and *D. pseudoobscura* (Harshman and Prout 1990), the sheep blowfly, *Lucilia cuprina* (Foster and Maddern 1985), and the onion fly, *Hylemya antiqua* (van Heemert 1977). A capture system for compound chromosomes has been constructed for *A. albimanus* but never applied (Kaiser et al. 1982). Compound chromosomes are still not known in anopheline mosquitoes. Field applications were restricted to *Drosophila* and *Lucilia* and both applications were unsuccessful. The field trials might have failed for reasons other than an inherent flaw of compound chromosomes, however. Recently, Dernburg et al. (1996) found that *Drosophila* females selectively lose sperm-bearing compound chromosomes; this probably will restrict future use of this strategy to species that mate only once during their lifetimes.

Compound chromosomes do have potential as a genetic mechanism to drive a population replacement (Barclay 1984, Fitz-Earle and Barclay 1984, Barclay and Fitz-Earle 1988). However, strategies for employing them are not being pursued at the moment, either in modeling or in applied genetics.

10.2.3.3.3 Cytoplasmic factors.   In all animal species, the cytoplasmic genomes and factors are maternally inherited. Mature sperm are so densely packed that they no longer can carry any cytoplasm or cytoplasmic organelles. Even in the rare cases where unmodified paternal mitochondria are introduced into the egg cell, they do not survive. The maternal cytoplasm may carry mitochondria that increase the fitness of its carriers. This cytotype might spread and replace a feral population. To employ such a strategy would require a method to increase the fitness of mitochondria. Such a method is not available.

Cytoplasmic factors would also require a means to carry and express the transgene(s) in the cytoplasm. Theoretically, this could be a non-pathogenic, transovarially inherited virus. Two candidates have been suggested for insects: the sigma virus (a rhabdovirus) and Matsu virus (an unclassified RNA virus). The sigma virus is limited to *Drosophila*. Fly populations show a complex pattern of widespread viral resistance (Fleuriet 1996, Wayne et al. 1996). The maternal inheritance is less than 100%, which will cause a rapid segregation of the cytoplasmic expression system and the cytoplasmic drive. The second candidate, Matsu virus, is only known from populations of *Culex quinquefasciatus* on the islands of Matsu (in China) and Hawaii (Braig, unpublished). The morphology of the virion of Matsu is unlike any previously described virus (Vazeille-Falcoz et al. 1992, Shroyer and Rosen 1983). Both viruses are genetically not accessible. Thus, cytoplasmic inherited viral expression systems for transgenes are not an option for the foreseeable future.

Many insects show high rates of maternal inheritance of nutritive bacterial symbionts. An example of a vector insect that does so is the tsetse fly, which harbors three different bacteria; *Wolbachia pipientis* (a transovarially transmitted bacterium), *Wigglesworthia glossinidia* (a primary symbiont), and *Sodalis glossinidius* (a secondary symbiont) (Aksoy 2000). Unfortunately, both nutritive symbionts are not transovarially transmitted and thus might at one point in their life cycle, for a short moment, be considered to be in contact with the outside world. Unless it can be proved that non-transovarially transmitted bacteria are nevertheless clonal and do not at any point interact with the natural environment, it cannot be assumed that during transmission from mother to offspring extracellular, nutritive bacteria would be stably linked to any intracellular cytoplasmic drive. Further, *Anopheles* mosquitoes do not harbor nutritive symbionts.

The potential of nuclear-based sex ratio distorters as drives has already been discussed. Biased sex ratios can also be induced by intracellular cytoplasmic and extracellular bacteria (Hurst 1993b, Hurst et al. 1997, Jiggins 2000c). These bacteria could theoretically be used as a form of secondary sex drive if the problem of linking transgene and bacteria could be solved. Even so, there are no microbial sex ratio distorters in *Anopheles* or any other mosquitoes.

Wolbachia-induced cytoplasmic incompatibility.   Even before genetic control of vector insects was widely discussed, the first successful, local eradication of a vector mosquito using cytoplasmic incompatibility (CI) had already been reported. Laven (1967, 1971) released just enough CI males one winter in a small village in Burma to suppress the local *C. pipiens* population and bring it close to eradication.

Wolbachia pipientis is an obligate intracellular, rickettsial symbiont in the ovaries and testes of presumably more than 15% of all known insect species (Werren and O'Neill 1997, Stouthamer et al. 1999, Bourtzis and Braig 1999). The range of *Wolbachia* hosts stretches from flies and mosquitoes to parasitoid wasps, butterflies, moths, beetles, weevils, bugs, planthoppers, crickets, and termites. In a small survey on Neotropical insects by the Smithsonian Tropical Research Institute in Panama, 26 of 154 collected species (17%) carried *Wolbachia* (Werren et al. 1995). Surveys in Britain yielded between 16 and 22% of the species infected and 19% of North American insect species tested positive for *Wolbachia* (Werren and Windsor 2000). These data suggest an equilibrium in the global distribution of *Wolbachia* infections. However, whether there is also an equilibrium in time or whether we are witnessing a worldwide wave of increasing *Wolbachia* infection frequencies remains to be seen. An unexpectedly high detection rate of 25 of 50 different species was recently reported for Indo-Australian ants (Wenseleers et al. 1998).

Most of our knowledge about the distribution of *Wolbachia* is very sporadic and mainly limited to *Drosophila* and a few insect species of medical and agricultural importance. In most cases, *Wolbachia* has only been detected by PCR. The actual biology and the influence of *Wolbachia* on its host, especially its host's mode of reproduction, is known in only a handful of exemplary cases. Based on an estimate that 15% of all insect species carry *Wolbachia*, it may be the most widespread bacterium in the animal kingdom.

Endosymbiontic *Wolbachia* can have dramatic effects on insect populations by altering their mode of reproduction. Alterations range from cytoplasmic incompatibility in, for example, Diptera and Orthoptera (Hoffmann and Turelli 1997), through parasitism of another strain's cytoplasmic incompatibility in Diptera (Bourtzis et al. 1998), partheno-genesis in Hymenoptera (Stouthamer 1997), male killing in Coleoptera and Lepidoptera (Hurst et al. 1997, 1999), alteration of sperm precedence in Coleoptera (Wade and Chang 1995), increase of fecundity in Hymenoptera (Vavre et al. 1999), and enhancement of male fertility in Diptera (Hariri et al. 1998). *Wolbachia* has been well documented in terrestrial isopods (wood lice and pill bugs) where the infection can lead to cytoplasmic incompat-ibility and functional feminization of genetic males (Rigaud 1997, Bouchon et al. 1998). In predatory and spider mites, *Wolbachia* can cause cytoplasmic incompatibility (Breeuwer and Jacobs 1996, Johanowicz and Hoy 1998). Cytoplasmic incompatibility in haplo-diploid insects and mites results in all male or male-biased offspring as a result of the haploidiza-tion of the fertilized egg, with little or no mortality of the zygote. Through a genome-sequencing project, *Wolbachia* was confirmed in filarial nematodes, although the presence of unidentified bacteria had been known for many years (McCall et al. 1999). Further work revealed that the majority of human and animal filarial worm species harbor *Wolbachia* as a symbiont (Bandi et al. 1998). Antibiotic treatment results suggest that the removal of *Wolbachia* might arrest the development of juveniles and cause infertility in adult worms, killing both microfilariae and adult worms after prolonged treatment (Hoerauf et al. 1999, Taylor and Hoerauf 1999, Langworthy et al. 2000). The latest addition to the litany of *Wolbachia*-induced phenotypes comes from the detection of hybrid breakdown subsequent to matings between populations of the two-spotted mite *Tetranychus urticae* (Vala et al. 2000). Hybrid breakdown shows itself as increased male mortality in the second filial generation. Although one might be tempted to read the list of *Wolbachia*-induced pheno-types as a cabinet of curiosities in animal reproduction, with at least 15% of all insect species presumably infected, such curiosities might be more widespread than expected.

In many insects, *Wolbachia* are not limited to the reproductive organs. Depending on species, *Wolbachia* has been detected, often at high concentrations, in many different somatic tissues including midgut, hemolymph, and salivary gland tissues, all of which are of special interest in disease transmission in vector insects (Dobson et al. 1999). *Wolbachia* cannot be cultured *in vitro*. It can be cultured in a tissue culture system. However, it is not known whether or not *Wolbachia* in tissue culture is expressing the molecules responsible for cytoplasmic incompatibility (O'Neill et al. 1997).

Wolbachia is transovarially transmitted to 95 to 100% of the offspring. The incomplete transmission is compensated for through cytoplasmic incompatibility. The incompatibility is transmitted through sperm. The level of incompatibility depends on both the *Wolbachia* strain and the host; incompatibility can range between 0 and 100%. Increasing age of the males, repeated copulation, high temperature, stress in the form of high larval densities, and insufficient nutrition have all been correlated with lower bacterial densities and reduced incompatibility.

An advantage of a cytoplasmic or microbial driving system is that in such a system the risk of losing the drive is several times higher than the risk of losing the transgene. Losing a transgene would generate an empty drive, which would directly interfere with the drive of the transgene. An empty drive may burden its carrier with less genetic load, thereby increasing its fitness and presumably outcompeting the drive of the transgene. A transgene without a drive would not interfere at all with the drive of the complete system. Losing *Wolbachia* or having less than perfect transovarial transmission would not constitute a problem in terms of replacing a population.

Wolbachia is transmitted only maternally. But, on an evolutionary timescale, there must be extensive horizontal, intertaxon transmission. *Wolbachia* is obligatory intracellular, and an infectious form of *Wolbachia* has not yet been described. Consequently, the mechanism of horizontal transfer has been a mystery. Very recently, Nota and colleagues (2000) detected that populations from China and Japan of the small brown planthopper (*Laodelphax striatellus*), the whitebacked planthopper (*Sogatella furcifera*), and the strepsipteran endoparasite (*Elenchus japonicus*), were carrying identical strains of *Wolbachia*. This suggests that Strepsiptera (twisted wings or stylops) are instrumental in the horizontal transmission of *Wolbachia* between planthoppers.

Cytoplasmic incompatibility is best known in flies and mosquitoes. It is expressed when the sperm from an infected male enters an uninfected egg or an egg infected with a different strain of *Wolbachia*. Incompatible crosses result in early embryonic death. During spermatogenesis, *Wolbachia* imprints the sperm with the help of a strain-specific modification factor. *Wolbachia* itself is not present in mature sperm cells as a result of the lack of cytoplasm. The sperm imprint is rescued through a *Wolbachia* factor in the newly fertilized egg cell. The infection renders an advantage to the infected female over the uninfected female. This advantage can enable infected insects to replace naive populations (Turelli and Hoffmann 1991, 1995, Hoffmann and Turelli 1997). *Wolbachia* may be used to spread genetically engineered insects repeatedly into natural populations, even if they are already *Wolbachia* infected (Sinkins et al. 1995a, Merçot and Poinsot 1998). *Wolbachia*-induced cytoplasmic incompatibility may be applied to disease vectors that are incapable of transmitting pathogens or to beneficial insects rendered, for example, resistant to insecticides (Curtis 1994, Curtis and Sinkins 1998).

Very little is known about the molecular biology of *Wolbachia* (Sasaki et al. 1998, Braig et al. 1998). Cytological studies have shown that *Wolbachia* disrupts the kinetics of the decondensation/recondensation reactions of the pronuclei during fertilization. The sperm chromatin condensation is delayed in infected embryos, such that condensed maternal chromosomes and entangled prophase-like paternal fibers coalesce in the equatorial plane of the first metaphase spindle. At anaphase, the maternal chromosomes migrate to opposite poles of the spindle, whereas the paternal chromatin lag behind in the midzone of the spindle. This leads to death through the formation of embryos with aneuploid or haploid nuclei. Centrosome duplication and migration appear to be unaffected. *Wolbachia* is associated with centrosomal organized microtubules through cell division. The anaphase onset of maternal chromosomes is unaffected by the improper alignment of the paternal complement. The first metaphase spindle of the zygote consists of twin bundles of microtubules each holding one parental complement, suggesting that each half spindle regulates the timing of anaphase onset of its own chromosome set (Kose and Karr 1995, Lassy and Karr 1996, Callaini et al. 1996, 1997).

Currently, *Wolbachia*-infected vinegar flies *Drosophila simulans* are replacing the naive population in the Central Valley in Southern California and spreading north at a rate of 100 km/year. The *Wolbachia*-infected small brown planthopper, *Laodelphax stratiatellus*, is spreading in China and is migrating from mainland Asia over the East China Sea into Japan. Further data suggest that the *Wolbachia*-infected tsetse fly, *Glossina austeni*, is spreading at this moment in Africa (Hoshizaki et al. 1995, Cheng et al. 2000). One analysis indicates that worldwide mainland populations of the Asian tiger mosquito, *Aedes albopictus*, are double-infected with two different strains of *Wolbachia*, whereas populations on remote islands are only single-infected, thus suggesting that double-infected populations are currently replacing single-infected populations (Sinkins et al. 1995a). The fact that no population of the mosquito *Culex quinquefasciatus/pipiens* has been identified that is not *Wolbachia* infected indicates that *Wolbachia* infections may have worldwide consequences (Curtis 1992). Whereas unidirectional cytoplasmic incompatibility in certain hosts has been shown to be able to eliminate and replace natural populations completely, bi-directional cytoplasmic incompatibility is predicted to have the opposite effect. Bi-directional incompatibility should create a reproductive barrier within a population and might contribute to sympatric reproductive isolation and speciation (Hurst and Schilthuizen 1998, Werren 1998). Bi-directional incompatibility could theoretically also be utilized as a form of negative heterosis or hybrid sterility. Its hosts could only replace a local population if they were released in such numbers that they constituted the local majority.

Many mosquito species carry *Wolbachia* and express cytoplasmic incompatibility. Two prominent examples are the Asian tiger mosquito, *Aedes albopictus*, most populations of which carry a double infection, and the house mosquito, *C. pipiens/quinquefasciatus* (Sinkins et al. 1995b, Rousset et al. 1991). *Wolbachia* can be introduced into new hosts through microinjection and new crossing types can be generated *de novo* (Braig et al. 1994, Clancy and Hoffmann 1997). The artificial transfer into a new host can be very labor intensive and uncertain because there is no way to predict whether the chosen host will be capable of supporting a *Wolbachia* infection. In *Drosophila*, stable artificial double and triple infections have been generated (Sinkins et al. 1995a, Rousset et al. 1999). The incompatibility expressed in these strains is nearly additive. The double infection was capable of replacing a single infection in cage experiments (Sinkins et al. 1995a). Merçot and Poinsot (1998) estimate that a naturally bi-infected *D. simulans* strain from New Caledonia is converting at a rate of 3% to one of its constituent single infections. A triple infection will replace populations with double infections, with single infections, or without infections. This property of different *Wolbachia* strains would seem ideal for repeated population sweeps. Yet, none of the mosquitoes vectoring malaria has been found to carry *Wolbachia*. Attempts to microinject *Wolbachia* into eggs of *Anopheles gambiae* and *A. stephensi* have not been successful (Sinkins 1996).

The linkage of cytoplasmic drive and transgene still constitutes the biggest problem. Two options emerge for solving the problem of linking transgenes with *Wolbachia* as the drive. One option would take advantage of the fact that *Wolbachia* is an expression system in addition to a drive. A transformation of *Wolbachia* has not yet been achieved, but that should be only a matter of time. *Wolbachia* is always surrounded by a host membrane, and hence the transgene product could be transported by a type III or type IV transporter system into the host cytoplasm. Theoretically, this is possible; however, it would be a first for any intracellular bacterium. Much more difficult to realize would be any requirement of developmental regulation of the transgene. And, finally, there is another barrier to linkage: several post-translational modifications of the transgene product are out of reach in a bacterial expression system. The other option looks simple by comparison. If one were able to identify the factors by which *Wolbachia* causes and prevents or rescues cytoplasmic incompatibility, the most efficient utilization of cytoplasmic incompatibility might be possible. The ultimate objective would be to transform mosquitoes with these genes, inducing cytoplasmic incompatibility without the need of the bacterium. One model study showed that nuclear-coded cytoplasmic incompatibility genes (now on the chromosomes of the transformed insects) are still capable of spreading their host through a naive population (Sinkins et al. 1997, Turelli and Hoffmann 1999). Ideally, cytoplasmic incompatibility and transgenes could be linked together, for example, on an artificial supernumerary chromosome.

The very first field application of *Wolbachia*-induced incompatibility resulted in the successful eradication of a natural population of *C. pipiens* in the village Okpo, 10 miles north of Rangoon, Burma (Laven 1967). The daily release of relatively small numbers of cytoplasmic incompatible males during only one winter resulted in 100% inviable egg rafts. After the big success with cytoplasmic incompatible males in Burma, Laven and Aslamkhan (1970) proposed the combination of cytoplasmic incompatibility with a Y chromosome–linked translocation in what they called an integrated strain. Three years later, Curtis et al. (1982) released such a strain in two villages near New Delhi, India. Although up to 40,000 males were released per day per village, incompatibility never exceeded 68% and eventually the population-suppressive effect declined as well. Immigrating females, which were already mated, were responsible for the failure of this eradication attempt. A 3-km-wide buffer zone surrounding one of the villages, which was kept free of any breeding places with the help of larvicides, had no impact on the movement of adults. This example also shows that, regardless of the technique, where the limitations of the sterile male release become evident, the release of cytoplasmic incompatible females might have surmounted the obstacles in replacing the naive population and, most importantly, in fighting immigrating wild-type females.

Recent examples of natural horizontal transfer of *Wolbachia* between grasshoppers (Noda et al. 2000) and between parasitoid wasps (Huigens et al. 2000) raise questions about how well *Wolbachia* bacteria as a drive could be isolated in target species. Von der Schulenburg and others (2000) recently reconstructed the phylogenies for the *Wolbachia* genes *ftsZ* and *wsp* and concluded that they are mutually incompatible, suggesting extensive recombination between *Wolbachia* strains. However, there is no indication yet about the timescale of recombination. Until this question has been addressed, the use of *Wolbachia* as an expression system will seem less attractive.

Wolbachia-*induced parthenogenesis.*   In meiotic drive systems, we have seen that chromosomal sex ratio distorters can constitute strong drive mechanisms. An extreme form of sex ratio distortion is parthenogenesis, creating all-female broods. The molecular mechanisms of chromosomal parthenogenesis are totally unknown and inaccessible to manipulation. The investigation of cytoplasmic, microbial-based parthenogenesis systems might accelerate the possibility of an application.

Examples of obligate parthenogenesis can be found in almost every invertebrate order (e.g., Suomalainen et al. 1987, Hughes 1989). The association of *Wolbachia* with parthenogenesis has been intensively studied in parasitoid wasps (Hymenoptera) (Stouthamer 1997). Recently, the horizontal transfer of *Wolbachia* through microinjection from the parthenogenic species *Trichogramma pretiosum* to a naive, uninfected, sexually reproducing species, *T. dendrolimi*, succeeded in the induction of partial thelytoky (Grenier et al. 1998). Most invertebrate orders and almost all insect orders contain scattered examples of thelytokous parthenogenesis. Some *Wolbachia*-infected, parthenogenic parasitoid wasps cannot return to sexual reproduction after the elimination of *Wolbachia*. Unfortunately, parthenogenesis has not been described in mosquitoes.

Recently, a natural horizontal transfer was suggested for a parasitoid wasp (Heath et al. 1999). Further, Stouthamer's group (Huigens et al. 2000) found evidence of an unexpectedly high frequency of horizontal transfer between parthenogenesis-inducing parasitoid wasp larvae sharing a common host. Although that transfer was within the same species, it increases the possibility of interspecific transmission.

Parthenogenic females do not interact with a wild population. Their spread depends completely on increased reproductive fitness of the parthenogenic populations. Any genetic load affecting the overall fitness or any change in the environment affecting adaptation will threaten the population replacement. Fixation cannot be assured.

10.2.4   *Research priorities for genetic driving mechanisms*

Meaningful discussion of driving mechanisms requires detailed knowledge of the host species' population biology and the fitness costs of the proposed transgenes to the host. Transgenes potentially useful in the control of one of the major human vector diseases have not yet been identified. At present, even if genes were to be identified that were capable of rendering anopheline mosquitoes refractory to malaria transmission or even if naturally refractory anopheline mosquitoes were to become available though a mass-rearing program, no genetic driving force would be available to enable replacement of the natural vector population with the new strain. There is no mechanism available at the moment to link a refractory transgene or a natural phenotype to a genetic driving mechanism.

It is a shame that the promises of early successes in genetic control of vector insects (Laven 1967, 1971, Curtis et al. 1982) fell victim to the Cold War and were cut short by public relation disasters in India (e.g., Curtis and von Borstel 1978) and Pakistan (e.g., Jayaraman 1982, Davidson et al. 1982) just at a time when those projects were about to be placed on a solid footing based on field data. Countless victims of vector-borne diseases have paid the price during the last 30 years. Field applications of genetic control methods are still suffering the aftermath.

If it turns out that most, if not all, natural phenotypes of resistance are recessive for all practical purposes, that will leave only negative heterosis or a cytoplasmic system as a potential driving mechanism.

Random mutations sooner or later will affect every transgene and lead eventually to its inactivation. It is essential that the linkage of genetic drive and transgene be of such a nature that, even after the transgene becomes inactive, the linkage cannot be broken. The biggest threat would be an empty driver going wild. Genetic population replacement projects will need the same detailed knowledge about the genetics and behavior of the target population that has been characteristic of successful SIT projects. Those projects have clearly demonstrated that the dynamics of natural populations will not allow cutting a single corner when investigating and implementing a control program.

We regard supernumerary chromosomes and biased gene conversion as the two most-exciting areas for long-term research into developing ideal drive mechanisms to modify insect populations.

To further the development of cytoplasmic drive mechanisms for vector control, it will be essential that a driving mechanism be made available for anopheline mosquitoes and that a way be found to link drive and nuclear genotype with each other. Identification of the molecular mechanisms applied by *Wolbachia* to induce cytoplasmic incompatibility in its hosts is pivotal for cloning this system into *Anopheles* mosquitoes. It is becoming increasingly unlikely that a natural *Wolbachia* infection will be detected in anophelines, and it is equally unlikely that a stable interspecific transfer of *Wolbachia* into an *Anopheles* species will be achieved (Sinkins 1996). The development of interspecific transfer and of a transformation system for *Wolbachia*, as recently suggested by Sinkins and O'Neill (2000), will add valuable tools for studying the basic biology of *Wolbachia*, but such developments are secondary compared with the need to obtain a working population replacement method for *Anopheles* species. In light of this, it is surprising to read that a knowledge or understanding of the mode of action of a genetic system able to modify insect populations is no longer considered either essential or a research priority for use of that system in pest control (Sinkins and O'Neill 2000). Nevertheless, any contribution is important and likely to be fruitful in augmenting the palette of methods that might help reduce the millions of deaths each year inflicted by vector-borne diseases.

10.3   Factors influencing the spread rate of genetic constructs in natural insect populations

As discussed above, introduced transgenic populations are not likely to replace existing populations through competitive displacement. Some gene-driving mechanisms are needed. If transposable elements are used as a force to drive genes into natural populations, the spread of the genetic construct in natural populations will depend on:

1. The probability of a non-infected gamete becoming infected with a transposon when paired with an infected gamete, i.e., the transposition rate;
2. The magnitude of the negative fitness effect of the genetic construct on the host; and
3. The gene flow among host populations.

If *Wolbachia*-induced cytoplasmic incompatibility is used to drive transgenes into populations, then the initial frequency of the transformed hosts must exceed a threshold, and the threshold density will be determined by the maternal transmission rate of *Wolbachia*, the relative fitness of the transformants, and the strategies used to introduce transgenes (Turelli and Hoffmann 1999). Ginzburg et al. (1984) used a deterministic, one-locus, two-allele model and examined the conditions necessary for a newly introduced transposon to invade and spread in a panmictic, sexually reproducing population. If fitness loss of the dysgenic individual is independent of sex, as is the case with the majority of transposable elements, then the condition for the transposable elements to spread in a population is

$$\beta > \frac{S}{1 - S} \tag{10.1}$$

where β is transposition rate and S is fitness loss of the dysgenic individual. The equation implies that if the fitness loss of the dysgenic individual is over 0.5, the transposon cannot be spread even if the transposition rate is 100%. For a transposon with low transposition rate to spread in a population, hybrid fitness depression must be small. The conditions for a transposon to invade a population are the same for genetic constructs in which the transposon is coupled with a target gene. However, host fitness depression caused by the genetic construct may be different from that caused by the transposon alone if the target gene is associated with a fitness effect on the host.

Malaria places enormous burdens on humans in tropical and subtropical countries. Current strategies for malaria control involve treatment of infected individuals with anti-malarial drugs to kill the parasites and vector control to reduce human–vector contacts via residual spraying and the use of insecticide-impregnated bednets. The emergence of insecticide resistance in mosquito vectors and anti-malarial drug resistance in *Plasmodium* has significantly reduced the viability of many malaria control programs. Recently, a novel malaria control strategy based on genetic disruption of vector competence has been proposed (James 1992, Collins and Besansky 1994). Refractoriness, the inability of malaria parasites to develop in the mosquito midgut, is a common, naturally occurring parasite-inhibiting mechanism. Refractory genes are, therefore, promising candidates for mosquito genetic transformation. Research efforts have been initiated to identify and clone the refractory genes (Severson et al. 1995). The biggest challenge to the success of the transgenic mosquito approach is finding a method of driving parasite-inhibiting genes into natural populations. Two issues are clearly relevant to the spread of the refractory genes regardless of the type of the gene-driving mechanism: (1) fitness costs and benefits associated with mosquito refractoriness to malaria parasites and (2) gene flow of natural mosquito populations.

10.3.1   Fitness benefits/costs associated with refractoriness to malaria parasites

Sporogonic development of the malaria parasite in mosquito hosts involves a sequential series of developmental steps, including gametocyte fertilization, ookinete formation, penetration through the peritrophic matrix and midgut epithelium to form oocysts, and production and migration of sporozoites to the salivary glands. Failure to complete any of these steps inhibits parasite development and thereby prevents transmission to the vertebrate host.

10.3.1.1   Refractory mechanisms

These refractory mechanisms may or may not cause changes in vector fitness. At least four parasite-inhibiting mechanisms have been recognized in mosquitoes. These include parasite failure to exflagellate in the midgut of mosquitoes (Nijhout 1979); melanotic encapsulation of ookinetes by mosquito midgut cells (Collins et al. 1986, Paskewitz and Riehle 1994, Zheng et al. 1997); failure of oocyst development in mosquito midgut, termed *refractoriness* (Kilama and Craig 1969, Thathy et al. 1994, Severson et al. 1995); and a salivary gland barrier that prevents sporozoite penetration or reduces survivorship of sporozoites (Rosenberg 1985). As stated above, because refractory genes are promising targets, the discussion below will focus on the refractory phenotype.

Mosquito refractoriness to malaria parasites is genetically determined. Using the yellow fever mosquito (*Aedes aegypti*) and avian malaria parasite (*P. gallinaceum*), Kilama and Craig (1969) crossed mosquito populations with different levels of susceptibility and found that the refractoriness is determined by a single dominant gene. Severson et al. (1995) used quantitative trait loci (QTL) analysis with the aid of molecular markers and identified two loci conferring refractoriness. The major QTL was located on chromosome 2 and explained 49 to 65% of the variance in infection intensity, and the minor QTL was found on chromosome 3 and accounted for 10 to 14% of the variance (Severson et al. 1995). Both QTLs exhibit partial dominance effects on susceptibility. That is, heterozygotes derived from a susceptible and a refractory parent are also quite refractory. However, epistasis between the two QTLs was not significant.

The physiological mechanisms of naturally occurring refractoriness are not well understood. Previous studies have identified various factors limiting *Plasmodium* parasite development in the mosquitoes, including damage to ookinetes by the digestive enzymes in the mosquito midgut (Gass and Yeates 1979, Feldmann et al. 1990), peritrophic matrix and midgut epithelial cells acting as barriers to ookinete penetration (Billingsley and Rudin 1992, Shahabuddin et al. 1993, Vernick et al. 1995). A study examining ookinete formation and blood meal digestive enzymes in two *A. aegypti* populations highly refractory to *P. gallinaceum* and one *A. aegypti* population highly susceptible to *P. gallinaceum* suggested that the failure of oocyst development in the refractory mosquito population is not caused by ookinete damage by the proteolytic enzymes (Yan, unpublished data). Ultrastructural studies on ookinete invasion of midgut epithelium cells in the refractory population suggested that ookinetes extend to the basal lamina of the midgut, but fail to develop further. Melanotic encapsulation of ookinetes was not observed. Therefore, mosquito refractoriness likely results from a midgut barrier.

10.3.1.2   Distribution of Plasmodium *refractory genes in natural populations*

In nature, even in the most efficient malaria vector species, only a proportion of the mosquitoes is refractory to malaria parasites after exposure to a *Plasmodium*-infected blood meal. That is, malaria parasites cannot develop into oocysts in those mosquitoes. Kilama (1973) studied susceptibility of seven African *A. aegypti* populations to *P. gallinaceum*, and found that 10 to 30% of the mosquitoes did not have oocysts. Toure et al. (1998) exposed

F_1 *Anopheles gambiae* mosquitoes from field-collected females to *P. falciparum*-infected human blood, and found that >50% of the mosquitoes were not infected with *Plasmodium* oocysts, although ookinetes were observed in all mosquitoes. Parasites are intimately dependent upon their hosts for survival and often have deleterious effects on host reproductive success because of the physical damage they cause during invasion and migration, release of toxins, and usurpation of host resources. Thus, parasite-imposed selection could lead a host population to the point of complete resistance or to a point where additive genetic variance is depleted. This is obviously inconsistent with the empirical observation that host populations in nature exhibit considerable genetic variation in resistance to parasitism.

Three hypotheses were proposed to explain polymorphism of refractory genes in nature (Yan et al. 1997). The first hypothesis is that genetic variation in host susceptibility may not be due to selection by the parasite. Rather, it is simply a result of founder effects. That is, if parasite infection does not reduce host fitness and if refractoriness is not associated with fitness costs, then refractoriness is selectively neutral, and genetic variability in refractoriness is the result of genetic drift or other ecological constraints. The second hypothesis is that the observed genetic variation in susceptibility is a transient state prior to complete refractoriness or susceptibility. If there are no negative fitness effects associated with refractoriness, parasite-imposed selection could ultimately result in fixation for refractory alleles. On the other hand, when the parasite exerts no selection and the fitness costs of refractoriness are high, a polymorphic population could rapidly become completely susceptible. The third hypothesis is that genetic variation in susceptibility is subject to stabilizing selection, resulting in intermediate levels of refractoriness. If infection by a parasite reduces host fitness, but refractoriness is costly for the host to produce and maintain, stabilizing selection for an intermediate level of refractoriness could occur. Thus, information on the negative fitness effects of parasites on hosts and the fitness costs associated with refractoriness is critical for understanding the evolution of refractory genes.

10.3.1.3   *Fitness benefits and costs associated with refractoriness*

Malaria parasites are virulent to their vertebrate hosts, but generally benign to the mosquito vectors. Several studies (Hacker and Kilama 1974, Freier and Friedman 1976) have demonstrated reduced fecundity in mosquitoes exposed to vertebrate hosts infected with malaria parasites. It is unclear to what degree this effect can be attributed to the malaria parasite itself because blood from an infected animal is also of reduced nutritive quality to the mosquito. Under laboratory conditions, the survivorship of *Plasmodium gallinaceum*-infected *Aedes aegypti* mosquitoes was not different from that of uninfected mosquitoes (Yan et al. 1997). Thus, the direct fitness effect of malaria parasites on the mosquito vectors is small.

Fitness costs associated with refractoriness may result from self-inflicted toxicity, disruption of previous functions for biological pathways that are involved in refractoriness, diversion of resources from fitness-enhancing functions, or other pleiotropic effects of genes related to refractoriness (Simms and Triplett 1994). For example, the melanotic encapsulation of parasites by mosquitoes requires tyrosine as a precursor, and tyrosine is important for egg chorion, cuticular tanning, and wound healing (Li et al. 1994). Indeed, reduced ovary size and protein content and delayed oviposition have been observed in mosquitoes actively encapsulating parasites (Ferdig et al. 1993). Although active immune response against the malaria parasites is apparently not involved in the refractory phenotype described above, it would be interesting to know whether naturally occurring refractoriness is associated with reduced fitness.

Yan et al. (1997) compared the reproductive success and survivorship of two *A. aegypti* populations, refractory and susceptible to *P. gallinaceum*. The two populations were selected from a common stock population through inbreeding (inbreeding coefficient = 0.5; Thathy et al. 1994). Fitness components studied include fecundity, adult survivorship and egg-to-adult developmental time, blood-meal size, adult body size, and survivorship. The refractory population had a significantly shorter larva-to-adult developmental time and a smaller body size, took a smaller blood meal, and subsequently laid fewer eggs than the susceptible population. The mean longevity of the refractory population was nearly half that of the susceptible population. Thus, the refractory population exhibited substantially reduced fitness compared with the susceptible population under the laboratory conditions.

The study by Yan et al. (1997) only established a phenotypic correlation between refractoriness and reduced fitness. Three hypotheses could explain the reduced fitness observed in the refractory population. First, fitness variation between the two populations may simply reflect founder effect, or variation in inbreeding depression. Second, the reduced fitness in the refractory population may be caused by uncharacterized genes tightly linked to genes determining refractoriness. Third, genes conferring refractoriness may have direct pleiotropic effects on other factors determining fitness. Further studies to determine the genetic basis of reduced fitness observed in the refractory population are critical for understanding the distribution of the refractory genes in nature and for designing rational strategies for driving the refractory genes into natural mosquito populations.

10.3.1.4   The rate of genetic construct spread

Refractory genes cannot be spread to natural vector populations without a driving mechanism because malaria parasites confer little selective advantage to refractory mosquitoes. Further, the prevalence of malaria parasites in natural mosquito populations, even in areas with high vertebrate disease transmission, is generally low (Githeko et al. 1993, Paul et al. 1995). If transposable elements are used as gene-driving mechanisms, the rate of genetic construct spread will be a function of transposition rate and the fitness loss in dysgenic individuals, assuming that recombination between transposon and targeted genes is negligible (Ginzburg et al. 1984). The higher the transposition rate and the lower the fitness loss, the faster the construct is spread. Computer simulation suggests that the genetic constructs will be spread at low frequency by transposons and fixed rapidly (within 100 generations; Ribeiro and Kidwell 1994). If *Wolbachia*-induced cytoplasmic incompatibility is used as a gene-driving mechanism, fitness decrease of the transformants will determine the initial threshold frequency of the transformed hosts to be released; above that threshold the transgene can be spread and fixed in the population (Turelli and Hoffmann 1999). Once the initial transgene frequency is above the critical threshold density, the transgenes can be spread and fixed within 60 to 100 generations (Turelli and Hoffmann 1999).

10.3.2   Mosquito population genetic structure

Knowledge of mosquito population genetic structure provides important information on the spatial scale required for the spread of *Plasmodium*-refractory genes. For example, for populations with extensive gene flow, genetically engineered mosquitoes can be released in a few locations and gene exchange in nature will facilitate the spread of the refractory genes. However, for populations experiencing little or no gene exchange, gene introduction will be required for each population. The discussion below focuses on *Anopheles gambiae*, the major malaria vector in Africa.

Various genetic markers have been used to assess mosquito population genetic structure, including chromosomal inversions, isozymes, microsatellites, rDNA intergenic spacers, and mitochondrial DNA sequences. Microsatellite markers are commonly used because they are abundant, co-dominant, and highly polymorphic. In addition, they employ a PCR-based genotyping method and can be readily used for preserved specimens. For *A. gambiae*, more than 130 microsatellite markers spanning the three chromosomes have been genetically mapped, and the cytogenetic locations of many markers have been determined by *in situ* hybridization (Zheng et al. 1996). Microsatellite and mitochondrial DNA sequence data suggest extensive gene flow in *A. gambiae* in areas without obvious geographical barriers. For example, in western Kenya, mosquitoes collected from villages that were 50 km apart did not exhibit significant genetic differentiation, suggesting that the minimum area in western Kenya associated with a deme is >50 km in diameter (Lehmann et al. 1997).

Climate conditions can cause discontinuity in mosquito species distribution. *Anopheles gambiae* prefers moist areas, whereas *A. arabiensis* is more common in dry areas (Lindsay et al. 1998). Thus, *A. gambiae* does not inhabit the arid Great Rift Valley, but is the predominant species in coastal Kenya and in the basin regions of Lake Victoria where rainfall is abundant. Therefore, the Rift Valley could act as a gene flow barrier for *A. gambiae*. Empirical evidence supports this hypothesis. Little genetic differentiation was found among populations on each side of the rift, but high differentiation was detected among populations across the rift (Lehmann et al. 1999). Environmental conditions may also affect the distribution of mosquito genetic variants. For example, in Mali, western Africa, three chromosomal forms of *A. gambiae* were recognized based on characteristics of chromosome 2R inversions: Bamako, Mopti, and Savanna. The Mopti form is dominant in all ecological zones in Mali, whereas the Savanna and Bamako forms prevail in humid savannas. Rainfall is correlated positively with the frequency of Savanna form and negatively with the frequency of Mopti form (Toure et al. 1994). The magnitude of genetic differentiation between the three chromosome forms depends on the genome region in question. Loci on chromosome 2 exhibit a higher degree of differentiation than loci on chromosomes X and 3 (Lanzaro et al. 1998), suggesting that gene flow is not equal across the genome. As this is the case, population genetic studies should consider the constraints of evolutionary and ecological forces.

It is worth noting that different genetic markers lead to incongruent results in studies of mosquito population genetic structure. For example, rDNA intergenic spacer markers indicated that *A. gambiae* populations <10 km apart in western Kenya exhibited significant genetic differentiation (McLain et al. 1989), while microsatellite and mitochondrial DNA markers did not detect any genetic differentiation for populations >50 km apart (Lehmann et al. 1997). When microsatellite markers were used to study populations with different spatial scale, contradicting results were also observed. For example, little genetic differentiation was detected for *A. gambiae* populations >6000 km apart between eastern and western Africa (Lehmann et al. 1996a), whereas large differentiation was found for populations <900 km apart between western and eastern Kenya (Lehmann et al. 1999). Little genetic differentiation for *A. gambiae* across the African continent is surprising because there are many potential gene flow barriers, such as large arid areas in central Africa and high-elevation mountains in eastern Democratic Republic of the Congo. Such incongruent results may result from potential problems with microsatellite markers (Walton et al. 1998). First, constraints on microsatellite allele sizes (Lehmann et al. 1996b) and high mutation rate of microsatellites can lead to underestimation of genetic differentiation between populations. Second, allelic homoplasy caused by base substitution and insertion/deletion in the region flanking the tandem repeats could conceal the genetic differences between populations because microsatellite alleles of same size do not necessarily have the same number of repeat units (Grimaldi and Crouau-Roy 1997, Orti et al. 1997, Walton et al. 1998). The problem of allelic homoplasy can also occur in comparisons between populations over large geographical ranges. Third, approximation in conversion from allele size to the number of repeat units can lead to serious underestimation of population genetic differentiation. Thus, it is necessary to carefully screen microsatellite loci to minimize potential problems. Results from studies of mosquito population genetic structure derived from microsatellite markers should be confirmed with other molecular markers in the same genome region. Finally, it is worth remembering that gene flow inferred from population genetic structure reflects the effect of historical migration and recent population expansion. The spread of the introduced parasite-inhibiting genes, on the other hand, depends on current and future gene flow. Thus, direct measurement of gene flow using a mark–release–recapture method should provide valuable information for designing rational strategies to spread parasite-inhibiting genes into populations.

10.3.3   Viability of transgenic mosquito approach for malaria control

The debates surrounding use of genetic disruption of mosquito vector competence to control malaria transmission have raged since the concept was first proposed a decade ago (James 1992, Curtis 1994, Spielman 1994a). Over the past several years, remarkable progress has been made in the development of mosquito germ-line transformation and in the identification of parasite-inhibiting molecules. For example, *Aedes aegypti* mosquitoes and *Anopheles gambiae* cell lines were successfully transformed with the *Hermes* element (Jasinskiene et al. 1998, Zhao and Eggleston 1998), and the *Minos* transposable element bearing an exogenous gene was efficiently integrated into the genome of *A. stephensi* (Catteruccia et al. 2000a, 2000b). Genetic linkage maps have been constructed for various mosquito species (Severson et al. 1993, Zheng et al. 1996), and genes conferring mosquito refractoriness and resistance to malaria parasites have been mapped (Severson et al. 1995, Zheng et al. 1997). Further, the about-to-be-launched *A. gambiae* genome-sequencing project will greatly facilitate identification and cloning of parasite-inhibiting genes. There is no doubt of the feasibility of cloning parasite-inhibiting genes and developing efficient transformation tools for *A. gambiae*. However, despite all the progress and in addition to unresolved social and ethical issues, there remain important scientific questions that must be addressed prior to a mosquito release program.

1. Can parasite-inhibiting genes be spread and fixed in natural mosquito populations?
2. Can reduced vector competence be translated into reduction in transmission intensity and, ultimately, reduction in mortality and morbidity caused by malaria?
3. Will the release program increase overall vector population density and increase transmission of other vector-borne diseases?
4. Can *Plasmodium* parasites evade the introduced parasite-inhibiting mechanisms?

The fate of introduced parasite-inhibiting genes is determined not only by the gene-driving mechanisms and the magnitude of fitness loss in dysgenic individuals, but also by various ecological and abiotic conditions. For example, the climate in Africa can generally be divided into rainy and dry seasons. Dry seasons vary from 2 to 3 months to more than 6 months. Larval development sites of mosquitoes are reduced during the dry period, and mosquito density is very low and often not detectable (Charlwood et al. 1995). Anopheline mosquito populations build up rapidly and reach peak density shortly after the onset of the rainy season (Joshi et al. 1975, Mbogo et al. 1995). If the survivorship of mosquitoes carrying the genetic constructs during the dry season is lower than mosquitoes in the wild, then the spread of parasite-inhibiting genes will be slowed. Another problem arises because multiple malaria vector species occur sympatrically in many areas in sub-Saharan Africa. Although they vary in vector competence and biting behaviors (Coluzzi

---

et al. 1979, Collins and Paskwitz 1996), they all are efficient malaria vectors. Thus, any release program would need to tackle more than one species and would need to consider the ecological associations between the mosquito species and their overall effects on malaria transmission.

The ultimate goal of the transgenic mosquito approach is to reduce malaria prevalence, morbidity, and mortality. Theoretically, if all mosquitoes are refractory to malaria parasites, the parasite life cycle will be disrupted. However, before the refractory genes reach fixation in a population, it is important to evaluate whether releasing transgenic mosquitoes may bring adverse effects on public health (Spielman 1994a; see also Spielman et al., Chapter 11 of this volume). Parasite transmission force is far more sensitive to mosquito survivorship than vector competence (the ability to transmit parasites). That is, transmission force characterized by mosquito vectorial capacity is linearly proportional to the density of competent vectors, but exponentially proportional to vector survivorship (Macdonald 1957). If the survivorship of hybrids between native mosquitoes and laboratory-reared transgenic mosquitoes exceeds the survivorship of the native mosquitoes (Turelli and Ginzburg 1983) and the hybrid is not completely refractory to malaria parasites, then the release program will increase malaria transmission. Increase in hybrid survivorship will also pose increased risks for other vector-borne diseases such as lymphatic filariasis.

The transgenic mosquito approach aims to reduce malaria transmission intensity through population replacement with parasite-inhibiting genes. To what levels should malaria transmission intensity be reduced? The relationship between transmission intensity and malaria prevalence is not clear-cut. For example, malaria transmission intensity, as measured by entomological inoculation rates (EIRs), is generally much lower on the coast of Kenya than in western Kenya; however, malaria prevalence is similar in both places (Githeko et al. 1993, 1996, Mbogo et al. 1995). Detailed analysis suggested that malaria prevalence and EIR varied considerably among villages within a district, and that there was no correlation between the two variables (Mbogo et al. 1995). There is growing evidence from studies throughout Africa that EIR must be reduced to levels as low as one infective bite per year from current levels of tens or hundreds of infective bites per year in order to achieve any meaningful reductions in malaria prevalence (Trape and Rogier 1996, Beier et al. 1999). Population replacement by parasite-inhibiting genes must therefore reduce vector competence to an extremely low level if it is to have a positive consequence on public health.

Finally, the transgenic mosquito approach must consider species diversity and genetic variability of malaria parasites. In most areas in sub-Saharan Africa, two or more *Plasmodium* species exist, and *P. falciparum* is the dominant species. If the parasite-inhibiting mechanism is specific to *P. falciparum*, then suppression of *P. falciparum* populations by the release program may lead to resurgence of other *Plasmodium* species. Further, *P. falciparum* population always consists of multiple genotypes (e.g., Babiker et al. 1995, Robert et al. 1996, Rich et al. 1997), thus introduced refractory mechanisms must be able to combat all genotypes. From an evolutionary perspective, new mechanisms able to evade the introduced parasite-inhibiting mechanisms can be expected to evolve inevitably, just as anti-malarial drug resistance evolved. The success of the transgenic mosquito approach depends on our ability to slow that evolution. All the above issues emphasize the importance of conducting field-based evaluations of the risk and viability of the transgenic mosquito approach for malaria control.

10.4   Conclusions

Genetic manipulation of vector competence is a promising alternative approach to vector-borne disease control. Genetically engineered insects present a desperately needed addition to a tool chest that only contains insecticides and anti-parasite drugs. Efficient mechanisms must be designed to drive genetic constructs into natural vector populations. Without gene-driving mechanisms, replacing a natural insect population permanently with a different genotype is extremely difficult and almost unachievable. An ideal gene-driving mechanism should be:

1. Strong enough to much more than compensate for any loss of fertility and competitiveness caused by transgenic payload drive;
2. Able to achieve fixation of its load rapidly in a population;
3. Able to carry several large transgenes and regulatory elements at the same time;
4. Stable in its linkage to its load until fixation has been accomplished;
5. Easily applicable to many local strains or species of vectors;
6. Capable of more than one application in the same population and capable of repeated population sweeps with different loads; and
7. Resistant to significant levels of immigration, especially if eventual isolation infrastructures break down over time.

A successful driving strategy might encompass small local replacement attempts and ways to isolate or subdivide a population to facilitate small replacements. We regard supernumerary chromosomes and biased gene conversion as the two most-exciting areas for long-term research aimed at developing ideal drive mechanisms to modify insect populations.

A number of factors influence the spread of the genetic constructs in nature. Most importantly, these include:

1. Transposition rate for transposons and transmission rate for *Wolbachia* parasites;
2. Magnitude of negative fitness effect of the genetic construct on the host; and
3. Gene flow among host populations.

The higher the transposition/transmission rates and the lower the fitness loss, the faster the genetic construct will be spread if the initial frequency of the transformed hosts exceeds the threshold density. In the case of vector-borne disease control, genetic association between parasite-inhibiting genes and host fitness must be determined for selection of appropriate gene-driving mechanisms and predictions of the spread rate of the genetic constructs. Knowledge of insect population genetic structure can provide important information about the spatial scale required for the spread of parasite-inhibiting genes. Finally and most importantly, the benefits and risks of the particular transgenic mosquito approach for vector-borne disease control must be carefully evaluated prior to any transgene release.

Acknowledgments

We thank Prof. Andrew Spielman, Harvard University, for bringing the two authors together for a special symposium on Public Health Promise of Genetically Modified Vectors during a conference of the Society for Vector Ecology. H.R.B. is pleased to acknowledge funding by the Royal Society (20965), the Biotechnology and Biological Sciences Research Council (5/S11854), and the Natural Environment Research Council (GR3/13199), U.K. G.Y. is supported by grants from the National Institutes of Health (U19 AI45511 and D43 TW001505) and the UNDP/WORLD BANK/WHO Special Programme for Research and Training in Tropical Diseases (TDR 980690).

Literature cited

Akashi, H., Kliman, R. M., and Eyre-Walker, A., Mutation pressure, natural selection, and the evolution of base composition in *Drosophila, Genetica*, 102, 49, 1998.

Aksoy, S., Tsetse — a haven for microorganisms, *Parasitol. Today*, 16, 114, 2000.

Anderson, R. M. and May, R. M., *Infectious Diseases of Humans — Dynamics and Control*, Oxford University Press, Oxford, U.K., 1991.

Anonymous, *The New World Screwworm Eradication Programme North Africa 1988–1992*, Food and Agricultural Organization, United Nations, Rome, 1992.

Anonymous, *FAO/IAEA International Conference on Area-Wide Control of Insect Pests Integrating the Sterile Insect and Related Nuclear and Other Techniques*, Food and Agricultural Organization, United Nations, Rome, 1998a.

Anonymous, *Fifth International Symposium on Fruit Flies of Economic Importance*, Organizational Committee, Pendang, Malasyia, 1998b.

Anxolabéhère, D. and Periquet, G., P-homologous sequences in Diptera are not restricted to the Drosophilidae family, *Genet. Iber.*, 39, 211, 1987.

Araujo, S. M. S. R., Pompolo, S. G., Dergam, J. A. S., and Campos, L. A. O., The B chromosome system of *Trypoxylon (Trypargilum) albitarse* (Hymenoptera, Sphecidae), 1. banding analysis, *Cytobios*, 101, 7, 2000.

Arnold, M. L., Contreras, N., and Shaw, D. D., Biased gene conversion and asymmetrical introgression between subspecies, *Chromosoma*, 96, 368, 1988.

Ashburner, M., Transformation, in *Drosophila — A Laboratory Handbook*, Cold Spring Harbor Laboratory Press, Cold Spring Harbor, NY, 1989a, 969.

Ashburner, M., Hybrid dysgenesis and related phenomena, in *Drosophila — A Laboratory Handbook*, Cold Spring Harbor Laboratory Press, Cold Spring Harbor, NY, 1989b, 1017.

Ashburner, M., Hoy, M. A., and Peloquin, J. J., Prospects for the genetic transformation of arthropods, *Insect Mol. Biol.*, 7, 201, 1998.

Asman, S. M., Nelson, R. L., McDonald, P. T., Milby, M. M., Reeves, W. C., White, K. D., and Fine, P. E. M., Pilot release of a sex-linked multiple translocation into a *Culex tarsalis* population in Kern County, CA, *Mosq. News*, 39, 248, 1979.

Asman, S. M., Zalom, F. G., and Meyer, R. P., A field release of irradiated male *Culex tarsalis* in California, *Proc. Calif. Mosq. Vector Control Assoc.*, 48, 64, 1980.

Asman, S. M., McDonald, P. T., and Prout, T., Field studies of genetic control systems for mosquitoes, *Annu. Rev. Entomol.*, 26, 289, 1981.

Atkinson, P. W. and O'Brochta, D. A., Genetic transformation of non-drosophilid insects by transposable elements, *Ann. Entomol. Soc. Am.*, 92, 930, 1999.

Babiker, H. A., Satti, G., and Walliker, D., Genetic changes in the population of *Plasmodium falciparum* in a Sudanese village over a three-year period, *Am. J. Trop. Med. Hyg.*, 53, 7, 1995.

Baker, R. H., Reisen, W. K., Sakai, R. K., Hayes, C. G., Aslamkhan, M., Saifuddin, U. T., Mahmood, F., Perveen, A., and Javed, S., Field assessment of mating competitiveness of male *Culex tritaeniorhynchus* carrying a complex chromosomal aberration, *Ann. Entomol. Soc. Am.*, 72, 751, 1979.

Baker, R. H., Sakai, R. K., Rathor, H. R., Raana, K., Azra, K., Niaz, S., and Reisen, W. K., *Anopheles culicifacies* mating behavior and competitiveness in nature of males carrying a complex chromosomal aberration, *Ann. Entomol. Soc. Am.*, 73, 581, 1980.

Bandi, C., Anderson, T. J. C., Genchi, C., and Blaxter, M. L., Phylogeny of *Wolbachia* in filarial nematodes, *Proc. R. Soc. London B Biol. Sci.*, 265, 2407, 1998.

Barclay, H. J., The estimation of fitness in iteroparous species during population replacement experiments, *Can. J. Genet. Cytol.*, 26, 91, 1984.

Barclay, H. J. and Fitz-Earle, M., Maximum-likelihood estimates of fitness in iteroparous species population replacement experiments using compound chromosomes, *Genome*, 30, 83, 1988.

Barr, A. R., Divided-eye, a sex-linked mutation in *Culex pipiens* L., *J. Med. Entomol.*, 6, 393, 1969.

Bauser, C. A., Elick, T. A., and Fraser, M. J., Characterization of *hitchhiker*, a transposon insertion frequently associated with baculovirus FP mutants derived upon passage in the TN-368 cell line, *Virology*, 216, 235, 1996.

Bazin, C. and Higuet, D., Lack of correlation between dysgenic traits in the *hobo* system of hybrid dysgenesis in *Drosophila melanogaster*, *Genet. Res.*, 67, 219, 1996.

Beard, C. B., Durvasula, R. V., and Richards, F. F., Bacterial symbiont transformation in Chagas disease vectors, in *Insect Transgenesis: Methods and Applications*, Handler, A. M. and James, A. J., Eds., CRC Press, Boca Raton, FL, 2000, 289.

Beckendorf, S. K. and Hoy, M. A., Genetic improvement of arthropod natural enemies through selection, hybridization or genetic engineering techniques, in *Biological Control in Agricultural IPM Systems*, Hoy, M. A. and Herzog, D. C., Eds., Academic Press, Orlando, FL, 1985, 167.

Beeman, R. W. and Friesen, K. S., Properties and natural occurrence of maternal-effect selfish genes (*Medea* factors) in the red flour beetle, *Tribolium castaneum*, *Heredity*, 82, 529, 1999.

Beeman, R. W., Friesen, K. S., and Denell, R. E., Maternal-effect selfish genes in flour beetles, *Science*, 256, 89, 1992.

Beier, J. C., Killeen, G. F., and Githure, J. I., Short report: entomologic inoculation rates and *Plasmodium falciparum* malaria prevalence in Africa, *Am. J. Trop. Med. Hyg.*, 61, 109, 1999.

Belling, J., A study of semi-sterility, *J. Hered.*, 5, 65, 1914.

Benedict, M. O., McNitt, L. M., Cornel, A. J., and Collins, F. H., A new marker, *black*, a useful recombination suppressor, *In(2)2*, and a balanced lethal for chromosome 2 of the mosquito *Anopheles gambiae*, *Am. J. Trop. Med. Hyg.*, 61, 618, 1999.

Berg, D. E. and Howe, M. M., *Mobile DNA*, American Society for Microbiology, Washington, D.C., 1989.

Berghammer, A. J., Klingler, M., and Wimmer, E. A., A universal marker for transgenic insects, *Nature*, 402, 370, 1999.

Biemont, C. and Cizeron, G., Distribution of transposable elements in *Drosophila* species, *Genetica*, 105, 43, 1999.

Billingsley, P. F. and Rudin, W., The role of the mosquito peritrophic membrane in bloodmeal digestion and infectivity of *Plasmodium* species, *J. Parasitol.*, 78, 430, 1992.

Bond, H. A., Craig, G. B., Jr., and Fay, R. W., Field mating and movement of *Aedes aegypti*, *Mosq. News*, 30, 394, 1970.

Bonnivard, E., Bazin, C., Denis, B., and Higuet, D., A scenario for the *hobo* transposable element invasion, deduced from the structure of natural populations of *Drosophila melanogaster* using tandem *TPE* repeats, *Genet. Res.*, 75, 13, 2000.

Bouchon, D., Rigaud, T., and Juchault, P., Evidence for widespread *Wolbachia* infection in isopod crustaceans: molecular identification and host feminization, *Proc. R. Soc. London B Biol. Sci.*, 265, 1081, 1998.

Bourtzis, K. and Braig, H. R., The many faces of *Wolbachia*, in *Rickettsiae and Rickettsial Diseases at the Turn of the Third Millennium*, Raoult, D. and Brouqui, P., Eds., Elsevier, Paris, 1999, 199.

Bourtzis, K., Dobson, S. L., Braig, H. R., and O'Neill, S. L., Rescuing *Wolbachia* have been overlooked, *Nature*, 391, 852, 1998.

Boussy, I. A., A *Drosophila* model of improving the fitness of translocations for genetic control, 1. Autosomal translocations with euchromatic breakpoints, *Theor. Appl. Genet.*, 76, 627, 1988.

Braig, H. R., Guzman, H., Tesh, R. B., and O'Neill, S. L., Replacement of the natural *Wolbachia* symbiont of *Drosophila simulans* with a mosquito counterpart, *Nature*, 367, 453, 1994.

Braig, H. R., Zhou, W., Dobson, S. L., and O'Neill, S. L., Cloning and characterization of a gene encoding the major surface protein of the bacterial endosymbiont *Wolbachia pipientis*, *J. Bacteriol.*, 180, 2373, 1998.

Breeuwer, J. A. J. and Jacobs, G., *Wolbachia*: intracellular manipulators of mite reproduction, *Exp. Appl. Acarol.*, 20, 421, 1996.

Brown, S. D. M. and Dover, G., Organization and evolutionary progress of a dispersed repetitive family of sequences in widely separated rodent genomes, *J. Mol. Biol.*, 150, 441, 1981.

Callaini, G., Riparbelli, M. G., Giordano, R., and Dallai, R., Mitotic defects associated with cytoplasmic incompatibility in *Drosophila simulans*, *J. Invertebr. Pathol.*, 67, 55, 1996.

Callaini, G., Dallai, R., and Riparbelli, M. G., *Wolbachia*-induced delay of paternal chromatin condensation does not prevent maternal chromosomes from entering anaphase in incompatible crosses of *Drosophila simulans*, *J. Cell Sci.*, 110, 271–280, 1997.

Camacho, J. P. M., Sharbel, T. F., and Beukeboom, L. W., B-chromosome evolution, *Philos. Trans. R. Soc. London B.*, 355, 163, 2000.

Cantelo, W. W. and Childress, D., Laboratory and field studies with a compound chromosome strain of *Drosophila melanogaster*, *Theor. Appl. Genet.*, 45, 1, 1974.

Capillon, C. and Atlan, A., Evolution of driving X chromosomes and resistance factors in experimental populations of *Drosophila simulans*, *Evolution*, 53, 506, 1999.

Caprio, M. A., Hoy, M. A., and Tabashnik, B. E., A model for implementing a genetically improved strain of the parasitoid *Trioxys pallidus* Haliday (Hymenoptera: Aphidiidae), *Am. Entomol.*, 34, 232, 1991.

Carareto, C. M. A., Kim, W., Wojciechowski, M. F., O'Grady, P., Prokchorova, A. V., Silva, J. C., and Kidwell, M. G., Testing transposable elements as genetic drive mechanisms using *Drosophila P* element constructs as a model system, *Genetica*, 101, 13, 1997.

Carvalho, A. B. and Vaz, S. C., Are *Drosophila SR* drive chromosomes always balanced? *Heredity*, 83, 221, 1999.

Carvalho, A. B., Sampaio, M. C., Varandas, F. R., and Klaczko, L. B., An experimental demonstration of Fisher's principle: evolution of sexual proportion by natural selection, *Genetics*, 148, 719, 1998.

Catteruccia, F., Nolan, T., Loukeris, T. G., Blass, C., Savakis, C., Kafatos, F. C., and Crisanti, A., Stable germline transformation of the malaria mosquito *Anopheles stephensi*, *Nature*, 405, 959, 2000a.

Catteruccia, F., Nolan, T., Blass, C., Müller, H. M., Crisanti, A., Kafatos, F. C., and Loukeris, T. G., Toward *Anopheles* transformation: *Minos* element activity in anopheline cells and embryos, *Proc. Natl. Acad. Sci. U.S.A.*, 97 (5), 2157, 2000b.

Cayol, J. P. and Zaraï, M., Field releases of two genetic sexing strains of the Mediterranean fruit fly (*Ceratitis capitata* Wied.) in two isolated oases of Tozeur governorate, Tunisia, *J. Appl. Entomol.*, 123, 613, 1999.

Cazemajor, M., Joly, D., and Montchamp-Moreau, C., Sex-ratio meiotic drive in *Drosophila simulans* is related to equational nondisjunction of the Y chromosome, *Genetics*, 154, 229, 2000.

Charlwood, J. D., Kihonda, J., Sama, S., Billingsley, P. F., Hadji, H., Verhave, J. P., Lyimo, E., Luttikhuizen, P. C., and Smith, T., The rise and fall of *Anopheles arabiensis* (Diptera: Culicidae) in a Tanzanian village, *Bull. Entomol. Res.*, 85, 37, 1995.

Cheng, Q., Ruel, T. D., Zhou, W., Moloo, S. K., Majiwa, P., O'Neill, S. L., and Aksoy, S., Tissue distribution and prevalence of *Wolbachia* infections in tsetse flies, *Glossina* spp., *Med. Vet. Entomol.*, 14, 44, 2000.

Christophides, G. K., Savakis, C., Mintzas, A. C., and Komitopoulou, K., Expression and function of the *Drosophila melanogaster* ADH in male *Ceratitis capitata* adults: a potential strategy for medfly genetic sexing based on gene-transfer technology, *Insect Mol. Biol.*, 10, 249, 2001.

Clancy, D. J. and Hoffmann, A. A., Behavior of *Wolbachia* endosymbionts from *Drosophila simulans* in *D. serrata*, a novel host, *Am. Nat.*, 149, 975, 1997.

Clark, J. B., Maddison, W. P., and Kidwell, M. G., Phylogenetic analysis supports horizontal transfer of *P* transposable elements, *Mol. Biol. Evol.*, 11, 40, 1994.

Coates, C. J., Jasinskiene, N., Miyashiro, L., and James, A. A., Mariner transposition and transformation of the yellow fever mosquito, *Aedes aegypti*, *Proc. Natl. Acad. Sci. U.S.A.*, 95, 3748, 1998.

Coleman, P. G., Goodman, C. A., and Mills, A., Rebound mortality and the cost-effectiveness of malaria control: potential impact of increased mortality in late childhood following the introduction of insecticide treated nets, *Trop. Med. Int. Health*, 4, 175, 1999.

Collins, F. H. and Besansky, N. J., Vector biology and the control of malaria in Africa, *Science*, 264, 1874, 1994.

Collins, F. H. and Paskewitz, S. M., A review of the use of ribosomal DNA (rDNA) to differentiate among cryptic *Anopheles* species, *Insect Mol. Biol.*, 5, 1, 1996.

Collins, F. H., Sakai, R. K., Vernick, K. D., Paskewitz, S., Seeley, D. C., Miller, L. H., Collins, W. E., Campbell, C. C., and Gwadz, R. W., Genetic selection of a *Plasmodium*-refractory strain of the malaria vector *Anopheles gambiae*, *Science*, 234, 607, 1986.

Coluzzi, M., Sabatini, A., Petrarca, V., and Di Deco, M. A., Chromosomal differentiation and adaptation to human environments in the *Anopheles gambiae* complex, *Trans. R. Soc. Trop. Med. Hyg.*, 73, 483, 1979.

Cousserans, J. and Guille, G., Expérience de lutte génétique contra *Culex pipiens* dans la region de Montpellier. Synthèse de quatre années d'observations, *Bull. Biolog.*, 108, 253, 1974.

Coyne, J. A., Meyers, W., Crittenden, A. P., and Sniegowski, P., The fertility effects of pericentric inversions in *Drosophila melanogaster*, *Genetics*, 134, 487, 1993.

Craig, G. B., Hickey, W. A., and Vandehey, R. C., An inherited male-producing factor in *Aedes aegypti*, *Science*, 123, 1887, 1960.

Cuisance, D., Politzar, H., Clair, H. M., Sellin, E., and Taze, Y., Impact des lachers des males steriles sur les niveaux de deux populations sauvages de *Glossina palpalis gambiensis* en Haute-Volta (sources de la Volta Noire), *Rev. Elev. Med. Vet. Pays Trop.*, 31, 315, 1978.

Curtis, C. F., Possible use of translocations to fix desirable genes in insect pest populations, *Nature*, 218, 368, 1968.

Curtis, C. F., Male-linked translocations and the control of insect pest populations, *Experientia*, 31, 1139, 1975.

Curtis, C. F., Genetic control of insect pests: growth industry or lead balloon? *Biol. J. Linnean Soc.*, 26, 359, 1985.

Curtis, C. F., Selfish genes in mosquitoes, *Nature*, 357, 450, 1992.

Curtis, C. F., The case for malaria control through the genetic manipulation of its vectors, *Parasitol. Today*, 10, 371, 1994.

Curtis, C. F. and Graves, P. M., Methods for replacement of malaria vector populations, *J. Trop. Med. Hyg.*, 91, 43, 1988.

Curtis, C. F. and Sinkins, S. P., *Wolbachia* as a possible means of driving genes into populations, *Parasitology*, 116 (Suppl.), S111, 1998.

Curtis, C. F. and von Borstel, R. C., Allegations against Indian research unit refuted, *Nature*, 273, 96, 1978.

Curtis, C. F., Grover, K. K., Suguna, S. G., Uppal, D. K., Dietz, K., Agarwal, H. V., and Kazmi, S. J., Comparative field cage tests of the population suppressing efficiency of 3 genetic control systems for *Aedes aegypti*, *Heredity*, 36, 11, 1976.

Curtis, C. F., Brooks, G. D., Ansari, M. A., Grover, K. K., Krishnamurthy, B. S., Rajagopalan, P. K., Sharma, L. S., Sharma, V. P., Singh, D., Singh, K. R. P., and Yasuno, M., A field trial on control of *Culex quinquefasciatus* by release of males of a strain integrating cytoplasmic incompatibility and a translocation, *Entomol. Exp. Appl.*, 31, 181, 1982.

Dame, D. A., Woodard, D. B., Ford, H. R., and Weidhaas, D. E., Field behavior of sexually sterile *Anopheles quadrimaculatus* males, *Mosq. News*, 24, 6, 1964.

Dame, D. A., Lowe, R. E., and Williamson, D. L., Assessment of released sterile *Anopheles albimanus* and *Glossina morsitans morsitans*, in *Cytogenetics and Genetics of Vectors, Proceedings of the XVI International Congress of Entomology*, Kyoto, 1980, Pal, R., Kitzmiller, J. B., and Kanda, T., Eds., Elsevier, Amsterdam, 231, 1981.

Daniels, S. B., McCarron, M., Love, C., Clark, S. H., and Chovnick, A., Dysgenesis induced instability of *rosy* locus transformation in *Drosophila melanogaster*: analysis of excision events and the selective recovery of control element deletions, *Genetics*, 109, 95, 1985.

Daniels, S. B., Peterson, K. P., Strausbaugh, L. D., Kidwell, M. G., and Chovnick, A., Evidence for horizontal transfer of the P transposable element between *Drosophila* species, *Genetics*, 124, 339, 1990a.

Daniels, S. B., Chovnick, A., and Boussy, I. A., Distribution of *hobo* transposable elements in the genus *Drosophila*, *Mol. Biol. Evol.*, 7, 589, 1990b.

Davidson, G., *The Genetic Control of Insect Pests*, Academic Press, New York, 1974.

Davidson, G., Odetoyinbo, J. A., Colussa, B., and Coz, J., A field attempt to assess the mating competitiveness of sterile males produced by crossing 2 member species of the *Anopheles gambiae* complex, *Bull. World Health Org.*, 42, 55, 1970.

Davidson, G., Curtis, C. F., White, G. B., and Rawlings, P., Mosquito war games, *Nature*, 296, 700, 1982.

Davis, S. A., Catchpole, E. A., and Pech, R. P., Models for the introgression of a transgene into a wild population within a stochastic environment, with applications to pest control, *Ecol. Model.*, 119, 267, 1999.

Davis, S. A., Catchpole, E. A., and Fulford, G. R., Periodic triggering of an inducible gene for control of a wild population, *Theor. Popul. Biol.*, 58, 95, 2000.

de La Roche Saint André, C. and Bregliano, J.-C., Evidence for a multistep control in transposition of I factor in *Drosophila melanogaster*, *Genetics*, 148, 1875, 1998.

Dernburg, A. F., Daily, D. R., Yook, K. J., Corbin, J. A., Sedat, J. W., and Sullivan, W., Selective loss of sperm bearing a compound chromosome in the *Drosophila* female, *Genetics*, 143, 1629, 1996.

DeVault, J. D., Hughes, K. J., Leopold, R. A., Johnson, O. A., and Narang, S. K., Gene transfer into corn earworm (*Helicoverpa zea*) embryos, *Genome Res.*, 6, 571, 1996.

Dobson, S. L. and Tanouye, M. A., The paternal sex ratio chromosome induces chromosome loss independently of *Wolbachia* in the wasp *Nasonia vitripennis*, *Dev. Genes Evol.*, 206, 207, 1996.

Dobson, S. L. and Tanouye, M. A., Evidence for a genomic imprinting sex determination mechanism in *Nasonia vitripennis* (Hymenoptera; Chalcidoidea), *Genetics*, 149, 233, 1998a.

Dobson, S. L. and Tanouye, M. A., Interspecific movement of the paternal sex ratio chromosome, *Heredity*, 81, 261, 1998b.

Dobson, S. L., Bourtzis, K., Braig, H. R., Jones, B. F., Zhou, W., Rousset, F., and O'Neill, S. L., *Wolbachia* infections are distributed throughout insect somatic and germ line tissue, *Insect Biochem. Mol. Biol.*, 29, 153, 1999.

Dover, G. A., Molecular drive: a cohesive mode of species evolution, *Nature*, 229, 111, 1982.

Dyck, V. A., Juma, K. G., Msangi, A. R., Saleh, K. M., Kiwia, N., Vreysen, M. J. B., Parker, A. G., Hendrichs, J., and Feldmann, U., Eradication of the tsetse fly *Glossina austeni* by the Sterile Insect Technique (SIT) in Zanzibar: could South Africa be next? in *Insects in African Economy and Environment*, Robertson, H. G., Ed., Entomological Society of Southern Africa, Pretoria, 1997, 168.

Economopoulos, A. P., Quality control and SIT field testing with genetic sexing Mediterranean fruit fly males, in *Fruit Fly Pests: A World Assessment of Their Biology and Management*, McPheron, B. A. and Steck, G. J., Eds., St. Lucie Press, Delray Beach, FL, 1996, 385.

Edwards, O. R. and Hoy, M. A., Random amplified polymorphic DNA markers to monitor laboratory-selected, pesticide-resistant *Trioxys pallidus* (Hymenoptera: Aphidiidae) after release into three California walnut orchards, *Environ. Entomol.*, 24, 487, 1995.

Eichner, M. and Kiszewski, A., Ribeiro and Kidwell's transposon model, *J. Med. Entomol.*, 32, 1, 1995.

Ellis, B. A., Mills, J. N., Childs, J. E., Muzzini, M. C., McKee, K. T., Enria, D. A., and Glass, G. E., Structure and floristics of habitats associated with five rodent species in an agroecosystem in Central Argentina, *J. Zool.*, 243, 437, 1997.

Enkerli, J., Reed, H., Briley, A., Bhatt, G., and Covert, S. F., Physical map of a conditionally dispensable chromosome in *Nectria haematococca* mating population VI and location of chromosome breakpoints, *Genetics*, 155, 1083, 2000.

Eskelinen, O., On the population fluctuations and structure of the wood lemming *Myopus schisticolor*, *Z. Saugetierkd.*, 62, 293, 1997.

Fawcett, D. H., Liste, C. K., Kellet, E., and Finnegan, D. J., Transposable elements controlling I-R hybrid dysgenesis in *Drosophila melanogaster* are similar to mammalian LINEs, *Cell*, 47, 1007, 1986.

Feldmann, A. M., Billingsley, P. F., and Savelkoul, E., Bloodmeal digestion by strains of *Anopheles stephensi* liston (Diptera: Culicidae) of differing susceptibility to *Plasmodium falciparum*, *Parasitology*, 101, 193, 1990.

Feldmann, A. M., van Gemert, G.-J., van de Vegte-Bolmer, M. G., and Jansen, R. C., Genetics of refractoriness to *Plasmodium falciparum* in the mosquito *Anopheles stepheni*, *Med. Vet. Entomol.*, 12, 302, 1998.

Ferdig, M. T., Beerntsen, B. T., Spray, F. J., Li, J., and Christensen, M. M., Reproductive costs associated with resistance in a mosquito-filirial worm system, *Am. J. Trop. Med. Hyg.*, 49, 756, 1993.

Fitz-Earle, M. and Barclay, H. J., Maximum-likelihood fitness estimates from population replacement experiments with compound autosomes and compound-free arm combinations, *Jpn. J. Genet.*, 59, 612, 1984.

Fitz-Earle, M. and Holm, D. G., Exploring the potential of compound free arm combinations of chromosome 2 in *Drosophila melanogaster* for insect control and the survival to pupae of whole arm trisomies, *Genetics*, 89, 499, 1978.

Fitz-Earle, M. and Suzuki, D. T., Conditional mutations for the control of insect populations, in *Sterility Principle for Insect Control*, International Atomic Energy Agency, Vienna, 1975, 365.

Fleuriet, A., Polymorphism of the *Drosophila melanogaster* — Sigma virus system, *J. Evol. Biol.*, 9, 471, 1996.

Foster, G. G., The use of bridging systems to increase genetic variability in compound chromosome strains for genetic control of *Lucilia cuprina*, *Theor. Appl. Genet.*, 63, 295, 1982.

Foster, G. G., Simulation of genetic control — homozygous viable pericentric inversions in field female killing systems, *Theor. Appl. Genet.*, 82, 368, 1991.

Foster, G. G. and Maddern, R. H., Segregation and pairing of compound fifth chromosomes in *Lucilia cuprina* males, *Genet. Res.*, 46, 149, 1985.

Foster, G. G. and Smith, P. H., Genetic control of *Lucilia cuprina*, analysis of field trial data using simulation techniques, *Theor. Appl. Genet.*, 82, 33, 1991.

Foster, G. G., Maddern, R. H., and Mills, A. T., Genetic instability in mass rearing colonies of a sex-linked translocation strain of *Lucilia cuprina* (Diptera: Calliphoridae) during a field trial of genetic control, *Theor. Appl. Genet.*, 58, 169, 1980.

Foster, G. G., Vogt, W. G., and Woodburn, T. L., Genetic analysis of field trials of sex-linked translocation strains for genetic control of the Australian sheep blowfly *Lucilia cuprina*, *Aust. J. Biol. Sci.*, 38, 275, 1985a.

Foster, G. G., Maddern, R. H., Helman, R. A., and Reed, E. M., Field trial of a compound chromosome strain for genetic control of the sheep blowfly *Lucilia cuprina, Theor. Appl. Genet.,* 70, 13, 1985b.

Foster, G. G., Weller, G. L., and Bedo, D. G., Homozygous-viable pericentric inversions for genetic control of *Lucilia cuprina, Theor. Appl. Genet.,* 82, 681, 1991a.

Foster, G. G., Weller, G. L., and Clarke, G. M., Male crossing over and genetic sexing systems in the Australian sheep blowfly, *Lucilia cuprina, Heredity,* 67, 365, 1991b.

Foster, J. E. and Gallun, R. L., Populations of the Eastern races of the Hessian fly controlled by release of dominant avirulent Great Plains race, *Ann. Entomol. Soc. Am.,* 65, 750, 1972.

Freier, J. E. and Friedman, S., Effect of host infection with *Plasmodium gallinaceum* on the reproductive capacity of *Aedes aegypti, J. Invertebr. Pathol.,* 28, 161, 1976.

Fryxell, K. J. and Miller, T. A., Autocidal biological control: a general strategy for insect control based on genetic transformation with a highly conserved gene, *J. Econ. Entomol.,* 88, 1221, 1995.

Galvin, T. J. and Wyss, J. H., Screwworm eradication program in Central America, *Ann. N.Y. Acad. Sci.,* 791, 233, 1996.

Gass, R. F. and Yeates, R. A., *In vitro* damage of cultured ookinetes of *Plasmodium gallinaceum* by digestive proteinases from susceptible *Aedes aegypti, Acta Trop.,* 36, 243, 1979.

Ginzburg, L. R., Bingham, P. M., and Yoo, S., On the theory of speciation induced by transposable elements, *Genetics,* 107, 331, 1984.

Githeko, A. K., Service, M. W., Mbogo, C. M., Atieli, F. K., and Juma, F. O., *Plasmodium falciparum* sporozoite and entomological inoculation rates at the Ahero rice irrigation scheme and the Miwani sugar-belt in western Kenya, *Ann. Trop. Med. Parasitol.,* 87, 379, 1993.

Githeko, A. K., Service, M. W., Mbogo, C. M., and Atieli, F. K., Resting behaviour, ecology and genetics of malaria vectors in large scale agricultural areas of Western Kenya, *Parasitologia,* 38, 481, 1996.

Good, A. G., Meister, G. A., Brook, H. W., Grigliatti, T. A., and Hickey, D. A., Rapid spread of transposable *P* elements in experimental populations of *Drosophila melanogaster, Genetics,* 122, 387, 1989.

Gordon, R. M., Notes on the bionomics of *Stegomyia calopus* in Brazil, II, *Ann. Trop. Med. Parasitol.,* 16, 425, 1922.

Grant, G. H., Snow, W., and Vargas, M., A screwworm eradication programme for Jamaica and other Caribbean nations, in *FAO/IAEA International Conference on Area-Wide Control of Insect Pests Integrating the Sterile Insect and Related Nuclear and Other Techniques,* Pengang, Malaysia, 28 May–2 June, 1998, 21.

Green, M. M., Mapping a *Drosophila melanogaster* "controlling element" by interallelic crossing over, *Genetics,* 61, 423, 1969a.

Green, M. M., Controlling element mediated transpositions of the *white* gene in *Drosophila melanogaster, Genetics,* 61, 429, 1969b.

Grenier, S., Pintureau, B., Heddi, A., Lassblière, F., Jager, C., Louis, C., and Khatchadourian, C., Successful horizontal transfer of *Wolbachia* symbionts between *Trichogramma* wasps, *Proc. R. Soc. London B Biol. Sci.,* 265, 1441, 1998.

Grigliatti, T. A., Transposons — gene tagging and mutagenesis, in *Drosophila — A Practical Approach,* Roberts, D. B., Ed., Oxford University Press, Oxford, U.K., 1998, 84.

Grimaldi, M. C. and Crouau-Roy, B., Microsatellite allelic homoplasy due to variable flanking sequences, *J. Mol. Evol.,* 44, 336, 1997.

Grossniklaus, U., Vielle-Calzada, J.-P., Hoeppner, M. A., and Gagliano, W. B., Maternal control of embryogenesis by *MEDEA,* a *polycomb* group gene in *Arabidopsis, Science,* 280, 446, 1998.

Grover, K. K., Suguna, S. G., Uppal, D. K., Singh, K. R. P., Ansari, M. A., Curtis, C. F., Singh, D., Sharma, V. P., and Panicker, K. N., Field experiments on the competitiveness of males carrying genetic control systems for *Aedes aegypti, Entomol. Exp. Appl.,* 20, 8, 1976.

Gubler, D. J., Competitive displacement of *Aedes (Stegomyia) polynesiensis* Marks by *Aedes (Stegomyia) albopictus* Skuse in laboratory populations, *J. Med. Entomol.,* 7, 229, 1970.

Gubler, D. J., Resurgent vector-borne diseases as a global health problem, *Emerg. Infect. Dis.,* 4, 442–50, 1998.

Hacker, C. S. and Kilama, W. L., The relationship between *Plasmodium gallinaceum* density and the fecundity of *Aedes aegypti, J. Invertebr. Pathol.,* 23, 101, 1974.

Hagemann, S., Haring, E., and Pinsker, W., Repeated horizontal transfer of *P* transposons between *Scaptomyza pallida* and *Drosophila bifasciata, Genetica,* 98, 43, 1996.

Hamilton, W. D., Extraordinary sex ratios, *Science*, 156, 477, 1967.

Handler, A. M. and Harrell II, R. A, Germline transformation of *Drosophila melanogaster* with the *piggyBac* transposon vector, *Insect Mol. Biol.*, 8, 449, 1999.

Handler, A. M. and Harrell II, R. A., Transformation of the Caribbean fruit fly, *Anastrepha suspensa*, with a *piggyBac* vector marked with polyubiquitin-regulated GFP, *Insect Biochem. Mol. Biol.*, 31, 199, 2001.

Handler, A. M. and James, A. J., Eds., *Insect Transgenesis: Methods and Applications*, CRC Press, Boca Raton, FL, 2000.

Handler, A. M. and McCombs, S. D., The *piggyBac* transposon mediates germ-line transformation in the Oriental fruit fly and closely related elements exist in its genome, *Insect Mol. Biol.*, 9, 605, 2000.

Handler, A. M., McCombs, S. D., Fraser, M. J., and Saul, S. H., The lepidopteran transposon vector, *piggyBac*, mediates germ-line transformation in the Mediterranean fruit fly, *Proc. Natl. Acad. Sci. U.S.A.*, 95, 7520, 1998.

Hariri, A. R., Werren, J. H., and Wilkinson, G. S., Distribution and reproductive effects of *Wolbachia* in stalk-eyed flies (Diptera, Diopsidae), *Heredity*, 81, 254, 1998.

Harshman, L. G. and Prout, T., Synthesis of an attached autosome C-3-RM in *Drosophila pseudo-obscura*, *Genetica*, 80, 87, 1990.

Hartl, D., Lohe, A. R., and Lozovskaya, E. R., Regulation of the transposable element *mariner*, *Genetica*, 100, 177, 1997.

Hastings, I. M., Selfish DNA as a method of pest control, *Philos. Trans. R. Soc. London B*, 344, 313, 1994.

Hatcher, M. J., Persistence of selfish genetic elements: population structure and conflict, *Trends Ecol. Evol.*, 15, 271, 2000.

Heath, B. D., Butcher, R. D. J., Whitfield, W. G. F., and Hubbard, S. F., Horizontal transfer of *Wolbachia* between phylogenetically distant insect species by a naturally occurring mechanism, *Curr. Biol.*, 9, 313, 1999.

Hediger, M., Niessen, M., Wimmer, E. A., Dubendorfer, A., and Bopp, D., Genetic transformation of the housefly *Musca domestica* with the lepidopteran derived transposon *piggyBac*, *Insect Mol. Biol.*, 10, 113, 2001.

Heinrich, J. J. and Scott, M. J., A repressible female-specific lethal genetic system for making transgenic insect strains suitable for a sterile-release program, *Proc. Natl. Acad. Sci. U.S.A.*, 97, 8229, 2000.

Hickey, W. A. and Craig, G. B., Genetic distortion of sex ratio in mosquito, *Aedes aegypti*, *Genetics*, 53, 1177, 1966.

Higuet, D., Merçot, H., Allouis, S., and Montchamp-Moreau, C., The relationship between structural variation and dysgenic properties of P elements in long-established P-transformed lines of *Drosophila simulans*, *Heredity*, 77, 9, 1996.

Hillis, D. M., Moritz, C., Porter, C. A., and Baker, R. J., Evidence for biased gene conversion in concerted evolution of ribosomal DNA, *Science*, 251, 308, 1991.

Hiraizumi, Y., Spontaneous recombination in *Drosophila melanogaster* males, *Proc. Natl. Acad. Sci. U.S.A.*, 68, 268, 1971.

Hoerauf, A., Nissen-Pahle, K., Schmetz, C., Henkle-Duhrsen, K., Blaxter, M. L., Buttner, D. W., Gallin, M. Y., Al-Qaoud, K. M., Lucius, R., and Fleischer, B., Tetracycline therapy targets intracellular bacteria in the filarial nematode *Litomosoides sigmodontis* and results in filarial infertility, *J. Clin. Invest.*, 103, 11, 1999.

Hoffmann, A. A. and Turelli, M., Cytoplasmic incompatibility in insects, in *Influential Passengers: Inherited Microorganisms and Arthropod Reproduction*, O'Neill, S. L., Hoffman, A. A., and Werren, J. H., Eds., Oxford University Press, Oxford, U.K., 1997, 42.

Hoshizaki, S. and Shimada, T., PCR-based detection of *Wolbachia*, cytoplasmic incompatibility microorganisms, infected in natural populations of *Laodelphax striatellus* (Homoptera: Delphacidae) in central Japan: has the distribution of *Wolbachia* spread recently? *Insect Mol. Biol.*, 4, 237, 1995.

Houck, M. A., Clark, J. B., Peterson, K. R., and Kidwell, M. G., Possible horizontal transfer of *Drosophila* genes by the mite *Proctolaelaps regalis*, *Science*, 253, 1125, 1991.

Hoy, M. A., Genetic improvement of arthropod natural enemies: becoming a conventional tactic? in *New Directions in Biological Control: Alternatives for Suppressing Agricultural Pests and Diseases*, Baker, R. R. and Dunn, P. E., Eds., Alan R. Liss, New York, 1990a, 405.

Hoy, M. A., Pesticide resistance in arthropod natural enemies: variability and selection responses, in *Pesticide Resistance in Arthropods*, Roush, R. T. and Tabashnik, B. E., Eds., Chapman & Hall, New York, 1990b, 203.

Hoy, M. A., Use of parasites and predators in the biological control of arthropod pests: emerging technologies and challenges, in *Entomology Serving Society: Emerging Technologies and Challenges*, Vinson, S. B. and Metcalf, R. L., Eds., Entomological Society of America, Lanham, MD, 1991, 272.

Hoy, M. A., Impact of risk analyses on pest management programs employing transgenic arthropods, *Parasitol. Today*, 11, 229, 1995.

Hoy, M. A., Challenges for the safe use of transgenic arthropod natural enemies for pest management programs, *Phytoprotection*, 79 (Suppl.), 139, 1998.

Hoy, M. A., Deploying transgenic arthropods in pest management programs: risks and realities, in *Insect Transgenesis: Methods and Applications*, Handler, A. M. and James, A. J., Eds., CRC Press, Boca Raton, FL, 2000, 335.

Hughes, R. N., *A Functional Biology of Clonal Animals*, Chapman & Hall, London, 1989.

Huigens, M. E., Luck, R. F., Klaassen, R. H. G., Maas, M. F. P. M., Timmermans, M. J. T. N., and Stouthamer, R., Infectious parthenogenesis, *Nature*, 405, 178, 2000.

Hurst, G. D. D. and Schilthuizen, M., Selfish genetic elements and speciation, *Heredity*, 80, 2, 1998.

Hurst, G. D. D., Hurst, L. D., and Majerus, M. E. N., Cytoplasmic sex-ratio distorters, in *Influential Passengers: Inherited Microorganisms and Arthropod Reproduction*, O'Neill, S. L., Hoffman, A. A., and Werren, J. H., Eds., Oxford University Press, Oxford, U.K., 1997, 125.

Hurst, G. D. D., Jiggins, F. M., von der Schulenburg, J. H. G., Bertrand, D., West, S. A., Goriacheva, I. I., Zakharov, I. A., Werren, J. H., Stouthamer, R., and Majerus, M. E. N., Male-killing *Wolbachia* in two species of insect, *Proc. R. Soc. London B Biol. Sci.*, 266, 735, 1999.

Hurst, L. D., *scat+* is a selfish gene analogous to *Medea* of *Tribolium castaneum*, *Cell*, 75, 407–408, 1993a.

Hurst, L. D., The incidences, mechanisms and evolution of cytoplasmic sex ratio distorters in animals, *Biol. Rev. Cambridge Philos. Soc.*, 68, 121–193, 1993b.

Ishihara, M., Persistence of abnormal females that produce only female progeny with occasional recovery to normal females in Lepidoptera, *Res. Popul. Ecol.*, 36, 261, 1994.

James, A. A., Mosquito molecular genetics: the hands that feed bite back, *Science*, 257, 37, 1992.

James, A. A., Control of disease transmission through genetic modification of mosquitoes, in *Insect Transgenesis: Methods and Applications*, Handler, A. M. and James, A. J., Eds., CRC Press, Boca Raton, FL, 2000, 319.

Jasinskiene, N., Coates, C. J., Benedict, M. Q., Cornel, A. J., Rafferty, C. S., James, A. A., and Collins, F. H., Stable transformation of the yellow fever mosquito, *Aedes aegypti*, with the *Hermes* element from the housefly, *Proc. Natl. Acad. Sci. U.S.A.*, 95, 3743, 1998.

Jayaraman, K. S., Foreign labs shut, *Nature*, 296, 104, 1982.

Jehle, J. A., Nickel, A., Vlak, J. M., and Backhaus, H., Horizontal escape of the novel *Tc1*-like lepidopteran transposon *TCp3.2* into *Cydia pomonella* granulovirus, *J. Mol. Evol.*, 46, 215, 1998.

Jeyaprakash, A., Lopez, G., and Hoy, M. A., Extrachromosomal plasmid DNA transmission and amplification in *Metaseiulus occidentalis* (Acari: Phytoseiidae) transformants generated by maternal microinjection, *Ann. Entomol. Soc. Am.*, 91, 730, 1998.

Jiggins, F. M., Hurst, G. D. D., and Majerus, M. E. N., How common are meiotically driving sex chromosomes in insects? *Am. Nat.*, 154, 481, 1999.

Jiggins, F. M., Hurst, G. D. D., and Majerus, M. E. N., Sex-ratio-distorting *Wolbachia* causes sex-role reversal in its butterfly host, *Proc. R. Soc. London B Biol. Sci.*, 267, 69, 2000a.

Jiggins, F. M., Hurst, G. D. D., Dolman, C. E., and Majerus, M. E. N., High-prevalence male-killing *Wolbachia* in the butterfly *Acraea encedana*, *J. Evol. Biol.*, 13, 495, 2000b.

Jiggins, F. M., Hurst, G. D. D., Jiggins, C. D., von der Schulenburg, J. H. G., and Majerus, M. E. N., The butterfly *Danaus chrysippus* is infected by a male-killing *Spiroplasma* bacterium, *Parasitology*, 120, 439, 2000c.

Johanowicz, D. L. and Hoy, M. A., Experimental induction and termination of non-reciprocal reproductive incompatibilities in a parahaploid mite, *Entomol. Exp. Appl.*, 87, 51, 1998.

Johnson, B. W., Olson, K. E., Allen-Miura, T., Rayms-Keller, A., Carlson, J. O., Coates, C. J., Jasinskiene, N., James, A. A., Beaty, B. J., and Higgs, S., Inhibition of luciferase expression in transgenic *Aedes aegypti* mosquitoes by Sindbis virus expression of antisense luciferase RNA, *Proc. Natl. Acad. Sci. U.S.A.*, 96, 13399, 1999.

Jones, R. N., B chromosome drive, *Am. Nat.*, 137, 430, 1991.

Jones, R. N. and Rees, H., *B Chromosomes*, Academic Press, New York, 1982.

Joshi, G. P., Service, M. W., and Pradhan, G. D., A survey of species A and B of the *Anopheles gambiae* Giles complex in the Kisumu area of Kenya prior to insecticidal spraying with OMS-43 (fenitrothion), *Ann. Trop. Med. Parasitol.*, 69, 91, 1975.

Kaiser, P. E., Seawright, J. A., Benedict, M. Q., Narang, S., and Suguna, S. G., Radiation induced reciprocal translocations and inversions in *Anopheles albimanus, Can. J. Genet. Cytol.*, 24, 177, 1982.

Kaiser, P. E., Seawright, J. A., Benedict, M. Q., and Narang, S., Homozygous translocations in *Anopheles albimanus, Theor. Appl. Genet.*, 65, 207, 1983.

Kakinohana, H., Kuba, H., Kohama, T., Kinjo, K., Taniguchi, M., Nakamori, H., Tanahara, A., and Sokei, Y., Eradication of the melon fly, *Bactrocera cucurbitae* Coquillett, by mass release of sterile flies in Okinawa Prefecture, Japan, *Jpn. Agric. Res. Q.*, 31, 91, 1997.

Kerremans, P. and Franz, G., Use of a temperature-sensitive lethal mutation strain of medfly (*Ceratitis capitata*) for the suppression of pest populations, *Theor. Appl. Genet.*, 90, 511, 1995a.

Kerremans, P. and Franz, G., Isolation and cytogenetic analyses of genetic sexing strains for the medfly (*Ceratitis capitata*), *Theor. Appl. Genet.*, 91, 255, 1995b.

Kidwell, M. G., Evolution of hybrid dysgenesis determinants in *Drosophila melanogaster, Proc. Natl. Acad. Sci. U.S.A.*, 80, 1655, 1983.

Kidwell, M. G., Regulatory aspects of the expression of P-M hybrid dysgenesis in *Drosophila*, in *Eukaryotic Transposable Elements as Mutagens*, Lambert, M. E., McDonald, J. F., and Weinstein, I. B., Eds., Cold Spring Harbor Laboratory Press, Cold Spring Harbor, NY, 1987, 183.

Kidwell, M. G., Lateral transfer in natural populations of eukaryotes, *Annu. Rev. Genet.*, 27, 235, 1993.

Kidwell, M. G. and Ribeiro, J. M. C., Can transposable elements be used to drive disease refractoriness genes into vector populations? *Parasitol. Today*, 8, 325, 1992.

Kidwell, M. G., Kidwell, J. F., and Sved, J. A., Hybrid dysgenesis in *Drosophila melanogaster* — a syndrome of aberrant traits including mutation sterility and male recombination, *Genetics*, 86, 813, 1977.

Kidwell, M. G., Novy, J. B., and Feeley, S. M., Rapid unidirectional change of hybrid dysgenesis potential in *Drosophila, J. Hered.*, 72, 32, 1981.

Kilama, W. L., Distribution of a gene for susceptibility to *Plasmodium gallinaceum* in populations of *Aedes aegypti* (L.), *J. Parasitol.*, 59, 920, 1973.

Kilama, W. L. and Craig, G. B., Monofactorial inheritance of susceptibility to *Plasmodium gallinaceum* in *Aedes aegypti, Ann. Trop. Med. Parasitol.*, 63, 419, 1969.

Kim, S. S., Seawright, J. A., and Kaiser, P. E., A genetic sexing strain of *Anopheles quadrimaculatus* species A, *J. Am. Mosq. Assoc.*, 3, 50, 1987.

Kiszewski, A. E. and Spielman, A., Spatially explicit model of transposon-based genetic drive mechanisms for displaying fluctuating populations of anopheline vector mosquitoes, *J. Med. Entomol.*, 35, 584, 1998.

Kiyasu, P. K. and Kidwell, G. M., Hybrid dysgenesis in *Drosophila melanogaster*: the evolution of mixed P and M populations maintained at high temperature, *Genet. Res.*, 44, 251, 1984.

Klassen, W., Knipling, E. F., and McGuire, J. U., Jr., The potential for insect-population suppression by dominant conditional lethal traits, *Ann. Entomol. Soc. Am.*, 63, 238, 1970.

Kokoza, V., Ahmed, A., Cho, W.-L., Jasinskiene, N., James, A. A., and Raikhel, A., Engineering blood meal-activated systemic immunity in the yellow fever mosquito, *Aedes aegypti, Proc. Natl. Acad. Sci. U.S.A.*, 97, 9144, 2000.

Kose, H. and Karr, T. L., Organization of *Wolbachia pipientis* in the *Drosophila* fertilized egg and embryo revealed by an anti-*Wolbachia* monoclonal antibody, *Mech. Dev.*, 51, 275, 1995.

Kostiainen, T. and Hoy, M. A., Genetic improvement of *Amblyseius finlandicus* (Acari, Phytoseiidae) — laboratory selection for resistance to azinphosmethyl and dimethoate, *Exp. Appl. Acarol.*, 18, 469, 1994.

Kostiainen, T. and Hoy, M. A., Laboratory evaluation of a laboratory selected organophosphate-resistant strain of *Amblyseius finlandicus* (Acari, Phytoseiidae) for possible use in Finnish apple orchards, *Biocontrol Sci. Technol.*, 5, 297, 1995.

Krafsur, E. S., Sterile insect technique for suppressing and eradicating insect population: 55 years and counting, *J. Agric. Entomol.*, 15, 303, 1998.

Krafsur, E. S. and Davidson, G., Production of semisterile mutants in the *Anopheles gambiae s.l.* complex, *Iowa State J. Res.*, 62, 85, 1987.

Krafsur, E. S. and Lindquist, D. A., Did the sterile insect technique or weather eradicate screwworms (Diptera: Calliphoridae) from Libya? *J. Med. Entomol.*, 33, 877, 1996.

Krishnamurthy, B. S., Ray, S. N., and Joshi, G. C., A note on preliminary field studies of the use of irradiated males for reduction of *C. fatigans* Wied. populations, *Indian J. Malariol.*, 16, 365, 1962.

Lamb, B. C. and Helmi, S., The effects of gene conversion control factors on conversion-induced changes in allele frequencies in populations and on linkage disequilibrium, *Genetica*, 79, 167, 1989.

Lande, R. and Wilkinson, G. S., Models of sex-ratio meiotic drive and sexual selection in stalk-eyed flies, *Genet. Res.*, 74, 245, 1999.

Langworthy, N. G., Renz, A., Mackenstedt, U., Henkle-Duhrsen, K., Bronsvoort, M. B. D., Tanya, V. N., Donnelly, M. J., and Trees, A. J., Macrofilaricidal activity of tetracycline against the filarial nematode *Onchocerca ochengi*: elimination of *Wolbachia* precedes worm death and suggests a dependent relationship, *Proc. R. Soc. London B Biol. Sci.*, 267, 1063, 2000.

Lanzaro, G. C., Toure, Y. T., Carnahan, J., Zheng, L., Dolo, G., Traore, S., Petrarca, V., Vernick, K. D., and Taylor, C. E., Complexities in the genetic structure of *Anopheles gambiae* populations in west Africa as revealed by microsatellite DNA analysis, *Proc. Natl. Acad. Sci. U.S.A.*, 95, 14260, 1998.

Lassy, C. W. and Karr, T. L., Cytological analysis of fertilization and early embryonic development in incompatible crosses of *Drosophila simulans*, *Mech. Dev.*, 57, 47, 1996.

Laven, H., Eradication of *Culex pipiens fatigans* through cytoplasmic incompatibility, *Nature*, 216, 383, 1967.

Laven, H., Genetische Methoden zur Schädlingsbekämpfung, *Anz. Schaedlingsk.*, 41, 1, 1968.

Laven, H., Eradicating mosquitoes using translocations, *Nature*, 221, 958, 1969.

Laven, H., Une expérience de lutte génétique contre *Culex pipiens fatigans*, *Ann. Parasitol. Hum. Comp.*, 46, 117, 1971.

Laven, H. and Aslamkhan, M., Control of *Culex pipiens pipiens* and *C. p. fatigans* with integrated genetical systems, *Pak. J. Sci.*, 22, 303, 1970.

Laven, H., Cousserans, J., and Guille, G., Inherited semisterility for control of harmful insects: III. A first field experiment, *Experientia*, 27, 1355, 1971.

Laven, H., Cousserans, J., and Guille, G., Eradicating mosquitoes using translocations: a first field experiment, *Nature*, 236, 456, 1972.

Lee, S. H., Clark, J. B., and Kidwell, M. G., A P element-homologous sequence in the house fly, *Musca domestica*, *Insect Mol. Biol.*, 8, 491, 1999.

Lehmann, T., Hawley, W. A., Kamau, L., Fontenille, D., Simard, F., and Collins, F. H., Genetic differentiation of *Anopheles gambiae* populations from East and West Africa: comparison of microsatellite and allozyme loci, *Heredity*, 77, 192, 1996a.

Lehmann, T., Hawley, W. A., and Collins, F. H., An evaluation of evolutionary constraints on microsatellite loci using null alleles, *Genetics*, 144, 1155, 1996b.

Lehmann, T., Besansky, N. J., Hawley, W. A., Fahey, T. G., Kamau, L., and Collins, F. H., Microgeographic structure of *Anopheles gambiae* in western Kenya based on mtDNA and microsatellite loci, *Mol. Ecol.*, 6, 243, 1997.

Lehmann, T., Hawley, W. A., Grebert, H., Danga, M., Atieli, F., and Collins, F. H., The Rift Valley complex as a barrier to gene flow for *Anopheles gambiae* in Kenya, *J. Hered.*, 90, 613, 1999.

Lewallen, L. L., Chapman, H. C., and Wilder, W. H., Chemosterilant application to an isolated population of *Culex tarsalis*, *Mosq. News*, 25, 16, 1965.

Li, J. B. and Hoy, M. A., Adaptability and efficacy of transgenic and wild-type *Metaseiulus occidentalis* (Acari: Phytoseiidae) compared as part of a risk assessment, *Exp. Appl. Acarol.*, 20, 563, 1996.

Li, J., Zhao, X., and Christensen, B. M., Dopachrome conversion activity in *Aedes aegypti*: significance during melanotic encapsulation of parasites and cuticular tanning, *Insect Biochem. Mol. Biol.*, 24, 1043, 1994.

Lidholm, D.-A., Lohe, A. R., and Hartl, D. L., The transposable element *mariner* mediates germline transformation in *Drosophila melanogaster*, *Genetics*, 134, 859, 1993.

Lindquist, D. A., Atoms for pest control, *Int. A.E.A. Bull.*, 26, 22, 1984.

Lindsay, S. W., Parson, L., and Thomas, C. J., Mapping the ranges and relative abundance of the two principal African malaria vectors, *Anopheles gambiae* sensu stricto and *An. arabiensis*, using climate data, *Proc. R. Soc. London B Biol. Sci.*, 265, 847, 1998.

Liu, W. S., Eriksson, L., and Fredga, K., XY sex reversal in the wood lemming is associated with deletion of *Xp(21-23)* as revealed by chromosome microdissection and fluorescence *in situ* hybridization, *Chromosome Res.*, 6, 379, 1998.

Lofgren, C. S., Dame, D. A., Breeland, S. G., Weidhaas, D. E., Jeffery, G., Kaiser, R., Ford, H. R., Boston, M. D., and Baldwin, K. F., Release of chemosterilized males for the control of *Anopheles albimanus* in El Salvador: III. Field methods and population control, *Am. J. Trop. Med. Hyg.*, 23, 288, 1974.

Lohe, A. R. and Hartl, D. L, Reduced germline mobility of a mariner vector containing exogenous DNA: effect of size or site? *Genetics*, 143, 1299, 1996a.

Lohe, A. R. and Hartl, D. L., Germline transformation of *Drosophila virilis* with the transposable element *mariner*, *Genetics*, 143, 365, 1996b.

Lohe, A. R., Moriyama, E. N., Lidholm, D.-A., and Hartl, D. L., Horizontal transmission, vertical inactivation, and stochastic loss of *mariner*-like transposable elements, *Mol. Biol. Evol.*, 12, 62, 1995.

Lombardo, F., Di Cristina, M., Spanos, L., Louis, C., Coluzzi, M., and Arca, B., Promoter sequences of the putative *Anopheles gambiae* apyrase confer salivary gland expression in *Drosophila melanogaster*, *J. Biol. Chem.*, 275, 23,861, 2000.

Lorimer, N., Long-term survival of introduced genes in a natural population of *Aedes aegypti* (Diptera, Culicidae), *Bull. Entomol. Res.*, 71, 129–132, 1981.

Lorimer, N., Lounibos, L. P., and Petersen, J. L., Field trials with a translocation homozygote in *Aedes aegypti* for population replacement, *J. Econ. Entomol.*, 69, 405, 1976.

Loukeris, T. G., Arca, B., Livadaras, I., Dialektaki, G., and Savakis, C., Introduction of the transposable element *Minos* into the germ-line of *Drosophila melanogaster*, *Proc. Natl. Acad. Sci. U.S.A.*, 92, 9485, 1995a.

Loukeris, T. G., Livadaras, I., Arca, B., Zabalou, S., and Savakis, C., Gene transfer into the medfly, *Ceratitis capitata*, with a *Drosophila hydei* transposable element, *Science*, 270, 2002, 1995b.

Lowe, R. E., Ford, H. R., Cameron, A. L., Smittle, B. J., Dame, D. A., Patterson, R. S., and Weidhaas, D. E., Competitiveness of sterile male *Culex pipiens quinquefasciatus* released into a natural population, *Mosq. News*, 34, 447, 1974.

Lozovskaya, E. R., Nurminsky, D. I., Hartl, D. L, and Sullivan, D. T., Germline transformation of *Drosophila virilis* mediated by the transposable element *hobo*, *Genetics*, 142, 173, 1996.

Lyttle, T. W., Segregation distorters, *Annu. Rev. Genet.*, 25, 511, 1991.

Macdonald, G., *The Epidemiology and Control of Malaria*, Oxford University Press, Oxford, U.K., 1957.

Mackay, T. F. C., Lyman, R. F., and Jackson, M. S., Effects of *P* element insertions on quantitative traits in *Drosophila melanogaster*, *Genetics*, 130, 315, 1992.

Magaudda, P. L., Sacca, G., and Guarniera, D., Sterile male method supplemented by insecticides for the control of *Musca domestica* on the island of Vulcano, Italy, *Ann. Ist. Super. Sanita*, 5, 29, 1969.

Matsubara, T., Beeman, R. W., Besansky, N. J., Mukabayire, O., Higgs, S., James, A. A., and Burns, J. C., Pantropic retroviral vectors integrate and express in cells of the malaria mosquito, *Anopheles gambiae*, *Proc. Natl. Acad. Sci. U.S.A.*, 93, 6181, 1996.

Mbogo, C. N., Snow, R. W., Khamala, C. P., Kabiru, E. W., Ouma, J. H., Githure, J. I., Marsh, K., and Beier, J. C., Relationships between *Plasmodium falciparum* transmission by vector populations and the incidence of severe disease at nine sites on the Kenyan coast, *Am. J. Trop. Med. Hyg.*, 52, 201, 1995.

McCall, J. W., Jun, J. J., and Bandi, C., *Wolbachia* and the antifilarial properties of tetracycline. An untold story, *Ital. J. Zool.*, 66, 7, 1999.

McDermott, G. J. and Hoy, M. A., Persistence and containment of *Metaseiulus occidentalis* (Acari: Phytoseiidae) in Florida: risk assessment for possible releases of transgenic strains, *Fla. Entomol.*, 80, 42, 1997.

McDonald, P. T., Hausermann, W., and Lorimer, N., Sterility introduced by release of genetically altered males to a domestic population of *Aedes aegypti* at the Kenya coast, *Am. J. Trop. Med. Hyg.*, 26, 553, 1977.

Mckenzie, J. A., The release of a compound chromosome stock in a vineyard cellular population of *Drosophila melanogaster*, *Genetics*, 82, 685, 1976.

McLain, D. K., Collins, F. H., Brandling-Bennett, A. D., and Were, J. B., Microgeographic variation in rDNA intergenic spacers of *Anopheles gambiae* in western Kenya, *Heredity*, 62, 257, 1989.

Meister, G. A. and Grigliatti, T. A., Rapid spread of a *P* element/*Adh* gene construct through experimental populations of *Drosophila melanogaster*, *Genome*, 36, 1169, 1993.

Merçot, H. and Poinsot, D., *Wolbachia* transmission in a naturally bi-infected *Drosophila simulans* strain from New-Caledonia, *Entomol. Exp. Appl.*, 86, 97, 1998.

Meyer, N., *History of the Mexico–United States Screwworm Eradication Program*, Vantage Press, New York, 1994.

Miao, V. P., Covert, S. F., and Vanetten, H. D., A fungal gene for antibiotic resistance on a dispensable (B) chromosome, *Science*, 254, 1773, 1991.

Michel, K., Stamenova, A., Pinkerton, A. C., Franz, G., Robinson, A. S., Gariou-Papalexiou, A., Zacharopoulou, A., O'Brochta, D. A., and Atkinson, P. W., Hermes-mediated germ-line transformation of the Mediterranean fruit fly *Ceratitis capitata*, *Insect Mol. Biol.*, 10, 155, 2001.

Miller, D. W. and Miller, L. K., A virus mutant with an insertion of a *copia*-like transposable element, *Nature*, 299, 562, 1982.

Mora-Garcia, S. and Goodrich, J., Genomic imprinting: seeds of conflict, *Curr. Biol.*, 10, R71, 2000.

Moreira, L. A., Edwards, M. J., Adhami, F., Jasinskiene, N., James, A. A., and Jacobs-Lorena, M., Robust gut-specific gene expression in transgenic *Aedes aegypti* mosquitoes, *Proc. Natl. Acad. Sci. U.S.A.*, 97, 10895, 2000.

Morgan, P. B., Wagoner, D. E., and Fye, R. L., Genetic manipulation used against a field population of house flies. III. Males and females bearing a heterozygous translocation; release begun after initial seasonal peak population level reached, *Environ. Entomol.*, 2, 779, 1973.

Morlan, H. B., McCray, E. M., Jr., and Kilpatrick, J. W., Field tests with sexually sterile males for control of *Aedes aegypti*, *Mosq. News*, 22, 295, 1962.

Nchinda, T. C., Malaria: a reemerging disease in Africa, *Emerg. Infect. Dis.*, 4, 398, 1998.

Nijhout, M. M., *Plasmodium gallinaceum*: exflagellation stimulated by a mosquito factor, *Exp. Parasitol.*, 48, 75, 1979.

Nota, H., Miyoshi, T., Zhang, Q., Watanabe, K., Deng, K., and Hoshizaki, S., A plausible interspecies transmission of *Wolbachia* among planthoppers and their strepsipteran endoparasite, in *First International* Wolbachia *Conference*, Kolymbari, June 7–12, 2000.

Nur, U., Werren, J. H., Eickbush, D. G., Burke, W. D., and Eickbush, T. H., A "selfish" B chromosome that enhances its transmission by eliminating the paternal genome, *Science*, 240, 512, 1988.

O'Brochta, D. A., Warren, W. D., Saville, K. J., and Atkinson, P. W., *Hermes*, a functional non-drosophilid insect gene vector from *Musca domestica*, *Genetics*, 142, 907, 1996.

O'Brochta, D. A., Atkinson, P. W., and Lehane, M. J., Transformation of *Stomoxys calcitrans* with a *Hermes* gene vector, *Insect Mol. Biol.*, 9, 531, 2000.

O'Neill, S. L., Pettigrew, M. M., Sinkins, S. P., Braig, H. R., Andreadis, T. G., and Tesh, R. B., *In vitro* cultivation of *Wolbachia pipientis* in an *Aedes albopictus* cell line, *Insect Mol. Biol.*, 6, 33, 1997.

Orti, G., Pearse, D. E., and Avise, J. C., Phylogenetic assessment of length variation at a microsatellite locus, *Proc. Natl. Acad. Sci. U.S.A.*, 94, 10745, 1997.

Owusu-Daaku, K. O., Wood, R. J., and Butler, R. D., Selected lines of *Aedes aegypti* with persistently distorted sex ratios, *Heredity*, 79, 388, 1997.

Pal, R. and Whitten, M. J., *Genetic Control of Insects*, Elsevier/North-Holland, Amsterdam, 1974.

Pascual, L. and Periquet, G., Distribution of *hobo* transposable elements in natural populations of *Drosophila melanogaster*, *Mol. Biol. Evol.*, 8, 282, 1991.

Paskewitz, S. and Riehle, M. A., Response of *Plasmodium* refractory and susceptible strains of *Anopheles gambiae* to inoculated Sephadex beads, *Dev. Comp. Immunol.*, 18, 369, 1994.

Patterson, R. S., Weidhaas, D. E., Ford, H. R., and Lofgren, C. S., Suppression and elimination of an island population of *Culex pipiens quinquefasciatus* with sterile males, *Science*, 168, 1368, 1970.

Patterson, R. S., Sharma, V. P., Singh, K. R. P., La Brecque, G. C., Seetheram, P. L., and Grover, K. K., Use of radiosterilized males to control indigenous populations of *Culex pipiens quinquefasciatus*, laboratory and field studies, *Mosq. News*, 35, 1, 1975.

Patterson, R. S., Lowe, R. E., Smittle, B. J., Dame, D. A., Boston, M. D., and Cameron, A. L., Release of radiosterilized males to control *Culex pipiens quinquefasciatus* (Diptera: Culicidae), laboratory and field studies, *J. Med. Entomol.*, 14, 299, 1977.

Patz, J. A., Martens, W. J. M., Focks, D. A., and Jetten, T. H., Dengue fever epidemic potential as projected by general circulation models of global climate change, *Environ. Health Perspect.*, 106, 147, 1998.

Paul, R. E. L., Packer, M. J., Walmsley, M., Lagog, M., Ranford-Cartwright, L. C., Paru, R., and Day, K. P., Mating patterns in malaria parasite populations of Papua New Guinea, *Science*, 269, 1709, 1995.

Pearson, A. M. and Wood, R. J., Combining the meiotic drive gene *D* and the translocation *T-1* in the mosquito *Aedes aegypti*. 2. Recombination, *Genetica*, 54, 79, 1980.

Peloquin, J. J., Thibault, S. T., Staten, R., and Miller, T. A., Germ-line transformation of pink bollworm (Lepidoptera: Gelechiidae) mediated by the *piggyBac* transposable element, *Insect Mol. Biol.*, 9, 323, 2000.

Peters, L. L. and Barker, J. E., Novel inheritance of the murine severe combined anemia and thrombocytopenia (*scat*) phenotype, *Cell*, 74, 135, 1993.

Petersen, J. L., Lounibos, L. P., and Lorimer, N., Field trials of double translocation heterozygote males for genetic control of *Aedes aegypti* (L.) (Diptera: Culicidae), *Bull. Entomol. Res.*, 67, 313, 1977.

Petrov, D. A., Schutzman, J. L., Hartl, D. L., and Lozovskaya, E. R., Diverse transposable elements are mobilized in hybrid dysgenesis in *Drosophila virilis*, *Proc. Natl. Acad. Sci. U.S.A.*, 92, 8050, 1995.

Pettigrew, M. M. and O'Neill, S. L., Control of vector-borne disease by genetic manipulation of insect populations — technological requirements and research priorities, *Aust. J. Entomol.*, 36, 309, 1997.

Pfeifer, T. A. and Grigliatti, T. A., Future perspectives on insect pest management: engineering the pest, *J. Invertebr. Pathol.*, 67, 109, 1996.

Picard, G. and L'Heritier, P., A maternally inherited factor inducing sterility in *Drosophila melanogaster*, *Dros. Inf. Serv.*, 46, 54, 1971.

Pigozzi, M. I. and Solari, A. J., Germ cell restriction and regular transmission of an accessory chromosome that mimics a sex body in the zebra finch, *Taeniopygia guttata*, *Chromosome Res.*, 6, 105, 1998.

Pinkerton, A. C. et al., Green fluorescent protein as a genetic marker in transgenic *Aedes aegypti*, *Insect Mol. Biol.*, 9, 1, 2000.

Plasterk, R. H. A., Izsvak, Z., and Ivics, Z., Resident aliens — the *Tc1/mariner* superfamily of transposable elements, *Trends Genet.*, 15, 326, 1999.

Pleshkova, G. N., Radiation-induced inversions and reciprocal translocations in *Anopheles atroparvus*, *Genetika*, 20, 2011, 1984.

Pointier, J. P., Invading freshwater gastropods: some conflicting aspects for public health, *Malacologia*, 41, 403, 1999.

Pointier, J. P. and Augustin, D., Biological control and invading freshwater snails. A case study, *C. R. Acad. Sci. III Vie*, 322, 1093, 1999.

Politzar, H. and Cuisance, D., An integrated campaign against riverine tstetse *Glossina palpalis gambiensis* and *Glossina tachinoides* by trapping and the release of sterile males, *Insect Sci. Appl.*, 5, 439, 1984.

Politzar, H., Cuisance, D., Lafaye, A., Clair, M., Taze, Y., and Sellin, E., Field trial of genetic control by sterile male release, longevity and dispersion of irradiated males of *Glossina palpalis gambiensis*, Upper Volta, *Ann. Soc. Belge Méd. Trop.*, 59, 59, 1979.

Powell, J. R., *Progress and Prospects in Evolutionary Biology — The* Drosophila *Model*, Oxford University Press, New York, 1997.

Presnail, J., Jeyaprakash, A., Li, J., and Hoy, M. A., Genetic analysis of four lines of *Metaseiulus occidentalis* (Acari: Phytoseiidae) transformed by maternal microinjection, *Ann. Entomol. Soc. Am.*, 90, 237, 1997.

Rai, K. S., Genetic control of vectors, in *The Biology of Disease Vectors*, Beaty, B. J. and Marquardt, W. C., Eds., University Press of Colorado, Niwot, 1996, 564.

Rai, K. S. and Asman, S. M., Possible application of a reciprocal translocation for genetic control of the mosquito, *Aedes aegypti*, in *Proceedings of the XII International Congress of Genetics*, Vol. 1, Tokyo, Japan, Science Council of Japan, Tokyo, 1968, 164.

Rai, K. S., Grover, K. K., and Suguna, S. G., Genetic manipulation of *Aedes aegypti*: incorporation and maintenance of a genetic marker and chromosomal translocation in natural populations, *Bull. World Health Org.*, 48, 49, 1973.

Rand, D. M. and Harrison, R. G., Mitochondrial DNA transmission in crickets, *Genetics*, 114, 955, 1986.

Reed, K. M. and Werren, J. H., Induction of paternal genome loss by the paternal sex ratio chromosome and cytoplasmic incompatibility bacteria (*Wolbachia*) — a comparative study of early embryonic events, *Mol. Reprod. Dev.*, 40, 408, 1995.

Regner, L. P., Abdelhay, E., Rohde, C., Rodrigues, J. J. S., and Valente, V. L. S., Temperature-dependent gonadal hybrid dysgenesis in *Drosophila willistoni*, *Genet. Mol. Biol.*, 22, 205, 1999.

Reisen, W. K., Sakai, R. K., Baker, R. H., Rathor, H. R., Raana, K., Azra, K., and Niaz, S., Field competitiveness of *Culex tritaeniorhynchus* Giles males carrying a complex chromosomal aberration: a second experiment, *Ann. Entomol. Soc. Am.*, 73, 479, 1980.

Reisen, W. K., Baker, R. H., Sakai, R. K., Mahmood, F., Rathor, H. R., Ranna, K., and Toqir, G., *Anopheles culicifacies* mating behavior and competitiveness in nature of chemosterilized males carrying a genetic sexing system, *Ann. Entomol. Soc. Am.*, 74, 395, 1981b.

Reisen, W. K., Asman, S. M., Milby, M. M., Bock, M. E., Stoddard, P. E., Meyer, R. P., and Reeves, W. C., Attempted suppression of a semi-isolated population of *Culex tarsalis* by release of irradiated males, *Mosq. News*, 41, 736, 1981a.

Reisen, W. K., Milby, M. M., Asman, S. M., Bock, M. E., Meyer, R. P., McDonald, P. T., and Reeves, W. C., Attempted suppression of a semi-isolated population by the release of irradiated males: a second experiment using males from a recently colonized strain, *Mosq. News*, 42, 565, 1982.

Reisen, W. K., Bock, M. E., Milby, M. M., and Reeves, W. C., Attempted insertion of a recessive autosomal gene into a semi-isolated population of *Culex tarsalis* (Diptera: Culicidae), *J. Med. Entomol.*, 22, 250, 1985.

Ribeiro, J. M. C. and Kidwell, M. G., Transposable elements as population drive mechanisms: specification of critical parameter values, *J. Med. Entomol.*, 31, 10, 1994.

Ribeiro, J. M. C. and Kidwell, M. G., Ribero and Kidwell's transposon model — reply, *J. Med. Entomol.*, 32, 1, 1995.

Rich, S. M., Hudson, R. R., and Ayala, F. J., *Plasmodium falciparum* antigenic diversity — evidence of clonal population structure, *Proc. Natl. Acad. Sci. U.S.A.*, 94, 13040, 1997.

Rigaud, T., Inherited microorganisms and sex determination of arthropod hosts, in *Influential Passengers: Inherited Microorganisms and Arthropod Reproduction*, O'Neill, S. L., Hoffman, A. A., and Werren, J. H., Eds., Oxford University Press, Oxford, U.K., 1997, 81.

Robert, F., Ntoumi, F., Angel, G., Candito, D., Rogier, C., Fandeur, T., Sarthou, J. L., and Mercereau-Puijalon, O., Extensive genetic diversity of *Plasmodium falciparum* isolates collected from patients with severe malaria in Dakar, Senegal, *Trans. R. Soc. Trop. Med. Hyg.*, 90, 704, 1996.

Robertson, H. M., The *mariner* transposable element is widespread in insects, *Nature*, 362, 241, 1993.

Robertson, H. M., Preston, C. R., Phillis, R. W., Johnson-Schlitz, D., Benz, W. K., and Engels, W. R., A stable genomic source of P element transposase in *Drosophila melanogaster*, *Genetics*, 118, 461, 1988.

Robinson, A. S., Progress in the use of chromosomal translocations for the control of insect pests, *Biol. Rev.*, 51, 1, 1976.

Robinson, A. S. and Franz, G., The application of transgenic insect technology in the sterile insect technique, in *Insect Transgenesis: Methods and Applications*, Handler, A. M. and James, A. J., Eds., CRC Press, Boca Raton, FL, 2000, 307.

Robinson, A. S. and Van Lap, P., Cytological linkage and insecticide studies on a genetic sexing line in *Anopheles stephensi* Liston, *Heredity*, 58, 95, 1987.

Robinson, K. O., Ferguson, H. J., Cobey, S., Vaessin, H., and Smith, B. H., Sperm-mediated transformation of the honey bee, *Apis mellifera*, *Insect Mol. Biol.*, 9, 625, 2000.

Rosen, L., Rozeboom, L. E., Reeves, W. C., Saugrain, J., and Gubler, D. R., A field trial of competitive displacement of *Aedes polynesiensis* by *Aedes albopictus* on a Pacific atoll, *Am. J. Trop. Med. Hyg.*, 25, 906, 1976.

Rosenberg, R., Inability of *Plasmodium knowlesi* sporozoites to invade *Anopheles freeborni* salivary glands, *Am. J. Trop. Med. Hyg.*, 34, 687, 1985.

Ross, M. H., Keil, C. B., and Cochran, D. G., The release of sterile males into natural populations of the German cockroach *Blattella germanica*, *Entomol. Exp. Appl.*, 30, 241, 1981.

Rousset, F., Raymond, M., and Kjellberg, F., Cytoplasmic incompatibilities in the mosquito *Culex pipiens*: how to explain a cytotype polymorphism? *J. Evol. Biol.*, 4, 69, 1991.

Rousset, F., Braig, H. R., and O'Neill, S. L., A stable triple *Wolbachia* infection in *Drosophila* with nearly additive incompatibility effects, *Heredity*, 82, 620, 1999.

Rozeboom, L. E., Relative densities of freely breeding populations of *Aedes (S.) polynesiensis* Marks and *Ae. (S.) albopictus* Skuse, *Am. J. Trop. Med. Hyg.*, 20, 356, 1971.

Rubin, M. C. and Spradling, A. C., Genetic transformation of *Drosophila* with transposable element vectors, *Science*, 218, 348, 1982.

Sacca, G., Magaudda, P. L., and Guarniera, D., A trial for the integrated (chemical and biological) control of *Musca domestica* in the Lipari Islands (preliminary note), *Riv. Parassitol.*, 28, 295, 1967.

Sakai, R. K. and Mahmood, F., Homozygous chromosomal aberrations in *Anopheles stephensi*, *J. Hered.*, 76, 230, 1985.

Sandler, L. and Golic, K., Segregation distortion in *Drosophila*, *Trends Genet.*, 1, 181, 1985.

Sandler, L. and Novitski, E., Meiotic drive as an evolutionary force, *Am. Nat.*, 91, 105, 1957.

Sasaki, T., Braig, H. R., and O'Neill, S. L., Analysis of *Wolbachia* protein synthesis in *Drosophila in vivo*, *Insect Mol. Biol.*, 7, 101, 1998.

Seawright, J. A. and Cockburn, A. F., Genetic control of mosquitoes and the stable fly in Florida, in *Pest Management in the Subtropics: Integrated Pest Management: A Florida Perspective*, Rosen, D., Bennett, F. D., and Capinera, J. L., Eds., Intercept, Andover, U.K., 1996, 259.

Seawright, J. A., Kaiser, P. E., Dame, D. A., and Willis, N. L., Field competitiveness of males of *Aedes aegypti* heterozygous for a translocation, *Mosq. News*, 35, 30, 1975.

Seawright, J. A., Kaiser, P. E., Willis, N. L., and Dame, D. A., Field competitiveness of double translocation heterozygote males of *Aedes aegypti*, *J. Med. Entomol.*, 13, 208, 1976.

Seawright, J. A., Kaiser, P. E., and Dame, D. A., Mating competitiveness of chemosterilized males of *Aedes aegypti* in field test, *Mosq. News*, 37, 615, 1977.

Seawright, J. A., Kaiser, P. E., Suguna, S. G., and Focks, D. A., Genetic sexing strains of *Anopheles albimanus*, *Mosq. News*, 41, 107, 1981.

Seleme, M.-d.-C., Busseau, I., Malinsky, S., Bucheton, A., and Teninges, D., High-frequency retrotransposition of a marked I factor in *Drosophila melanogaster* correlates with a dynamic expression pattern of the ORF1 protein in the cytoplasm of oocytes, *Genetics*, 151, 761, 1999.

Serebrovskii, A. S., On the possibility of a new method for the control of insect pests [in Russian], *Zool. Zh.*, 19, 618, 1940. English translation in Panel on Application of the Sterile-Male Technique for the Eradication or Control of Harmful Species of Insects, Sterile-Male Technique for Eradication or Control of Harmful Insects, Proceedings, Joint FAO/IAEA, Division of Atomic Energy in Food and Agriculture, Vienna, 27–31 May 1968, STI/PUB/224, International Atomic Energy Agency, Vienna, 1969, 123.

Severson, D. W., Mori, A., Zhang, Y., and Christensen, B. M., Linkage map for *Aedes aegypti* using restriction fragment length polymorphisms, *J. Hered.*, 84, 241, 1993.

Severson, D. W., Thathy, V., Mori, A., Zhang, Y., and Christensen, B. M., Restriction fragment length polymorphism mapping of quantitative trait loci for malaria parasite susceptibility in the mosquito *Aedes aegypti*, *Genetics*, 139, 1711, 1995.

Shahabuddin, M., Toyoshima, T., Aikawa, M., and Kaslow, D. C., Transmission-blocking activity of a chitinase inhibitor and activation of malarial parasite chitinase by mosquito protease, *Proc. R. Soc. London B. Biol. Sci.*, 90, 4266, 1993.

Sharma, M. I. D., Ed., Special Issue on Genetic Control of Mosquitoes, *J. Communicable Dis.*, 6(2), 1974.

Sharma, V. P., Patterson, R. S., La Brecque, G. C., and Singh, K. R. P., Three field release trials with chemosterilized *Culex pipiens fatigans* Wied. in Delhi villages, *J. Communicable Dis.*, 8, 18, 1976.

Shaw, M. W. and Hewitt, G. M., B chromosomes, selfish DNA and theoretical models: where next? in *Oxford Surveys in Evolutionary Biology*, Vol. 7, Futuyma, D. and Antonovics, J., Eds., Oxford University Press, Oxford, U.K., 1990, 197.

Shroyer, D. A. and Rosen, L., Extrachromosomal inheritance of carbon dioxide sensitivity in the mosquito *Culex quinquefasciatus*, *Genetics*, 104, 649, 1983.

Simms, E. L. and Triplett, J., Costs and benefits of plant responses to disease: resistance and tolerance, *Evolution*, 48, 1973, 1994.

Sinkins, S. P., *Wolbachia* as a Potential Gene Drive System, Ph.D. dissertation, University of London, 1996.

Sinkins, S. P. and O'Neill, S. L., *Wolbachia* as a vehicle to modify insect populations, in *Insect Transgenesis: Methods and Applications*, Handler, A. M. and James, A. J., Eds., CRC Press, Boca Raton, FL, 2000, 271.

Sinkins, S. P., Braig, H. R., and O'Neill, S. L., *Wolbachia* superinfections and the expression of cytoplasmic incompatibility, *Proc. R. Soc. London B Biol. Sci.*, 261, 325, 1995a.

Sinkins, S. P., Braig, H. R., and O'Neill, S. L., *Wolbachia pipientis*: bacterial densities and unidirectional cytoplasmic incompatibility between infected populations of *Aedes albopictus*, *Exp. Parasitol.*, 81, 284, 1995b.

Sinkins, S. P., Curtis, C. F., and O'Neill, S. L., The potential application of inherited symbiont systems to pest control, in *Influential Passengers: Inherited Microorganisms and Arthropod Reproduction*, O'Neill, S. L., Hoffman, A. A., and Werren, J. H., Eds., Oxford University Press, Oxford, U.K., 1997, 155–208.

Skinner, S. W., Maternally-inherited sex ratio in the parasitoid wasp *Nasonia vitripennis*, *Science*, 215, 1135, 1982.

Slatkin, M., Interchromosomal biased gene conversion, mutation and selection in a multigene family, *Genetics*, 112, 681, 1986.

Smit, A. F. A., Interspersed repeats and other mementos of transposable elements in mammalian genomes, *Curr. Opin. Genet. Dev.*, 9, 657, 1999.

Smith, N. G. C., The dynamics of maternal-effect selfish genetic elements, *J. Theor. Biol.*, 191, 173, 1998.

Smith, P. H. and Morton, R., Assessment of the field performance of compound chromosome strains compared to laboratory-reared wild-type strains in *Lucilia cuprina* (Diptera: Calliphoridae), *Bull. Entomol. Res.*, 75, 233, 1985.

Smith, R. H., Induced conditional lethal mutations for the control of insect populations, in *Sterility Principle for Insect Control*, International Atomic Energy Agency, Vienna, 1971, 453.

Smith, R. H. and von Borstel, R. C., Genetic control of insect populations, *Science*, 178, 1164, 1972.

Snow, R. W., The burden of malaria: understanding the balance between immunity, public health and control, *J. Med. Microbiol.*, 49, 1053, 2000.

Snow, R. W. and Marsh, K., Will reducing *Plasmodium falciparum* transmission alter malaria mortality among African children? *Parasitol. Today*, 11, 188, 1995.

Snow, R. W. and Marsh, K., New insights into the epidemiology of malaria relevant for disease control, *Br. Med. Bull.*, 54, 293, 1998.

Solomon, M. G., Cross, J. V., Fitzgerald, J. D., Campbell, C. A. M., Jolly, R. L., Olszak, R. W., Niemczyk, E., and Vogt, H., Biocontrol of pests of apples and pears in northern and central Europe — 3. Predators, *Biocontrol Sci. Technol.*, 10, 91, 2000.

Somboon, P., Prapanthadara, L., and Suwonkerd, W., Selection of *Anopheles dirus* for refractoriness and susceptibility to *Plasmodium yoelii nigeriensis*, *Med. Vet. Entomol.*, 13, 355, 1999.

Spielman, A., Why entomological anti-malaria research should not focus on transgenic mosquitoes, *Parasitol. Today*, 10, 374, 1994a.

Spielman, A., Research priorities for managing the transmission of vector-borne disease, *Prev. Med.*, 23, 693, 1994b.

Spielman, A., Kitron, U., and Pollack, R. J., Time limitation and the role of research in the worldwide attempt to eradicate malaria, *J. Med. Entomol.*, 30, 6, 1993.

Spollen, K. M. and Hoy, M. A., Genetic improvement of an arthropod natural enemy: relative fitness of a carbaryl-resistant strain of the California red scale parasite *Aphytis melinus* DeBach, *Biol. Control*, 2, 87, 1992.

Spradbery, J. P., Tozer, R. S., Robb, J. M., and Cassells, P., The screw-worm fly *Chrysomya bezziana* Villeneuve (Diptera: Calliphoridae) in a sterile insect release trial in Papua New Guinea, *Res. Popul. Ecol.*, 31, 353, 1989.

Stacey, S. N., Lansman, R. A., Brock, H. W., and Grigliatti, T. A., Distribution and conservation of mobile elements in the genus *Drosophila*, *Mol. Biol. Evol.*, 3, 522, 1986.

Steffens, R. J., Methodology of translocation production and stability of translocations in the Mediterranean fruit fly, *Ceratitis capitata* Wied. (Diptera: Tephritidae), *Z. Angew. Entomol.*, 95, 181, 1983a.

Steffens, R. J., Combination of radiation-induced and translocation sterility for genetic control of fruit flies, *Entomol. Exp. Appl.*, 33, 253, 1983b.

Steiner, L. F., Harris, E. J., Mitchell, W. C., Fujimoto, M. S., and Christenson, L. D., Melon fly eradication by overflooding with sterile flies, *J. Econ. Entomol.*, 58, 519, 1965.

Stouthamer, R., *Wolbachia* induced parthenogenesis, in *Influential Passengers: Inherited Microorganisms and Arthropod Reproduction*, O'Neill, S. L., Hoffman, A. A., and Werren, J. H., Eds., Oxford University Press, Oxford, U.K., 1997, 102.

Stouthamer, R., Breeuwer, J. A. J., and Hurst, G. D. D., *Wolbachia pipientis*: microbial manipulator of arthropod reproduction, *Annu. Rev. Microbiol.*, 53, 71, 1999.

Suguna, S. G., Wood, R. J., Curtis, C. F., Whitelaw, A., and Kazmi, S. J., Resistance to meiotic drive at the M^D locus in an Indian wild population of *Aedis aegypti*, *Genet. Res.*, 29, 123–132, 1977a.

Suguna, S. G., Kazmi, S. J., and Curtis, C. F., Sex-ratio distorter translocation homozygotes in *Aedes aegypti*, *Genetica*, 47, 125, 1977b.

Suomalainen, E., Saura, A., and Lokki, J., *Cytology and Evolution in Parthenogenesis*, CRC Press, Boca Raton, FL, 1987.

Sved, J. A., Hybrid dysgenesis in *Drosophila melanogaster* — a possible explanation in terms of spatial organization of chromosomes, *Aust. J. Biol. Sci.*, 29, 375, 1976.

Sweeny, T. L. and Barr, A. R., Sex ratio distortion caused by meiotic drive in the mosquito *Culex pipiens*, *Genetics*, 88, 427, 1978.

Sweeny, T. L., Guptavanij, P., and Barr, A. R., Abnormal salivary gland puff associated with meiotic drive in mosquitoes (Diptera, Culicidae), *J. Med. Entomol.*, 24, 623, 1987.

Takken, W., Oladunmade, M. A., Dengwat, L., Feldmann, H. U., Onah, J. A., Tenabe, S. O., and Hamann, H. J., The eradication of *Glossina palpalis papalis* (Diptera: Glossinidae) using traps, insecticide-impregnated targets and the sterile insect technique in Central Nigeria, *Bull. Entomol. Res.*, 76, 275, 1986.

Tamura, T., Thibert, C., Royer, C., Kanda, T., Abraham, E., Kamba, M., Komoto, N., Thomas, J. L., Mauchamp, B., Chavancy, G., Shirk, P., Fraser, M., Prudhomme, J. C., and Couble, P., Germline transformation of the silkworm *Bombyx mori* L. using a *piggyBac* transposon-derived vector, *Nat. Biotechnol.*, 18, 81, 2000.

Taylor, M. J. and Hoerauf, A., *Wolbachia* bacteria of filarial nematodes, *Parasitol. Today*, 15, 437, 1999.

Thathy, V., Severson, D. W., and Christensen, B. M., Reinterpretation of the genetics of susceptibility of *Aedes aegypti* to *Plasmodium gallinaceum*, *J. Parasitol.*, 80, 705, 1994.

Thibault, S. T., Luu, H. T., Vann, N., and Miller, T. A., Precise excision and transposition of *piggyBac* in pink bollworm embryos, *Insect Mol. Biol.*, 8, 119, 2000.

Thistlewood, H. M. A., Pree, D. J., and Crawford, L. A., Selection and genetic analysis of permethrin resistance in *Amblyseius fallacis* (Garman) (Acari: Phytoseiidae) from Ontario apple orchards, *Exp. Appl. Acarol.*, 19, 707, 1995.

Thomson, M. S. and Beeman, R. W., Assisted suicide of a selfish gene, *J. Hered.*, 90, 191, 1999.

Thomson, M. S., Friesen, K. S., Denell, R. E., and Beeman, R. W., A hybrid incompatibility factor in *Tribolium castaneum*, *J. Hered.*, 86, 6, 1995.

Toure, Y. T., Petrarca, V., Traore, S. F., Coulibaly, A., Maiga, H. M., Sankare, O., Sow, M., Di Deco, M. A., and Coluzzi, M., Ecological genetic studies in the chromosomal form Mopti of *Anopheles gambiae* s.str. in Mali, west Africa, *Genetica*, 94, 213, 1994.

Toure, Y. T., Doumbo, O., Toure, A., Bagayoko, M., Diallo, M., Dolo, A., Vernick, K. D., Keister, D. B., Muratova, O., and Kaslow, D. C., Gametocyte infectivity by direct mosquito feeds in an area of seasonal malaria transmission: implications for Bancoumana, Mali as a transmission-blocking vaccine site, *Am. J. Trop. Med. Hyg.*, 59, 481, 1998.

Trape, J. F. and Rogier, C., Combating malaria morbidity and mortality by reducing transmission, *Parasitol. Today*, 12, 236, 1996.

Tremblay, A., Jasin, M., and Chartrand, P., A double-strand break in a chromosomal LINE element can be repaired by gene conversion with various endogenous LINE elements in mouse cells, *Mol. Cell Biol.*, 20, 54, 2000.

Turelli, M. and Ginzburg, L. R., Should individual fitness increase with heterozygosity? *Genetics*, 104, 191, 1983.

Turelli, M. and Hoffmann, A. A., Rapid spread of an inherited incompatibility factor in California, USA, *Drosophila*, *Nature*, 353, 440, 1991.

Turelli, M. and Hoffmann, A. A., Cytoplasmic incompatibility in *Drosophila simulans*: dynamics and parameter estimates from natural populations, *Genetics*, 140, 1319, 1995.

Turelli, M. and Hoffmann, A. A., Microbe-induced cytoplasmic incompatibility as a mechanism for introducing transgenes into arthropod populations, *Insect Mol. Biol.*, 8, 243, 1999.

Uyenoyama, M. and Nei, M., Quantitative models of hybrid dysgenesis: rapid evolution under transposition, extrachromosomal inheritance and fertility selection, *Theor. Popul. Biol.*, 27, 176, 1985.

Vala, F., Breeuwer, J. A. J., and Sabelis, M. W., *Wolbachia*-induced "hybrid breakdown" in the two-spotted spider mite *Tetranychus urticae* Koch, presented at First International *Wolbachia* Conference, Kolymbari, June 7–12, 2000.

Vanderplank, F. L., Hybridization between *Glossina* species and suggested new methods for control of certain species of tsetse, *Nature*, 154, 607, 1944.

Vanderplank, F. L., Experiments in the hybridization of tsetse flies (*Glossina*, Diptera) and the possibility of a new method of control, *Trans. R. Entomol. Soc.*, 98, 1, 1947.

van Heemert, C., Synthesis of compound chromosomes from a pericentric inversion in the onion fly *Hylemya antiqua*, *Nature*, 266, 445, 1977.

Vavre, F., Girin, C., and Bouletreau, M., Phylogenetic status of a fecundity-enhancing *Wolbachia* that does not induce thelytoky in *Trichogramma*, *Insect Mol. Biol.*, 8, 6, 1999.

Vazeille-Falcoz, M., Ohayon, H., Gounon, P., and Rosen, L., Unusual morphology of a virus which produces carbon dioxide sensitivity in mosquitoes, *Virus Res.*, 24, 235, 1992.

Vernick, K. D., Fujioka, H., Seeley, D. C., Tandler, B., Aikawa, M., and Miller, L. H., *Plasmodium gallinaceum*: a refractory mechanism of ookinete killing in the mosquito, *Anopheles gambiae*, *Exp. Parasitol.*, 80, 583, 1995.

Vielle-Calzada, J. P., Thomas, J., Spillane, C., Coluccio, A., Hoeppner, M. A., and Grossniklaus, U., Maintenance of genomic imprinting at the *Arabidopsis medea* locus requires zygotic *DDM1* activity, *Genes Dev.*, 13, 2971, 1999.

Vogt, W. G., Woodburn, T. L., and Foster, G. G., Ecological analysis of field trials conducted to assess the potential of sex-linked translocation strains for genetic control of the Australian sheep blowfly *Lucilia cuprina*, *Aust. J. Biol. Sci.*, 38, 259, 1985.

Von der Schulenburg, J. H. G., Hurst, G. D. D., Huigens, T. M. E., van Meer, M. M. M., Jiggins, F. M., and Majerus, M. E. N., Molecular evolution and phylogenetic utility of *Wolbachia ftsZ* and *wsp* gene sequences with special reference to the origin of male-killing, *Mol. Biol. Evol.*, 17, 584, 2000.

Wade, M. J. and Beeman, R. W., The population dynamics of maternal-effect selfish genes, *Genetics*, 138, 1309, 1994.

Wade, M. J. and Chang, N. W., Increased male fertility in *Tribolium confusum* beetles after infection with the intracellular parasite *Wolbachia*, *Nature*, 373, 72, 1995.

Wagoner, D. E., Morgan, P. B., Labrecque, G. C., and Johnson, O. A., Genetic manipulation used against a field population of house flies: I. Males bearing a heterozygous translocation, *Environ. Entomol.*, 2, 128, 1973.

Walsh, J. B., Interaction of selection and biased gene conversion in a multigene family, *Proc. Natl. Acad. Sci. U.S.A.*, 82, 152, 1985.

Walsh, J. B., Selection and biased gene conversion in a multigene family: consequences of interallelic bias and threshold selection, *Genetics*, 112, 699, 1986.

Walsh, J. B., Intracellular selection conversion bias and the expected substitution rate of organelle genes, *Genetics*, 130, 939, 1992.

Walton, C., Thelwell, N. J., Priestman, A., and Butlin, R. K., The use of microsatellites to study gene flow in natural populations of *Anopheles* malaria vectors in Africa: potential and pitfalls, *J. Am. Mosq. Control Assoc.*, 14, 266, 1998.

Wang, W., Swevers, L., and Iatrou, K., *Mariner* (*Mos1*) transposase and genomic integration of foreign gene sequences in *Bombyx mori* cells, *Insect Mol. Biol.*, 9, 145, 2000.

Wayne, M. L., Contamine, D., and Kreitman, M., Molecular population genetics of *ref(2)P*, a locus which confers viral resistance in *Drosophila*, *Mol. Biol. Evol.*, 13, 191, 1996.

Weidhaas, D. E., Schmidt, C. H., and Seabrook, E. L., Field studies of the release of sterile males for the control of *Anopheles quadrimaculatus*, *Mosq. News*, 22, 283, 1962.

Wenseleers, T., Ito, F., van Borm, S., Huybrechts, R., Volckaert, F., and Billen, J., Widespread occurrence of the microorganism *Wolbachia* in ants, *Proc. R. Soc. London B Biol. Sci.*, 265, 1447, 1998.

Werren, J. H., *Wolbachia* and speciation, in *Endless Forms: Species and Speciation*, Howard, D. J. and Berlocher, S. H., Eds., Oxford University Press, Oxford, U.K., 1998.

Werren, J. H. and O'Neill, S. L., The evolution of heritable symbionts, in *Influential Passengers: Inherited Microorganisms and Arthropod Reproduction*, O'Neill, S. L., Hoffman, A. A., and Werren, J. H., Eds., Oxford University Press, Oxford, U.K., 1997, 1–41.

Werren, J. H. and Windsor, D. M., *Wolbachia* infection frequencies in insects: evidence of a global equilibrium? *Proc. R. Soc. London B Biol. Sci.*, 267, 1277, 2000.

Werren, J. H., Nur, U., and Eickbush, D., An extrachromosomal factor causing loss of paternal chromosomes, *Nature*, 327, 75, 1987.

Werren, J. H., Windsor, D., and Guo, L. R., Distribution of *Wolbachia* among Neotropical arthropods, *Proc. R. Soc. London B Biol. Sci.*, 262, 197, 1995.

Whitten, M. J., Use of chromosome rearrangements for mosquito control, in *Sterility Principle for Insect Control or Eradication*, International Atomic Energy Agency Symposium, Athens, Greece, 399, 1970.

Whitten, M. J. and Foster, G. G., Genetical methods of pest control, *Annu. Rev. Entomol.*, 20, 461, 1975.

Wilkinson, G. S., Presgraves, D. C., and Crymes, L., Male eye span in stalk-eyed flies indicates genetic quality by meiotic drive suppression, *Nature*, 391, 276, 1998.

Wilson, E. B., The supernumerary chromosomes of hemiptera, *Science*, 26, 870, 1907.

Wood, R. J., Biological and genetical studies on sex ratio in DDT resistant and susceptible strains of *Aedes aegypti* (Linn.), *Genet. Agrar.*, 13, 287, 1961.

Wood, R. J. and Newton, M. E., Sex-ratio distortion caused by meiotic drive in mosquitoes, *Am. Nat.*, 137, 379, 1991.

Wood, R. J., Cook, L. M., Hamilton, A., and Whitelaw, A., Transporting the marker gene *re* (red eye) into a laboratory cage population of *Aedes aegypti* (Diptera: Culicidae) using meiotic drive at the M^D locus, *J. Med. Entomol.*, 14, 461, 1977.

Wyss, J. H., Screwworm eradication in the Americas — overview, in *FAO/IAEA International Conference on Area-Wide Control of Insect Pests Integrating the Sterile Insect and Related Nuclear and Other Techniques*, 28 May–2 June, Penang, Malaysia, 1998, 20.

Yamamoto, K., Kausano, K., Takahashi, N. K., Yoshikura, H., and Kobayashi, I., Gene conversion in the *Escherichia coli RecF* pathway — a successive half-crossing-over model, *Mol. Gen. Genet.*, 234, 1, 1992.

Yan, G., Severson, D. W., and Christensen, B. M., Costs and benefits of mosquito refractoriness to malaria parasites: implication for genetic diversity of mosquitoes and genetic control of malaria, *Evolution*, 51, 441, 1997.

Yasuno, M., Macdonald, W. W., Curtis, C. F., Grover, K. K., Rajagopalan, P. K., Sharma, L. S., Sharma, V. P., Singh, D., and Singh, K. R. P., A control experiment with chemosterilized male *Culex pipiens fatigans* Wied. in a village near Delhi, India, surrounded by a breeding-free zone, *Jpn. J. Sanit. Zool.*, 29, 325, 1978.

Yoshiyama, M., Honda, H., and Kimura, K., Successful transformation of the housefly, *Musca domestica* (Diptera: Muscidae) with the transposable element, *mariner, Appl. Entomol. Zool.*, 35, 321, 2000.

Young, C. J., Notes on the bionomics of *Stegomyia calopus* Meigen in Brazil, I. *Ann. Trop. Med. Parasitol.*, 16, 389, 1922.

Young, O. P., Ingebritsen, S. P., and Foudin, A. S., Regulation of transgenic arthropods and other invertebrates in the United States, in *Insect Transgenesis: Methods and Applications*, Handler, A. M. and James, A. J., Eds., CRC Press, Boca Raton, FL, 2000, 369.

Zhao, Y. G. and Eggleston, P., Stable transformation of an *Anopheles gambiae* cell line mediated by the *Hermes* mobile genetic element, *Insect Biochem. Mol. Biol.*, 28, 213, 1998.

Zheng, L., Benedict, M. Q., Cornel, A. J., Collins, F. H., and Kafatos, F. C., An integrated genetic map of the African human malaria vector mosquito, *Anopheles gambiae, Genetics*, 143, 941, 1996.

Zheng, L., Cornel, A. J., Wang, R., Erfle, H., Voss, H., Ansorge, W., Kafatos, F. C., and Collins, F. H., Quantitative trait loci for refractoriness of *Anopheles gambiae* to *Plasmodium cynomolgi* B., *Science*, 276, 425, 1997.

Zwolinski, S. A. and Lamb, B. C., Non-locus-specific polygenes giving responses to selection for gene conversion frequencies in *Ascobolus immersus, Genetics*, 140, 1277, 1995.

chapter eleven

Ecological and community considerations in engineering arthropods to suppress vector-borne disease

Andrew Spielman, John C. Beier, and Anthony E. Kiszewski

Contents

11.1 Introduction

Advances in molecular biology have encouraged major research efforts devoted to improving human health by reducing the ability of natural populations of vector arthropods to transmit certain pathogens. In 1986, a well-attended symposium on this subject was held at the national meeting of the American Society of Tropical Medicine and Hygiene. The speakers participating in this earliest of formal discussions on the subject agreed that the risk of vector-borne disease might be reduced if a genetic "construct" could be developed that would block development of certain pathogens in the vector arthropod and if that construct could be linked to a genetic "drive mechanism" that would cause a disproportionate portion of the descendants of the released arthropods to carry the construct. Malaria was the primary disease discussed at the symposium, and the main construct under consideration was a gene or combination of genes that destroyed one of the developmental stages of the malaria pathogen in the vector. At the time, the newly discovered global sweep by the P-element in natural populations of *Drosophila melanogaster* (Spradling and Rubin 1986) inspired the participants to identify transposable elements as the most feasible drive mechanism for the proposed public health intervention against *Anopheles gambiae*, the main African vector mosquito. The strategy proposed at that early symposium more recently has been extended to the *Aedes aegypti* mosquitoes that transmit dengue virus (Olson et al. 1996). The enduring spirit of optimism that began in the 1980s now causes a large share of the public health entomology research budget to be invested in the genetics of vector competence, transposable elements, and the structure of vector populations (Spielman 1994).

Although, clearly, the deliberate release of hematophagous arthropods into a site rife with vector-borne human infections would be designed to improve human health, such a release may also threaten well-being, both in the short and the long term. The released organisms or the first few generations of their descendants might themselves directly cause human annoyance or transmit agents of disease. More fundamentally, the disease burden might be exacerbated until the desired health effects were accomplished, but only temporarily. Another possible consequence is a resurgence of the disease should the manipulation not be sustained in the population. Accordingly, the discussion that follows examines the need for regulatory oversight that would govern the release of genetically modified vector arthropods. In particular, we identify potential unintended consequences of such releases and recommend a rationale for endorsing such releases.

11.2   Ethics of releasing reared vector arthropods

Unique ethical problems accompany any release of hematophagous arthropods because local residents who may be affected by the experiment must register their "informed consent." Presumably, each such "experimental subject" must be provided with relevant details of the research protocol, must agree to participate, must be monitored throughout the course of the experiment, must be able to withdraw from the experiment at will, and ultimately must be apprised of the results of the work. Conventional pharmaceutical or vaccine trials are levied against a closely defined group of people, every one of whom would be identified and interviewed. Phase 1 trials generally are conducted under laboratory conditions and involve a few tens of people; Phase 2 trials are conducted in endemic field sites and involve a few hundred subjects; and Phase 3 trials may include populations comprising many thousands of people. Entomological experiments, on the other hand, permit no such distinctions. Released vector arthropods may attack anyone in their vicinity, including transient visitors, and their attacks may continue for weeks. Once released, free-ranging arthropods cannot be recalled.

Debates over the implementation of public health interventions tend to pit conservative advocates of the "Precautionary Principle" against the more permissive advocates of the "principle of minimal risk." Few non-contained entomological releases could take place if a precautionary proof of safety were prerequisite. Certain kinds of entomological exposures may be so contained and apparently innocuous that little debate over their deployment is justified. A "vector competence" experiment using laboratory-reared mosquitoes, for example, would pose few problems. In those operations, pathogen-free insects would be caged against an infected person's body upon which they would be permitted to feed. Eventually, the course of infection in these insects would be evaluated. Olfactometer experiments similarly would be subject to conventional rules of ethics. Institutional Review Boards (IRBs) could review such limited "releases" as though they were conventional Phase 1 drug or vaccine trials. Even experimental exposure of human subjects to ticks would present few ethical problems, despite the distastefulness of these organisms. Any possibility of inherited infection (= transovarial transmission, etc.) must be, of course, rigorously excluded. Although such "contained exposures" to vector arthropods would encounter few ethical obstacles, a "noncontained release" would be far more problematic.

11.2.1   Marked organisms

Mosquitoes have long been used in "mark–release–recapture" experiments that were designed to explore any of a variety of phenomena, but mainly dealing with dispersion behavior and longevity. Service (1993) listed hundreds of these experiments, including one that released 3 million hematophagous mosquitoes in 1951. Although no long-term damage has been reported, the regulatory climate has changed (Aultman et al. 2000). Any release of hematophagous female mosquitoes must now be reviewed and approved by IRBs or by other regulatory authorities. The heart of the difficulty lies in obtaining informed consent. Investigators who plan to release any hematophagous or potentially pestiferous insect might first inform and obtain the consent of all people who would find themselves in the vicinity of the release. Provision for withdrawing from such an experiment might be required, and such a provision could not be honored unless residents were willing to relocate. Even a single objection by an individual to the release of a potential pest might disqualify the entire experiment.

11.2.2   Infertile organisms

A large series of "genetic control" experiments were executed during the 1960s and 1970s, each aiming to suppress the density of one or another population of vector or pest insects. These experiments used strategies based on the release of laboratory-reared organisms exploiting sterile males, cytoplasmic incompatibility, sterile hybrid males, translocation heterozygotes, conditional lethals, meiotic drive, and compound chromosomes (Pal and LaChance 1974). The objectives of these experiments included: "1. eradication or suppression to an acceptable level, i.e., below the critical density for disease transmission; 2. long-term suppression without eradication (which might involve a commitment to continued releases); or 3. replacement of the target strain by an introduced strain that has low fertility or inability to transmit disease, conditional lethals, or some other desirable characteristics."

The late Edward Knipling's (1955) seminal demonstration that screwworm flies can be eliminated by "the use of sexually sterile males" stimulated diverse attempts designed to displace or modify various vector populations. The successful attack on those flies required the release of about 20 irradiated, factory-reared males for each female fly present at the site. Some 10 billion irradiated screwworm flies were released each year along the Texas–New Mexico border (Smith and von Borstel 1972). These insects are particularly vulnerable to such an approach because they tend to distribute themselves uniformly over the landscape and to mate randomly. Males can readily be distinguished from the much-larger female flies and, of course, are entirely innocuous — they lay no eggs. The ultimate objective of this highly successful operation was to limit the northern distribution of the "endemic population to the Isthmus of Panama, where continuous small releases would keep the North American continent free of the pest" (Smith and von Borstel 1972). Although screwworm flies originally were grown in whale meat, an environmentally friendly vegetable diet now serves that purpose. Few ethical issues arise directly from such a release.

Laven's (1967) "Okpo experiment" was the first attempt to register similar gains against a vector mosquito. That sophisticated attack on the tropical house mosquito (*Culex pipiens quinquefasciatus*) in a village in Burma was based on the principle of "cytoplasmic incompatibility," i.e., failure of egg embryonation due to infection by a rickettsial symbiont. Every aquatic breeding site in that isolated community was identified, and the abundance of the larval and pupal stages of the mosquitoes in each was estimated. About as many male pupae deriving from an incompatible laboratory colony were placed in each water container each day as were present there naturally. Fertile eggs, thereafter, became increasingly scarce. Great care had to be taken to release no females of the incompatible strain; disastrous population replacement might then have occurred. Female pupae were excluded by examining each pupa microscopically. The apparently successful outcome of this experiment, however, remains in doubt because of the absence of comparison treatments; the experiment was uncontrolled. The Okpo experimental release presented few ethical issues because only non-hematophagous male mosquitoes were released.

Quite the opposite was the case, however, when World Health Organization (WHO) personnel attempted to apply sterile male technology in India against the tropical house mosquito. Intense controversy surrounded the fiasco, which is remembered as the "Delhi experiment." Politically motivated critics accused the WHO of practicing U.S. Central Intelligence Agency–inspired biological warfare against the Indian people (Curtis and von Borstel 1978). The Indian Parliamentary Public Accounts Committee concluded that these irradiated mosquitoes would sterilize male human residents at the release site. Alarm had been generated by the massive manner in which about half a million sterilized male mosquitoes were released daily from prominent, truck-mounted cages. The intense political response to the releases caused the program to be discontinued. The outcome was a lesson teaching that, even in cases with few objective problems to burden a release, the attitudes of people residing at the release site may be exceedingly costly.

A more objective problem accompanied the landmark "Lake Apastepeque experiment" in El Salvador. This trial, sponsored by the U.S. Department of Agriculture and the Centers for Disease Control, aimed to reduce the fertility of *Anopheles albimanus*, the dominant vector of malaria in the region (Lofgren et al. 1974). Chemosterilized laboratory-reared mosquitoes of both sexes were released into the study site. As pupae, the mosquitoes were sorted by sex on the basis of size. Unfortunately, because this distinction tends to be ambiguous, nearly 700,000 females accompanied the 4.4 million males that were released over a 10-month period into a 15-km^2 malaria-endemic agricultural community of about 1000 people (Breeland et al. 1974). To protect the released mosquitoes, no public health applications of insecticide were permitted in the vicinity of the release site throughout the 2-year duration of this experiment. Instead, additional supplies of antimalaria drugs were distributed there. The authors of the study noted that "the release of large numbers of females could be a [health] problem if their longevity is sufficient to permit them to become malaria vectors. This possibility must be investigated." No such investigation was recorded. The Lake Apastepeque experiment was a great success; the fertility of the target population was reduced sharply, and their abundance transiently declined. The immediate impact of this experiment on the well-being of the residents of the site, however, remains uncertain.

Releases such as that at Lake Apastepeque were conducted in a far more permissive environment than exists today. Problems deriving from the "release ratio," the proportion of released vector or pest insects compared with those in the ambient situation, encountered in the implementation of "genetic control" strategies, illustrate issues that ultimately may impede future release of genetically modified arthropods.

11.2.3   Exotic organisms

A novel attempt to modify the genetic composition of an established pest or vector population involved the release of tropical *Aedes albopictus* mosquitoes in a suburban site in the north-central United States (Hanson et al. 1993). The aim was to reduce the hibernal survival of a recently established infestation of a temperate zone variety of these mosquitoes by releasing males derived from a laboratory colony of similar but diapause-incapable mosquitoes that originated in tropical Malaysia. Approximately 40,000 adult male mosquitoes were released. Few female mosquitoes would have been released. An unusually detailed section dealing with "safety" in the project's report discusses the remote possibility that Japanese B encephalitis virus might somehow be introduced into North America as a result of this experiment (Hanson et al. 1993). We understand that the safety discussion was a reaction to severe criticism from the regulatory authorities. Eventually, the authorities were satisfied that this release of male mosquitoes posed no danger to the residents of the study site and that no ethical issues were evident in the situation. Although this first attempt at "genetic engineering" did proceed and apparently succeed, diverse obstacles to any release should be expected.

11.3   Diversity of transgenic strategies

Prospects for creating transgenic organisms for release against vector arthropods generally focus on attempts to modify the vector genome such that the organism cannot support the development of a pathogen. The most likely general target in the case of *Plasmodium falciparum* malaria is an oocyst melanization trait in Anopheles gambiae mosquitoes that is regulated at two major genetic loci (Vernik et al. 1989). Genes from alien kinds of organisms serve similarly as promising bases for devising intervention strategies. Promoter sequences that regulate gene expression generally must be present in the construct (James et al. 1991, Mueller et al. 1993). Alternatively, bacterial symbionts may serve as vehicles for expressing foreign genes in vector arthropods. Those arthropods that feed exclusively on blood generally depend on "nutritionally mutualistic" symbionts in the gut to digest their blood meal (Beard et al. 1998). Other kinds of microbial symbionts are present in some mosquitoes and in certain other flies. Mycobacteria genes, for example, have been introduced stably into the coryneform bacteria that are symbiotic in the guts of the kissing bugs that transmit Chagas' disease, and this promises to serve as a basis for public health interventions.

Yet another form of genetic manipulation, which only indirectly involves vector arthropods, focuses on the food that the candidate vector arthropod may ingest. Algae, for example, may be transformed to express *Bacillus thuringiensis israelensis* toxins to destroy larval mosquitoes that ingest them. This protein also may be expressed in the pollen of corn that has been transformed for similar purposes (Yihdego et al., in press). The primary anopheline vectors of malaria in Africa feed abundantly on pollen. The release of such "insecticidal" GEOs (genetically engineered organisms) also requires regulatory oversight because their use may select against conventional insecticides that could otherwise remain effective. Although diverse constructs, including arthropods, microbial symbionts, crop plants, and algae, are being developed for intervening against vector-borne infection, no practical system is yet in operation.

11.4   Drive mechanisms

Genetic drive mechanisms are required to disseminate a GEO rapidly and thoroughly through a wild population. Certain non-Mendelian genetic phenomena may transfer genes most effectively between germ lines or favor the proliferation of particular reproductive combinations in a population. Favorable traits that could be linked to such entities therefore would be "driven" through a wild population. Patently unacceptable alternatives include sustained inundative releases of the modified vector or use of particular conditional lethals, such as a gene conferring insecticide insusceptibility. Conditional lethals might be more acceptable when used in a sexing scheme that ensures against the presence of females in the release population. In conjunction with direct drive mechanisms, such conditional lethals would serve as a precaution in the event that the diversity of drive mechanisms is limited.

The bulk of efforts seeking to design a transgenic vector with reduced competence has focused on methods for linking genetic constructs to a transposable element. Such transposons are parasite-like segments of DNA that reproduce within genomes. Retrotransposons function similarly but use an RNA intermediate. Certain types carry a gene for a transposase that facilitates copying and reinsertion of the transposon into a new site within the genome. Others must rely on their hosts for expression of transposition enzymes. A single kind of transposon may occupy multiple sites within a genome. A transposon named *hobo*, for example, occupies some 66 sites within the genome of *Drosophila simulans* populations throughout the world (Vieira et al. 1999). Replication frequency appears to be regulated by interactions between the host and the transposon (Hartl et al. 1997a, 1997b). A single host may carry numerous transposons within its genome. A global survey of *D. melanogaster* revealed that some populations contain more than 1000 copies of about 30 different transposons and retrotransposons. *Mariner*-like elements are unique in not requiring transmission through the germ line; they also can transpose within somatic cells. Such profligate transposition throughout evolutionary history has provided about 42% of the genetic material of the human genome, much of it residing in the less functional heterochromatin (Smit 1999). Fruit flies carry between a 5 and a 15% load of errors, and certain studies suggest that flies carrying few errors remain far from saturated. The great diversity of transposon varieties and the ability of genomes to function under heavy loads would appear to facilitate artificial introductions.

The stability of transposon-based constructs remains in question. Although copy number would be expected to increase after a release, it remains unclear how many copies would accumulate before reaching an equilibrium and how long they might persist intact. Degraded and inactive transposons predominate among those detected in genomes thus far. Although molecular geneticists have succeeded in transfecting mosquitoes with transposons in the laboratory with a 4% success rate (Coates et al. 1998), it is not clear how stable their constructs might be, how many generations a transgenic mosquito might remain incompetent, or how readily these elements would be incorporated into field populations. Considerable sequence diversity characterizes the 297 elements of *D. melanogaster*, including frequent internal rearrangement (Dominguez and Albornoz 1999). Frequent frame shifts, substitutions, insertions, and deletions, which can disrupt their open reading frame and render them nonfunctional, characterize other transposons in nature. A 2×10^{-5} mutation rate has been estimated for certain of these elements in *D. melanogaster*. Particular non-LTR retrotransposons often produce "dead-on-arrival" copies because of the frequent loss of fragments at their 5′ end (Petrov et al. 1996). Upon reaching a tenuous equilibrium, transposon copy numbers generally decline incrementally as a result of stochastic loss. Transposons have a limited life span and therefore are inherently unstable. Their instability must be considered when designing interventions based on these drive mechanisms.

Transgenic releases should have in place a series of different transposable element constructs before any proposed introduction commences. This permits a backup release in the event that poor saturation, impaired fitness, or instability compromises the effectiveness of the previously released construct before the trait of interest becomes fixed in the wild population. Otherwise, a wild population inevitably would rebound after several generations because of its reproductive fitness advantage over transposon bearers. Fitness decreases as transposon copies increase in the genome. Cumulative fitness deficits may overwhelm the ability of an organism to cope with its environment. Each genetic construct should occupy a distinctly different set of transposition sites, and be demonstrated to be capable of transfecting its host in serial combination. Even with a backup, some minimum measure of confidence in the stability of each construct must be demonstrated. Releasing one unstable construct after another would be pointless.

Until recently, reliable methods for transforming mosquitoes have been elusive. A *Science Research News* report in 1993 commented that "most vector biologists [are] pinning their hopes on a genetic manipulation technique, [that] the race is on to find a transposable element that functions in *An[opheles] gambiae*," and *minos* appeared to be "one leading candidate" (Aldhous 1993). The pace was set when, nearly a decade later, the *minos* transposon was used to insert a stably inherited fluorescent marker into *A. stephensi* mosquitoes (Cattaruccia et al. 2000), and *hermes* was used to insert a defensin gene into *Aedes aegypti* (Kokoza et al. 2000). Neither accomplishment approached the ultimate requirement for a usable drive mechanism because separate promoter mechanisms were required. The power of these newly developed techniques, however, is likely to accelerate the pace of this perceived race to identify a suitable transposable element.

11.5 Theoretical considerations

Several other considerations arise in the context of designing successful interventions using GEOs. Although the following factors have not been documented as problematic, they nevertheless are areas of potential concern.

11.5.1 Competence

Most GEO interventions that have been proposed seek to reduce the suitability of the target vector population as a host for the pathogen. Although this strategy generally assumes that the pathogen develops and multiplies freely in natural vector populations, no estimates of the prevalence of malaria competence in nature appear to have been published, nor are estimates available that would lead us to anticipate the effect of a change in competence on human health. Indeed, competence is a weak element in the classical model of "vectorial capacity" (Garrett-Jones and Shidrawi 1969). Although vector longevity or narrowness of host range contribute exponentially or geometrically to the force of transmission, competence contributes only linearly. Thus, indoor applications of residual insecticide that reduce the likelihood that an adult mosquito would survive to become infectious would interrupt transmission far more powerfully than would any reduction in competence.

11.5.2 One-pathogen strategies

An arthropod vector of one infection is likely to transmit other infections, as well, and this presents special ethical problems. Anopheline vector mosquitoes, for example, may transmit at least six human pathogens: *P. falciparum*, *P. vivax*, *P. malariae*, *P. ovale*, O'nyong nyong virus, and *Wuchereria bancrofti*. In addition, numerous apomictic or partially panmictic vector populations may be present in the same site. A release directed against one pathogen in only one of these vector populations, therefore, may have little effect on the overall force of transmission of these pathogens. Such an outcome may be complex. In the event that a given release results in an increase in vector abundance, human health may bear an added burden.

11.5.3   Herd immunity

Any temporary improvement in health tends to prejudice the future well-being of a population, a relationship that holds particularly true in the case of malaria. Where the force of transmission is intense, infection may be virtually universal. Episodes of patent disease, however, tend to be infrequent in adults. Although few indigenous adults express malaria-related symptoms, virtually all visitors from non-endemic sites would become incapacitated and suffer life-threatening episodes of illness. Such immune protection of perpetually infected residents dissipates when transmission ceases. This loss of herd immunity, as would follow transient elimination of infection in a site, may be mirrored by a devastating outbreak of disease in the event that transmission resumes. Reversal of a major public health gain also would be burdened by a series of social disruptions. An indigenous population adapts psychologically to frequent illness, an attitude that rapidly becomes reversed once the burden is relieved. Resumption of transmission in this "newly virgin" population, of course, would provoke outrage. In the event that an intervention were supported by outside funding, over time the donor community might become "fatigued" and make renewed funding difficult. Resumed transmission of malaria after a period of relief would produce damage that far exceeded the transient benefit that might have resulted from the period of relief.

11.5.4 Maintaining the population density of GEOs

Additional problems must be considered in the likely event that the density of the population of released organisms must be maintained at a specified level, relative to that of the population of native vectors. Toward this end, the density of the released population must be monitored and steps taken to ensure that its density remains at the specified level. Peridomestic insecticide use might have to be curtailed, as in the case of the Lake Apastepeque experiment. Similarly, the use of bednets or screening might be incompatible with the objectives of the release. Indeed, success might hinge on the availability of sites that are suitable for the breeding of the released arthropod, and peridomestic artificial breeding sites might be required. Even if the modified vector were completely incompetent as a vector of any human infection, presumably it would still be anthropophagous and such a pest cannot be nurtured.

11.6   Temporal–spatial relationships

Transposon-driven releases may be subject to strong environmental forces that create considerable uncertainty regarding the successful fixation of an accompanying construct, especially early in a release and when conditions for perpetuation of the released arthropods are marginal. Because vector populations are distributed patchily and are interrupted seasonally, a transgenic release should not be expected to progress smoothly, but rather in fits and starts. The instability of such progression toward fixation would tend to increase as conditions for the perpetuation of released transposon-bearing vectors become less optimal. The reproductive and ecological fitness of vectors bearing transgenic constructs is critical in determining the rate and stability of ultimate fixation.

The dissemination of transposon-driven constructs in a population may be limited by particular characteristics of the transformed organisms and by their interactions with wild populations. Under ideal conditions with unrestricted, panmictic breeding and global dispersal, certain mathematical models indicate that a released construct may become fixed in about 30 generations as long as overall fitness is at least half that of wild types (Ribeiro and Kidwell 1994). Organisms that mate but once in their lifetime remain relatively close to discontinuous breeding sites (such as those of *An. gambiae* s.l.) and undergo seasonal constriction of their populations, and this may require greater fitness and much more time to become fixed. Certain models suggest that under such conditions, even constructs that preserve fitness at about 80% of that in wild-type mosquitoes would require many hundreds of generations to achieve fixation (Kiszewski and Spielman 1998).

The unstable progress toward fixation may endanger health in areas hyperendemic for malaria. The ebb and flow of transposon-bearing vector populations may create situations in which exposure to malaria disappears for a few years and then returns. This may compromise protective immunity in certain local populations, leading to epidemics with high rates of morbidity and mortality. Even where fixation progresses more stably, pockets of transmission may become interspersed with refugia, thereby exposing members of protected communities to infection when they travel relatively short distances from their homes.

Measures designed to enhance the ability of a transgenic organism to overtake a wild population may have limitations. Achieving highly favorable release ratios (1:1 or greater) will accelerate fixation but may require intensive interventions against wild populations prior to release. Stewardship strategies, including repeated release of transgenic organisms, may have a similar effect. Mathematical models suggest that, while the average time until fixation may be reduced considerably by such practices, considerable uncertainty and variability in outcomes remains, primarily as a result of stochastic instabilities associated with the early stages of a release (Kiszewski and Spielman 1998).

11.7   Epidemiological objectives

Although efforts to produce transgenic malaria-incompetent *Anopheles gambiae* mosquitoes have been pursued for more than a decade now, the health objective of such efforts has not precisely been defined. Simply put, the question remains: How would reduced vector competence translate into reduced public health burden? The question is most appropriately asked for malaria in sub-Saharan Africa because the burden imposed by vector-borne disease is greatest there and because the *A. gambiae* complex has been the most frequent subject of studies to reduce vector competence. Although these mosquitoes are exceptionally long-lived and human biting, their competence as malaria vectors is restrained because they tend to develop far fewer oocysts than do certain vector mosquitoes native to North America or India. The magnitude of the sporozoite inoculum, presumably, would reflect oocyst load, and pathogenesis would, presumably, relate to this quantum of infection. The proportion of field-derived mosquitoes that become infective after feeding on an infectious person, however, has not yet been determined. We lack a model that would help relate increments of reduced malaria burden with reduced vector competence.

Risk of malaria infection is estimated most readily by calculating an entomological inoculation rate (EIR) as the product of the mosquito biting rate times the proportion of mosquitoes with their salivary glands infected by sporozoites. The EIR measures the average number of infective bites per person per unit time. Annual EIRs throughout sub-Saharan Africa range from <1 to more than 1000, depending upon the environment and climate supporting vector populations (Beier et al. 1999). The frequency of symptomatic infection as well as the severity of symptoms in Africa relate directly to EIR (Mbogo et al. 1993). Interventions that reduce malaria competence in the vector population, therefore, would translate directly into a similar increment of improved health.

These considerations suggest that an effective anti-vector intervention would be likely to reduce malaria incidence but may improve human health only marginally because severe disease occurs even when the EIR exceeds 1 (Mbogo et al. 1993). Irrespective of natural immunity, severe malaria episodes frequently follow only a single infective bite. Malaria prevalence frequently exceeds 40 to 60% in African sites characterized by an EIR < 1. An EIR ≤ 1 would seem a likely minimal goal for antimalaria interventions. To be worthwhile, a GEO release against malaria must promise to reduce risk to at least one sporozoite inoculum per person per year.

Considerations relating to the structure of vector populations suggest difficulties that may obstruct proposed releases of transgenic malaria-refractory mosquitoes in real-world African communities. Many such sites are infested by three potent vector species, *A. gambiae* and *A. arabiensis* of the *A. gambiae* complex, and *A. funestus* (Powell et al. 1999). The *A. gambiae* s.l. mosquitoes populating individual African sites are likely to include such genetically isolated demes as those designated as Mopti, Bamako, and Savanna (Favia et al. 1997). If specimens of an *A. gambiae* GEO deme were to be released in such a site, prevalence would continue unabated if any of the resident non-GEO populations were to continue to transmit at EIRs that exceed one infective bite per year. In much of Africa, the proportion of EIRs due to non-*A. gambiae* mosquitoes ranges from about 10 to 90%. The species composition within projected release sites will, therefore, be a key determinant of the public health success of GEOs.

The requirement for sustainable gains introduces complexity into the planning of any intervention against a vector-borne disease. To be avoided, of course, is an intervention that increases the disease burden or produces an outcome that cannot be sustained or one that impedes subsequent interventions.

11.8   Regulatory requirements

The deployment of transgenic arthropods is regulated in the United States by the Biotechnology Permits Unit of the Animal and Plant Health Inspection Service (APHIS) of the Department of Agriculture operating under the provisions of the Federal Plant Pest Act and the Plant Quarantine Act. A description of these functions is displayed on the Internet at the following address:

www.aphis.usda.gov:80/bbep/bp/arthropod/

On that Web site, APHIS indicates that, "The fundamental risk assessment to be addressed is: Will the genetic alteration modify ecologically or environmentally relevant properties of the organism? Specific potential or perceived risks associated with the release of a transgenic arthropod could include the displacement of native populations, a change in host or prey utilization or ecological distribution, the transfer of exogenous DNA to other organisms, or, if one of the characteristics of the transgenic arthropod was increased resistance to herbicides or pesticides, subsequent usage of such chemicals." The description, however, includes no discussion of criteria specifically relating to human health. In commenting on this lacuna, Beard and colleagues (1998) suggest that the release of numerous genetically modified vectors may actually increase risk of human disease if numerous vectors are involved in transmission.

The complex interactions and unintended consequences that might derive from transgenic releases call for a careful approach to their planning and execution. Thorough field trials generally should precede large-scale interventions, except when the intervention methodology shows too little promise in laboratory experimentation to warrant such a test. Ease of containment must be considered in field trial design to limit any deleterious

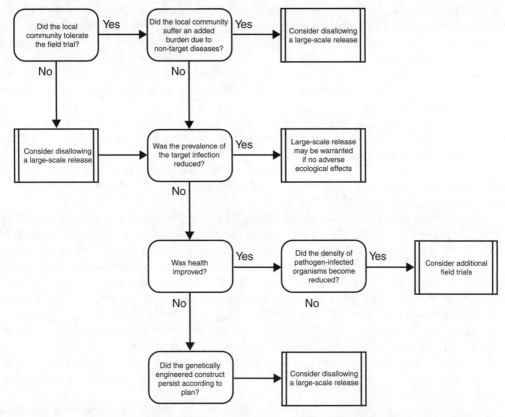

Figure 11.1   Decision tree for judging whether field trials of GEO releases are warranted based on the characteristics of the genetic construct, its drive mechanism, and the vector organism (Modified from Scientists' Working Group, *Manual for Assessing Ecological and Human Health Effects of Genetically Engineered Organisms*, Part Two: *Flowcharts and Worksheets*, The Edmonds Institute, Edmonds, WA, 1998.)

effects, in case a release causes unanticipated complications. Isolated populations, such as occur on an island, would facilitate limiting the spread of a disadvantageous release. The likelihood of unintended consequences warrants a careful, stepped approach in designing release strategies.

To determine whether a preliminary field trial is warranted, one must first consider the pathogen-transmitting competence of a transgenic vector (see decision tree in Figure 11.1). There is little point in proceeding farther if tangible benefits cannot be realized. If drive mechanisms are required, some demonstration of their effectiveness must be provided experimentally. The stability of a drive mechanism is also critical in determining the feasibility of a release. Unstable construct–drive mechanism linkages can prove dangerous when they decay after a period of interrupted transmission. Immunological lapses in the population during this period of protection would render it susceptible to explosive outbreaks that would ensue when wild-type vectors returned. This threat proves less critical if alternative drive mechanisms are available that would allow protection to be restored through a new release. The detailed molecular characteristics of a genetic construct may, therefore, prove important in the field.

The scale required of a release also may provide an indication of its desirability. If transgenic vectors must outnumber wild types in order to drive replacement and if transgenic vectors are as pestiferous as the wild ones, it may not be ethical to impose such discomfort on the residents of a release site, and such residents may impose limits on the scale of further releases. Many vectors are capable of transmitting more than one type of infection. Massive releases can decrease one infectious disease burden while promoting all others associated with a vector. Even if the disease whose incidence is reduced is associated with greater morbidity and mortality than all other local diseases combined, such an outcome may not necessarily be welcomed by the affected communities. If transgenic vectors provide little annoyance and if they do not transmit other infections that are unaffected by the genetic construct, then these points become moot.

To determine whether large-scale releases should follow after field trials, the response of local communities to the field trials must be closely monitored (Figure 11.2). Precedent indicates that certain communities may react violently even to relatively innocuous releases of sterile male insects. The biological success of a trial intervention is meaningless unless affected communities are able to perceive that a GEO release is desirable.

Trials should monitor the infectious disease burdens suffered by a community. Any unanticipated enhancement of burdens revealed by trials would provide further grounds for discontinuing a release program. Trials also must demonstrate that the prevalence of the target pathogen is reduced. Otherwise, a large-scale release would not confer sufficient benefits to the population to warrant the release. Reduction of prevalence, however, may not always provide a proportionate increase in level of health. In such cases, the criterion for continuation must be a decrease in overall morbidity or infection rates. Once evidence for a health benefit can be shown, its sustainability must be considered. Some indication of the stability of genetic constructs may be derived from field trials, although the time span for the work may be too limited to detect any potential for instability. As a result, investigators may be limited to theoretical considerations regarding sustainability.

If the design of a transgenic release meets each of the considerations discussed above, then a large-scale field trial may indeed be warranted, particularly if pilot field trials have already been performed and no evidence of adverse ecological effects has been identified. A limited set of field trials, however, may not provide sufficient confidence to persuade biologists, government officials, and the general public to form a consensus toward proceeding with an intervention. Although absolute confidence in outcomes may not be required, ambiguous or weak outcomes may indicate a need for additional testing. Furthermore, these decision trees consider only the biological and ethical rationality of GEO releases. Analyses of cost-effectiveness and operational feasibility also deserve serious consideration.

11.9   Conclusions

The release of genetically modified arthropods that would otherwise serve as vectors of human pathogens carries special burdens and requires regulation. In general, such a release should not be permitted unless:

1. The released organisms do not annoy local residents more than do any ambient vector organisms.
2. The release results in no increase in abundance of hematophagous arthropods.
3. The release requires no reduction in ongoing health-promoting activities.
4. The force of transmission of microbes other than the target pathogen would not increase.
5. The release does not compromise future interventions against the target disease.
6. Any improved state of health of people living in the release site is sustainable.

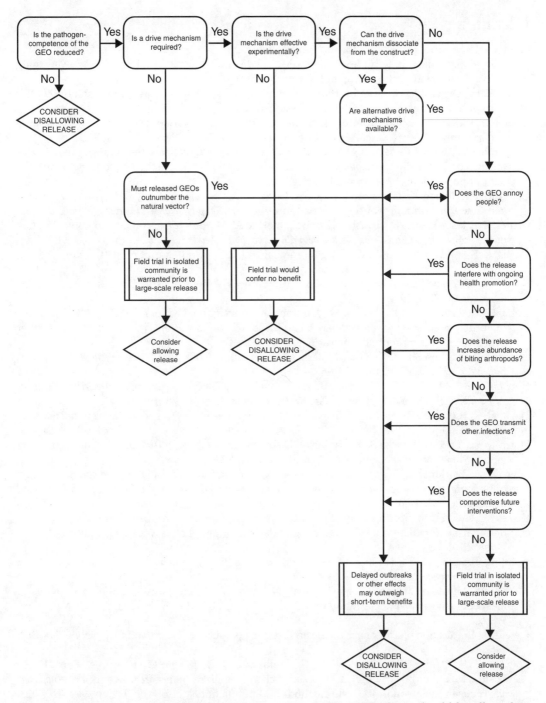

Figure 11.2   Decision tree for judging whether a large-scale GEO release should be allowed to proceed, based on the outcomes of isolated field trials. (Modified from Scientists' Working Group, *Manual for Assessing Ecological and Human Health Effects of Genetically Engineered Organisms*, Part Two: *Flowcharts and Worksheets*, The Edmonds Institute, Edmonds, WA, 1998).

Literature cited

Aldhous, P., Malaria: focus on mosquito genes, *Science*, 261, 546, 1993.

Aultman, K. S., Walker, E. D., Gifford, F., Severson, D. W., Beard, C. B., and Scott, T. W., Managing risks of arthropod vector research, *Science,* 288, 2321, 2000.

Beard, C. H., Durvasula, R. V., and Richards, F. H., Bacterial symbiosis in arthropods and the control of disease transmission, *Emerg. Infect. Dis.,*4, 581, 1998.

Beier, J. C., Killeen, G. F., and Githure, J. J., Entomological inoculation rates and *Plasmodium falciparum* malaria prevalence in Africa, *Am. J. Trop. Med. Hyg.,* 61, 109, 1999.

Breeland, S. G., Jeffery, G. M., Lofgren, C. S., and Weidhaas, D. E., Release of chemosterilized males for the control of *Anopheles albimanus* in El Salvador: I. Characteristics of the test site and the natural population, *Am. J. Trop. Med. Hyg.,* 23, 274, 1974.

Cattaruccia F., Nolan, T., Loukeris, T. G., Blass, C., Savakis, C., Kafotos, F. C., and Crisanti, A., Stable germline transformation of the malaria mosquito *Anopheles stephensi, Nature,* 405, 958, 2000.

Coates, C. J., Jasinskeine, N., Miyashiro, L., and James, A. A., Mariner transposition and transformation of the yellow fever mosquito, *Aedes aegypti., Proc. Natl. Acad. Sci. U.S.A.,* 95, 3748, 1998.

Curtis, C. F. and von Borstel, R. C., Allegations against Indian research unit refuted, *Nature,* 273, 96, 1978.

Dominguez, A. and Albornoz, J., Structural instability of 297 element in *Drosophila melanogaster, Genetica,* 105, 239, 1999.

Favia, G., della Torre, A., Bagayoko, M., Lanfrancotti, A., Sagnon, N., Toure, Y. T., and Coluzzi, M., Molecular identification of sympatric chromosomal forms of *Anopheles gambiae* and further evidence of their reproductive isolation, *Insect Mol. Biol.,* 6, 377, 1997.

Garrett-Jones, C. and Shidrawi, G. R., Malaria vectorial capacity of a population of *Anopheles gambiae, Bull. World Health. Org.,* 40, 531, 1969.

Hanson, S. M., Mutebi, J.-P., Craig, G. B., and Novak, R. J., Reducing the overwintering ability of *Aedes albopictus* by male release, *J. Am. Mosq. Control. Assoc.,* 9, 78, 1993.

Hartl, D. L., Lohe, A. R., and Lozovskaya, E. R., Regulation of the transposable element mariner, *Genetica,* 100, 177, 1997a.

Hartl, D. L., Lozovskaya, E. R., Nurminsky, D. I., and Lohe, A. R., What restricts the activity of mariner-like transposable elements? *Trends Gen.,* 13, 197, 1997b.

James, A. A., Blackmer, K., Marinotti, O., Ghosn, C. R., and Racciopi, J. V., Isolation and characterization of the gene expressing the major salivary gland protein of the female mosquito *Aedes aegypti, Mol. Biochem. Parisitol.,* 44, 245, 1991.

Kiszewski, A. E. and Spielman, A., Spatially explicit model of transposon-based genetic drive mechanisms for displacing fluctuating populations of anopheline vector mosquitoes, *J. Med. Entomol.,* 35, 584, 1998.

Knipling, E. F., Possibilities of insect control or eradication through the use of sexually sterile males, *J. Econ. Entomol.,* 48, 459, 1955.

Kokoza, V., Ahmed, A., Cho, W.-L., Jasinskeine, N., James, A. A., and Raikel, A., Engineering blood meal-activated systemic immunity in the yellow fever mosquito, *Aedes aegypti, Proc. Natl. Acad. Sci. U.S.A.,* 97, 9144, 2000.

Laven, H., Eradication of *Culex pipiens fatigans* through cytoplasmic incompatibility, *Nature,* 216, 383, 1967.

Lofgren, C. S., Dame, D. A., Breeland, S. G., Weidhaas, D. E., Jeffery, G., Kaiser, R., Ford, H. R., Boston, M. D., and Baldwin, K. F., Release of chemosterilized males for the control of *Anopheles albimanus* in El Salvador: III. Field methods and population control, *Am. J. Trop. Med. Hyg.,* 23, 288, 1974.

Mbogo, J., Snow, R. W., Kabiru, E. W., Ouma, J. H., Githure, J., Marsh, K., and Beier, J. C., Low level *Plasmodium falciparum* transmission and the incidence of severe malaria infections on the Kenyan coast, *Am. J. Trop. Med. Hyg.,* 49, 245, 1993.

Mueller, H. M., Crampton, J. M., della Torre, A., Sinden, R., and Chrisanti, A., Members of a trypsin gene family in *Anopheles gambiae* are induced in the gut by blood meal, *EMBO J.,* 12, 2891, 1993.

Olson, K. E., Higgs, S., Gaines, P. J., Powers, A. M., Davis, B. S., Ramrud, K. I., Carlson, J. O., Blair, C. D., and Beaty, B. J., Genetically engineered resistance to Dengue-2 virus transmission in mosquitoes, *Science*, 272, 884, 1996.

Pal, R. and LaChance, L. E., The operational feasibility of genetic methods for the control of insects of medical and veterinary importance, *Annu. Rev. Entomol.*, 19, 269, 1974.

Petrov, D. A., Lozovskaya, E. R., and Hartl, D. L., High intrinsic rate of DNA loss in *Drosophila*, *Nature*, 384, 346, 1996.

Powell, J. R., Petrarca, V., della Torre, A., Caccone, A., and Coluzzi, M., Population structure, speciation, and introgression in the *Anopheles gambiae* complex, *Parassitologia* (Rome), 41, 101, 1999.

Ribeiro, J. M. and Kidwell, M. G., Transposable elements as population drive mechanisms: specification of critical parameter values, *J. Med. Entomol.*, 31, 10, 1994.

Service, M.W., *Mosquito Ecology*, Elsevier Applied Science, London, 1993, 652.

Smit, A. F. A., Interspersed repeats and other mementos of transposable elements in mammalian genomes, *Curr. Opin. Genet.*, 9, 657, 1999.

Smith, R. H. and von Borstel, R. C., Genetic control of insect populations, *Science*, 178, 1164, 1972.

Spielman, A., Why entomological antimalaria research should not focus on transgenic mosquitoes, *Parasitol. Today*, 10, 374, 1994.

Spradling, A. C. and Rubin, G. M., Transposition of clonal P elements into *Drosophila* germ line chromosomes, *Science*, 218, 341, 1986.

Vernik, K. D., Collins, F. H., and Gwadz, R. W., A general system of resistance to *Plasmodium* infection in the vector *Anopheles gambiae* is controlled by two genetic loci, *Am. J. Trop. Med. Hyg.*, 40, 585, 1989.

Vieira, C., Lepetit, D., Dumont, S., and Biemont, C., Wake up of transposable elements following *Drosophila simulans* worldwide colonization, *Mol. Biol. Evol.*, 16, 1251, 1999.

Yihdego, Y., Pollack, R. J., and Spielman, A., Enhanced development in nature of larval *Anopheles arabiensis* mosquitoes feeding on maize pollen, *Am. J. Trop. Med. Hyg.*, in press.

chapter twelve

Environmental risks of genetically engineered vaccines

Terje Traavik

Contents

0-8493-0439-3/00/$0.00+$1.50
© 2002 by CRC Press LLC

12.1   Introduction: The past and the present

The bicentennial celebration of the first vaccination took place just a few years ago. In 1796, Edward Jenner injected cowpox virus into James Phipps, and later on challenged the boy with fully virulent human smallpox (variola) virus (Jenner 1798). The boy survived, and Jenner thus had protected him against one of the most dreaded human diseases of all time. The smallpox vaccination story ended in triumph over the variola virus, following a worldwide vaccination campaign (Fenner et al. 1988). In recognition of Edward Jenner's contribution, procedures aimed at protection against disease by pre-mobilization of the immune system were termed *vaccination*, derived from the Latin word for cow, *vacca*.

Vaccination is meant to provide the individual with immunological protection before infection takes place. If the vaccinated proportion of a population is high enough, disease symptoms of the individual, as well as transmission of the disease agent, may be prevented. Ultimately, this may result in eradication of the disease agent, as achieved with smallpox and hoped for with poliovirus (Bloom and Widdus 1998).

12.1.1   Prevention or treatment?

The reasons for giving priority to prevention and prophylaxis instead of therapy are stronger than ever. The therapeutic benefits offered by chemotherapeutics and antibiotics are reduced by the development of resistance in microorganisms, viruses, and cancer cells. Furthermore, for many infectious and neoplastic diseases, damage done before symptoms are recognized can result in lasting deleterious effects, death, or loss of livelihood, even if efficient therapy is available.

Although a number of vaccine strategies and successful vaccines were developed in the last century, vaccines are still unavailable for a number of diseases that theoretically may be prevented by vaccination (Ellis 1999). Through progress in molecular immunology and genetic engineering, opportunities to produce vaccines for a number of purposes and target organisms (mammals, birds, fish, plants) have been dramatically improved. The new strategies may give rise to vaccines that are pure, provide efficient and long-lasting immunological responses at a low price, and have decreased potentials for unpredicted side effects in vaccinated individuals. On the other hand, theoretical and unpredicted harmful effects and hazards to the environment and specific ecosystems may turn out to be more severe than current worst-case scenarios (Traavik 1999b). From that perspective, it is highly disturbing to find so little effort dedicated to risk-associated research and so much vaccine-relevant research dominated by commercial interests.

The definition of *safety* in vaccinology is very narrow compared with the putative risks and hazards vaccine use may imply. Primarily, safety research studies unintended and unwanted side effects in the vaccinated individuals themselves. Second, such research is often directed at non-target effects on unvaccinated individuals within the same species. Few efforts have been dedicated to examining unintended and non-target effects across species or wider biological borders. Because the narrow vision of safety is associated with a narrow set of research strategies, many potential hazards and harms related to various vaccine categories may remain unexamined, or even unimagined, until one or more of them actually occurs. Very few research reports concerning environmental or ecological effects of genetically engineered vaccines have been published. On the other hand, examples of scientists defending — some with almost religious zeal — the total innocuousness of vaccines, without taking environmental and non-target effects into consideration, are numerous (Danner 1997).

12.1.2   Vaccines and vaccination

Vaccination is intended to prevent or limit the effects of disease. The ideal vaccine provides an optimal mobilization of the adaptive immune system with no unwanted side effects.

Vaccination may take place many years before exposure to the pathogen, and hence immunological memory is a critical element.

Vaccination may be used to protect against infectious diseases and cancer (Ellis 1999, Jaffee 1999). Recently, vaccines have also been used to induce infertility in humans and in domestic and wild animals. Free-ranging wild animals are reservoirs of important human and domestic animal pathogens, i.e., rabies virus, hantaviruses, Ebola/Marburg viruses, Nipah/Hendai viruses, and a number of arboviruses. Hence, vaccination of the reservoir animals may prevent disease in humans and domestic animals. Other free-ranging mammalian species are considered "pest animals" in the context of food and feed production. Vaccine-enforced infertility is becoming an alternative for control and reduction of such pest populations.

Most vaccines have been of the "whole disease agent" type. Bacterial cells or virus particles, inactivated or "live," were used for immunization after varying degrees of purification. Live vaccine agents are able to infect those vaccinated, but their disease-eliciting abilities are attenuated through propagation in unnatural host cells or organisms (review by Hilleman 1998).

The safest vaccines are those that only consist of an essential polypeptide (antigen) derived from a pathogen. The polypeptide should contain antigenic epitopes that elicit protective immunity against the disease agent upon uptake in the organism. However, immunization with antigens alone often elicits weak or no immunity, and the responses may be restricted to antibodies and Th (T-helper cells). Such responses often fail to eliminate cancer cells or virus-infected cells. Cancer cells or virus-infected cells may be efficiently killed by cytotoxic T lymphocytes (CTLs). CTLs are mobilized when "foreign" peptides are present on the surface of cells. "Live" viral vaccines efficiently mobilize both specific antibody production and CTLs, and the same is the case for naked DNA/RNA vaccines that are taken up by and have their genes expressed intracellularly. It must be noted, however, that a number of promising strategies are now being pursued to obtain more adequate immune responses with polypeptide-based vaccines (review by Raychaudri and Rock 1998).

Delivery of vaccines may take place by injection, inhalation, or ingestion. Free-ranging animals are offered vaccine in baits spread from airplanes or helicopters. Fish are bathed in or sprayed with vaccine-containing solutions. All vaccine delivery strategies hold the potential for vaccine-containing material to end up in unintended locations, biotopes, and ecosystems. Hence, escape of viruses, microorganisms, transgenic organisms, and bioactive molecules, i.e., proteins, DNA, or RNA, may take place.

12.2   "Risk" and "the Precautionary Principle"

Risk is defined as the probability that a hazardous event will take place multiplied by the consequences arising *if* it does take place. With regard to development and commercialization of genetically engineered nucleic acids, organisms, and viruses, we often are neither able to define probability of unintended events nor to know the consequences of those events. The present state of ignorance calls for invoking the "Precautionary Principle," which implies that responsibility for future generations and the environment is to be combined with the anthropocentric needs of the present (Myhr and Traavik 1999). Used in the context of genetically engineered vaccines, the principle might read: "Environmental and health policies must be aimed at predicting, preventing and attacking the causes of environmental or health hazards. When there is reason to suspect threats of serious, irreversible damage, lack of scientific evidence should not be used as a basis for postponement of preventive measures" (revised after Cameron and Abouchar 1991).

The potential value of the Precautionary Principle both for risk management and for generation of risk-associated research can hardly be overestimated. A good example of this follows. In the last decade, researchers have been eager to make plants resistant to viral infections by inserting virus genes in the plant genome. If, for example, the gene that codes for the coat protein for the cowpea chlorotic mottle virus (CCMV) is inserted in plants, the plants become resistant to both CCMV and several other related viruses. When such transgenic plants are infected with other viruses, new, recombinant progeny, which have had their host specificity and other biological properties changed, may arise (Greene and Allison 1994). After this became apparent, Greene and Allison (1996) demonstrated that a targeted trimming of the viral transgene seemed to eliminate development of viral recombinants. Their work illustrates the value of invoking the Precautionary Principle: to allow sufficient time to identify risk-imposing mechanisms and to look for means to prevent them.

12.3   Genetically engineered vaccines: strategies and potential hazards

It must be emphasized that the risk factors and hazards discussed here are hypothetical, based on theoretical considerations. There is, however, a valid and definite reason for this. A vast number of research articles concerning vaccine safety as related to adverse immunological and other reactions in vaccinated individuals have been published. But to date scientific literature concerning environmental and ecological risks is very sparse and exclusively related to the use of live virus vector and deletion mutant vaccines (for review, see Traavik 1999b).

12.3.1   Subunit and recombinant vaccines

So-called subunit and recombinant vaccines seem relatively safe, from both an environmental and a human health point of view. Subunit vaccines are based on technologies ranging from chemical purification of components of the *in vitro* propagated pathogen to recombinant DNA techniques producing a single viral or bacterial protein.

A DNA sequence coding for the immunogen(s) of a given infectious agent or cancer cell type may be translated into amino acid sequences. Corresponding peptides may then be synthesized in the laboratory. When properly administered, the peptides may elicit protective immune responses directed against the infectious agent/cancer cell. The disadvantages of such approaches are that the protective immune responses elicited have been too weak and the production of vaccines too expensive. Intensive research and developmental activity, however, is devoted to this field, especially to the combination of pure immunogenic peptides and potent and safe adjuvants (reviewed by Raychaudhuri and Rock 1998).

Recombinant vaccine production occurs under contained conditions. Recombinant DNA is made by isolation of DNA fragment(s) coding for the immunogen(s) of an infectious agent/cancer cell, followed by the insertion of the fragment(s) into vector DNA molecules (i.e., plasmids or viruses), which can replicate and conduct protein expression within bacterial, yeast, insect, or mammalian cells. The immunogen(s) may then be completely purified by modern separation techniques. When essential antigenic epitopes are determined by conformation, this approach may yield better results than corresponding synthetic vaccines. Otherwise, the theoretical drawbacks are the same: the vaccines tend to provide good antibody responses, but weak T-cell activation. A number of recombinant vaccines have already been marketed.

Subunit and recombinant vaccines simply contain pure proteins (immunogens, antigens). Potential hazards associated with unintended releases of such vaccines will be toxic, allergic, and other unwanted immunological reactions in animal or human individuals within the release area. However, the DNA constructs that are used to produce recombinant vaccines, or the cell cultures used to have them expressed, conceivably may escape from laboratories/manufacturing units. Risk assessment would then involve any potential hazards associated with the accidental release of a genetically modified nucleic acid or organism (see Section 12.3.4 and subheadings in Section 12.4, and reviews in Traavik 1999a, 1999b).

12.3.2   Genetically engineered viruses and microorganisms as vaccines

For a number of viruses, and recently also for some bacteria, "disease-causing" genes have been identified and characterized. Some viruses have been genetically engineered, by introducing deletions or specific point mutations, without destroying the ability of the virus to infect its target host species by a natural route. The genetically modified virus may then activate and stimulate the immune system as in a natural infection, but without or with substantially weaker symptoms. Very efficient and long-lasting protection may be obtained since vaccination results in optimal stimulation of both B and T lymphocytes. Local immunity on the epithelial surfaces used as portals of entry during natural infection may also be obtained.

The most common strategy for engineering a vaccine is, however, to insert the DNA fragment(s) coding for an immunogen(s) into the genome of a "non-dangerous" virus or bacterium, the vector. The insertion is performed so that the vector is still infectious ("live"). Following vaccination, the inserted vaccine gene, as well as the genes of the vector, will be expressed, and the gene products (proteins) will be present within the infected host cells. Highly efficient activation of both B and T lymphocytes may be obtained. In addition, significant local immunity may be evoked if the vector is able to infect the epithelial surface that the infectious agent in question is regularly using as its portal of entrance. Some vectors, i.e., bacteria and viruses within the family Poxviridae, permit insertion and expression of genes (at the cDNA level) from a number of infectious agents without losing their infectivity. It will hence be possible to immunize against a number of infectious diseases with one vaccine shot, aerosol inhalation, or capsule ingestion.

12.3.2.1   Genetically engineered viruses in general

Two different types of live, genetically engineered viral vaccines are being developed. Gene-deleted viruses are used to prevent the disease caused by the very same viruses. Virus vector vaccines are used to carry and express genes from other infectious agents into the organism.

In general, different strains of the same viral species may have different virulence, as well as host-cell or host-species tropism. Even genetic differences at the single point mutation level may result in virus strains with aberrant phenotypic characteristics (Tyler and Fields 1996). Thus, for GEVs (genetically engineered viruses), it is conceivable that unintended phenotypic characteristics with unwanted ecological consequences may be established, in addition to the intended modification(s). In the short run, this may not become evident unless very comprehensive and carefully planned experiments are conducted. In many instances, fully adequate experiments are totally precluded by the complexity and the regular or occasional variations of the recipient ecosystem.

Ideally, before any GEV is released into a new location/ecosystem, a number of crucial questions should be answered, for example:

1. Can the virus engage in genetic recombination, or acquire new genetic material by other means? If so, will the hybrid offspring have changed their host preferences and virulence characteristics?
2. Can other viruses, present within the ecosystem, interfere with the released virus or its offspring?

3. Can insects or migrating birds/animals act as vectors for the released virus or its offspring, and disseminate the viruses beyond the intended release areas?
4. For how long can the virus and its offspring survive outside host organisms under realistic environmental and climatic conditions?
5. Are the virus and its offspring genetically stable over time?
6. Can the virus or its offspring establish long-lasting, clinically mute, persistent or latent infections in naturally accessible host organisms?
7. Can the virus or its offspring activate or aggravate naturally occurring latent or persistent virus infections?

Generally, we cannot predict how a gene modification, i.e., point mutation(s), deletions, insertions, may affect the transmission of a virus within and between different species (Tyler and Fields 1996, Mulder 1997). We can minimize such potential risk factors by well-planned "microcosm" experiments that include ecosystem imitation, host organisms, vaccine viruses, and naturally occurring relatives. If the GEV is breaking assumed species barriers, replication in new hosts may lead to emergence and selection of secondary, spontaneous genetic changes and mutations, which in turn may influence transmission abilities, host preferences, and virulence. Such changes may take place and accumulate over time and may not be detected during short-term experiments. Unintended spread of a GEV hence implies unpredictable hazards.

A given GEV vaccine may be advantageous in every imaginable way in the intended vaccine species, but it may be detrimental on the whole because of its effects on unintended non-target species, and hence on ecosystems. This may relate not only to acute disease symptoms, but also to persistent and latent infections that interfere with reproduction and behaviors. Genetic differences between virus strains and geographical variants of the same animal species may influence the effects of the GEV infection. Over time, new or spontaneous genetic changes in the GEVs may modulate the interplay with host species in new and unpredicted ways. Furthermore, it is possible that a given GEV or its constituent(s) may access genomes in related, naturally occurring viruses, or in the DNA of host cells. Our knowledge concerning naturally occurring relatives of present and candidate vaccine viruses is very limited. Theoretically, different GEVs may exchange nucleic acids and genetic information with related viruses, by different mechanisms (i.e., recombination, reassortment). This may result in hybrid, recombinant viruses with unpredictable characteristics. If a GEV or a hybrid is able to integrate its DNA into host cell chromosomes of some species, this may have dramatic biological, and hence ecological, effects on either a short- or long-term timescale. Gene expression patterns and functions of host cells may become influenced by the integrated DNA or by viral or chimeric proteins translated from integrated DNA.

GEV particles that are broken down in the environment will release naked nucleic acids. Horizontal gene transfer of GEV genomes, or parts thereof, is then a potential hazard in parallel with any other genetically engineered nucleic acid (for review, see Nielsen et al. 1998, Traavik 1999a, 1999b).

12.3.2.2   Gene-deleted viruses
Gene-deleted viruses were initially considered a promising strategy for vaccines. Recent experience with an AIDS-related vaccine, however, has raised serious concerns about both target and non-target effects of such vaccines. A supposedly safe vaccine, made by deleting several genes from the simian immunodeficiency virus (SIV), caused AIDS in infant and adult macaques (Baba et al. 1999).

To illustrate some of the theoretical hazards, and the lack of key knowledge for risk assessments, gene-deleted pseudorabies virus vaccine will be briefly discussed.

Pseudorabies virus (PRV; synonyms Aujeszky's disease virus and herpesvirus suis type 1) causes neurological disorders in pigs, as well as in a number of other domestic and wildlife animal species. The disease in pigs results in vast economic losses as well as other practical problems, e.g., creation of trade barriers between countries. The epizootiology of PRV is complicated. Healthy virus carriers and shedders, latent infections that may be reactivated with or without accompanying clinical symptoms, vertical transmission from sow to offspring, etc. must be taken into consideration (Stegeman et al. 1994). Control of the disease and eradication of PRV, with or without the use of vaccines, consequently has high priority worldwide.

The genetically engineered vaccine virus for PRV was developed by genetic engineering of a live, attenuated PRV strain. The vaccine virus is a multi-deletion mutant. The main deletion removes gE and part of the so-called 11K protein-gene. The glycoprotein gE seems to be an essential protein in transneuronal spread of PRV (Jacobs 1994). The gE-negative PRV replicates in peripheral tissues, infects first-order neurons, and spreads toward the central nervous system via both olfactory and trigeminal routes (review by Mulder et al. 1997). There may be considerable differences in the effect of gE-deletions among different neuron circuits in the pig, and also between pigs and other permissive animal species (Standish et al. 1994).

In eradication campaigns, gE-deleted vaccine strains have been used in conjunction with a serological test that specifically detects antibodies against gE. This makes it possible to identify vaccinated individuals that have become infected with wild-type PRV (Kit 1994). The vaccines that have been introduced so far, including the gene-deleted ones, are unable to stop transmission of wild-type PRV strains, and the vaccine strain itself may spread to contact pigs, setting the stage for single- or multi-step recombinational events (reviewed in Traavik 1999b).

Non-target effects are another concern. It is well known that a number of domestic animals and wildlife species are susceptible to PRV infection and may contract dramatic and even fatal forms of the disease (Kimman et al. 1991, Kit 1994). There are striking differences in virulence for the same wild-type PRV in different host species, and for different PRV strains in the same host species. The molecular basis for these aberrations is unknown (Christensen and Lomniczi 1993, Bouma et al. 1996). There are indications of human infections during outbreaks in pigs or cattle (Anusz et al. 1992). Wild boars may be natural PRV reservoirs, and can be experimentally infected with porcine PRV (Oslage et al. 1994). No systematic studies of PRV susceptibility have been published for other wildlife animal species, nor are there any published reports about natural occurrences of PRV or PRV-like viruses in wildlife species. Whether PRV deletion mutants are able to infect wildlife species is also unknown. Such knowledge is essential for risk assessment concerning recombination between PRV vaccines and naturally occurring viruses. Warning signs have been noted. Sheep housed together with pigs started to die from PRV infection. Pigs were vaccinated by the 783 gE-/*tk*-vaccine and sheep with the live attenuated Bartha vaccine. In spite of that effort, sheep continued to die. Both vaccine viruses, as well as wild-type PRV, were detected in the sheep, the latter in extremely small amounts (Jacobs et al. 1997). Jacobs et al. (1997) suggested that the wild-type virus was the killer, but this was by no means proved, nor were recombinations between the three involved viruses investigated.

In 1992, gE-deleted PRV vaccines were authorized as the preferred choice by an expert group (Pensaert et al. 1992). The experts concentrated on target effects. Demands for documentation concerning non-target effects were not expressed until 1997 (Mulder et al. 1997). The European Pharmacopoeia has formulated safety requirements for live PRV vaccines (Kimman 1992). These requirements do *not* include information about hazards that are important from environmental and ecological points of view.

There are other important questions concerning latency and gene-deleted PRV that have not at the moment been answered in a satisfactory manner (see Traavik 1999b for detailed discussion). Several experimental studies have demonstrated that co-innoculated, modified live PRV vaccine strains could recombine *in vivo* to create virulent recombinant strains (Mulder et al. 1997). The potential risks represented by recombination events have not been satisfactorily clarified for the gene-deleted PRV vaccine, or for any of the other gene-deleted vaccine candidates.

12.3.2.3   Live virus vector vaccines

12.3.2.3.1   Poxviruses.   Both orthopox and avipoxviruses are subjects of extensive evaluation as live genetically modified vectors for use in vaccination against contagious diseases and cancers, and, recently, also against fertility in pest animals. The viruses belong to Poxviridae, a large virus family with known representatives infecting many vertebrates and a number of insects. It is safe to assume that many family members are still unknown. A member of the genus *Orthopoxvirus*, vaccinia virus (VV), was the basis for the successful eradication campaign against smallpox, caused by the closely related variola virus. The origin of VV is uncertain, but it may have developed by repeated recombinations between different orthopoxviruses. VV has a very broad spectrum of susceptible host animals, making dissemination within ecosystems and across species borders a real possibility (Moss 1996, Fenner 1996).

Parts of the poxvirus genomes may be removed by recombinant DNA techniques and replaced by foreign DNA, without destroying the ability of the viruses to infect host cells. GEVs with foreign gene inserts up to approximately 30 kbp may infect inoculated individuals, replicate, and express the proteins that the foreign gene(s) are coding for. The inoculated individual may then mount an immune response against that foreign gene product. If the foreign gene is taken from another virus (i.e., rabies virus), this immunity may protect, partly or totally, against a later natural infection. The vaccines consist of live, infectious virus that are shed from vaccinated individuals. Human beings or domestic animals usually are not placed in isolation or quarantine after vaccination. Consequently, not only wildlife vaccinations, but all practical applications of VV-based vaccines, may result in environmental release of GEV.

VV has a number of important advantages as a vaccine vector but also a number of proven and theoretical disadvantages. Unmodified VV may cause serious, generalized infections in some individuals with immunosuppressive disorders or treatments. The potential host spectrum of VV is very broad. Laboratory animals, pigs, cattle, camel, and monkey species are susceptible, but that list is certainly not complete. There is a high number of closely related viral species, and high degrees of sequence homology across species borders. VV has great potential for engaging in recombinations. VV particles have a high degree of environmental resistance (Fenner 1996, Moss 1996). This makes possible dissemination and transmission by insects, migratory birds and animals, domestic and pet animal trade, etc. A number of these disadvantages are admittedly theoretical, having modest scientific support, mainly because research that might prove or disapprove them has yet to be carried out (review by Traavik 1999b).

To vaccinate reservoir animals in Europe (e.g., red foxes) and North America (e.g., raccoons) recombinant VV/rabies vaccines have been released over vast land areas (Brochier et al. 1991, Anderson 1991, Pastoret et al. 1996). The rabies vaccine is but the first of a series of poxvirus-based GEVs that are on the horizon. A number of these, for medical and veterinary purposes, are in various stages of regulatory approval in the United States, the European Union, etc.

Risk assessment and monitoring were performed before, during, and after release of the rabies vaccine in Europe. The investigators, however, have been met with harsh criticism from different quarters. The most serious criticism was that the investigations focused on target effects of vaccination, and that the studies of ecological non-target effects were too limited both with respect to extent and penetration (Kaplan 1989, McNally 1994).

Ecological non-target effects are difficult to assess because even carefully designed model studies will not directly reflect the real ecosystem conditions, which themselves are dependent on locally variable parameters. However, some warning signs of non-target effects have already been observed:

- During the human smallpox eradication campaign, VV found a new host species and established itself in a new reservoir, namely, the buffalo (Dumbell and Richardson 1993).
- It is a common experience that gene inserts *may* change the virulence and host preferences of viruses (Mulder et al. 1997).
- MRV (malignant rabbit virus) seems to be a recombinant between SFV (Shope fibroma virus) and myxoma virus, created by mixed infection in wild rabbits. MRV causes an invasive malignant disease and profound immunosuppression in adult rabbits, much more serious than the disease caused by any of the parental viruses (Strayer et al. 1983). MRV has received more than 90% of its DNA from one parent (myxoma virus) in a coupled recombination and transposition event (Block et al. 1985). The MRV story exemplifies the unpredictability of virus recombinants with regard to biological characteristics and virulence.
- A recombinant field isolate of capripoxvirus has also been detected (Gershon et al. 1989). The new virus was the result of recombination between a capripoxvirus vaccine strain and a naturally occurring virus strain.

If a genetically modified orthopoxvirus infects an individual, animal or human, that already carries another orthopoxvirus, a hybrid progeny virus with unpredictable pathogenicity and altered host range might be the outcome. At the moment, this risk is impossible to assess because systematic studies of naturally occurring orthopoxviruses are lacking.

Investigations in the author's laboratory (Sandvik and Tryland 1996, Sandvik et al. 1998, Tryland et al. 1998a, 1998b, 1998c, 1998d, 1998e, Hansen et al. 1999) have demonstrated that in Norway orthopoxvirus(es) closely related to VV are widely distributed in terms of geography, ecosystems, and host animal species. Approximately 20% of shrews and small rodents belonging to eight species carry orthopoxvirus(es) in one or more of their organs (lungs, kidneys, liver, spleen) at a given point in time; a similar proportion has specific antibodies as a sign of past infections. The DNA sequences of two different genes (*tk* and *atip*) demonstrate that the Norwegian viruses are so closely related to VV that recombinations and hybrid offspring in doubly infected animals are very real prospects. Experiments to clarify these possibilities are being performed by encouraging mixed infections in an authentic orthopoxvirus host animal — the bank vole, *Clethrionomys glareolus* — and in an artificial host (the common laboratory mouse strain Balb/c).

Avian poxviruses offer an alternative to VV vectors. The bird viruses possess many of the desirable characteristics of VV (Baxby and Paoletti 1992), and recombinant avipoxvirus vaccines have induced protective immune responses to their inserted foreign gene products in several mammalian species (Taylor et al. 1991, Fries et al. 1996). The established wisdom is that avipoxviruses do not replicate, and are hence safe for use in mammalian hosts (Baxby and Paoletti 1992, Fries et al. 1996). However, returning to the early papers from which the citations originate (Taylor and Paoletti 1988, Taylor et al. 1988), one does not find any definite foundation for this generalization. A very restricted number of avipoxviruses (fowlpox, canarypox, and pigeon pox) have been used to infect a very restricted number of mammalian species and cell lines of mammalian origin. Non-target wildlife mammals have not been investigated, nor has the prospect of recombinations between genetically engineered avipox vectors and naturally occurring avipoxvirus species been investigated. The genetic diversity of avipoxviruses in different bird species is not well researched (see Traavik 1999b). There is an urgent need for penetrating studies designed to elucidate the biology, ecology, and genetics of avipoxviruses before they are used as vaccine vectors under uncontained conditions.

12.3.2.3.2   Adenoviruses.   Recombinant adenoviruses may be constructed by either insertion or replacement of viral sequences. Depending on the chosen strategy, the resulting vectors are replication competent or defective. The size of DNA that can be packaged in virus particles represents a limit to the insertion of foreign genetic material into adenovirus vectors (Bett et al. 1993).

Most adenoviruses seem to have rather restricted host ranges, although this has not been examined systematically in most cases. To circumvent this host species restriction, recombinant adenoviruses containing host-range mutations that allow human adenoviruses to infect non-permissive host cells were constructed (Cheng et al. 1992, Caravakyri et al. 1993). Although, obviously, this may increase the putative ecological and health risks of the resultant vectors, that prospect has not been investigated. This seems all the more unfortunate given that adenovirus excretion from the pharynx was demonstrated following enteric inoculation in human volunteers (Schwartz et al. 1974). Environmentally resistant viruses, like adenoviruses, may be spread over amazingly large areas in the aerosols created from the respiratory organs of virus-excreting individuals.

12.3.2.3.3   Other live virus vaccine vectors.   Recently, considerable research effort has focused on developing live recombinant vectors from a number of other viruses. Such vectors have been designed to accommodate entire genes, as well as parts of genes encoding a protective antigenic epitope (reviewed by Sheppard and Boursnell 1998). Herpesviruses, poliovirus and other picornaviruses, influenzaviruses, alphaviruses, and other positive-stranded RNA viruses (reviewed by Sheppard and Boursnell 1998, Liu 1998) are in different stages of development as vaccine vectors. They all have attractive prospects, but they also present different and uncertain environmental and health risks.

12.3.3   Bacterial vectors

Much recent attention has focused on the use of several bacterial species for delivery of a variety of antigens. Among these are *Salmonella* species; BCG (Bacille Calmette Guérin); *Streptococcus* and *Staphylococcus* species; *Listeria monocytogenes*; and enteric bacteria such as *Escherichia coli, Vibrio cholerae, Yersinia enterocolitica*, and *Shigella flexneri* (reviewed by Jones 1998). As recombinant vaccine vectors, any of these organisms may be delivered parenterally. Most importantly, they may be delivered on mucosal surfaces, potentially evoking local immune responses. Furthermore, these organisms may be used for the expression of multiple antigens. Hence, they are important in the continuing search for combined vaccines (reviewed by Jones 1998).

From some points of view, genetically engineered bacterial vectors have produced conflicting and unpredictable results. In addition, since most of these vectors are designed for mucosal administration, the issue of oral tolerance must be addressed. How long a given recombinant organism persists within the host is also a matter of concern (Jones 1998). Most attenuated pathogens remain in the host for a few days to weeks. However, it is not known whether the introduction of foreign genes may change the relationships between vector and host. The unpredictability of gene expression levels is another cause for concern. During studies of genetically engineered *Salmonella/Leishmania major* vectors in mice, it was revealed that a *low* antigen dose regimen was necessary to establish stable, protective CTL immune responses in susceptible mice, whereas *high* doses elicited only antibody responses and exacerbated the disease (Bretscher et al. 1992).

It has recently been reported that genetically modified bacteria may transfer their transgene efficiently to indigenous bacteria in the mammalian gut (MacKenzie 1999). This possibility has not been investigated for the bacteria that are now being genetically engineered as oral vaccines.

At the moment, intensive efforts are under way to create effective DNA vaccines for oral use (Lowrie 1998). A number of studies have underscored the fact that genetically engineered and recombinant DNA molecules may be absorbed from the intestines and taken up in the mammalian organism (Courvalin et al. 1995, Sizemore et al. 1997, Darji et al. 1997, Jones 1998). Indeed, the efficiency of the oral DNA vaccines indicates that there may be a normal process whereby large segments of intact DNA trapped in particles move from the gut, through Peyer patches (PP), into the nuclei of antigen-presenting cells (APC) (Lowrie 1998) and perhaps other cells in the organism (Schubbert et al. 1997, Doerfler and Schubbert 1998). If genetically engineered bacterial vaccine vectors access the environment, their DNA may be subject to horizontal gene transfer by the same processes as any other DNA (for review, see Nielsen et al. 1998, Traavik 1999a).

12.3.4 DNA vaccines

Viruses consist of protective membranes and protein shells in addition to the genome (RNA or DNA). Hence vaccination with whole viral particles always burdens the individual with a number of proteins besides the immunogen(s), which are the basis for protective immunity. These extra proteins are introduced whether the virus is "live" or dead, or attenuated, or genetically modified, or is a live vector. The chances of unwanted reactions in the vaccinated individual are augmented, simply because of the extra burden of irrelevant proteins.

A DNA vaccine, on the other hand, may be engineered to express just immunologically relevant protein(s). In its simplest form, it consists of a bacterial plasmid with a strong viral promoter, the gene of interest, and a polyadenylation/transcriptional termination sequence. The plasmid is grown in bacteria (*E. coli*), purified, dissolved in a saline solution, and then simply injected into the host. The DNA is taken up by, and the encoded antigen expressed in, host cells. The DNA vaccines are administered by direct intramuscular injection. Use of a "gene gun" delivering gold beads onto which DNA has been precipitated is also under evaluation (Liu 1998).

It was discovered by chance that naked DNA vectors could express genes in intact animal organisms (Wolff et al. 1990). In the wake of the initial experiments, it also became evident that expression of inserted genes from naked plasmids gave a very efficient immunological presentation of immunogens (Ulmer et al. 1993). With the optimal DNA constructs, delivery systems, and adjuvants, very efficient immunization against a number of disease agents was achieved in laboratory animals. Antigens are presented to the immune system in a way that imitates a natural viral infection and, therefore, provide the type of immunity that is attained by a natural viral infection.

A number of other potential advantages of DNA vaccines have been advanced (reviews by Donnelly et al. 1997, 1998, Liu 1998). It is believed that, presently, a number of DNA vaccines will be developed and become very widely used. In recent years, the DNA vaccination prospect has created great hope for immunization against such notorious killers as HIV (Letvin 1998), various cancer forms (Benton and Kennedy 1998), and malaria (Hoffman et al. 1997).

It has not been conclusively determined whether, and under which circumstances, plasmid vectors may become integrated into the chromosomes of target or non-target cells. The integration tendency will probably vary for the same plasmid in different target cells and for different plasmids in the same type of cell. Plasmid DNA integrations may theoretically contribute to generation of tumors and other pathological conditions. That is a real concern for skeptics, but not for believers (Donnelly et al. 1997, 1998). There is a growing debate over the potential harmful effects of random insertions into cellular genomes and of unintended viral recombinants (Brower 1998, Jane et al. 1998, Putnam 1998, Ho et al. 1998). The problems have been highlighted by observations made during gene therapy trials in animals and humans (Boyce 1998). Furthermore, uptake of naked DNA by sperm cells of marine organisms and mammals has been demonstrated and transgenic animals have been created. Indeed, the use of sperm to deliver therapeutic genes is under consideration (Spadafora 1998).

12.3.5   RNA vaccines

The viral genomic RNA of so-called positive-strand RNA viruses is infectious when introduced into permissive host cells. RNA from such viruses has been transcribed *in vitro* into cDNA clones, and this was also done for the 11-kb-long TBE (tick-borne encephaliti) virus RNA. Full-length cDNA derived from TBE virus was transcribed *in vitro* into infectious RNA that was coated on gold microcarrier particles and biolistically shot into the abdominal skin of mice by a gene gun. Like intact TBE virus, RNA derived from a fully virulent TBE virus strain efficiently killed the mice. However, when RNA from an attenuated TBE virus strain was used, the mice were fully protected against a later challenge from the virulent virus (Mandl et al. 1998).

The use of *in vitro* synthesized RNA corresponding to the genome of an attenuated virus seems to combine the main advantages of conventional live virus vaccines with those of DNA immunization without some of the drawbacks of either approach. Pure RNA may be produced without the need for cell cultures. The production does not require manipulation of infectious virus. Propagation of live, attenuated viruses, RNA viruses in particular, may result in reversion to virulence. This necessitates extensive safety testing of each vaccine production lot. For RNA vaccines there is no risk of chromosomal integration of foreign genetic material. However, the potential hazards of RNA recombination have not been properly clarified (see Section 12.4.2).

12.3.6   Edible vaccines produced by transgenic plants

In 1990, a group of philanthropic organizations led by the World Health Organization (WHO) launched the Children's Vaccine Initiative. The goals were vaccines that were safe, heat stable, orally administered, and widely accessible. These goals led to the idea of producing subunit vaccines in edible tissues of transgenic crop plants (Mason et al. 1992). Transgenic, edible crop plants as production and delivery systems for subunit vaccines have a number of attractive advantages (Mason et al. 1992, Mor et al. 1998). Production is as cheap as agriculture. Distribution is as convenient as marketing fresh products. Vaccine administration is as simple and safe as eating.

Various crop plants were genetically engineered to express subunits of infectious agents (Mor et al. 1998). A common experience, however, was that only low levels of antigens were detected in the plants. Various strategies were then employed to increase expression levels (reviewed by Mor et al. 1998). But high expression sometimes proved detrimental to the plants, because of the harmful effects of some of the antigenic proteins. Site-directed mutagenesis and inducible rather than constitutive promoters have been proposed as means to alleviate the toxicity of foreign proteins.

Feeding with transgenic potato tubers expressing subunits of bacterial enterotoxins provided partial protection against diarrhea to mice later challenged with the intact toxins (Mason et al. 1998). The mouse experiments eventually led to testing of an "edible vaccine" in humans. These tests represented the first human clinical trial of a plant-derived vaccine (Tacket et al. 1998). Volunteers who consumed 50 or 100 mg potato tubers in these tests developed specific mucosal and systemic immune responses.

Other mucosal immunogens are being evaluated for production in transgenic plants, including antigens from pathogens of the respiratory and urogenital tracts. Bananas, maize, and soybeans have emerged as attractive vaccine crop plants (Mor et al. 1998).

The most serious arguments against large-scale, commercial use of first-generation GEPs (genetically engineered plants), and hence against "edible vaccines," are based on the difficulty of knowing *where* in the recipient cell chromosomes insertion of vector DNA occurs. The consequences of insertion in eukaryotic chromosomes may vary considerably according to the precise insertional location (Doerfler et al. 1997). Hence, unintended changes in expression levels of the inserted transgenes as well as changes in endogenous genes of recipient organisms and their expression levels may be seen. Some of the greater concern about DNA vector insertion is related to the fact that the recipient organism receives a new promoter/enhancer. These elements govern the gene expression levels of the attached transgenes, but, after insertion, they may also influence gene expression and methylation patterns in the recipient chromosome(s) over long distances, up- and down-stream, from the insertion site. Promoter/enhancers function in response to signals received from both the internal or external environment of the organism. For a GEO (genetically engineered organism), this creates uncertainty with regard to the expression level of the inserted foreign gene(s); expression of the organism's own genes; influence of geographical, climatic, chemical (i.e., xenobiotics), and ecological changes in the environment; transfer of vector sequences within the chromosomes of the organism and vertical and/or horizontal gene transfer to other organisms; and activation of endogeneous mobile genetic elements in the GEO itself. The uncertainty created by variability in expression levels of plant transgenes is an acknowledged problem (Mor et al. 1998). Such variation has been observed in different plants of the same transgenic line, and even within the same plant. Appreciable variation in transgenic expression levels has been demonstrated, for example, between different tubers of the same potato plant (Mor et al. 1998).

So-called genetic pollution from GEPs, exerted by cross-pollination and horizontal gene transfer, is a real possibility (for review, see Kidwell 1993, Nielsen et al. 1998, Traavik 1999a, 1999b) and presents us with a variety of complex and unpredictable health, environmental, and socioeconomic problems. The issue has been given added salience by the demonstration that ingested DNA, under some circumstances, may be taken up from the intestines of mice and vertically transmitted to offspring (Doerfler et al. 1997, Schubbert et al. 1997, Doerfler and Schubbert 1998). A number of unpredicted incidents have already occurred with GEPs (Mikkelsen et al. 1996, Chévre et al. 1997, Oger et al. 1997, Bergelson et al. 1998, see also Traavik 1999a, 1999b, and references therein).

12.3.7   Aquaculture and genetically engineered vaccines

The growing demand for fish and shellfish products for human and animal consumption, combined with the continuous decline of wild fishery resources, has made aquaculture the fastest-growing segment of animal farming in many parts of the world (Hanfman 1993). As with other intensive farming operations, however, this rapid growth has been accompanied by a proliferation of profit-limiting infectious diseases. Pathogens now constitute the most important cause of economic loss for fish and shellfish farmers, destroying 10% of all cultured animals (Leong and Fryer 1993).

The potential environmental problems associated with use of live vector vaccines and DNA/RNA vaccines in aquaculture are obviously even more serious than in terrestrial ecosystems. Particles, microorganisms, viruses, and DNA may be distributed over vast areas, distances, and phylae, as a result of the relative lack of physical and physiological barriers. For such reasons, live attenuated or genetically engineered vector vaccines have not been introduced, or seriously considered. However, various adaptations of DNA vectors to fish vaccinology have been attempted (Heppel et al. 1998, Lorenzen et al. 1999).

Strong expression of reporter genes in fish injected with plasmid DNA has been recorded and protective immunity following DNA vaccine injection in fish has been demonstrated (see Traavik 1999b, and references therein). Both antibody and cell-mediated responses were reported. Consequently, development of DNA vaccines for infectious fish diseases recalcitrant to traditional strategies is now being seriously considered (Heppell et al. 1998). Such vaccines would have several attractive advantages: low cost, ease of production and quality control, heat stability, an identical production process for different vaccines, and the possibility of producing multivalent vaccines by use of plasmid cocktails. Further, intramuscular rather than intraperitoneal injection would facilitate use of fully automated devices and avoid some growth-retarding pathological changes that occur with intraperitoneal injection (Poppe and Breck 1997).

In view of what is known about the aquatic ecosystems as well as about DNA vaccines, and considering the general lack of knowledge about the environmental consequences of DNA vaccine use, including non-target effects and the possibility and consequences of horizontal transfer, it would seem strongly advisable to argue against further application and commercialization of DNA vaccines for fish. A number of basic questions should be answered first. After intramuscular DNA plasmid injections in fish, reporter gene expression was detected in gills (Heppell et al. 1998). It was unclear whether this was due to primary DNA transfection of migrating cells passing through the injected muscle or to diffusion of injected DNA and secondary transfection of cells at distant sites. Whatever the mechanism, the data suggest the existence of a route for potential exposure of the environment to genetically engineered DNA. The high level of reporter gene expression exhibited over prolonged periods adds to the concerns already generated.

12.4   Unpredictable risks and hazards related to genetically engineered vaccines

12.4.1   Horizontal gene transfer

Major genomic rearrangements, such as duplications, deletions, translocations, or insertions (integrations), are important for evolution. Duplications may provide additional copies of genes, which may accumulate mutations and thereby offer opportunities for further evolution. Translocations and deletions can fuse genes, thereby creating new proteins with new functional domains; they can also change the surroundings of one gene, thereby helping it to be influenced by new regulatory mechanisms. Insertions of foreign DNA into a genome may help to overcome the need for repeated evolution of similar functions in different organisms. However, genomic rearrangements may also have serious damaging effects when they occur in the wrong places, at the wrong time, or on an abnormal scale (Doerfler et al. 1997). Rearrangements may cause cell growth aberrations, and the extent of rearrangements often increases during development of malignant tumors (Croce 1987). Fetal death, developmental defects, metabolic illnesses, and hereditary disorders such as

Duchenne and Becker's muscular dystrophy may be due to such rearrangements (Bakker et al. 1987).

Genomic rearrangements may result from *legitimate* recombination, i.e., those that take place between long, homologous sequences. However, they can also arise as a consequence of *illegitimate* recombinations between sequences with little or no homology. In transgenic mice (Dellaire and Chartrand 1998) and in plants (Gorbunova and Levy 1997) illegitimate recombination leads to the random and unpredictable integration of transgenes in recipient chromosomes. Such recombination is probably universal, since it has been found in all organisms examined for its presence (see Traavik 1999a, and references therein).

It is crucial to identify the various factors that affect illegitimate recombination and determine their significance in relevant model systems. So far, the majority of such model systems have been developed for microorganisms, presumably in the hope that they will provide insight into the phenomenon of illegitimate recombination in any type of cell. Recent studies strongly indicate that illegitimate recombination is a major driving force in the evolution of plants, animals, and microorganisms, as well as a major cause of disease (see Traavik 1999a, and references therein). Unfortunately, at the moment, we cannot claim more than a very rudimentary understanding of the phenomenon.

Horizontal (lateral) gene transfer is defined as non-sexual transfer of genetic information between genomes. Horizontal transfer is thus distinct from the ordinary form of gene transfer, which takes place vertically from parent to offspring. There is now good evidence that horizontal transfer takes place naturally for both genomic (usually non-mobile) sequences and sequences derived from transposable genetic elements or mobile introns. Documented cases exist of the transfer of genomic sequences from eucaryotes to procaryotes, from procaryotes to eucaryotes, between procaryotes, and between eucaryotes (see Traavik, 1999a, 1999b, and references therein).

For any given gene construct or GEO able to access an environment, the state of our present knowledge does not allow pre-assessment of either the probability or the consequences of horizontal gene transfer. Hence, genuine risk assessments are impossible at this time. Only extensive research into the mechanisms of horizontal gene transfer and its implications for an ecosystem can change this situation.

Thus far, the focus has been on horizontal transfer of entire genes, but for some bacterial species, it has been shown that far shorter elements are stably transferred and can give rise to mosaic genes (Heinemann and Roughan 2000). There are strong indications that this also occurs in eucaryotic organisms, exemplified by cytochrome *c* in plants and β-globins in mammals (Heinemann and Roughan 2000). Theoretically, shorter DNA sequences may contain control elements for expression of genes (e.g., transcription factor binding motifs of promoters or enhancers), which may change the amount of some gene products in the recipient, perhaps with substantial biological consequences. Conservation of regulatory elements between phyla can sometimes result in low levels of gene expression under unlikely circumstances (Lowrie 1998). That is, the nature of the actual DNA used may play a role. At this stage, we simply do not know how plasmid DNA moves from a phagolysosomal vesicle into the nucleus of an APC, or any other cell type, in a functional state (Lowrie 1998). Again, the kind of DNA may play a role.

There are limitations to the kind of DNA that can be transferred, but we do not know what kind of mechanisms sort DNA for transfer and so are unable to pre-assess whether a particular plasmid or other genetic construction will be transferred horizontally, when it will be transferred, and where it will end up. Furthermore, as stated by Nielsen et al. (1998): "Transfer frequencies should not be confounded with the likelihood of environmental implications, since the frequency of horizontal gene transfer is probably only marginally important compared with the selective force acting on the outcome." We know very little about selective forces in different ecosystems.

12.4.1.1   Protection and persistence of DNA in nature

One of the inherent curses of experimental resesearch is that nature only seldom views it as a priority to reproduce or mimic "normal" laboratory conditions! Until a few years ago, it was generally believed that any free, naked DNA in the environment or in multicellular organisms would be very rapidly broken down and hence have no biological effects. Recently, a number of published reports have demonstrated that various types of DNA may persist in the environment for extended periods under various conditions. Nevertheless, we still have no way of knowing in advance what will happen under the actual natural conditions in which a particular, genetically modified vaccine may be used (recent reviews: Nielsen et al. 1998, Traavik 1999a, 1999b, Heinemann and Roughan 2000).

12.4.1.2   Uptake of foreign nucleic acids in the mammalian organism

Mammalian cells can take up foreign DNA in a manner that permits biological activity. This is, of course, precisely the basis for transfections of cell cultures and genetic modifications of plants and animals, for gene therapy and for DNA vaccination.

Oligo- and polynucleotides cannot diffuse through the lipid membranes of living cells. In some eukaryotic cells, it has been shown that nucleic acids can be taken up by endocytosis, which is mediated by nucleic acid–specific receptors (Vlassov et al. 1994). Similar mechanisms may be active in bacteria as well (Lorenz and Wackernagel 1994). Following uptake, the nucleotides find a way of escaping from the endosomes in eucaryotic cells and reach nucleic acids that are located in both the cytoplasm and the nucleus (Vlassov et al. 1994). Bacteria normally remove foreign DNA by restriction enzymes that distinguish foreign DNA from their own; however, under certain circumstances, this mechanism fails (for recent review, see Nielsen et al. 1998, Traavik 1999a). The biological and evolutionary importance of the above-described mechanisms are not known and we have no knowledge about the difference between nucleic acids that are taken up in biologically active form and those that are broken down. Nor do we know whether and, if so, which environmental conditions can increase or reduce the expression of the nucleic acid receptors, or whether this change, if it occurs, can affect the uptake and the further handling of nucleic acids in the cells.

The fate of nucleic acids in the gastrointestinal tract was studied in ruminants and rats in the 1970s and 1980s. The limited sensitivity of the methods available at that time meant that lack of discoveries could not be used to deny the possibility of either uptake of biologically active DNA from the intestinal tract or dispersal to the surroundings through the feces. More recent work has demonstrated that ingested DNA under some circumstances may be taken up from the intestines of mice, inserted into chromosomes, and vertically transmitted to offspring (Doerfler et al. 1997, Schubbert et al. 1997, Doerfler and Schubbert 1998). The authors assume that other types of DNA might behave in the same manner, but they add that this must be investigated experimentally. Furthermore, the efficiency of the oral DNA vaccines indicates that there may be a normal process whereby large segments of intact DNA move from the gut, through PPs, into the nuclei of APCs (Lowrie 1998).

A number of recent reports have demonstrated that naked DNA may be taken up in a variety of ways, some unexpected. Naked viral DNA may initiate a full-blown viral infection following intravenous injection (Fredriksen et al. 1994). Similarly, efficient expression of DNA intravenously delivered in rat muscle has been demonstrated (Budker et al. 1998). High levels of foreign gene expression were observed in the liver cells of rats, mice, and dogs when naked plasmid DNA was injected into blood vessels supplying the liver (Zhang et al. 1997). Naked DNA has been integrated into cellular chromosomes and expressed in human and pig skin (Hengge et al. 1995). Unexpected side effects, such as myositis, appeared when plasmids carrying the gene for a tRNA-activating enzyme were injected into the bloodstream of mice (Blechynden et al. 1997). The ability of naked DNA to penetrate intact skin has been known for years, e.g., within weeks of applying cloned DNA, including a human oncogene, to the skin of mice, tumors developed in endothelial cells lining blood vessels (Brown 1990). More recently, gene therapy based on cutaneous application of DNA is being seriously considered (Khavari 1997).

That nucleic acids are taken up and have biological activity is obviously not a common phenomenon. Throughout the history of evolution, animals, including humans, have received foreign DNA from other animals and plants through uptake of nutrients and breathing of air. The problem is that, yet again, in the case of a few, perhaps rare, combinations of nucleic acids and circumstances, nucleic acids will be able to be taken up from the mucous membranes (Doerfler et al. 1997, Schubbert et al. 1997, Doerfler and Schubbert 1998). However, we have no knowledge of the sequences, structures, or environmental factors that can contribute to such an event. Nor, at the present time, can we predict what type of DNA will avoid rapid breakdown in the organism and which environmental factors might contribute to such a breakdown.

12.4.1.3   Interactions with xenobiotics?

Xenobiotics are generally considered to be compounds that are alien in the biosphere. However, by such a narrow definition, metals, some pesticides, and many organic chemicals would not be considered xenobiotics because they are also found naturally in ecosystems. The narrow definition does not take human activity into account, which may lead us to ignore concentrations of compounds that can result in damaging effects. The essential element phosphorus is a good example. It is not usually a xenobiotic compound, but in large amounts it can create major environmental problems. Consequently, the following definition is suggested: xenobiotics are compounds that people release into nature in concentrations that create undesirable impacts.

Many xenobiotics, e.g., herbicides, pesticides, heavy metals, emissions from industry and burning of fossil hydrocarbons, etc., are chemically inert and hydrophobic. Hydrophobism means that the xenobiotic can easily enter organisms by diffusing through biological membranes, will be difficult to separate in urine and gall, and can accumulate in certain areas of the cell, including the phosphorus-lipid double layer of membranes, where it may be able to disturb normal cellular functions (Lundgren and DePierre 1990).

Very little work has been done to study how xenobiotics may interfere with horizontal gene transfer under natural conditions or in microcosms and other types of controlled environments. However, a vast literature concerning other effects of xenobiotics suggests that such effects may exist (see Traavik 1999a, and references therein).

Different xenobiotics have properties and biological activities that enable us to posit two possible scenarios concerning the fate of naked DNA in an ecosystem.

1. Some xenobiotics, such as radioactive substances, some industrial chemicals, and pesticides, can act as mutagens. Mutagens may change the sequence or structure of naked DNA that accesses an environment in which mutagens are found. This, in turn, may affect DNA uptake in cells and organisms, horizontal transfer, and long-term establishment in the ecosystems in ways that are not predictable at this time. It has been reported, for example, that minor changes in a DNA sequence may alter the host spectrum of a transferable genetic element (Kipling and Kearsey 1990).
2. Some xenobiotics can affect cell membrane and/or intracellular functions in ways that influence the ability of cells to uptake and horizontally transfer naked DNA. Xenobiotics may affect the structure of cell membranes, the content of both surface

receptors and transport channels, intracellular signal conversion, and gene expression. For example, xenobiotics that mimic hormones or affect the local conditions in the organ systems of mammals (e.g., respiratory passages) may change the possibilities for both uptake and establishment of foreign nucleic acids in those mammals.

Some xenobiotics may follow both scenarios. We neither know the cumulative effect of such events on any ecosystem, nor do we know the combined effects of xenobiotics following different scenarios and polluting the same environment. Finally, we have no knowledge of how such situations will affect DNA uptake and dispersal in the environment.

12.4.2　RNA recombination

Recombination is common in RNA viruses, and may be more important than accumulation of point mutations for significant evolutionary change and speciation of such viruses (see review by Miller and Koev 1998). Viruses in taxonomical groupings as distant as the α-viruses, coronaviruses, and luteoviruses contain closely related genes, suggesting recent recombination events in their evolution. The implications of the RNA recombination concept for use of self-expressing vaccines ought to be self-evident. If an RNA replicase and one foreign RNA species are present in a cell or organism, totally new species of viral RNA or mRNA may arise. Understanding viral RNA recombination is obviously important for the safe deployment of all kinds of vaccines that are genetically self-expressing, i.e., all kinds of "live" vaccines, edible vaccines, and DNA and RNA vaccines. From the point of view of risk assessment, the details of viral RNA recombination should have been clarified before GEVs and edible vaccines produced by transgenic crops were placed into general use.

The mechanisms for RNA recombination are not known in all instances, but most evidence supports a copy-choice process (Nagy and Simon 1997, Nagy et al. 1998). In such a process, an RNA replicase switches from copying an RNA donor template to copying an acceptor template without releasing the nascent RNA strand, resulting in a hybrid RNA molecule. It is conceivable, although not proved, that more than two different RNA molecules, present in the same cell at the same time, might be involved the process.

12.5　Concluding remarks

From an ecological and environmental point of view, many first-generation live GEVs are inherently unpredictable, potentially dangerous, and should not be placed into widespread use until a number of putative problems have been clarified. At the moment, it is not possible either to assess or to manage the potential environmental risks involved. Most likely, we have not yet even conceived of all the theoretical risks. Taking the Precautionary Principle into consideration, it would seem that many live GEV strategies should not be rushed into common use in medicine, veterinary medicine, or fish farming. At the same time as we must be careful to weigh the environmental risks, we must be careful to ensure that only serious, scientifically proven risks prevent us from saving lives and food resources with whatever means are available. For quite understandable and ethical reasons, this is a field where long-term, theoretical problems are likely to yield to more dramatic, short-term goals.

To make reliable risk assessments and perform sensible risk management with regard to GEVs in particular, much pertinent knowledge is lacking. The prerequisite for obtaining such knowledge is science and scientists dedicated to relevant projects and research areas. It is the responsibility of national governments and international authorities to make funding available for such research. Such funding is not the responsibility of producers and manufacturers. Risk-associated research should be publicly funded in order to keep it totally independent and above reproach.

Finally, it should be kept in mind that although vaccinology may seem the "Holy Grail" of medicine, other ways of preventing infectious diseases in humans and animals must not be ignored. Many of the most burdensome infectious agents of humanity and our domesticated animals are caused by pathogens that have reservoirs and are circulating among wild animals. By increasing our knowledge about these reservoirs and the transmission routes within and between indigenous ecosystems, we may be able to break the chain of transmission or keep our activities out of dangerous ecosystems. There is a void in knowledge about the ecological interactions and implications of many important pathogens. This field is constrained by the very confidence we place in vaccines; it is effectively an orphan of science.

Literature cited

Anderson, R. M., Rabies. Immunization in the field, *Nature*, 354, 502, 1991.

Anusz, Z. et al., Is Aujeszky's disease a zoonosis? *Przegl. Epidemiol.*, 46, 181, 1992.

Baba, T. W. et al., Live attenuated, multiply deleted simian immunodeficiency virus causes AIDS in infant and adult macaques, *Nat. Med.*, 5, 194, 1999.

Bakker, E. et al., Germline mosaicism and Duchenne muscular dystrophy mutations, *Nature*, 329, 554, 1987.

Baxby, D. and Paoletti, E., Potential use of non-replicating vectors as recombinant vaccines, *Vaccine*, 10, 8, 1992.

Bergelson, J., Purrington, C. B., and Wichmann, G., Promiscuity in transgenic plants, *Nature*, 395, 25, 1998.

Benton, P. A. and Kennedy, R. C., DNA vaccine strategies for the treatment of cancer, *Curr. Topics. Microbiol. Immunol.*, 226, 1, 1998.

Bett, A. J., Prevec, L., and Graham, F. L., Packaging capacity and stability of human adenovirus type 5 vectors, *J. Virol.*, 67, 5911, 1993.

Blechynden, L. M. et al., Myositis induced by naked DNA immunization with the gene for histidyl-tRNA transferase, *Hum. Gene Ther.*, 8, 1469–1480, 1997.

Block, W., Upton, C., and McFadden, G., Tumorigenic poxviruses: genomic organization of malignant rabbit virus, a recombinant between Shope fibroma virus and myxoma virus, *Virology*, 140, 113, 1985.

Bloom, B. R. and Widdus, R., Vaccine visions and their global impact, *Nat. Med.*, 4, 480, 1998.

Bouma, A., De Jong, M. C., and Kimman, T. G., Transmission of two pseudorabies virus strains that differ in virulence and virus excretion in groups of vaccinated pigs, *Am. J. Vet. Res.*, 57, 43, 1996.

Boyce, N., Suffer the children. The great taboo of gene therapy may have been broken, *New Sci.*, 14, 7, 1998.

Bretscher, P. A. et al., Establishment of stable, cell-mediated immunity that makes "susceptible" mice resistant to *Leishmania major*, *Science*, 257, 539, 1992.

Brochier, B. et al., Large-scale eradication of rabies using recombinant vaccinia-rabies, *Nature*, 354, 520, 1991.

Brower, V., Naked DNA vaccines come of age, *Nat. Biotechnol.*, 16, 1304, 1998.

Brown, P., Naked DNA raises cancer fears for researchers, *New Sci.*, 128, 17, 1990.

Budker, V. et al., The efficient expression of intravenously delivered DNA in rat muscle, *Gene Ther.*, 5, 272, 1998.

Cameron, J. and Abouchar, J., The precautionary principle, *Boston Coll. Int. Comp. Law Rev.*, 14, 1, 1991.

Caravokyri, C., Pringle, C. R., and Leppard, K. N., Human adenovirus type 5 recombinants expressing simian immunodeficiency virus macaque strain gag antigens, *J. Gen. Virol.*, 74, 2819, 1993.

Cheng, S. M. et al., Coexpression of the simian immunodeficiency virus Env and Rev proteins by a recombinant human adenovirus host range mutant, *J. Virol.*, 66, 6721, 1992.

Chévre, A. M. et al., Gene flow from transgenic crops, *Nature*, 389, 924, 1997.

Christensen, L. S. and Lomniczi, B., High frequency intergenomic recombination of suid herpesvirus 1 (SHV-1, Aujeszky's disease), *Arch. Virol.*, 132, 37, 1993.

Courvalin, P. et al., Gene transfer from bacteria to mammalian cells, *C. R. Acad. Sci. III*, 318, 1207, 1995.

Croce, C. M., Role of chromosome translocations in human neoplasia, *Cell*, 49, 155, 1987.

Danner, K., Acceptability of bio-engineered vaccines, *Comp. Immunol. Microbiol. Infect. Dis.*, 20, 3, 1997.

Darji, A. et al., Oral somatic transgene vaccination using attenuated *S. typhimurium*, *Cell*, 91, 765, 1997.

Dellaire, G. and Chartrand, P., Direct evidence that transgene integration is random in murine cells, implying that naturally occurring double-strand breaks may be distributed similarly within the genome, *Radiat. Res.*, 149, 325, 1998.

Doerfler, W. and Schubbert, R., Uptake of foreign DNA from the environment: the gastrointestinal tract and the placenta as portals of entry, *Wien. Klin. Wochenschr.*, 110, 40, 1998.

Doerfler, W. et al., Integration of foreign DNA and its consequences in mammalian systems, *Trends Biotechnol.*, 15, 297, 1997.

Donnelly, J. J. et al., DNA vaccines, *Annu. Rev. Immunol.*, 15, 617, 1997.

Donnelly, J. J., Ulmer, J. B., and Liu, M. A., DNA vaccines, *Dev. Biol. Stand.*, 95, 43, 1998.

Dumbell, K. and Richardson, M., Virological investigations of specimens from buffaloes affected by buffalopox in Maharashtra State, India between 1985 and 1987, *Arch. Virol.*, 128, 257, 1993.

Ellis, R. W., New technologies for making vaccines, *Vaccine*, 17, 1596, 1999.

Fenner, F., Poxviruses, in *Fields Virology*, Vol. 2, 3rd ed., Fields, B. N. et al., Eds., Lippincott-Raven, Philadelphia, 1996, 2673.

Fenner F. et al., *Smallpox and Its Eradication*, World Health Organization, Geneva, 1988.

Fredriksen, K. et al., On the biological origin of anti-double-stranded (ds) DNA antibodies: systemic lupus erythematosus-related anti-dsDNA antibodies are induced by polyomavirus BK in lupus-prone (NZBxNZW) F1 hybrids, but not in normal mice, *Eur. J. Immunol.*, 24, 66, 1994.

Fries, L. F. et al., Human safety and immunogenicity of a canarypox-rabies glycoprotein recombinant vaccine: an alternative poxvirus vector system, *Vaccine*, 14, 428, 1996.

Gershon, P. D. et al., Poxvirus genetic recombination during natural virus transmission, *J. Gen. Virol.*, 70, 485, 1989.

Gorbunova, V. and Levy, A. A., Non-homologous DNA end joining in plant cells is associated with deletions and filler DNA insertions, *Nucl. Acids Res.*, 25, 4650, 1997.

Greene, A. E. and Allison, R F., Recombination between viral RNA and transgenic plant transcripts, *Science*, 263, 1423, 1994.

Greene, A. E. and Allison, R. F., Deletions in the 3′ untranslated region of cowpea chlorotic mottle virus transgene reduced recovery of recombinant viruses in transgenic plants, *Virology*, 225, 231, 1996.

Hanfman, D. T., The Status and Potential of Aquaculture in the United States: An Overview and Bibliography, U.S. Department of Agriculture, National Agricultural Library, Washington, D.C., 1993.

Hansen, H. et al., Comparison of thymidine kinase and A-type inclusion protein gene sequences from Norwegian and Swedish cowpox virus isolates, *Acta Pathol. Microbiol. Scand.*, 107, 667, 1999.

Heinemann, J. and Roughan, P., New hypotheses on the material nature of horizontally mobile genes, *Ann. N.Y. Acad. Sci.*, 906, 169, 2000.

Hengge, U. R., Walker, P. S. and Vogel, J. C., Expression of naked DNA in human, pig and mouse skin, *J. Clin. Invest.*, 97, 2911, 1996.

Heppell, J. et al., Development of DNA vaccines for fish: vector design, intramuscular injection and antigen expression using viral haemorrhagic septicaemia virus genes as model, *Fish Shellfish Immunol.*, 8, 271, 1998.

Hilleman, M. R., A simplified vaccinologists' vaccinology and the pursuit of a vaccine against AIDS, *Vaccine*, 16, 778, 1998.

Ho, M.-W. et al., Gene technology and gene ecology of infectious diseases, *Microb. Ecol. Health Dis.*, 10, 33, 1998.

Hoffman, S. L. et al., Toward clinical trials of DNA vaccines against malaria, *Immunol. Cell Biol.*, 75, 376, 1997.

Jacobs, L., Glycoprotein E of pseudorabies virus and homologous proteins in other alphaherpesvirinae, *Arch. Virol.*, 137, 209, 1994.

Jacobs, L. et al., Detection of wild-type Aujeszky's disease virus by polymerase chain reaction in sheep vaccinated with a modified live vaccine strain, *Res. Vet. Sci.*, 62, 271, 1997.

Jaffee, E. M., Immunotherapy of cancer, *Ann. N.Y. Acad. Sci.*, 886, 67, 1999.

Jane, S. M., Cunningham, J. M., and Vanin, E. F., Vector development: a major obstacle in human gene therapy, *Ann. Med.*, 30, 413, 1998.

Jenner, E., An inquiry into the causes and effects of the variolae vaccinae, a disease discovered in some of the western counties of England, particularly Gloucestershire, and known by the name of cow pox, London, 1798. Reprinted in Camac, C. N. B., Ed., *Classics of Medicine and Surgery*, Dover, New York, 1959, 213.

Jones, D. H. et al., Poly (DL-lactide-co-glycolide)-encapsulated plasmid DNA elicits systemic and mucosal antibody responses to encoded protein after oral administration, *Vaccine*, 15, 814, 1997.

Jones, K. F., Bacterial vectors for vaccine delivery, in *Designer Vaccines — Principles for Successful Prophylaxis*, Hughes, H. P. A. and Campos, M., Eds., CRC Press, Boca Raton, FL, 1998, 151.

Kaplan, C., Vaccinia virus: a suitable vehicle for recombinant vaccines? *Arch. Virol.*, 106, 127, 1989.

Khavari, P. A., Therapeutic gene delivery to the skin, *Mol. Med. Today*, 3, 533, 1997.

Kidwell, M. G., Lateral transfer in natural populations of eukaryotes, *Annu. Rev. Genet.*, 27, 235, 1993.

Kimman, T. G., Acceptability of Aujeszky's disease vaccines, *Dev. Biol. Stand.*, 79, 129, 1992.

Kimman, T. G. et al., Aujeszky's disease in horses fulfills Koch's postulates, *Vet. Rec.*, 128, 103, 1991.

Kipling, D. and Kearsey, S. E., Reversion of autonomously replicating sequence mutations in *Saccharomyces cerevisiae*: creation of a eucaryotic replication origin within procaryotic vector DNA, *Mol. Cell. Biol.*, 10, 265, 1990.

Kit, S., Pseudorabies virus, in *Encyclopedia of Virology*, Webster, R. G. and Granoff, A., Eds., Academic Press, London, 1994, 1173.

Leong, J. C. and Fryer, J. L., Viral vaccines for aquaculture, *Annu. Rev. Fish Dis.*, 3, 225, 1993.

Letvin, N. L., Progress in the development of an HIV-1 vaccine, *Science*, 280, 1875, 1998.

Liu, M. A., Vaccine developments, *Nat. Med.*, 4, 515, 1998.

Lorenz, M. G. and Wackernagel, W., Bacterial gene transfer by natural genetic transformation in the environment, *Microbiol. Rev.*, 58, 563, 1994.

Lorenzen, N. et al., Genetic vaccination of rainbow trout against viral haemorrhagic septicaemia virus: small amounts of plasmid DNA protect against a heterologous serotype, *Virus Res.*, 63, 19, 1999.

Lowrie, D. B., DNA vaccination exploits normal biology, *Nat. Med.*, 4, 147, 1998.

Lundgren, B. and DePierre, J. W., The metabolism of xenobiotics and its relationship to toxicity/genotoxicity: studies with human lymphocytes, *Acta Physiol. Scand. Suppl.*, 592, 49, 1990.

Mandl, C. W. et al., *In vitro*-synthesized infectious RNA as an attenuated live vaccine in a flavivirus model, *Nat. Med.*, 4, 1438, 1998.

Mason, H. S., Lam, D. M., and Arntzen, C. J., Expression of hepatitis B surface antigen in transgenic plants, *Proc. Natl. Acad. Sci. U.S.A.*, 89, 11745, 1992.

Mason, H. S. et al., Edible vaccine protects mice against *Escherichia coli* heat-labile enterotoxin (LT): potatoes expressing a synthetic LT-B gene, *Vaccine*, 16, 1336, 1998.

McNally, R., Genetic madness. The European rabies eradication programme, *Ecologist*, 24, 207, 1994.

Mikkelsen, T. R., Anderson, B., and Bagger Jörgensen, R., The risk of crop transgene spread, *Nature*, 380, 31, 1996.

Miller, W. A. and Koev, G., Getting a handle on RNA virus recombination, *Trends Microbiol.*, 6, 421, 1998.

Mor, T. S., Gomez-Lim, M. A., and Palmer, K. E., Perspective: edible vaccines — a concept coming of age, *Trends Microbiol.*, 6, 449, 1998.

Moss, B., Poxviridae and their replication, in *Fields Virology*, Vol. 2, 3rd ed., Fields, B. N. et al., Eds., Lippincott-Raven, Philadelphia, 1996, 2637.

Mulder, W. A. M. et al., Pseudorabies virus infections in pigs. Role of viral proteins in virulence, pathogenesis and transmission, *Vet. Res.*, 28, 1, 1997.

Myhr, A. I. and Traavik, T., The precautionary principle applied to deliberate release of genetically modified organisms (GMOs), *Microb. Ecol. Health Dis.*, 11, 65, 1999.

Nagy, P. D. and Simon, A. E., New insights into the mechanisms of RNA recombination, *Virology*, 235, 1, 1997.

Nagy, P. D. et al., Dissecting RNA recombination in vitro: role of RNA sequences and the viral replicase, *EMBO J.*, 17, 2392, 1998.

Nielsen, K. M. et al., Horizontal gene transfer from transgenic plants to terrestrial bacteria — a rare event? *FEMS Microbiol. Rev.*, 22, 79, 1998.

Oger, P., Petit, A., and Dessaux, Y., Genetically engineered plants to terrestrial bacteria — a rare event? *FEMS Microbiol. Rev.*, 22, 79, 1998.

Oslage, U. et al., Prevalence of antibodies against the viruses of European swine fever, Aujeszky's disease and "porcine reproductive and respiratory syndrome" in wild boars in the federal states Sachsen-Anhalt and Brandenburg, *Dtsch. Tierärztl. Wochenschr.*, 101, 33, 1994.

Pastoret, P. P. et al., Stability of recombinant vaccinia-rabies vaccine in veterinary use, *Dev. Biol. Stand.*, 87, 245, 1996.

Pensaert, M., et al., Round table on control of Aujeszky's disease and vaccine development based on molecular biology, *Vet. Microbiol.*, 33, 53, 1992.

Poppe, T. T. and Breck, O., Pathology of Atlantic *Salmo salar* intraperitoneally immunized with oil-adjuvated vaccine. A case report, *Dis. Aquat. Organ.*, 29, 219, 1997.

Puman, L., Debate grows on safety of gene-therapy vectors, *Lancet*, 351, 808, 1998.

Raychaudhuri, S. and Rock, K. L., Fully mobilizing host defense: building better vaccines, *Nat. Biotechnol.*, 16, 1025, 1998.

Sandvik, T. and Tryland, M., Orthopoxviruses in Norway, Ph.D. dissertation, University of Tromsö, 1996.

Sandvik, T. et al., Naturally occurring orthopoxviruses: potential for recombination with vaccine viruses, *J. Clin. Microbiol.*, 36, 2542, 1998.

Schubbert, R. et al., Foreign (M13) DNA ingested by mice reaches peripheral leukocytes, spleen, and liver via the intestinal wall mucosa and can be covalently linked to mouse DNA, *Proc. Natl. Acad. Sci. U.S.A.*, 94, 961, 1997.

Schwartz, A. R., Togo, Y., and Hornick, R. B., Clinical evaluation of live oral types 1, 2, and 5 adenovirus vaccines, *Am. Rev. Respir. Dis.*, 109, 233, 1974.

Sheppard, M. and Boursnell, M., Viral vectors for viral vaccine delivery, in *Designer Vaccines — Principles for Successful Prophylaxis*, Hughes, H.P. A. and Campos, M., Eds., CRC Press, Boca Raton, FL, 1997, 125.

Sizemore, D. R., Branstrom, A. A., and Sadoff, J. C., Attenuated bacteria as a DNA delivery vehicle for DNA-mediated immunization, *Vaccine*, 15, 804, 1997.

Spadafora, C., Sperm cells and foreign DNA: a controversial relation, *BioEssays*, 20, 955, 1998.

Standish, A., Enquist, L. W., and Schwaber, J. S., Innervation of the heart and its central medullary origin defined by viral tracing, *Science*, 263, 232, 1994.

Stegeman, J. A. et al., Intensive regional vaccination with a gI-deleted vaccine markedly reduces pseudorabies virus infections, *Vaccine*, 12, 527, 1994.

Strayer, D. S. et al., Malignant rabbit fibroma virus: observations on the culture and histopathologic characteristics of a new virus-induced rabbit tumor, *J. Natl. Cancer Inst.*, 71, 91, 1983.

Tacket, C. O. et al., Immunogenicity in humans of a recombinant bacterial antigen delivered in a transgenic potato, *Nat. Med.*, 4, 607, 1998.

Taylor, J. and Paoletti, E., Fowlpox virus as a vector in non-avian species, *Vaccine*, 6, 466, 1988.

Taylor, J. et al., Efficacy studies on a canarypox-rabies recombinant virus, *Vaccine*, 9, 190, 1991.

Taylor, J. et al., Recombinant fowlpox virus inducing protective immunity in non-avian species, *Vaccine*, 6, 497, 1988.

Traavik, T., Too early may be too late, Report to the Directorate of Nature Management, Trondheim, Norway, 1999a.

Traavik, T., An orphan in science: environmental risks of genetically engineered vaccines, Report to the Directorate of Nature Management, Trondheim, Norway, 1999b.

Tryland, M. et al., Clinical cowpox cases in Norway, *Scand. J. Infect. Dis.*, 30, 301, 1998a.

Tryland, M. et al., Characteristics of four cowpox virus isolates from Norway and Sweden, *APMIS*, 106, 623, 1998b.

Tryland, M. et al., Serosurvey for orthopoxviruses in rodents and shrews from Norway, *J. Wildl. Dis.*, 34, 240, 1998c.

Tryland, M. et al., Antibodies against orthopoxviruses in wild carnivores from Fennoscandia, *J. Wildl. Dis.*, 34, 443, 1998d.

Tryland, M. et al., Antibodies to orthopoxviruses in domestic cats in Norway, *Vet. Rec.*, 143, 105, 1998e.

Tyler, K. L. and Fields, B. N., Pathogenesis of viral infections, in *Fields Virology*, Vol. 2, 3rd ed., Fields, B. N. et al., Eds., Lippincott-Raven, Philadelphia, 1996, 173.

Ulmer, J. B. et al., Heterologous protection against influenza by injection of DNA encoding a viral protein, *Science*, 259, 1745, 1993.

Vlassov, V. V., Balakireva, L. A., and Yakubov, L. A., Transport of oligonucleotides across natural and model membranes, *Biochim. Biophys. Acta*, 1197, 95, 1994.

Wolff, J. A. et al., Direct gene transfer into mouse muscle in vivo, *Science*, 247, 1465, 1990.

Zhang, G. et al., Expression of naked plasmid DNA injected into the afferent and efferent vessels of rodent and dog livers, *Hum. Gene Ther.*, 8, 1763, 1997.

chapter thirteen

Methods to assess ecological risks of transgenic fish releases

William M. Muir and Richard D. Howard

Contents

0-8493-0439-3/00/$0.00+$1.50
© 2002 by CRC Press LLC

13.1   Introduction

Fish, as other organisms, can be made transgenic for any number of different genes from any number of species; however, for a transgenic fish to have commercial value, the transformation must impart some economic advantage. Fast-growing fish have several economic advantages, including lower fixed cost per fish and increased feed efficiency (Lakra and Das 1998, Dunham 1999). Thus, fish transgenic for genes that impart faster growth are of special interest to commercial producers.

13.1.1   Production of transgenic fish

It is now possible to make fish of nearly any species transgenic for growth hormone (GH) expression using genetic sequences from many other organisms. Recent discoveries of DNA sequences for GH genes in several species of fish (Devlin 1993, Pandian et al. 1999), allow the artificial reproduction of these DNA sequences. The sequences are then fused to various promoters, which regulate expression of the gene, to make a transgene construct. The construct is inserted into single-cell embryos via microinjection or other techniques. The construct becomes incorporated into the germ-line DNA of the host for only a small number of injected fish embryos, however. When those fish develop to sexual maturity, the construct, also known as a transgene, can be passed on to their offspring in the normal Mendelian mode of inheritance. As such, it is now possible to make nearly any fish species transgenic for GH using genes from nearly any other species.

A special category of genetically modified organisms (GEOs) are transgenic for their own genes but are fused to a promoter in a construct that overexpresses the gene. These GEOs are sometimes called isogenic or autogenic, meaning same- or self-gene, respectively. Devlin et al. (1994a, 1994b, 1995a, 1995b) and Du et al. (1992) have produced various lines of transgenic and autogenic salmonids (*Oncorhynchus* spp.) for GH which are, on average, 10 to 15 times larger than non-transgenic controls using the same or similar construct. Transgenic arctic charr (*Salvelinus alpinus*) lines have been produced that grow 14 times faster than wild-type controls (Pitkanen et al. 1999) and tilapia (*Oreochromis niloticus*) lines have been created that grow three times faster than wild-type (Rahman and Maclean 1999). Several other species have also been transformed with GH, producing growth enhancements up to 100% of that of wild-type fish (reviewed by Fletcher and Davies 1991, Houdebine and Chourrout 1991, Pandian and Marian 1994, Chen et al. 1995, Hew et al. 1995, Iyengar et al. 1996, Devlin 1997, Sin 1997, Sin et al. 1997, Maclean 1998, Pitkanen et al. 1999).

13.1.2   Need for critical evaluation

Globally, aquaculture yields are equal to wild fish harvests of some species (e.g., salmon) and greater than wild harvests of other species (e.g., carp *Cyprinus carpio* in China). The ongoing expansion of aquaculture and its proven ability to generate foreign exchange and enhance local food security are likely to increase investment in the development of transgenic fish. Some agencies in developing countries are especially interested in this technology as a way of increasing food production (Dunham 1999). Currently, an application is before the U.S. Food and Drug Administration (FDA) seeking approval to produce GH-transgenic Atlantic salmon commercially in net pens floating in the Atlantic Ocean.

Fish may become the first transgenic animals to be produced in the wild and be made available to consumers.

Adoption of transgenic technology depends on consumer acceptance of its products, however. Consumers and environmentalists currently question the safety of genetic engineering in food and agriculture. The topic of GEOs is a present focus of the World Trade Organization due to biosafety concerns and other issues (Vogt and Parish 1999, Halvorson and Quezada 1999). Ministers from more than 130 countries met in Montreal, Canada in January 2000 under the aegis of the Convention of Biological Diversity, to discuss GEOs in international trade. A biosafety protocol, the Cartagena Biosafety Protocol, was agreed upon, partly in response to widespread concerns that GEOs pose environmental risk. The protocol allows countries to block imports of GEOs on a "precautionary" basis in the absence of sufficient scientific evidence about their safety. This development, along with the current commercial research and development of transgenic salmon in New Zealand and Canada and the expected production of transgenic tilapia in Cuba (de la Fuente et al. 1996; 1997, Guillen et al. 1999) and other parts of the world (Dunham 1999), underscores the need to develop rigorous methodologies to estimate environmental risk. Although commercial production of transgenic fish is imminent, it is clear that the first step in risk assessment, hazard identification, is not complete. Scientific investigations have not yet identified all the possible mechanisms by which transgenic fishes might influence ecosystems, let alone quantified the likelihood of each of those impacts. Of the utmost importance is a thorough comparison of GEOs with wild-type fish with respect to critical aspects of their life histories.

Current standards do not specify which criteria are critical for determination of environmental risk. For example, Guillen et al. (1999) reported experiments "performed to evaluate the behavior of transgenic tilapia in comparison to wild tilapia as a way to assess the environmental impact of introducing transgenic tilapia into Cuban aquaculture." Based on feeding motivation, social dominance status, and adaptation to seawater, they concluded that under "the conditions found in Cuba, [there are] no environmental implications for the introduction of this transgenic tilapia line." Thus, based only on a few factors that indirectly influence survival and perhaps mating success, they concluded that tilapia transgenic for GH (hereafter referred to as GH-tilapia) are safe for commercial production. Unfortunately, net fitness parameters, such as survival to sexual maturity, mating success, age at sexual maturity, fertility, fecundity, or longevity, were not investigated. Thus, their conclusions may be seriously flawed.

13.1.3   *Ecological impacts*

The escape or introduction of transgenic fish into natural communities is a major ecological concern. Transgenic individuals retain most of the characteristics of their wild-type counterparts while possessing some novel advantage. For example, a transgene for cold tolerance would allow fish with that gene to invade waters in colder climates while maintaining populations in currently used habitats. Similarly, fish transgenic for GH (GH-fish) may prey on different organisms, reach sexual maturity sooner, spawn at different times, and escape from their normal predators. As such, GH-fish could reproduce at a faster rate; their population may increase unchecked and adversely affect other species, and disrupt the ecosystem. As a consequence, transgenic organisms might threaten the survival of wild-type conspecifics as well as other species in a community (Tiedje et al. 1989, Hallerman and Kapuscinski 1990, 1992, 1993, 1995, Kapuscinski and Hallerman 1990, 1991, 1999, Devlin and Donaldson 1992, Devlin 1998, and Kapuscinski, Chapter 14 of this volume). The environmental risks at issue are, to some degree, comparable with those posed by the introduction of non-indigenous species. Exotic species sometimes adversely affect biotic communities in many ways, including the elimination of populations of other species (Drake and Mooney 1986, Lodge 1993, Bright 1996).

This chapter examines scientific methods used to assess environmental risks of transgenic animals, discusses limitations of those methods, and explores possible counterarguments that may limit environmental risk. In particular, we will focus on finfish. However, the methods we offer here are general and, with some modification, can be applied to any organism, plant or animal, that reproduces sexually. The particular ecological risk of interest in this chapter is the spread of transgenes into wild populations through interbreeding of GEOs with wild-type fish. In some places, we have used the phrase *ecological risk* to stand for transgene introgression into wild populations and any resultant hazard that may arise from the shift in life history characteristics (Table 13.1). It will be noted further, that the term *risk* is used here in the general sense rather than the technical one (involving both probability of occurrence times strength of hazard). In most cases, the probability of the occurrence of specific hazards has not been calculated.

13.2   Genetic mechanisms resulting in ecological risk

There are many perspectives from which to view possible ecological hazards of transgenic fish. In some aspects, release of transgenic fish causes ecological problems similar to those caused by release of domesticated fish; in other aspects, a transgenic fish release is very different. In the short term, transgenic fish could displace their wild-type conspecifics. In the long term, such displacement could cause extinction of their own and/or different species. Release of domesticated fish can also cause displacement of wild-type conspecifics, but the mechanisms of displacement and long-term environmental impact may differ depending on population size and other stochastic events.

There are also non-genetic mechanisms that result in ecological risk. For example, were large numbers of either sterile domesticated or transgenic fish to escape from net pens and compete with native populations for resources, there could be severe and lasting impacts on native species. In this chapter, we do not consider such events; we only consider genetic risks that result from interbreeding.

13.2.1   Co-adapted gene complexes

Fish native to a region are the result of evolution occurring over many thousands of generations. The fish have been naturally selected for optimum life history characteristics, such as rate of growth, age and size at sexual maturity, and clutch size. Deviation from these optima may result in lower overall fitness of the species. Once disrupted by introduced fish, domesticated or transgenic, natural selection may return the population to its previous optimum if the population is large enough and given enough time.

However, if co-adapted gene complexes (epistatic linkage relationships among genes) are disrupted, a return to the prior optimal state may not be possible. Evolution of co-adapted gene complexes is not well understood and is thought to be the result of species isolation and interspecies selection. If that is the case, the population would suffer from an overall long-term reduction in fitness. If the population is struggling for existence prior to an introduction event, the induced genetic load (subsequent to the introduction) may be sufficient to drive the population to extinction.

13.2.2   Introduction of maladapted genes

Even in the absence of co-adapted gene complexes, if the native population is small, and the number of introduced fish is large, or the introduction is continuous over time, the gene pool could be swamped by genes for traits that are poorly adapted to the environment and

Table 13.1 Fitness Component (FC) Advantages Observed in Natural Populations That Could Have Implications for Environmental Risk from GH Transgenic (TR) Fish[a]

Species	FC	Trend	Potential impact	Risk[b]	Ref.
Pacific salmon, Atlantic salmon, Pink salmon, Sockeye salmon, Chum salmon, Coho salmon, Japanese medaka	S	Large males have a mating advantage over small males	If size advantage of TR males over WT males persists through adult stage, TR males should have a mating advantage	I	de Gaudemar 1998, Fleming 1996, 1998, Moran et al. 1996, Howard et al. 1998
Atlantic salmon, Brown trout, Sockeye salmon, Coho salmon, Chum salmon, Chinook salmon, Pink salmon, Masu salmon, Rainbow trout, Cutthroat trout, Dolly Varden char, Brook char, Arctic char, Lake trout, Huchen	S	Variation in male size greater than that of females indicating use of alternative mating tactics (e.g., sneak copulations) by small males, which may provide them higher mating success than if they used the same tactic that large males use	If size advantage of TR males over WT males persists through adult stage, WT males could use the alternative mating tactic; the reverse would occur if individuals with higher growth rate prior to sexual maturity utilize the alternative tactic	I or D	Fleming 1998
Brown trout	EV	Large trout have reduced predator avoidance when foraging than small trout	Large fish have increased mortality due to predation	D	Johnsson 1993
Brown trout	EV	Large trout are socially dominant over small trout	Large fish should have greater access to needed resources such as food which could increase their survival and reproduction	I	Johnsson 1993
Pacific salmon, Coho salmon	EV	Higher early survival of young of larger females because large females can control better nest sites	If size advantage of TR females over WT females persists through adult stage, early survival advantage of TR young	I	de Gaudemar 1998
Amago salmon, Masu salmon, Chinook salmon, Atlantic salmon, Pacific salmon	A	Precocious sexual maturation of males with faster growth rates	Accelerated early growth rate of TR males should result in early age at maturity	I	de Gaudemar 1998
Atlantic salmon	AV	Larger males and females have reduced longevity	If size advantage of TR individuals persists through adult stage, TR should have lower survival as adults	D	Moran et al. 1996

Table 13.1 (continued)   Fitness Component (FC) Advantages Observed in Natural Populations That Could Have Implications for Environmental Risk from GH Transgenic (TR) Fish[a]

Species	FC	Trend	Potential impact	Risk[b]	Ref.
Atlantic salmon, Sockeye salmon, Japanese medaka	F	Larger females produce larger clutches and/or larger eggs	TR females may have higher fecundity than WT females	I	de Gaudemar 1998, Fleming, 1996, 1998
Atlantic salmon	R	Large anadromous males have high fertilization success because of more sperm/ejaculate; parr males invest more relative to size in sperm production and may fertilize 11 to 40% of eggs; larger parr males have size advantage in fertilizing eggs	TR males may have higher fertilization success than WT males	I	de Gaudemar 1998, Fleming 1996, 1998, Thomaz et al. 1997

[a] Fitness components: A = age at sexual maturity; AV = adult viability; EV = early viability; F = fecundity; S = sexual selection; R = fertility.

[b] Environmental risk: I = risk increased; D = risk decreased

more adapted alleles for better-adapted traits may be lost through random genetic drift. In that case, the species could become extinct as a result of the introduced genetic load. This phenomenon may be the reason Atlantic salmon are on the endangered species list.

The environmental risks discussed above were either a result of co-adapted gene complexes or finite population sizes and other stochastic events. The environmental hazard in the short term would be the displacement of the native species; the long-term risk would be extinction of the species and other species dependent on it. The definition of environmental hazard did not include any impacts from disease that might have been transmitted by domesticated or transgenic fish to natural populations.

13.2.3   Directional selection for the transgene

Another type of environmental hazard results directly from the introduction of the transgene itself. Species displacement in the short term and extinction in the long term are the potential hazards, but the mechanisms differ from those discussed above. In this case, deterministic forces drive the transgene into the population (increasing the transgene frequency), because of pleiotropic effects of the transgene on various fitness components. Pleiotropic effects occur when a gene affects multiple traits. Often, such an increase in transgene frequency is assumed to be unlikely because transgenic fish typically have some viability disadvantage (e.g., Knibb 1997, Dunham 1999). Muir et al. (1994, 1995, 1996) and Muir and Howard (1999a, 2001) provide the first comprehensive study to evaluate this assumption in transgenic fish. The objective of our study was to develop a framework for testing transgenic fish by showing how data on several life history parameters could be incorporated into a model to assess environmental risk.

13.2.4   Conflict between fitness parameters

The greatest environmental hazard predicted by our model is local extinction of natural populations after a transgenic introduction, a phenomenon referred to as the "Trojan gene" effect (Muir and Howard 1999a). For this effect to occur, transgenes must affect two fitness components in opposite ways: increasing male mating success while reducing viability of transgenic offspring (viability here is defined as survival to sexual maturity). Evolutionary biologists have long recognized that sexual selection could result in species extinction. For example, Lande (1981) noted that Fisher, Haldane, and Wright had all stated that "the fitness of a trait with respect to mating success can override its value for survival, creating a maladaptive evolution that may contribute to the extinction of species." However, transgenes that increase mating success at the expense of offspring viability can cause population extinction even more rapidly than that expected from sexual selection on naturally occurring genes that influence secondary sexual traits. In the latter case, the viability reduction is limited to males as a direct result of selection for sex-limited trait expression (i.e., only males are affected because only they express the trait that increases mortality); whereas, in the former case, a transgene reduces offspring viability of *both* sexes equally.

13.3   Arguments against genetic mechanisms resulting in ecological risk

Knibb (1997) presented several arguments based on both theory and data from laboratory populations in other species regarding the reason transgenic fish are not likely to present an environmental risk. Theoretical considerations of Lande (1983) are also relevant to this discussion and can be used to argue both for and against possible environmental risks of transgenic fish.

13.3.1   Probability of a transgene initiating a speciation event

Knibb (1997) stated that environmental risk exists whenever laboratory genetic changes have a non-negligible probability of the formation of new species. Knibb (1997) suggests that speciation events through reproductive- or ecological-isolating mechanisms are unlikely in transgenic stocks, because they differ from their wild-type counterparts by only a tiny fraction of the genome and these changes are too small to give rise to a new species. However, the amount of the genome altered in producing a transgenic line is not relevant; the amount of change in *phenotype* is, particularly if the phenotypic change involves a sexually selected character. The 14-fold increase in growth rate observed in some transgenic salmonids can hardly be considered a small change, even if the number of loci involved is miniscule. Given such a size advantage, in a species such as sockeye salmon in which each sex prefers to mate with large individuals of the opposite sex (de Gaudemar 1998, and references therein), both transgenic and wild-type individuals will favor the larger transgenics as mates, and the affected population will diverge rapidly in body size from conspecific populations that did not experience a transgenic introduction. By Knibb's (1997) definition, if a transgene results in a major impact on mating preference, a new species, or evolutionary line, could rapidly evolve and produce significant environmental disruption.

13.3.2   Population vs. individual fitness

Knibb (1997) concluded that there is a negligible probability that a genetic change produced in a laboratory could be adaptive in the wild. He based this inference on experimental studies that show that mutations usually reduce overall fitness in the laboratory and never increase fitness in nature. However, adaptations that produce some short-term advantage to individuals may often result in population extinction in the long term.

According to some estimates, 99% of all species that ever existed have become extinct (Mourao and Ayala 1971). No doubt, such extinctions occurred despite the relentless action of natural selection to maximize the short-term fitness of individuals. The paradox is that natural selection favoring traits that increase individual success is blind to the consequences of these traits on population persistence. As Williams (1966) argued convincingly, individual-level selection favoring such traits is considerably more powerful than population-level selection is in removing them. Mourao and Ayala (1971) point to the example of Dawson (1969), who presented convincing data in this regard. He observed a new eye mutation in one of ten replicated competition experiments between the closely related mealworms *Tribolium castaneum* and *T. confusum*. In that replicate, the gene frequency of the new mutant increased (as a result of individual or Darwinian fitness advantage), yet the population became extinct (as a result of population-level fitness decline). None of the other nine replicates resulted in extinction. Thus, although new mutations that increase short-term fitness can occur in sufficiently high frequency to be observed in a laboratory situation, the long-term consequences to the population could be missed, unless the experiment is designed to detect them. That is, existing populations are a biased sample in that they only include those mutations present in the population that have continued to survive. Similarly, sampling existing species is biased; we can only observe those species that have continued to survive the speciation process. What cannot be known is what we would have learned from those that failed.

Maximization of individual fitness can result in population extinction because of intense competition, as shown theoretically by Griffing (1967, 1976) and experimentally in *Drosophila melanogaster* (Mourao and Ayala 1971) and in poultry (Muir 1996, Craig and Muir 1996a, 1996b, 1997, Muir and Craig 1998). Results with poultry clearly show that selection for individual productivity can reduce group productivity through increased competition. Similar effects of selection for competitive ability have been observed between domesticated and wild-type fish (Fleming and Einum 1997). Thus, if a transgene has a major impact on competitive ability, it presents another mechanism by which the Trojan gene effect could result. Because a number of studies show that larger fish are more socially dominant (see Table 13.1), this is not a negligible environmental risk.

13.3.3   Risk of transgenic fish for GH limited by reaction norms

If bigger is better, why are all fish not the size of whales? Perhaps, one should ask, what limits body size in any organism? According to evolutionary theory, the maximum body size of any organism represents a balance between the benefits and costs of body size, given the organism's ecological relationships with its physical and biotic environment. Thus, the benefit of being slightly larger than the current maximum size may not be offset by the costs of increased size.

Body size is a highly polygenic trait in all organisms. Selection favoring an increase (or decrease) in body size in a natural population in any one generation involves the relative ecological and physiological success of alternative genotypes in the population. Such a process depends not only on the nature of the phenotypic effect of each genotype but also on the prevailing environmental conditions. It is usually assumed that a large number of loci underlie differences in a trait between coexisting genotypes and that each locus has a small effect on the trait. The cumulative effect of such natural selection across many generations could alter the allele frequencies at a large number of loci and even be a primary force of speciation (Fisher 1930).

Environmental conditions determine the direction and force of selection on genotypes. In addition, because the phenotype of any organism is a product of both its genes and its environment, the phenotypic response of an organism often varies with environmental conditions. When such plasticity is a continuous function of environmental variation, it is referred to as a norm of reaction (Stearns 1989). Thus, plasticity in traits affecting fitness components are likely to vary not only under different laboratory and natural conditions but also between the laboratory and natural environment. This phenotypic variation could well interact with variation in the intensity of selection acting on genotypes. For example, selection may be relaxed if population density is low (as would be the case, for example, in areas where fish populations are overfished commercially) because needed resources would be abundant for all fish genotypes and little or no resource competition would occur. Alternatively, if the nature of the habitat were to be altered substantially, as may be the case in artificial impoundments, the genotypes that prevail could be radically different from those prevailing in undisturbed natural habitats. Knibb (1997), whose discussion was restricted to undisturbed habitats, indirectly acknowledged this possibility.

Lande (1983) considered the relative evolutionary impact of mutations with large phenotypic effects as opposed to the cumulative effect of selection on a large number of polygenes each with a small effect. Data on recently evolved adaptations support the view that selection favoring mutations with major effects are usually confined to domesticated or artificially disturbed populations. In that regard, release of domesticated fish into natural populations automatically results in a disturbed natural population.

Knibb (1997) and others have argued that it is difficult to establish a new population in nature through the release of laboratory populations. Such arguments consider the consequences of small increments in body size and assume that the ecological niche of the organism is not changed if a mutant individual is only slightly larger. Transgenes, however, represent mega-mutations, which substantially and instantaneously increase the body size of a transgenic organism and thus could change many aspects of its niche. In theoretical terms, based on Wright's (1969, 1982) shifting balance theory, such a mega-mutation would involve a switch in the current adaptive peak of an organism to a different adaptive peak. An example of such a major mutation causing a shift in evolutionary trajectory was cited in Lande (1983), wherein it was noted that a major mutation for mimicry had occurred in the evolution of butterflies. Two primary predator avoidance attributes in butterflies are either to remain hidden (camouflage) or to closely resemble another species that is distasteful to birds (mimics). Intermediate individuals that are neither well camouflaged nor good mimics are likely to be eaten. Similarly, a GH-transgene could act as a mega-mutation that could, for example, change an organism from being a prey of one species to being a predator of another.

The consensus of Wright, Fisher, and others is that mutations with large effects are usually deleterious because of either their direct effect or their pleiotropic effect on other characters. This view results in the conclusion that the "probability that a random (undirected) mutation will improve adaptation decreases rapidly with increasing magnitude of its effect" (Lande 1983). However, unlike other macro-mutants, in which mutations with a large effect were random with regard to adaptation, some increase in adaptive ability is the point of producing transgenic organisms in the first place.

Lande (1983) shows that if a favorable macro-mutation is recessive, with negative pleiotropic effects, and the macro-mutant is initially rare, it may take so long to increase in frequency in the population that the new optimum phenotype will evolve more rapidly by polygenic changes and the major mutation will be lost. In contrast, Lande (see figure 3 in Lande 1983) shows that if the favorable macro-mutation is dominant, it is more likely to be fixed in the populations, even with negative pleiotropic effects and with the initial frequency of the macro-mutation orders of magnitude lower than the wild-type allele. This is the case for transgenes, of course, because by design they behave as dominant mutations.

13.4   Environmental risk assessment

The following discussion assumes a large population in which stochastic events are limited, and thus results are able to be predicted using deterministic models. The introduction of a transgene into a natural population through accidental or intentional release of transgenic organisms can then be characterized by the frequency of that gene in a population. Frequency of transgenes can be computed simply as the total number of transgene alleles in the population divided by twice the population size of a diploid population at the time of sampling. There are only three possible outcomes of such a transgene introduction. Through time (i.e., generations): (1) the frequency of the transgene may decrease and eventually the transgene is eliminated from the population, (2) the transgene may increase in frequency but the population size remains stable or increases resulting in a niche shift and/or disruption of predator-prey relationships, and (3) the transgene may increase in frequency but the affected population declines in size. In the latter case, the population could become extinct, potentially causing a cascade of undesirable consequences throughout the ecosystem, an outcome we refer to as the Trojan gene effect (Muir and Howard 1999a). With outcome 2, it is not necessary to determine the exact form of the niche shift and/or predator–prey relationship disruption. It is sufficient to conclude that if the transgene increases in frequency, transgenic individuals have some advantage over their naturally occurring wild-type conspecifics, and this advantage will be manifest in some way. Our methods of environmental risk assessment are only applicable to outcomes 2 and 3 as defined above, i.e., situations of transgene spread in native populations.

13.4.1   Transgenic animals most in need of environmental risk assessment

The only transgenic organisms that create environmental outcomes 2 and 3 above are those that can establish themselves in nature. For agriculturally important animals, environmental risk is mostly limited to those species that have feral counterparts, that is, transgenic animals that can become established in nature through intermating. Thus, environmental risk increases greatly if there are conspecific populations that exist in the wild within the dispersal distance of transgenic individuals. In most countries, the only agriculturally important animal species that can readily access feral counterparts are aquacultured species and aquatic fowl. In certain parts of the world where chickens have natural populations of wild relatives, transgenic chickens could also pose an environmental risk. However, if a transgene provides some significant novel reproductive or survival advantage, the organism might still spread as an exotic species, despite the absence of feral counterparts. Thus, all transgenic organisms must be evaluated for environmental risk.

In the case of transgenic fish, it would appear that there would be no risk (of spread of the transgenes) in regions that do not contain native populations of transgenic stock. This issue has been raised in support of production of GH-transgenic carp in Israel, tilapia in Cuba, and Atlantic salmon in the Pacific Ocean and elsewhere. There are really two issues to be taken into consideration, however: (1) introduction of exotic fish (domestic and/or transgenic) into new areas and (2) likelihood for spread of the transgenic stock to regions of the world where conspecifics exist naturally.

Introduction of exotic species has often resulted in ecosystem disruption. This is likely to be a warning for both transgenic and non-transgenic domesticated fish. Methods of risk assessment are not straightforward in such situations because the potential environmental hazard is the elimination of prey and competing species; such displacement may be unique to each region, and all prey and competing species may not even be known. Thus, the magnitude of such a hazard is difficult to characterize.

In the case of spread of transgenics to other regions where there are no wild relatives, the immediate concern of interbreeding with natural populations is absent, and thus the biosecurity of the transgenic fish may not seem to be an issue. There may be little precedent, then, to prevent someone from collecting transgenic fish and introducing them into other regions of the world where relatives do exist naturally. The transported transgenic fish then may cause extensive environmental damage.

In the context of risk assessment, we have to think of environmental risk in broader terms than our own borders. Each transgenic aquatic organism should be treated as if it had feral counterparts in the region of introduction.

13.4.2   Methods to assess environmental risk

There are four methods that have been used to assess the environmental risk of the release or escape of transgenic individuals into a wild population. The methods are based on (1) inference from natural variation, (2) direct observations on transgenic fish, (3) use of surrogate organisms to mimic the transgene effect, and (4) models to predict changes in gene frequency based on net fitness parameters of transgenic fish relative to native fish.

13.4.2.1   Natural variation

When a transgene of interest enhances a trait that naturally exists within the species, natural variation associated with that trait can be used to infer that impact of the transgene on various aspects of fitness. For example, to infer impact of a transgene for GH, natural variation in body size and its association with other traits may be good indicators of potential impact of the GH-transgene.

Observations on many fish species show that larger body size provides many advantages for securing limited resources and mates (see Table 13.1). From these data, one can infer that GH-fish might pose an environmental risk to natural populations of conspecifics. One can further infer the mechanisms through which those environmental risks might occur. A limitation of this method is that it may underestimate or obscure the true impact of the transgene because GH-transgenic fish often exceed natural size maxima. For example, transgenic salmonids can be 15 times larger than their same-aged, non-transgenic siblings (Du et al. 1992). Although use of natural size variation in these salmonids may be irrelevant for predicting the potential for disruption in predator–prey relationships caused by transgenic individuals, it may be very relevant for predicting the effects of transgenic organisms on other phenomena, such as social dominance, mating success, and fecundity.

Based on natural size variation, larger fish are usually more socially dominant (Johnsson 1993), mature at an earlier age (de Gaudemar 1998), produce more sperm (Fleming 1996, 1998, Thomaz et al. 1997, de Gaudemar 1998) or larger eggs and clutches (Fleming 1996, 1998, de Gaudemar 1998), have higher mating success (Fleming 1996, 1998, Moran et al. 1996, de Gaudemar 1998, Howard et al. 1998), attract larger mates (de Gaudemar 1998), and produce larger juveniles that are competitively superior and have higher survival (de Gaudemar 1998). All these advantages enhance various aspects of fitness; however, larger fish may also have reduced predator-avoidance behavior (Johnsson 1993) and a lower likelihood of surviving to mate in subsequent seasons due to injuries sustained while maintaining social dominance (Moran et al. 1996).

13.4.2.2   Direct observations of transgenic fish

Direct measurements on the transgenic fish of interest is the optimal method of assessing its performance relative to wild-type individuals and hence potential environmental risk. However, different studies of the transgenic fish have not consistently measured the same traits, and most studies have often failed to incorporate all, or even most, aspects of fitness that are necessary to predict if the transgene enhances the phenotype sufficiently to pose an environmental risk. Thus, existing data may not be sufficient to permit rigorous comparisons.

Nevertheless, available data on GH-transgenics indicate that the GH-transgene does affect different aspects of fitness in various species of fish in ways that could pose environmental risk (Table 13.2). Few studies have quantified viability differences between transgenic and wild-type fish directly, except Devlin et al. (1995a) who observed that the largest transgenic fish had growth abnormalities of the head and jaw. Fish with the highest early growth performance were affected the most and had difficulty eating. Growth of those fish could not be sustained relative to other transgenics and they died prior to maturation. Thus, the severity of morphological abnormalities was correlated with initial growth rate.

Indirect evidence for reduced viability has been observed in GH-transgenic fish of several species (see Table 13.2). Transgenic fish have reduced predator-avoidance responses that should reduce their viability in nature (Abrahams and Sutterlin 1999, Dunham et al. 1999). Transgenes appear to have no consistent effect across species on swimming speed (Farrell et al. 1997, Stevens et al. 1998, Abrahams and Sutterlin 1999), a trait that could influence an individual's ability to escape predators or attract mates.

Behavioral differences that could indirectly influence viability have been documented between transgenic and wild-type conspecifics; these include differential feeding rates (de la Fuente at al. 1997, Devlin et al. 1999, Guillen et al. 1999) and competitive ability either for resources (Devlin et al. 1995b, Fu et al. 1998; but see de la Fuente et al. 1997, Guillen et al. 1999) or social status. However, comparisons were sometimes biased by limiting competition to size-matched rather than age-matched fish (e.g., Farrell et al. 1997, Guillen et al. 1999). While this criterion isolates the effect of a transgene on some aspect of performance, juvenile transgenic individuals are often compared with adult wild-type individuals of similar size. As Farrell et al. (1997) speculated, "the poorer swimming performance of these growth-enhanced transgenic fish [may] arise either from an onto-genetic delay or from disruption of locomotor muscles and (or) their associated support systems." Similarly, age differences (hence experience levels) might explain the observed lower dominance status of GH-transgenic tilapia when compared with wild-type individuals of similar size (see Guillen et al. 1999).

Reproductive consequences of transgenes have been sorely neglected. As would be expected from an enhanced growth rate, transgenic individuals have been shown to attain sexual maturity at an earlier age in three species (see Table 13.2). Only one study has demonstrated an influence of transgenes on clutch size (Muir and Howard 2001). No studies have assessed the effect of transgenes on mating success; however, large male mating advantages are commonly observed in fish (see Table 13.1). Thus, transgenic individuals may also obtain a mating advantage if their size advantage persists through adulthood.

It is unfortunate that all the above fitness components have not been assessed in all transgenic lines. It is even more unfortunate to base an environmental risk assessment on one or a few fitness component measures. As shown below, fitness components can interact to produce environmental risk in ways that would not have been predicted if the effect of a transgene on each fitness component was considered separately. Thus, as is outlined in the next section, a more holistic approach is needed to predict environmental risk.

13.4.2.3   *Models to predict changes in gene frequency*
For a transgene that spreads by interbreeding with native populations followed by natural selection, the likelihood of transgene spread can be determined from expected gene frequency change. Prediction of changes in gene frequency is a two-step process: estimation of fitness parameters for alternative genotypes, followed by development of a model that incorporates these estimates and predicts the change in gene frequency.

Table 13.2 Implication for Ecological Risk of TR Fish Based on Individual Fitness Components[a]

Species	TR Const.	Trait	Finding	Potential impact	Risk	Ref.
			Age at sexual maturity			
Atlantic salmon	AFP-GH	Age at smolt	TR attain smolt status earlier than WT	Earlier sexual maturation should increase intrinsic rate of natural increase in TR	I	Saunders et al. 1998
Coho salmon	POnMT-GH1	Age at smolt	Precocious smoltification of TR (6 months prior to WT)	Earlier sexual maturation should increase intrinsic rate of natural increase in TR	I	Devlin et al. 1995a, 1994b
			Viability			
Japanese medaka	SGH-hGH	Early viability	Survival to day 3 of TR fry 70% that of WT	Reduced early viability of TR	D	Muir and Howard 1999a
Brown trout	GH injection	Predator avoidance	GH reduced predator avoidance response relative to WT juveniles	Reduced viability in GH juveniles due to greater chance of predation	D	Johnsson et al. 1996
	GH injection	Predator avoidance	Under field conditions, GH-treated and control fish did not differ in survival rates	Reduced predator avoidance of GH fish in the laboratory may not translate into lower survival in nature	I	Johnsson et. al. 1999
Atlantic salmon	OpAFP-GHc	Predator avoidance	TR alevins more likely to forage near predators than WT alevins are	Reduced viability in TR alevins due to greater predation risk	D	Abrahams and Sutterlin 1999
Rainbow trout	GH injected	Predator avoidance	GH-injected juveniles had reduced predator avoidance behavior than did WT juveniles	GH increases susceptibility to predation	R	Jönsson et al. 1996
Channel catfish	RSVLTR-csGH or RSVLTR-rtGH1	Predator avoidance	Predation on TR fingerlings greater than that on WT	Greater susceptibility of young TR to predators may reduce viability relative to WT	R	Dunham et al. 1999
Rainbow trout	GH injected	Feeding efficiency	Food conversion efficiency was higher in GH-treated fish than in WT fish	TR fish should have higher growth rates than WT	I	Johnsson and Björnsson 1994
Common carp	PMT-hGH	Feeding efficiency	Growth rate of TR higher than that of WT	TR grow faster both on poor and high-quality diets than WT	I	Fu et al. 1998
Tilapia	HCMV-tiGH	Feeding motivation	TR less motivated to feed than WT	TR are less competitive than WT	D	de la Fuente et al. 1997, Guillen et al. 1999

Table 13.2 (continued)   Implication for Ecological Risk of TR Fish Based on Individual Fitness Components[a]

Species	TR Const.	Trait	Finding	Potential impact	Risk	Ref.
Coho salmon	OnMT-sGH	Feeding motivation	TR consumed 2.9 times more food than WT fish of the same size but one year older	TR are more competitive than WT	I	Devlin et al. 1999
	OpAFP-GHc	Weight at alevin stage	TR 21% heavier than WT at alevin stage	TR grow faster than WT fish prior to first feeding	I	Devlin et al. 1995b
Atlantic salmon	AFP-sGH, OnMT-sGH	Swimming speed	Critical swimming speed did not differ between TR and WT fish	TR may have ability to escape predators similar to WT	N	Stevens et al. 1998
Coho salmon	OnMT-sGH	Swimming speed	Critical swimming speed of TR was half that of 1-year-older WT of similar size	TR may have reduced ability to escape predators than WT of the same size	D	Farrell et al. 1997
Atlantic salmon	OpAFP-GHc	Swimming speed	TR alevins had a three-fold speed advantage over WT	Greater swimming speed of TR may increase ability to obtain prey and escape predators	I	Abrahams and Sutterlin 1999
Tilapia	HCMV-tiGH	Dominance status	TR less competitive in obtaining food than WT	WT may be dominant to TR in obtaining food and mates	D	de la Fuente et al. 1997, Guillen et al. 1999
Rainbow trout	GH injected	Aggressive behavior	GH increased level of aggression but not social status relative to control in interactions between juvenile fish pairs matched in size	TR may not enhance social status	N	Jönsson et al. 1998
Atlantic salmon	AFP-sGH, OnMT-sGH	Gill morphology	Gill surface area in TR fish did not increase in proportion to oxygen uptake demands	Unknown	?	Stevens and Sutterlin 1999
Mating success						
Coho salmon	OpAFP-GHc	Weight at sexual maturity	At 24 months, the average weight of TR was approximately 6 times WT	Larger adult males may have a mating advantage	I	Devlin et al. 1995b
Tilapia	OpAFP-GHc	Weight at sexual maturity	4–7-month-old TR >3 times that of WT; at 25 months, one male was 7.5 times larger than its WT sib	Larger adult males may have a mating advantage	I	Rahman et al. 1998, Rahman and Maclean 1999
	HCMV-tiGH	Weight at sexual maturity	9-month-old TR were 81% heavier than WT	Larger fish at sexual maturity may have a mating advantage	I	Martinez et al. 1996

[a] Environmental risk: I = increases, D = decreases risk; N = neutral (impact was subjectively determined based on risk as discussed in text); WT = wild-type, TR = transgenic.

13.4.2.3.1 Fitness component estimation. There are certain critical "check points" in an organism's life history that can be used to assess its net fitness. From theoretical considerations and data from laboratory organisms, Prout (1971a) showed that the entire life cycle must be studied to determine overall fitness. Prout (1971a) also demonstrated that incomplete information would result in erroneous population predictions. Our research (Muir and Howard 1999a, 2001) also showed that to assess environmental risk, all fitness factors must be measured, because a disadvantage due to a transgene in one category, such as viability, can be offset by an advantage in another category, such as mating success. For example, reduced viability due to a transgene could be offset by (1) earlier sexual maturation (which would greatly reduce the generation interval, and hence reproductive rate, of a transgenic line), (2) enhanced ability of larger transgenic males to obtain mates; (3) increased egg production by females due to larger body size, and (4) reduced early mortality from cannibalism due to a faster growth rate (Muir and Howard 1999a, 2001).

Prout (1971a) also emphasized that it is not necessary to study every aspect of the entire life cycle. Instead, aspects of the life cycle could be grouped into a small number of fitness components. The problem is to determine the minimum number of such components. He suggested three components corresponding to sexual, fecundity, and viability (zygotic) selection. The various processes that influence viability fitness, such as predator avoidance, disease resistance, feeding efficiency, swimming speed, social dominance, and foraging ability, could all be subsumed in the viability measure. Similarly, factors that influence mating success, such as mate attraction, nest site acquisition and protection, and mate competition, could all be subsumed under the sexual fitness component. For environmental risk assessment, detailed knowledge of the factors that underlie fitness component differences between transgenic and wild-type individuals would not be necessary.

Prout's (1971a) procedures were designed to predict genotype frequencies in a population. Prout (1971b) later showed that such estimates could be incorporated into a recurrence equation to predict gene frequencies in experimental populations of fruit flies (*Drosophila melanogaster*). Results from his experiments showed that this method was reasonably good in predicting the dynamics of mutant allele frequencies in a population.

In the following three sections we (1) define the minimum number of fitness parameters necessary for accurate predictions, (2) show how to estimate the fitness parameters based on either static observation or dynamic methods as suggested by Prout (1971a), and, finally, (3) provide a recurrence equation using these estimates to predict population dynamics and gene frequency trajectories.

13.4.2.3.2 The minimum number of fitness components necessary for accurate predictions. Prout's (1971a) model was primarily designed for *D. melanogaster* populations and resolved fitness parameters into those associated with juvenile and adult stages; the model did not include variation in mating success, age at sexual maturity, and longevity. Prout (1971a, p. 145) states that for "an investigator who is about to study an unknown system, it is safest to assume a model of selection of maximal complexity." Toward this aim, Muir and Howard (1999a, 2001) concluded that estimates of six fitness components were needed for each genotype relative to wild-type in order to obtain an accurate model for predicting gene frequency change. These included:

1. Viability (survival to sexual maturity);
2. Longevity (adult survival);
3. Age at sexual maturation;
4. Fecundity (clutch or spawn size);
5. Male fertility; and
6. Mating success of both females and males.

Note that it is not necessary to know the processes that underlie any fitness component. For example, all that is needed to estimate the viability fitness component is the relative number of transgenics and non-transgenics that reach sexual maturity. It is not necessary to know whether any differences occur because of disease resistance, foraging ability, escape from predation, improper gill size, swimming ability, etc.

In this regard, Hallerman and Kapuscinski (1993) also discuss key parameters to measure to evaluate environmental risk, including changes in metabolic rate, tolerance to physical factors, behavior (including reproductive), resource use, and disease resistance. While these parameters describe some critical effects of a transgene on the phenotype and underlie the potential for risk, they are subcomponents of the net fitness parameters discussed above.

In an ideal situation, transgenic fish should be exposed to all factors that might be encountered in nature. However, this is not possible to accomplish in a laboratory situation and is a limitation of the method, a point that we will discuss in more detail later.

13.4.2.3.3   Static estimates of fitness components.   Static estimates of the six required fitness parameters can be obtained using the following experimental and observational approaches (Muir and Howard 1999a, 2001):

1. Assuming complete genetic dominance of the transgene, i.e., the effect of the transgene in the heterozygous state is the same as that in the homozygote, using crosses between heterozygous transgenic and wild-type fish, viability fitness can be assessed as the proportion of each genotype reaching sexual maturity. If dominance of the transgene is incomplete, it would also be necessary to cross transgenic heterozygotes with each other in a separate experiment. Deviations from a 1:1 ratio in the former crosses (or 1:2:1 in the latter) indicates the relative viability of each genotype.
2. Longevity may best be considered in terms of adult viability or survival after sexual maturity.
3. Age at sexual maturity can be determined by calculating the average onset of mating behavior and/or oviposition for each genotype.
4. Fecundity (clutch or spawn size) can be ascertained by calculating the average number of eggs in a clutch produced by each genotype.
5. Male fertility can be estimated by crossing transgenic and wild-type males with wild-type females and comparing percent fertile eggs via microscopic examination for first cleavage. Early assessment of fertilization success is necessary to avoid the confounding effects of differential viability.
6. Mating success of each genotype can be determined by staging mating trials that allow both mate competition and mate choice to occur and observing the mating frequency of each genotype (e.g., Hallerman and Kapuscinski 1993, Howard et al. 1998). Because the effect of the transgene on mating success can vary with its relative frequency, such trials should also be performed using different ratios of genotypes.

For a static estimate of mating success, the probability that a female of genotype j mates with a male of genotype k (U_{jk}) may depend on a number of factors including the frequency of other male and female genotypes present. For any given population, the probability that two random genotypes (one male and one female) will mate is the probability the genotypes meet, which is the product of the frequencies of the mating types $(P_j^m)(P_k^f)$, times the conditional probability they mate given that they meet. We define the conditional probability of mating given two genotypes meet as the affinity factor (A_{jk}). Thus, the probability of two genotypes mating is

$$U_{jk} = \frac{P_j^m P_k^f A_{ij}}{\sum_j \sum_k P_j^m P_k^f A_{ij}} \qquad (13.1)$$

Affinity and determination of frequency dependence can be estimated from a series of observational experiments in which the adult genotypes of each sex are placed together in known frequencies and all matings are recorded; these are termed mating sets. At least three mating sets should be examined in which the male and female genotype frequencies vary, for example:

Mating set	Male genotype frequencies		Female genotype frequencies	
	Transgenic	Wild-type	Transgenic	Wild-type
1	1/2	1/2	1/2	1/2
2	1/8	7/8	1/2	1/2
3	1/2	1/2	1/8	7/8

From these mating sets, observed frequencies of matings are recorded (Q_{jk}). The A_{jk} can be solved for because (P_j^m) and (P_k^f) were set by the experimenter for each mating set. The expected frequencies (U_{jk}) are set equal to the observed frequency (O_{jk}) to solve for the affinity factors. For example, if mating set 2 was used and the following hypothetical mating frequencies were observed

Genotype	Transgenic female	Wild-type female
Transgenic male	0.14	0.14
Wild-type male	0.48	0.24

the following set of equations would then be solved:

$$\frac{\frac{1}{16}A_{11}}{\frac{1}{16}A_{11} + \frac{1}{16}A_{12}\frac{7}{16}A_{21} + \frac{7}{16}A_{22}} = 0.14 \qquad (13.1a)$$

$$\frac{\frac{1}{16}A_{12}}{\frac{1}{16}A_{11} + \frac{1}{16}A_{12}\frac{7}{16}A_{21} + \frac{7}{16}A_{22}} = 0.14 \qquad (13.1b)$$

$$\frac{\frac{7}{16}A_{21}}{\frac{1}{16}A_{11} + \frac{1}{16}A_{12}\frac{7}{16}A_{21} + \frac{7}{16}A_{22}} = 0.48 \qquad (13.1c)$$

and

$$\frac{\frac{7}{16}A_{11}}{\frac{1}{16}A_{11} + \frac{1}{16}A_{12}\frac{7}{16}A_{21} + \frac{7}{16}A_{22}} = 0.24 \qquad (13.1d)$$

These constitute four sets of linear equations with four unknowns. Because only the affinity relative to one of the other genotype categories is needed, it is possible to set one factor equal to unity and reduce the number of equations to three. Solving these equations, the affinity factors equal: $A_{11} = 0.5$, $A_{12} = 0.5$, $A_{21} = 0.25$, $A_{22} = 0.125$, but relative to the frequency of a transgenic × transgenic mating they would be 1, 1, 0.5, and 0.25. These hypothetical results suggest that transgenic males have a competitive advantage over wild-type males in gaining access to females and/or wild-type females prefer to mate with transgenic males.

To examine frequency dependence, the actual, rather than relative, values must be estimated. The values of A_{jk} for each of the four mating sets are then combined and subjected to an analysis of variance with mating set, male genotype, and female genotype as main effects. A significant interaction between female and male genotype indicates that mating success depends on the relative frequency of both male and female genotypes present. If the interaction term is not significant, but the main effect of male genotype is significant, then male mating success is frequency dependent, and depends only on the frequency of other male genotypes present. If the interaction term is not significant, but the main effect of female genotype is significant, then female mating success is frequency dependent, and depends only on the frequency of other female genotypes present. If neither main effects nor interactions are significant, then matings are at random with respect to genotype.

If the main effects of male and female genotypes are significant but the interaction between them is not, the number of parameters for conditional mating success reduces from three to two, these are f_j (relative mating advantage of jth female genotype) and m_k (relative mating advantage of kth male genotype). In the absence of interaction, the probability of a genotype j female mating with a genotype k male is the product of the normalized marginal frequencies; that is,

$$U_{jk} = \left(\frac{f_j P_{jt}^f}{\sum\limits_{j=1}^{j=3} f_j P_{jt}^f} \right) \left(\frac{m_k P_{kt}^m}{\sum\limits_{k=1}^{k=3} m_k P_{kt}^m} \right) \tag{13.2}$$

Howard et al. (1998) observed that male Japanese medaka (*Oryzias latipes*) will mate with any receptive female, but must first contend with male–male competition to gain access to females. In addition, female mate choice occurs; that is, females are more likely to accept large males as mates and reject small males. However, female medaka do not compete with each other. Female–female mate competition is uncommon in fish and may require special circumstances (e.g., males provide parental care and a limited supply of high-quality males). The affinity factors in medaka for large and small males equaled 1.0 and 0.25, respectively, which is to say that large males had a fourfold mating advantage (Howard et al. 1998). Muir et al. (1996) and Muir and Howard (1999a, 1999b) used medaka as a model study organism to estimate some, but not all, of the above six fitness components. Medaka were convenient study organisms from which to obtain data on fitness components because they readily bred in the laboratory, were easily maintained, and attained sexual maturity in about 2 months. We produced a stock of transgenic medaka by inserting the human growth hormone gene, *hGH*, with a salmon promoter, *sGH* (Muir et al. 1996, Muir 1999). We found that viability of transgenic young was 70% that of wild-type, that the transgenic line had a 14% advantage in earlier sexual maturation, and that transgenic females produced 29% more eggs/spawn than wild-type females.

The main disadvantage of static estimation is accounting for successful matings. The simple observation of two individuals of different genotypes mating does not indicate if the mating successfully produced offspring. Low sperm count and short copulation duration may contribute to reduced fertility of a genotype; however, this problem can be resolved somewhat by examining the fertility component.

13.4.2.3.4 Dynamic estimates of fitness components. Dynamic estimation of fitness components is more relevant to situations where the individual components are difficult to measure (Prout 1971a). The estimation procedure involves breeding specific mixtures of adult genotypes and recording the proportions of offspring genotypes. The method is essentially a series of single-generation selection experiments where only one sex or the other is genetically variable. The estimates of fitness parameters are the values with which the known proportions among the parents must be weighted to produce the offspring in the observed numbers.

Prout's (1971a) experiments were designed primarily for estimating mating success; however, the same methodology can be used to estimate any fitness component. For example, Muir and Howard (2001) demonstrated how to estimate viability fitness without identifying offspring genotypes. With this method, a 2×2 factorial mating design of reciprocal male–female crosses between wild-type (WT) and transgenic (TR) fish is used. The number of zygotes from each mating is determined, and the numbers of resulting fish from each mating are recorded at any age. At any given age, the maximum likelihood estimate of viability for transgenic (homozygotes and heterozygotes combined) and wild-type offspring can be found by maximizing the following likelihood function:

$$L = \prod_{i=1}^{4} \binom{N_i}{A_i} (W_i)^{A_i} (1 - W_i)^{(N_i - A_i)} \tag{13.3}$$

where N_i is the number of fertile eggs produced by the *i*th mating class, A_i is the number of fish surviving from in the *i*th mating class, and W_i is the viability of each genotype of offspring, times the expected Mendelian proportion of each genotype, given the genotypes of the parents, defined as follows:

Mating (*i*)	Net expected viability of the *i*th mating class (W_i)
WT♂ × WT♀	v_1
WT♂ × TR♀	$(1/2)v_1 + (1/2)v_2$
TR♂ × WT♀	$(1/2)v_1 + (1/2)v_2$
TR♂ × TR♀	$(1/2)v_1 + (3/4)v_2$

where v_1 and v_2 are the viability fitness components of WT and TR fish, respectively, with the assumptions of complete dominance of the transgene and no maternal or paternal effects. The parameters v_1 and v_2 are found by maximizing the likelihood L, or equivalently, the log of the likelihood. For predicting changes in gene frequency, absolute viability of each genotype is not needed; rather one can find viability relative to that of WT, in which case, there is only one parameter to estimate.

Four mating sets were used above to estimate viability; however, by use of several such mating sets it is possible to estimate all net fitness parameters simultaneously. An example of an experimental design to estimate all components simultaneously is given in Table 13.3. For examination of frequency dependence, more mating sets with different frequencies would be needed, particularly of the rare genotype when all other genotypes are present.

Table 13.3 Listing of Mating Sets Necessary
to Estimate Net Fitness Components[a]

Mating set	Female mixture			Constant male[b]
	T/T	T/w	w/w	
1	1	1	0	T/w
2	1	2	1	T/w
3	0	1	1	T/w
4	0	0	1	T/w
5	1	1	0	w/w
6	1	2	1	w/w
7	0	1	1	w/w
8	0	1	0	w/w

Mating set	Male mixture			Constant female[b]
	T/T	T/w	w/w	
9	1	1	0	T/w
10	1	2	1	T/w
11	0	1	1	T/w
12	0	0	1	T/w
13	1	1	0	w/w
14	1	2	1	w/w
15	0	1	1	w/w
16	0	1	0	w/w

[a] Numbers in table are proportions of fish of each genotype.
w = wild-type, T = transgenic.

[b] Numbers of individuals of this sex equal the total number
of the opposite sex in each tank.

Prout's (1971a) estimates of fitness components from such mating sets were based on simple algebraic solutions of *expected* values. His estimates ignore the effects of differing amounts of information obtained for different mating sets due to numbers of observations and variances among those observations. The best estimate of fitness components results from using all the information provided by all matings along with relative information; such estimates are obtained by using maximum likelihood as follows.

For each of the i mating sets, the total number of eggs produced (by genotype if possible), age and number of each genotype at sexual maturity, and average age at death (longevity) of each genotype are recorded. Recording the total number of eggs produced is not absolutely necessary, but is useful for independent estimation of viability as shown previously, and if measured by genotype, there is one less parameter to solve for later. The ages at sexual maturity are defined as s_j^f and s_k^m for female and male genotypes j and k, respectively; average age at death is d_j^f and d_k^m for females and males of genotypes j and k, respectively; and average fecundity (spawn size) is c_j for the jth female genotype.

For any arbitrary mating set, such as those given in Table 13.3, the total number of offspring (N_{ij}) and number of transgenics (T_i) at sexual maturity is determined. The overall likelihood (L) of observing these frequencies is

$$L = \prod_{i=1}^{m} \binom{N_i}{T_i} (Q_i)^{T_i} (1 - Q_i)^{(N_i - T_i)} \qquad (13.4)$$

where Q_i and $1 - Q_i$ are the probabilities of observing TR and WT fish, respectively, at that age. The probability Q_i is defined as the product of the probability of a female of genotype j mating with a male of genotype k (U_{jk}), times the relative fecundity of a genotype j female (e_i), times the relative fertility of a genotype k male (r_i), times the probability that the offspring at conception will be transgenic given the genotypes of the parents are j and k (M_{Tjk}) (this is the Mendelian segregation factor), times the relative viability of TR offspring to sexual maturity (v_T), i.e.,

$$Q_i = \sum_{j=1}^{3} \sum_{k=1}^{3} [U_{jk} r_k e_j M_{Tjk} v_T] \tag{13.5}$$

With dominance of the transgene, the likelihood is a function of six parameters — three mating success parameters (U_{jk}), one relative fertility parameter (r_i), one relative fecundity parameter (e_i), and one relative viability parameter (v_T). If relative viability is found by observation of the total number of fertile eggs and survival as previously described, the number of parameters reduces to five. Further, if the number of eggs produced by each female genotype is known, the relative fecundity can be calculated directly from the data, leaving four parameters to estimate. The number of mating success parameters can be reduced to two if it is known that mate choice and mate competition affect only one sex, as discussed above in the section on static estimation.

The main disadvantage of dynamic estimation is the time required to produce a progeny generation. The main advantage of dynamic estimation is the greater accuracy with which mating success is measured; dynamic estimation actually measures the results of matings (the number of offspring produced).

13.4.2.3.5 Development of a model to predict gene frequency change. To determine the extent to which one fitness component offsets another, the parameters must be put into a mathematical model. Our model (Muir and Howard 1999a, 2001) combines all six factors, using population genetics theory, to predict changes in transgene frequency and population size. The expected genotype frequency change follows closely the derivation of dynamic estimates of fitness components. There are three main differences: (1) the frequency of parental genotypes is not known (except initially), (2) individuals in a natural population are distributed over an age distribution with overlapping generations, and (3) the absolute (rather than relative) fecundity and viability of each genotype must be estimated to determine the impact of the transgene on population size.

As shown by Muir and Howard (1999a, 2001), the recurrence equation begins with an arbitrary population composed of N_{ja}^f females of genotype j in the ath age class and N_{ka}^m males of genotype k in the ath age class. Age classes are normally measured in days but the frequency with which mating occurs and offspring are produced determines the actual number of age classes that must be monitored. For a species such as medaka that mates and produces offspring every day after attaining sexual maturity, age is measured in single days. Thus, the first age class is day-old fry. The total population size at time t is

$$N_t = \sum_{j=1}^{3} \sum_{a=1}^{d_j^f} N_{ja}^f + \sum_{k=1}^{3} \sum_{a=1}^{d_k^m} N_{ka}^m \tag{13.6}$$

The relative frequencies of sexually mature female and male genotypes j and k at time t are

$$P_{jt}^f = \frac{\displaystyle\sum_{a=s_j^f}^{d_j^f} N_{ja}^f}{\displaystyle\sum_{j=1}^{3}\sum_{a=s_j^f}^{d_j^f} N_{ja}^f} \tag{13.7}$$

$$P_{kt}^m = \frac{\displaystyle\sum_{a=s_k^m}^{d_k^m} N_{ka}^m}{\displaystyle\sum_{j=1}^{3}\sum_{a=s_k^m}^{d_k^m} N_{ka}^m} \tag{13.8}$$

Then, assuming an equal sex ratio at birth, the expected number of offspring of genotype i of each sex surviving to the first age class from all matings is

$$N_{i1}^f = N_{i1}^m = \frac{1}{2}\sum_{j=1}^{3}\sum_{k=1}^{3}\left[U_{jk}r_k\left(c_j \sum_{a=s_j^f}^{d_j^f} N_{ja}^f \right) M_{ijk}v_{ijk} \right] \tag{13.9}$$

where the net fitness parameters are as defined in static and dynamic estimation. These individuals compose the first age class, all other age classes advance by one, and depending on the longevity of the genotype, the oldest age class of each genotype dies off. Muir and Howard (2001) modeled mortality due to cannibalism and predation as a two-step process related to developmental rate. The validity of this approach depends on how viability was estimated in the static and dynamic steps. If predators and all stages of the life cycle were present during the estimation of those parameters, then the viability estimates already include those effects. Otherwise, those effects can be included in the model indirectly. Muir and Howard (2001) assumed that cannibalism and/or predation was size specific because predators were gape limited, i.e., their mouth size sets a limit on the size of prey they can consume (which depends on the age of the prey). With such a model, we divided mortality into that pertaining to juveniles and that pertaining to adults. We assumed that probability of death as a juvenile (g_1) or adult (g_2) on any given day was independent and constant. This assumption results in the classic negative exponential life expectancy curve. The number of each genotype and sex at time $t + 1$ adjusted for such mortality is

$$N_j^f = \left(\sum_{a=1}^{s_j^f-1} g_1 N_{ja}^f + \sum_{a=s_j^f}^{d_j^f} g_2 N_{ja}^f \right) \tag{13.10}$$

$$N_k^m = \left(\sum_{a=1}^{s_k^m-1} g_1 N_{ka}^m + \sum_{a=s_k^m}^{d_k^m} g_2 N_{ka}^m \right) \tag{13.11}$$

After mortality has occurred, the population size at time $t + 1$ is

$$N_{t+1} = \sum_{j=1}^{3} N_j^f + \sum_{k=1}^{3} N_k^m \qquad (13.12)$$

The relative frequency of the ith genotype at time $t + 1$ is

$$p_{i, t+1} = \frac{N_i^f + N_i^m}{N_{t+1}} \qquad (13.13)$$

We developed a computer program to keep constant accounting of all life stages and advancement through the stages to predict consequences of release of a few transgenic individuals into a large wild-type population.

Classic demographic analysis is based on tabulating age-specific survival and reproduction using life tables and schedules of fecundity. However, most demographic models are based on only one sex, usually females (Caswell 1989). These models assume the life cycles of the sexes are identical or the dynamics of the population are determined by one sex. Obviously, this assumption does not allow for differential mating success based on genotype. In the model developed above, if one assumes random mating and equal ages at sexual maturity, the equations are equivalent to those developed using the Leslie matrix approach (Caswell 1989). Caswell (1989) also shows that the Leslie matrix approach is equivalent to the Lotka model and the McKendrick–von Foerster equations for population growth. However, the matrix approach can allow for size-dependent vital rates and plastic growth. These effects were incorporated in our model by allowing age at sexual maturity to vary with genotype as influenced by factors such as body size.

An important aspect of demography is sensitivity analysis. Caswell (1989) shows that such an analysis can address questions such as the relative importance of each fitness component (vital rates) to population growth, errors in estimation, environmental perturbation, and intensity of selection. Unfortunately, to use his approach, the projection matrix must be reduced to eigenvalues and eigenvectors. With density or frequency dependence, as with a two-sex model and differential mating success, the projection matrix can no longer be reduced. Thus, it is not possible to obtain analytical results for a sensitivity analysis of eigenvalues. Nevertheless, this problem can be addressed by numerical methods, such as by varying each parameter by a constant percentage and examining its impact on projected growth.

Statistical properties (i.e., standard errors of estimates) of demographic models can be addressed in a number of ways. Because our estimates were found by maximum likelihood, asymptotic standard errors can be calculated by taking the partial derivative with respect to each parameter estimated. Alternatively, resampling methods such as bootstrapping can be used (Caswell 1989).

13.4.2.4 *Use of surrogate organisms to mimic the effect of a transgene*

Using methods similar to those we described above, Prout (1971b) showed that it was possible to predict gene frequency change successfully in a laboratory population of *Drosophila melanogaster*. However, Prout (1971b) noted that "one possible reason for this success is that it was possible to estimate the fitness components under conditions fairly similar to those of the rather simple population regime." It is questionable if we could ever measure fitness components of transgenic individuals under totally natural conditions. The inability to test transgenic fish under field conditions may result in important variables being overlooked; however, surrogate organisms could be used to obtain relevant data. For example, field data using GH-implanted trout revealed that the growth advantage associated with increased GH was much less pronounced in the wild than in the hatchery (Johnsson et al. 1999, 2000). These results suggest that increased growth of GH-transgenic fish may be restricted to conditions of non-limiting food supply. However, Johnsson et al. (1999) also showed that survival of GH-implanted trout did not differ from that of controls under field conditions with natural predation levels. They concluded that "escaped GH-manipulated fish may compete successfully with wild fish" despite differences observed in the laboratory between GH-treated and wild-type individuals in characteristics such as predator avoidance, foraging ability, and overwinter survival. Thus, use of GH-implanted fish as surrogates of GH-transgenic fish in field experiments may lend insights into relevant fitness parameters under natural conditions without risking the escape of transgenic individuals into nature.

13.5   Limitations imposed by the environment and constraints of the model

The methods described above to assess environmental risk have a number of limitations. Some limitations are imposed by the environment in which the parameters are estimated and others by the constraints of the model. The impact of each limitation is discussed below.

13.5.1   Limitations imposed by the environment

The most accurate method of testing environmental risk is from accidental or intentional release of a transgenic stock into nature. Unfortunately, this is the least desirable method, as the outcome might be adverse and irrevocable. Because transgenic organisms cannot be released into the wild, study of the potential environmental impact of such organisms must be inferred from inspection of causal components in a laboratory environment or through use of GH-implanted fish as surrogates of GH-transgenic fish. However, despite laboratory constraints that apply when studying a GH-transgenic fish, one can still test an important one-sided hypothesis: On one hand, if transgenic fish exhibit superior fitness in an optimum laboratory environment, in terms of growth, fecundity, competitive ability, etc., they might fare better in nature than indigenous fish but this may not necessarily be the case, given nature's less optimal conditions and greater complexity. On the other hand, if transgenic fish are shown to have a lower overall fitness in an artificial optimum environment, the transgenic fish very likely will not compete effectively or thrive in the wild where conditions are usually suboptimal.

13.5.2   Limitations imposed by the model

The model does not include effects due to antagonism between Darwinian (individual) and long-term (population) fitness, which could actually increase environmental risk (see discussion below), or effects of natural selection on expression of the transgene or modification of other traits in response to the transgene. Documenting the effect of natural selection on other traits in response to the transgene would require more experimental research. Incorporation of long-term population fitness is difficult to do, and will be addressed in a future paper. Transgene methylation could also turn the gene off in a portion of the population in advanced generations and further reduce environmental risk. The model also does not incorporate density-dependent effects on fecundity, growth rate, age at sexual maturity, or longevity. Such extensions can easily be incorporated in the model, but biological evidence must first be shown for such effects. We have begun to incorporate into our model density-dependent effects on early viability relative to the carrying capacity of the population. Thus far, we have found no qualitative difference in the predicted fate of the transgene.

13.6   Conclusions

The above discussion demonstrates mechanisms by which transgenic fish might flourish in natural communities in the short term, or even in the long term, because they obtain substantially greater benefits and/or face a different array of costs than would mutants that slightly exceed the previous maximum body size of wild-type relatives. Long-term persistence of a transgenic line seems unlikely but not impossible, and may represent the creation of a new species or evolutionary line capable of defining a novel niche. There is precedence for this in nature; mutations with large effects have been demonstrated. Thus, the possibility of such a mega-mutation, created in the laboratory and having a similar impact, cannot be ignored.

The methods discussed in Section 13.4 detail the critical parameters that should be measured to estimate the fate of a transgene in the population. If environmental risk, that is, the spread of the transgene is predicted, one should take appropriate measures to prevent release of fertile fish into the environment, as suggested by the "Precautionary Principle" of the Cartegena Biosafety Protocol recently adopted by more than 130 nations, the Performance Standard for Safely Conducting Research with Genetically Modified Fish and Shellfish (ABRAC 1995), and the Manual for Assessing Ecological and Human Health Effects of Genetically Engineered Organisms (Scientists' Working Group on Biosafety 1998).

Acknowledgments

This research was supported by Grants 93-33120-9468 and 97-39210-4997 from the USDA National Biological Impact Assessment Program. We thank Bob Devlin, Bruce Walsh, Jorgen Johnsson, Deborah Letourneau, and an anonymous reviewer for helpful comments.

Literature cited

ABRAC (Agricultural Biotechnology Research Advisory Committee, U.S. Department of Agriculture), Performance standards for safely conducting research with genetically modified fish and shellfish, available at http://www.nbiap.vt.edu/perfstands, 1995.
Abrahams, M. V. and Sutterlin, A., The foraging and antipredator behaviour of growth-enhanced transgenic Atlantic salmon, *Anim. Behav.*, 58, 933, 1999.
Bright, C., Understanding the threat of biological invasions, in *State of the World 1996. A World Watch Institute Report on Progress toward a Sustainable Society*, Starke, L., Ed., W. W. Norton, New York, 1996, 95.
Caswell, H., *Matrix Population Models: Construction, Analysis, and Interpretation*, Sinauer Associates, Sunderland, MA, 1989.
Chen, T. T., Lu, J. K., Shamblott, M. J., Cheng, C. M., Lin, C. M., Burns, J. C., Reimschuessel, R., Chatakondi, N., and Dunham, R. A., Transgenic fish — ideal models for basic research and biotechnological applications, *Zool. Stud.*, 34, 215, 1995.
Craig, J. V. and Muir, W. M., Group selection for adaptation to multiple-hen cages: beak-related mortality, feathering and body weight responses, *Poult. Sci.*, 75, 294, 1996a.
Craig, J. V. and Muir, W. M., Group selection for adaptation to multiple-hen cages: behavioral responses, *Poult. Sci.*, 75, 1145, 1996b.

Craig, J. V. and Muir, W. M., Genetic influences on the behavior of chickens associated with welfare and productivity, in *Genetics and the Behavior of Domestic Animals*, Grandin, T., Ed., Academic Press, San Diego, CA, 1997, 265.

Dawson, P., A conflict between Darwinian fitness and population fitness in *Tribolium* "competition" experiments, *Genetics*, 62, 413, 1969.

de Gaudemar, B., Sexual selection and breeding patterns: insights from salmonids (salmonidae), *Acta Biotheor.*, 46, 235, 1998.

de la Fuente, J. I., Hernandez, O., Martinez, R., Guellen, I., Estrada, M. P., and Lleonart, R., Generation, characterization and risk assessment of transgenic tilapia with accelerated growth, *Biotecnol. Apl.*, 13, 221, 1996.

de la Fuente, J. I., Martinez, R., Guellen, I., Estrada, M. P., and Lleonart, R., Gene transfer in tilapia for accelerated growth: from the laboratory to the consumer, in *Gene Transfer in Aquatic Organisms*, de la Fuente, J. I. and Castro., F. O., Eds., Landes Bioscience, Georgetown, TX, 1997, 83.

Devlin, R. H., Sequence of sockeye salmon type 1 and 2 growth hormone genes and the relationship of rainbow trout with Atlantic and Pacific salmon, *Can. J. Fish. Aquat. Sci.*, 50, 1738, 1993.

Devlin, R. H., Transgenic salmonids, in *Transgenic Animals Generation and Use*, Houdebine, L. M., Ed., Harwood Academic Publishers, Amsterdam, the Netherlands, 1997, 105.

Devlin, R. H. and Donaldson, E. M., Containment of genetically altered fish with emphasis on salmonids, in *Transgenic Fish*, Hew, C. L. and Hew, G. L., Eds., World Science, Singapore, 1992, 229.

Devlin, R. H., Yesaki, T. Y., Biagi, C. A., and Donaldson, E. M., Extraordinary salmon growth, *Nature*, 371, 209, 1994a.

Devlin, R. H., Yesaki, T. Y., Biagi, C., Donaldson, E. M., and Chan, W. K., Production and breeding of transgenic salmon, in *Proceedings of the 5th World Congress on Genetics Applied to Livestock Production*, Smith, C., Gavora, J. S., Benkel, B., Chesnais, J., Fairfull, W., Gibson, J. P., Kennedy, B. W., and Burnside, E. B., Eds., University of Guelph, Guelph, Ontario, Canada, 19, 1994b, 372.

Devlin, R. H., Yesaki, T. Y., Donaldson, E. M., Du, S. J., and Hew, C. L., Production of germline transgenic Pacific salmonids with dramatically increased growth performance, *Can. J. Fish. Aquat. Sci.*, 52, 1376, 1995a.

Devlin, R. H., Yesaki, T. Y., Donaldson, E. M., and Hew, C. L., Transmission and phenotypic effects of an antifreeze/GH gene construct in coho salmon (*Oncorhynchus kisutch*), *Aquaculture*, 137, 161, 1995b.

Devlin, R. H., Johnsson, J. I., Smailus, D. E., Biagi, C. A., Jonsson, E., and Bjornsson, B. Th., Increased ability to compete for food by growth hormone-transgenic coho salmon *Oncorhynchus kisutch* (Walbaum), *Aquacult. Res.*, 30, 479, 1999.

Drake, J. A. and Mooney, H. A., *Ecology of Biological Invasions of North America and Hawaii*, Springer-Verlag, New York, 1986.

Du, S. J., Gong, Z., Fletcher, G., Shears, M., King, M., Idler, D., and Hew, C., Growth enhancement in transgenic Atlantic salmon by the use of an "all fish" chimeric growth-hormone gene construct, *Bio/Technology*, 10, 176, 1992.

Dunham, R. A., Utilization of transgenic fish in developing countries: potential benefits and risks, *J. World Aquacult. Soc.*, 30, 1, 1999.

Dunham, R. A., Chitmanat, C., Nichols, A., Argue, B., Powers, D. A., and Chen, T. T., Predator avoidance of transgenic channel catfish containing salmonid growth hormone genes, *Mar. Biotechnol.*, 1, 545, 1999.

Farrell, A. P., Bennett, W., and Devlin, R. H., Growth-enhanced transgenic salmon can be inferior swimmers, *Can. J. Zool.*, 75, 335, 1997.

Fisher, R. A., *The Genetical Theory of Natural Selection*, Clarendon Press, Oxford, U.K., 1930.

Fleming, I. A., Reproductive strategies of Atlantic salmon: ecology and evolution, *Rev. Fish Biol. Fish.*, 6, 379, 1996.

Fleming, I. A., Pattern and variability in the breeding system of Atlantic salmon (*Salmo salar*), with comparisons to other salmonids, *Can. J. Fish. Aquat. Sci.*, 55(Suppl. 1), 59, 1998.

Fleming, I. A. and Einum, S., Experimental tests of genetic divergence of farmed from wild Atlantic salmon due to domestication, *ICES J. Mar. Sci.*, 54, 1051, 1997.

Fletcher, G. L. and Davies, P. L., Transgenic fish for aquaculture, *Genet. Eng.*, 13, 331, 1991.

Fu, C., Cui, Y., Hung, S. S. O., and Zhu, Z., Growth and feed utilization by F_4 human growth hormone transgenic carp fed diets with different protein levels, *J. Fish Biol.*, 53, 115, 1998.

Griffing, B., Selection in reference to biological groups. V. Analysis of full-sib groups, *Genetics*, 82, 703, 1976.

Griffing, B., Selection in reference to biological groups. I. Individual and group selection applied to populations of unordered groups, *Aust. J. Biol. Sci.*, 10, 127, 1967.

Guillen, I., Berlanga, J., Valenzuela, C. M., Morales, A., Toledo, J., Estrada, M. P., Puentes, P., Hayes, O., and de la Fuente, J., Safety evaluation of transgenic tilapia with accelerated growth, *Mar. Biotechnol.*, 1, 2, 1999.

Hallerman, E. M. and Kapuscinski, A. R., Incorporating risk assessment and risk management into public policies on genetically modified finfish and shellfish, *Aquaculture*, 137, 9, 1995.

Hallerman, E. M. and Kapuscinski, A. R., Transgenic fish and public policy: regulatory concerns, *Fisheries*, 15, 12, 1990.

Hallerman, E. M. and Kapuscinski, A. R., Ecological and regulatory uncertainties associated with transgenic fish, in *Transgenic Fish*, Hew, C. L. and Fletcher, G. L., Eds., World Science, Singapore, 1992, 209.

Hallerman, E. M. and Kapuscinski, A. R., Potential impacts of transgenic and genetically manipulated fish on natural populations: addressing the uncertainties through field testing, in *Genetic Conservation of Salmonid Fishes*, Cloud, J. G. and Thorgaard, G. H., Eds., Plenum Press, New York, 1993, 93.

Halvorson, H. O. and Quezada, F., Increasing public involvement in enriching our fish stocks through genetic enhancement, *Genet. Anal. Biomol. Eng.*, 15, 75, 1999.

Hew, C. L., Fletcher, G. L., and Davies, P. L., Transgenic salmon: tailoring the genome for food production, *J. Fish Biol.*, 47(Suppl. A), 1, 1995.

Houdebine, L. M. and Chourrout, D., Transgenesis in fish, *Experientia*, 47, 891, 1991.

Howard, R. D., Martens, R. S., Innes, S. A., Drnevich, J. M., and Hale, J., Mate choice and mate competition influence male body size in Japanese medaka, *Anim. Behav.*, 55, 1151, 1998.

Iyengar, A., Muller, F., and Maclean, N., Regulation and expression of transgenes in fish — a review, *Transgenic Res.*, 5, 147, 1996.

Johnsson, J. I., Big and brave — size selection affects foraging under risk of predation in juvenile rainbow-trout, *Oncorhynchus mykiss*, *Anim. Behav.*, 45, 1219, 1993.

Johnsson, J. I. and Bjornsson, B. T., Growth hormone increases growth rate, appetite and dominance in juvenile rainbow trout, *Oncorhynchus mykiss*, *Anim. Behav.*, 48, 177, 1994.

Johnsson, J. I., Petersson, E., Jönsson, E., Björnsson, B. T., and Jarvi, T., Domestication and growth hormone alter antipredator behaviour and growth patterns in juvenile brown trout, *Salmo trutta*, *Can. J. Fish. Aquat. Sci.*, 53, 1546, 1996.

Johnsson, J. I., Petersson, E., Jönsson, E., Jarvi, T., and Björnsson, B. T., Growth hormone-induced effects on mortality, energy status and growth: a field study on brown trout (*Salmo trutta*), *Funct. Ecol.*, 13, 514, 1999.

Johnsson, J. I., Jönsson, E., Petersson, E., Jarvi, T., and Björnsson, B. T., Fitness-related effects of growth investment in brown trout under field and hatchery conditions, *J. Fish Biol.*, 57, 326, 2000.

Jönsson, E., Johnsson, J. I., and Björnsson, B. T., Growth hormone increases predation exposure of rainbow trout, *Proc. R. Soc. London Ser. B*, 263, 647, 1996.

Jönsson, E., Johnsson, J. I., and Björnsson, B. T., Growth hormone increases aggressive behavior in juvenile rainbow trout, *Horm. Behav.*, 33, 9, 1998.

Kapuscinski, A. R. and Hallerman, E. M., Transgenic fish and public policy: anticipating environmental impacts of transgenic fish, *Fisheries*, 15, 2, 1990.

Kapuscinski, A. R. and Hallerman, E. M., Implications of introduction of transgenic fish into natural ecosystems, *Can. J. Fish. Aquat. Sci.*, 48, 99, 1991.

Kapuscinski, A. R., Nega, T., and Hallerman, E. M., Adaptive biosafety assessment and management regimes for aquatic genetically modified organisms in the environment, in *Toward Policies for Conservation and Sustainable Use of Aquatic Genetic Resources*, Pullin, R. S. V., Bartley, D. M., and Kooiman, J., Eds., ICLARM (International Center for Living Aquatic Resource Management) Conference Proceedings, 59, 225, 1999.

Knibb, W., Risk from genetically engineered and modified marine fish, *Transgenic Res.*, 6, 59, 1997.

Lakra, W. S. and Das, P., Genetic engineering in aquaculture, *Indian J. Anim. Sci.*, 68, 873, 1998.

Lande, R., Models of speciation by sexual selection on polygenic traits, *Proc. Natl. Acad. Sci. U.S.A.*, 76, 3721, 1981.

Lande, R., The response to selection on major and minor mutations affecting a metrical trait, *Heredity*, 50, 47, 1983.

Lodge, D. M., Biological invasions: lessons for ecology, *Trends Ecol. Evol.*, 8, 133, 1993.

Maclean, N., Regulation and exploitation of transgenes in fish, *Mutat. Res.*, 399, 255, 1998.

Martinez, R., Estrada, M. P., Berlanga, J., Guillen, I., Hernandez, O., Cabrera, E., Pimentel, R., Morales, R., Herrera, F., Morales, A., Pina, J. C., Abad, Z., Sanchez, V., Melamed, P., Lleonart, R., and de la Fuente, J., Growth enhancement in transgenic tilapia by ectopic expression of tilapia growth hormone, *Mol. Mar. Biol. Biotechnol.*, 5, 62, 1996.

Moran, P., Pendas, A. M., Beall, E., and Garcia-Vazquez, E., Genetic assessment of the reproductive success of Atlantic salmon precocious parr by means of VNTR loci, *Heredity*, 77, 655, 1996.

Mourao, C. A. and Ayala, F. A., Competitive fitness in experimental populations of *Drosophila willistoni*, *Genetics*, 42, 65, 1971.

Muir, W. M., Group selection for adaptation to multiple-hen cages: selection program and direct responses, *Poult. Sci.*, 75, 447, 1996.

Muir, W. M. and Craig, J. V., Improving animal well-being through genetic selection, *Poult. Sci.*, 77, 1781, 1998.

Muir, W. M. and Howard, R. D., Possible ecological risks of transgenic organism release when transgenes affect mating success: sexual selection and the Trojan gene hypothesis, *Proc. Natl. Acad. Sci. U.S.A.*, 96, 13853, 1999a.

Muir, W. M. and Howard, R. D., Effect of genetic background on transgene expression in medaka (*Oryzias latipes*) and models to assess environmental risk of GMOs, *Transgene Res.*, 8, 470, 1999b.

Muir, W. M. and Howard, R. D., Fitness components and ecological risks of transgenic release: a model using Japanese medaka (*Oryzias latipes*), *Am. Nat.*, 158, 1, 2001.

Muir, W. M., Howard, R. D., and Bidwell, C., Use of multigenerational studies to assess genetic stability, fitness, and competitive ability of transgenic Japanese medaka: I. Methodology, in *Proceedings of Biotechnology Risk Assessment Symposium*, Levin, M., Grim, C., and Angle, J., Eds., College Park, MD, 1994, 170.

Muir, W. M., Martens, R., Howard, R. D., and Bidwell, C., Use of multigenerational studies to assess genetic stability, fitness, and competitive ability of transgenic Japanese medaka: II. Development of transgenic medaka and mating preferences, in *Proceedings of Second International Conference on Risk Assessment Methodologies*, Levin, M., Grim, C., and Angle, J. S., Eds., Pensacola, FL, 1995, 140.

Muir, W. M., Howard, R. D., Martens, R. S., Schulte, S., and Bidwell, C. A., Use of multigenerational studies to assess genetic stability, fitness, and competitive ability of transgenic Japanese medaka: III. Results and predictions, in *Proceedings of 8th International Conference on Risk Assessment Methodologies*, College Park, MD, Levin, M., Grim, C., and Angle, J., Eds., 1996.

Pandian, T. J. and Marian, L. A., Problems and prospects of transgenic fish production, *Curr. Sci.*, 66, 635, 1994.

Pandian, T. J., Venugopal, T., and Koteeswaran, R., Problems and prospects of hormone, chromosome and gene manipulations in fish, *Curr. Sci.*, 76, 369, 1999.

Pitkänen, T. I., Krasnov, A., Teerijoki, H., and Mölsä, H., Transfer of growth hormone (GH) transgenes into Arctic charr (*Salvelinus alpinus* L.) I. Growth response to various GH constructs, *Genet. Anal. Biomol. Eng.*, 15, 91, 1999.

Prout, T., The relation between fitness components and population prediction in *Drosophila*. I: The estimation of fitness components, *Genetics*, 68, 127, 1971a.

Prout, T., The relation between fitness components and population prediction in *Drosophila*. II. Population prediction, *Genetics*, 68, 151, 1971b.

Rahman, M. A. and Maclean, N., Growth performance of transgenic tilapia containing an exogenous piscine growth hormone gene, *Aquaculture*, 173, 333, 1999.

Rahman, M. A., Mak, R., Ayad, H., Smith, A., and Maclean, N., Expression of a novel piscine growth hormone gene results in growth enhancement in transgenic tilapia (*Oreochromis niloticus*), *Transgenic Res.*, 7, 357, 1998.

Regal, P. J., Models of genetically engineered organisms and their ecological impact, *Recombinant DNA Tech. Bull.*, 10, 67, 1987.

Saunders, R. L., Fletcher, G. L., and Hew, C. L., Smolt development in growth hormone transgenic Atlantic salmon, *Aquaculture*, 168, 177, 1998.

Scientists' Working Group on Biosafety, *Manual for Assessing Ecological and Human Health Effects of Genetically Engineered Organisms.* Part 1: *Introductory Materials and Supporting Text for Flowcharts.* Part 2: *Flowcharts and Worksheets*, The Edmonds Institute, Edmonds, WA, 1998, http://www.Edmonds-Institute.org.

Sin, F. Y. T., Transgenic fish, *Rev. Fish Biol. Fish.*, 7, 417, 1997.

Sin, F. Y. T., Mukherjee, U. K., Walker, L., and Sin, I. L., The application of gene transfer techniques to marine resource management: recent advances, problems and future directions, *Hydrobiologia*, 352, 263, 1997.

Stearns, S. C., The evolutionary significance of phenotypic plasticity, *BioScience*, 39, 436, 1989.

Stevens, E. D. and Sutterlin, A., Gill morphometry in growth hormone transgenic Atlantic salmon, *Environ. Biol. Fish.*, 54, 405, 1999.

Stevens, E. D., Sutterlin, A., and Cook, T., Respiratory metabolism and swimming performance in growth hormone transgenic Atlantic salmon, *Can. J. Fish. Aquat. Sci.*, 55, 2028, 1998.

Thomaz, D., Beall, E., and Burke, T., Alternative reproductive tactics in Atlantic salmon: Factors affecting mature parr success, *Proc. R. Soc. London Ser. B Biol. Sci.*, 264, 219, 1997.

Tiedje, J. M., Colwell, R. K., Grossman, Y. L., Hodson, R. E., Lenski, R. E., Mack, R. N., and Regal, P. J., The planned introduction of genetically engineered organisms: ecological considerations and recommendations, *Ecology*, 70, 298, 1989.

Vogt, D. U. and Parish, M., Food biotechnology in the United States: science, regulation, and issues, CRS Report for Congress, Congressional Research Service, Library of Congress, Washington, D.C., 1999.

Williams, G. C., *Adaptation and Natural Selection: A Critique of Some Current Evolutionary Thought*, Princeton University Press, Princeton, NJ, 1966.

Wright, S., *Evolution and the Genetics of Populations*, Vol. 2, *The Theories of Gene Frequencies*, University of Chicago Press, Chicago, 1969.

Wright, S., The shifting balance theory of macroevolution, *Annu. Rev. Genet.*, 16, 1, 1982.

chapter fourteen

Controversies in designing useful ecological assessments of genetically engineered organisms

Anne R. Kapuscinski

Contents

14.1   Establishing a basis for biosafety

In 1989, the Ecological Society of America published a seminal paper on scientific questions about the ecological effects of genetically engineered organisms (GEOs) (Tiedje et al. 1989). The American Fisheries Society also issued a position statement that reviewed the status of scientific knowledge regarding possible ecological risks of transgenic fish and urged environmental caution, as well as improvements in regulatory policy, regarding commercial development of transgenic fish (Kapuscinski and Hallerman 1990). For most of the ensuing decade, the issues raised in these reports drew the attention of a relatively small number of people in academia, industry, government, and public interest groups. The general public — particularly the potential consumers of genetically engineered foods (GEFs) and other products — was largely unaware of concerns and questions about the biosafety of GEOs. But, as genetically engineered crops and other genetic engineering products were introduced on a large scale into the environment and the human food supply in the last few years, public attention grew in many countries. More and more diverse parties in the scientific, governmental, industrial, and public interest sectors now actively debate the merits and drawbacks of genetic technologies and strive to influence domestic and international policies governing GEOs. With this increased societal engagement with genetic engineering, the assessment of possible ecological and human health risks of GEOs became and remains a central focus of debate and negotiations at local, national, and international levels.

The importance of biosafety is recognized in the body of the Convention on Biological Diversity (CBD). Its Article 14 calls upon each contracting party to require environmental impact assessments of proposed projects "that are likely to have significant adverse effects on biological diversity with a view to avoiding or minimizing such effects." Furthermore, Article 19.3 calls for parties to consider the need for and modalities of a protocol setting out procedures in the field of safe transfer, handling, and use of living modified organisms that may have an adverse effect on biodiversity and its components. At the second meeting of the Conference of the Parties in November 1995, delegates established an Open-ended *Ad Hoc* Working Group on Biosafety to develop a biosafety protocol that would "take into account the principles enshrined in the Rio Declaration on Environment and Development, and in particular, the precautionary approach."

The Edmonds Institute, a public interest, non-profit organization, had pushed since its inception for a protocol that would be effective in protecting ecological and human health from risks posed by GEOs and their products. Having observed the inattention in CBD meetings to the breadth of biological sciences needed to inform risk assessment, the Edmonds Institute responded to the 1995 decision to negotiate a protocol by inviting an interdisciplinary group of biologists (from geneticists to ecologists) to develop a user-friendly and scientifically sound manual for assessing the ecological and human health effects of GEOs. I was a member of that group. We worked together over a 2-year period to publish such a manual (Scientists' Working Group on Biosafety 1998). Although aiming in the short term to provide scientific input to the negotiators of the biosafety protocol, our broader intent was to help consumers, scientists, regulators, policy makers, and biotechnology developers evaluate likely impacts of GEOs in a variety of settings and applications worldwide.

That guidelines for assessing and managing the biosafety of GEOs should be "science-based" has been the claim of parties around the world seeking approval for large-scale releases and transboundary trade of GEOs and a recurring theme in negotiations of a biosafety protocol under the CBD. A science-based assessment approach has also been an issue in international trade negotiations under the World Trade Organization (WTO) in order to avoid an exporting party from charging an importing party of having unfair barriers to trade. Yet some of the most prominently promoted guides for risk assessment were too vague in terms of scientific principles and methodologies to realistically assist a scientifically grounded process of risk assessment (see list in Ingham and Holmes 1995). The authors of the *Manual for Assessing Ecological and Human Health Effects of Genetically Engineered Organisms* (Scientists' Working Group on Biosafety 1998), hereafter abbreviated as the Manual, reclaimed the meaning of "science-based" by designing an assessment process of appropriate scope, covering ecology and other relevant life sciences (getting the right science), and of reliable content (getting the science right). The Manual went through a double-blind peer review whereby neither authors nor reviewers knew one another's identities. This accomplished a greater degree of anonymity than typically occurs in anonymous scientific peer review in which author names are revealed to the reviewers. This was done to assure focus on the scientific quality of its contents and reduce potential bias. People involved in biosafety issues within industry, government, and public interest groups in many countries have found the Manual to be based on good scientific foundations and have sought and used the Manual in various ways, from helping to shape national biosafety policies to evaluating specific proposals to import GEOs (Burrows 1999, 2000).[1] The Manual is available for free download on the Internet at www.edmonds-institute.org/manual.html.

The need for risk assessment and risk management procedures was a continuing topic of discussion during 4 years of arduous negotiations of the protocol, eventually named the Cartagena Protocol on Biosafety. The protocol becomes legally binding after 50 countries have ratified it. By September 2000, 74 countries and the European Community (a regional economic integration organization) had indicated their intent to ratify by signing the protocol. Although Articles 15 and Annex II of the final negotiated text of the Cartagena Protocol on Biosafety directly address risk assessment (Conference of the Parties to the Convention on Biological Diversity 2000), there remain some important obstacles to meaningful and adequate inclusion of contemporary ecological knowledge and methods of analysis in the assessment of GEOs involved in international trade under the terms of the protocol.

This chapter addresses four areas of controversy about what constitutes "science-based" biosafety assessment and decisions and describes how the design and content of the Manual dealt with these issues. It then describes and calls for adoption of "adaptive biosafety assessment and management," a learning-driven framework that brings together consideration of the latest science, important experiential knowledge, and democratic, cross-sectoral processes of decision making. I argue that adaptive biosafety regimes would be better suited than existing governance frameworks to address the areas of controversy regarding the environmental and human health safety of GEOs. The chapter finally points the genetic engineering industry toward the examples of industry-led safety programs in other engineering industries that produce complex systems. The maturation of these industries over the last century has involved gaining public trust by placing safety first from product design through post-market monitoring and incorporating government oversight and citizen participation in ways that reinforce industry responsibility and responsiveness.

14.1.1 Biosafety assessment: Questions and debates

Briefly, the spheres of controversy discussed in this chapter are the following:

1. What is the appropriate scope and specificity of risk assessment?
2. What kinds of information and knowledge are relevant for risk assessment? To what extent and how should risk assessment incorporate different kinds of knowledge?
3. To what extent should the Precautionary Principle guide risk assessment and management?

4. How should one consider scientific error and uncertainty in drawing biosafety conclusions?
5. Who should participate in the process of risk analysis and decision making?

By no means is this an exhaustive list of current controversies. It focuses rather on those controversies that lie at the direct intersection of ecological science and biosafety. The list omits important controversies related to socioeconomic impacts of large-scale adoption of GEOs. Examples of phenomena that will not be attended to in this chapter are the uneven social distribution of risks and benefits related to the products of genetic engineering; the impacts of large-scale adoption of GEOs on the structure of agriculture; the problems related to genetic engineering's potential displacement of traditional agricultural and natural resource practices that currently sustain the livelihoods of people in rural and coastal communities throughout the world; the problems related to genetic engineering's potential displacement of research on and application of ecological approaches (e.g., increasing the biodiversity of a cropping system) that are inexpensive for farmers and can solve agricultural and natural resource problems[2]; and the disparities among nations in the scientific, financial, political, or managerial resources and hence in the capacity to assess and maintain biosafety. To do justice to the broader set of socioeconomic and political concerns would have required a different mix of authors than the biologists comprising the Scientists' Working Group on Biosafety. The undertaking of an analysis of such social issues, however, presupposes the prior completion of an adequate scientific assessment of ecological and human health effects. The section of this chapter on adaptive biosafety assessment and management points to the large body of research documenting the inextricable and often unanticipated connections and feedbacks between social and ecological systems and the need for biosafety governance regimes to expect and explicitly address socioecological linkages.

14.1.2   Explanation of key terms and their usage

Definitions of a few key terms are in order before proceeding further. As used in the Manual and this chapter, a genetically engineered organism (GEO) is one that has been constructed by isolating nucleic acid molecules (the molecules that encode genetic information) from one organism and introducing these molecules or fragments thereof into another organism in a manner that makes them part of the permanent genetic makeup of the recipient, i.e., capable of being inherited by offspring. Included in this definition are those organisms constructed by transfer of subcellular organelles from one cell to another, followed by regeneration of an adult organism from the genetically altered cell, so long as the alteration can be transmitted to offspring.[3] A genetically engineered food (GEF) consists either of edible portions of a GEO (e.g., a transgenic potato or fish fillet) or processed foods containing one or more ingredients derived from a GEO (e.g., baked goods containing corn or soybean meal from a transgenic crop variety).

Biosafety, as used in this chapter, is best described by its goals. Drawing on the ideas from many prior fora, the Scientists' Working Group on Biosafety (1998) identified four major goals of biosafety:

1. To determine in advance when hazards to human health and natural systems will result if any particular GEO is released into the environment;
2. To anticipate when a given GEO or any of its products(s) will be harmful if it becomes part of human foods;
3. To discern whether a GEO actually will yield the benefit(s) it was designed to provide; and
4. To make as certain as possible that hazards will not arise when GEOs are transported, intentionally or unintentionally, among different ecosystems and nations.

Table 14.1 The Systematic Steps of Risk Assessment and Management, Essential but Not Sufficient Parts of an Analytic-Deliberative Framework of Risk Characterization and Decision Making

Step in risk assessment and management	Key question addressed at this step
Hazard identification	What could go wrong?
Risk analysis	How likely is the hazard?
	What would be the consequences of realization of the hazard and how severe are they?
	What is the risk assessment, i.e., a matrix of likelihood plotted against severity of consequence? Each cell of the matrix should be accompanied by a qualitative assessment of the response and level of assurance needed to reduce harm if the conditions of the cell were to occur.
	How certain is the knowledge used to identify the hazard, estimate its likelihood, and predict consequences?
Risk reduction planning and implementation	What can be done to reduce risk, either by reducing the likelihood or mitigating the consequences of hazard realization?
Risk tracking (monitoring)	How effective are the implemented measures for risk reduction?
	Are they as good, better, or worse than planned for?
	What follow-up/corrective action/intervention will be pursued if findings are unacceptable?
	Did the intervention adequately resolve the concern(s)?

The Manual addresses both ecological and human health hazards. It also addresses important connections between ecological and human health hazards. Consider, for example, proposals to commercialize a crop genetically engineered to produce a novel toxin or a pharmaceutical compound for human use. These would trigger the need to assess risks to non-target organisms in the agroecosystem while the crop is under cultivation as well as risks to human health if inappropriate parts of the plants entered the human food supply. The discussions in this chapter focus on ecological dimensions of biosafety. This is not meant to obscure the fact, however, that the four spheres of controversy listed above also apply to debates over appropriate methodologies for assessing human health effects of GEOs. An introduction to some of these methodological issues is available in Part One of the Manual on pages 96–105.

The terms *hazard* and *risk*, have distinct meanings in most formal systems of risk assessment, or risk characterization, and are so treated in this chapter. When discussing biosafety outside of more technical arenas, it is important to recognize that common parlance frequently uses these two terms interchangeably with little or no attention to the differences in meaning. As further explained in a section below, risk characterization is the outcome of an integrated analytic-deliberative process (Committee on Risk Characterization 1996). The formal distinction between hazard and risk is most relevant during the analytical phase of risk characterization. The analytical phase involves a series of systematic steps that logically build on each other (Committee on Risk Characterization 1996). Shared across the different taxonomies of risk characterization in different fields of technology assessment is this conception of systematic steps. The steps are along the lines of those laid out in Table 14.1 (Hann 2000, Rich 2000). The first step is hazard identification, whereby a hazard is defined as a possible adverse effect. This is followed by risk analysis, a multiple-step procedure designed to estimate the likelihood of occurrence of an identified hazard and the consequences and severity of its occurrence (e.g., how bad ecological harm might be). The subsequent steps involve developing and implementing plans to reduce risks as well as plans to monitor for effectiveness or breaches in risk management. These steps in risk characterization should be applied to the entire cycle of production and use of GEOs, from design of genetic constructs and early laboratory experiments through field tests, approved widespread use, and disposal or destruction of whole GEOs, their propagules, or other biologically active parts. These steps should also be used to address all possible scenarios of use and misuse of the GEO.

14.2   Areas of controversy

Decisions regarding the scope of concerns addressed and technical specificity of biosafety assessment guidelines tend to drive how these guidelines address other important areas of controversy. For example, a narrow scope might only consider ecological effects, whereas a broader scope would consider ecological and human health effects and the interactions between the two. Biosafety guidelines with little specificity might include a vague suggestion to compare the survivability of a GEO and its non-engineered parent organism. Guidelines with greater specificity might instruct the user to collect quantitative data, through scientifically reliable experiments, that compare the fitness parameters of the GEO and non-engineered parent organism. As the scope of the biosafety assessment process broadens and the specificity of assessment questions increases, the need for inter-disciplinary participation of a diversity of biologists, from molecular biologists to ecologists, in biosafety decision making becomes clearer.

14.2.1   Specificity and scope of biosafety assessment guides

Controversy over the scope and specificity of biosafety assessments involves several key questions that affect the design and implementation of biosafety programs: What are the relevant questions to pursue in order to identify relevant hazards? How much specificity is needed in laying out different questions and providing guidance on the types of information needed to address these questions? What constitutes an adequate, scientifically sound response to these questions?

At the start of negotiations of the Biosafety Protocol, a number of documents were available that aimed to provide scientific guidance on risk assessment (see list in Ingham and Holmes 1995). During negotiations of the Biosafety Protocol, parties pushing for unrestricted international trade of GEOs referred to one or more of these documents as exemplary of "science-based" guidelines. But when examined more closely and applied to the same specific case of a genetically engineered microorganism, *Klebsiella planticola*, engineered to produce ethanol from plant residues (Holmes et al. 1998), major scientific deficiencies in these documents became clear (Ingham and Holmes 1995). To quote from the 1995 Ingham and Holmes review, the documents examined established "general broad-brush guidelines which are too vague to adequately assess the risk posed by the test genetically engineered microorganism." To further illustrate the need for specificity, let me contrast two different approaches of biosafety guides, which I am calling the "minimalist" and the "full disclosure" approaches.

14.2.1.1   Example of a minimalist guide for ecological risk assessment

One of the most widely distributed examples of the minimalist approach is the document entitled, *International Technical Guidelines for Safety in Biotechnology*, issued by the United Nations Environment Program (Anonymous 1995) in anticipation of the negotiations of the Biosafety Protocol and referred to here as the Guidelines. The main section, titled "Assessment and Management of Risks," comprises one-and-a-half pages of generalities that read more like a table of contents for guidelines than the guidelines themselves. This section lists "key parameters" to consider, including the characteristics of the organism with novel traits, the intended use, scale of use, management procedures, and the potential receiving environment. Unfortunately, the text offers no systematic guidance on the relevant scientific questions to address, the design of ecologically relevant laboratory or confined field experiments to search for hazards and collect data that can inform risk estimation, and on the type and quality of scientific information needed to identify hazards and evaluate risks.

The main section of the Guidelines also refers the reader to a brief Annex listing examples of points to consider in risk assessment. The examples include a total of 11 lines presenting points about the receiving environment. This brief passage begins with "[t]he potential for an organism to cause harm is related to the environments into which it may be released and its interaction with other organisms. Relevant information can include" and then the text presents the following points to consider: site location and special features; proximity to humans and significant biota; flora, fauna, and ecosystems that could be affected; and potential gene flow. This list is a good start for an introductory paragraph, but does not provide sufficient guidance for decision making. Yet, the Guidelines promise to "provide assistance for identifying organisms whose characteristics may differ from those of the parent organisms … in ways that would suggest that additional scrutiny is appropriate." In short, the vagueness of the 1995 Guidelines makes them scientifically meaningless.

14.2.1.2 Example of a full-disclosure guide for ecological risk assessment

In contrast, the Manual took an approach that is consistent with the core of what it means to pursue some issue — in our case, biosafety assessment — in a manner that is science based. Philosopher of science David Miller (1999) recently stressed that the key attribute setting science apart from other ways of knowing is that "science is more than the sum of its hypotheses, its observations, and its experiments. From the point of view of rationality, science is above all its method — essentially the critical method of searching for errors."

In considering the relationship between ecological science and risk assessment of GEOs, there is an important corollary to Miller's statement: although biosafety itself is a goal that is not fully achievable, a biosafety assessment *process* that embraces a systematic, critical methodology can help to minimize the potential for harm. Miller's (1999) further exposition of the merits of "absolute skepticism" in scientific inquiry has important implications. Given the fact that uncertainty is involved, scientists have no choice but to decide upon two core questions when trying to inform policy decisions through the practice of science, in our case, biosafety assessment. First, should the biosafety assessment err on the side of protecting the environment and human health or on the side of protecting short-term financial interests? Second, how hard is one going to try to test the null hypothesis (i.e., that the GEO and non-engineered parental organism do not differ in their ecological effect)? I will return to both these questions in the section below on controversies regarding adherence to the Precautionary Principle.

To guide biosafety assessment of ecological and human health effects, the Manual provides clear and specific guidance for thinking systematically through a case-by-case assessment of a particular GEO or GEF. Figure 14.1 shows an overview of an interconnected series of flowcharts of questions that integrate knowledge from many different disciplines of biology. The various flowcharts included in the Manual help the user to evaluate whether a GEO or GEF poses specific ecological or human health hazards. If a hazard is identified, the text refers readers to the extensive literature on risk estimation where they can find guidance on the estimation of specific risks (e.g., Burgman et al. 1993, Committee on Risk Characterization 1996).

Different assessments will follow different paths through the flowcharts, so that each case addresses only a relevant subset of all the questions. For example, some users are

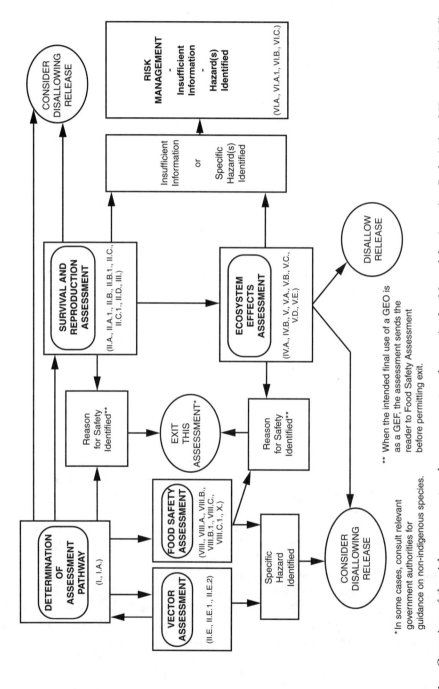

Figure 14.1   Overview of the risk assessment and management pathways in the *Manual for Assessing Ecological and Human Health Effects of Genetically Engineered Organisms*. The pathways guide case-specific assessment of possible genetic, ecological, and human health hazards associated with small- and large-scale uses of a particular GEO, including genetically engineered viruses, microbes, plants, or animals, or a particular GEF. Labels within boxes (e.g., I.A) identify the subordinate flowcharts in a given assessment pathway. (From Scientists' Working Group on Biosafety, *Manual for Assessing Ecological and Human Health Effects of Genetically Engineered Organisms. Part Two: Flowcharts and Worksheets*, The Edmonds Institute, Edmonds, WA, 1998. With permission.)

II.A.1. Impact of Deliberate Gene Changes

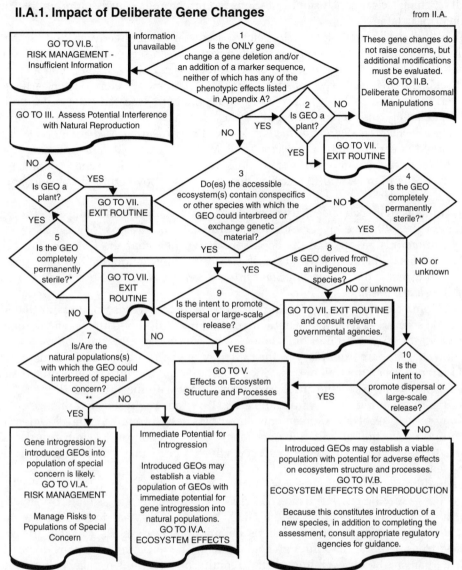

Figure 14.2 Chart II.A.1 begins the assessment of the ecological impacts of deliberate gene changes (e.g., gene transfer). (From Scientists' Working Group on Biosafety, *Manual for Assessing Ecological and Human Health Effects of Genetically Engineered Organisms. Part Two: Flowcharts and Worksheets*, The Edmonds Institute, Edmonds, WA, 1998. With permission.)

directed to Chart II.A.1, which is reproduced in Figure 14.2. Supporting text for each flowchart provides explanatory material for specific questions and statements, with citations of relevant scientific studies and a glossary of terms. The questions and supporting text together guide the user in assembling the scientific information needed to make an informed decision regarding release of the GEO in question.

The flowchart questions appear in the most parsimonious order. First come the questions for which it is relatively easy to gather necessary scientific information and that may permit reaching a clear conclusion. Questions requiring progressively more complex information appear later.

This "full-disclosure" style of biosafety assessment has already elicited strong and conflicting comments. Levin (1998) suggested that such "a series of highly detailed flowcharts [offers] less flexibility to regulators" and Balint et al. (1998) noted that a "primary drawback ... lies in its unnecessary specificity." These comments referred to an earlier guide restricted to fish and shellfish (Agricultural Biotechnology Research Advisory Committee 1995) that largely inspired the design of the Manual. This earlier document, entitled *Performance Standards for Safely Conducting Research with Genetically Modified Fish and Shellfish*, was adopted in early 1996 by the Secretary of Agriculture as *voluntary* guidelines for any research and development of fish and shellfish supported by USDA funds.[4] Because of the similarity of the Manual and this earlier document, the above criticisms would apply to the Manual as well.

Arguments that such specificity of guidance is unnecessary and inflexible for regulators (Levin 1998, Balint et al. 1998) failed to articulate the reasons for their views. Interestingly, numerous passages of the book containing their chapters (Zilinskas and Balint 1998) repeat the highly detailed and specific scientific information that had been provided as supporting text in the *Performance Standards for Safely Conducting Research with Genetically Modified Fish and Shellfish*.

van Dommelen (1999), arguing in the opposite direction, has pointed out that specificity is the scientific strength of what I am calling the "full-disclosure" style of guidance. In his view, the specific trail of questions and responses *make* the claim of risk or safety: "The flowcharts ... are exemplary in their attempt at explicitness. The resulting biosafety claim is an answer to the trail of prescribed questions and can thus be interpreted against the background of the specific questions." Concordant with the approach of the Scientists Working Group on Biosafety, van Dommelen (1999) has proposed that science can go a long way to help resolve disagreements about biosafety through the development of an agreed-upon collection of scientific questions. This collection of questions must make it possible to test all of the assumptions upon which a particular claim of safety or risk is based (see review of van Dommelen's book by Duvick 1999).

In guiding the reader through case-specific hazard identification, the flowchart questions address specific characteristics of the GEO, potentially accessible ecosystems, and potential ecological interactions of the GEO. When hazards are identified, readers are directed to appropriate recommendations for minimizing risks associated with research and field trials with the GEO through containment. These recommendations do not apply to microorganisms because it is impossible to confine microorganisms reliably in an outdoor setting. They do not apply to commercial-scale releases because containment is generally infeasible. When readers are considering commercial release of a GEO despite evidence of significant hazards, they might consider mitigation schemes. The Manual stresses that the development of risk management and mitigation is in its infancy at this time.

In providing in-depth assistance in assessing virtually any GEO in any ecosystem, the authors of the Manual had to guard against use of the Manual as a cookbook. The Manual offers a systematic way of thinking through risk assessment and management, not a static procedure to follow blindly. Fortunately, this message seems to have gotten across, based on comments from various users of the Manual from different countries who attended an April 2000 Advanced Workshop on Biosafety convened in Italy by the International Centre for Genetic Engineering and Biotechnology (ICGEB).[5]

The Manual is designed to stimulate users to think through in case-specific detail the following sorts of issues: What sorts of biological information are needed to identify ecological and human health hazards of a GEO? If hazards are identified, what kinds of qualitative and quantitative biological data are needed to inform risk estimation for application of the GEO in a particular setting? What laboratory, field, and computer simulation experiments and field studies do the preceding two questions require? Do the designs of these experiments incur an acceptable level of type II statistical error?[6] Is there enough information at hand to allow a scientifically defensible risk assessment? Unlike a cookbook, the Manual also does not prescribe one across-the-board level of acceptable risk for each identified hazard. The choice of acceptable level of risk is left to each community of people who might use the Manual to aid biosafety assessments and reach risk decisions in their own specific ecological and social contexts. The rationale for this design is further explained below in the discussion of who participates in risk analysis and decisions.

The Manual pushes for full disclosure and transparency in risk assessment by providing a worksheet that records the user's path through the flowchart questions and prompts the user to justify the rationale for key answers in the path. The completed worksheet facilitates independent scientific and public review of the risk assessment. Having the entire risk assessment process in the open — from the publicly available decision path of questions and answers to the completed worksheet of each project — makes it easy to revise appropriate parts of the Manual as new biosafety information comes to light.

The Manual has a broad enough scope to allow it to address present and future trends in the field trials, commercial release, and international trade of GEOs. It addresses ecological hazards of genetically engineered organisms of many taxa (e.g., microorganisms, terrestrial plants of all types, algae and other aquatic plants, and invertebrate and vertebrate animals) and human health hazards of foods derived from GEOs. Its questions were designed to assess potential hazards of small-scale and large-scale field trials as well as intentional and unintentional transboundary releases.

The supporting text of the Manual reminds readers that, in answering the Manual's various questions, "it is imperative that the user consider all ecosystems that are accessible to the GEO and that are suitable for survival of the GEO." This opposes the notion of "harmonization of ecosystems" advocated by many proponents of the international trade of GEOs. Such "harmonization" is implicit in proposals that would develop lists of GEOs that are considered safe for worldwide release and include GEOs that have undergone field tests in only one or a few types of ecosystems. From the perspective of the ecological and ecosystem sciences, it would be rare, if not impossible, to achieve a scientifically defensible claim of worldwide safety based on field tests in a limited number of the world's ecosystems.

14.2.2 Relevant information and scientific disciplines

There is disagreement over what scientific expertise is necessary for biosafety assessment and how to incorporate the knowledge of different scientific disciplines in an assessment. The Edmonds Institute chose to assemble an interdisciplinary team of biologists spanning scientific understanding from molecular to ecological levels of living systems. The members and affiliations of the Scientists Working Group on Biosafety are listed in Table 14.2. The members of this group interacted extensively with each other during and after two 4- to 5-day workshops; their more than 2-year-long interactions assured that interdisciplinary thinking went into the design of the structure and content of all the assessment paths in the Manual.[7] Such transdisciplinary collaboration is critically important to arriving at a scientifically accurate process for biosafety assessment. One or a few scientists, particularly if they are primarily molecular biologists, are unlikely to arrive at such a synthesis, even if they actively seek out scientific literature from different fields. The complexities and nuances involved in correctly integrating biological understanding from molecules to ecosystems are simply too challenging. The breadth and interdisciplinary interactions

Table 14.2   Members of the Scientists' Working Group on Biosafety

Name	Scientific affiliation	Institution
Dr. Mark Wheelis	Section of Microbiology	University of California at Davis
Dr. Andrew Spielman	Department of Immunology and Infectious Diseases	Harvard School of Public Health
Dr. Philip J. Regal	Department of Ecology, Evolution and Behavior	University of Minnesota
Dr. Deborah K. Letourneau	Department of Environmental Studies	University of California at Santa Cruz
Dr. Terrie Klinger	Friday Harbor Laboratories	University of Washington
Dr. Anne R. Kapuscinski	Department of Fisheries and Wildlife	University of Minnesota
Dr. Conrad Istock	(Emeritus of the Department of Ecology and Evolutionary Biology)	University of Arizona
Dr. Elaine Ingham	Department of Botany and Plant Pathology	Oregon State University
Dr. Norman Ellstrand	Department of Botany and Plant Sciences	University of California at Riverside
Dr. Pushpa M. Bhargava	Anveshna Consultancy Services	Hyderabad, India
Dr. Sharon Akabas	Institute of Human Nutrition	Columbia University

of the Scientists' Working Group on Biosafety are a model for how a country might best accomplish biosafety assessment of GEOs proposed for commercialization within its borders. It also serves as a model for local communities or other groups that may choose to implement sub-national review of proposals to introduce and commercialize GEOs in their area. The Manual encourages the formation of a "review planning team" of independent scientists (having a wide spectrum of expertise ranging from molecular to ecological and applied biological knowledge), scientists from the body proposing the commercial use, and scientists and administrators from appropriate government agencies.

In contrast, minimalist biosafety guides typically go no farther than listing fields that may be relevant. All too often, integrative and interdisciplinary thinking is not required and does not occur in the planning, execution, and review of biosafety assessments. Ecologically relevant analysis is often de-emphasized or omitted.

14.2.3   Adherence to the Precautionary Principle

A third major area of controversy involves the connection between the Precautionary Principle and biosafety assessment. Proponents of the biotechnology industry have been highly critical of the Precautionary Principle because they fear that it will be invoked to block international trade of GEOs and GEFs (Hegwood 1999, Anonymous 2000, Barnes et al. 2000, Cosbey 2000, Pollack 2000). This concern was pivotal in stalling negotiations of the International Biosafety Protocol in February 1999 and remained a central point of contention through the final negotiations in Montreal in January 2000. Articles 10.6 and 11.8 of the final text confirm the right of a party of import (i.e., a nation that has ratified the protocol) to apply precaution in deciding whether or not to allow the proposed importation of a GEO. However, the language is quite convoluted and has already generated divergent interpretations of what it will mean in practice. Some view the final text as a weak version of the Precautionary Principle, providing an importing nation the flexibility to weigh the importance of environmental risk against other factors (Barnes et al. 2000). Some industry and government leaders further argue that an importing nation's application of the Precautionary Principle is constrained by its obligations under the WTO, a point reinforced by one statement in the preamble and Article 2.4 of the protocol (Helmuth 2000, Hodgson 2000, Pollack 2000).[8] Many representatives of the European Union, developing countries, and public interest groups take a different view. They interpret the language of the protocol on precaution as a precedent-setting international victory, allowing a nation to bar import of a GEO when there is lack of scientific demonstration that it would be dangerous (Helmuth 2000, Pollack 2000, Walker 2000). This interpretation, in turn, has raised concerns among some biotechnology proponents that the text provides too much leeway to a party of import to reject a GMO product for arbitrary and unscientific reasons (Mahoney 2000, Pollack 2000).

The interpretation of the role of scientific information and lack thereof in biosafety decision making will hinge on the interpretation of the clause, "[l]ack of scientific certainty due to insufficient relevant scientific information and knowledge regarding the extent of the potential adverse effects of a living modified organism on the conservation and sustainable use of biological diversity in the Party of import, also taking into account risks to human health, shall not prevent that Party from taking a decision, as appropriate, with regard to the import of the living modified organism in question … in order to avoid or minimize such potential adverse effects." This clause raises a host of questions about the scientific criteria that will be used to make decisions about the sufficiency and relevance of scientific information and the overall quality of the risk assessment. In other words, Articles 10.6 and 11.8 have not resolved controversies over the issues at the interface of science and policy raised in this chapter.

The Precautionary Principle is a keystone of the Convention on Biological Diversity (1994); it states that "lack of full scientific certainty should not be used as a reason for postponing measures to avoid or minimize … a threat." The Manual is consistent with the convention, pointing out that adherence to the Precautionary Principle "requires shifting the burden of proof from those charged with post-release monitoring to those seeking approval for the release of new products" (Scientists' Working Group on Biosafety 1998, p. xi). No biosafety assessment process, no matter how scientifically rigorous, can *guarantee* that conclusions of safety will be sustained in the long run. The point here is that the developers of GEOs, rather than government or the general public, should bear the responsibility to *demonstrate* the safety of the proposed use of a given GEO against unambiguous safety benchmarks. This demonstration should be based on systematic, interdisciplinary, and transparent application of the latest scientific understanding in ecology and other relevant biological fields.

14.2.3.1 Examples of risk assessment questions that take a precautionary approach
The assessment paths in flowchart IV.B. (Figure 14.3) provide two examples of how the Precautionary Principle guided the design of the flowcharts. First, key questions with possible affirmative and negative answers also have an "unknown" option. The unknown response always directs the reader in the direction that errs on the side of caution. Question 1 on flowchart IV.B. addresses whether the abiotic characteristics of the accessible ecosystems would clearly prevent the GEO from reproducing in them. An affirmative answer, when supported by scientifically verifiable documentation, would provide reason to suggest the ecological safety of the proposed GEO release. An affirmative response thus leads the user toward the exit routine. (The exit routine leads to two possible conclusions: either that the assessment is complete or that the GEO will be used as a food source and now must proceed through the food safety assessment pathways.) However, a response of "unknown" to question 1 directs the user toward continuation of the ecological assessment rather than toward a chance to exit the assessment.

IV.B. Ecosystem Effects on Reproduction

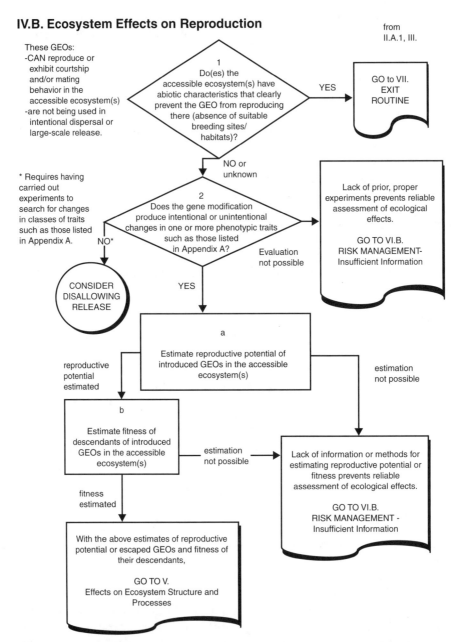

These GEOs:
-CAN reproduce or exhibit courtship and/or mating behavior in the accessible ecosystem(s)
-are not being used in intentional dispersal or large-scale release.

from
II.A.1, III.

1
Do(es) the accessible ecosystem(s) have abiotic characteristics that clearly prevent the GEO from reproducing there (absence of suitable breeding sites/ habitats)?

YES

GO to VII.
EXIT ROUTINE

NO or unknown

* Requires having carried out experiments to search for changes in classes of traits such as those listed in Appendix A.

NO*

2
Does the gene modification produce intentional or unintentional changes in one or more phenotypic traits such as those listed in Appendix A?

Evaluation not possible

Lack of prior, proper experiments prevents reliable assessment of ecological effects.

GO TO VI.B.
RISK MANAGEMENT-
Insufficient Information

CONSIDER DISALLOWING RELEASE

YES

a
Estimate reproductive potential of introduced GEOs in the accessible ecosystem(s)

reproductive potential estimated

estimation not possible

b
Estimate fitness of descendants of introduced GEOs in the accessible ecosystem(s)

estimation not possible

Lack of information or methods for estimating reproductive potential or fitness prevents reliable assessment of ecological effects.

GO TO VI.B.
RISK MANAGEMENT -
Insufficient Information

fitness estimated

With the above estimates of reproductive potential or escaped GEOs and fitness of their descendants,

GO TO V.
Effects on Ecosystem Structure and Processes

Figure 14.3   Chart IV.B guides assessment of effects of GEOs that can reproduce or exhibit mating behavior on the reproduction of wild relatives. Users assessing specific GEOs for which this is an issue will be directed to this chart along the ecological assessment pathway of the *Manual for Assessing Ecological and Human Health Effects of Genetically Engineered Organisms*, from Scientists Working Group on Biosafety (1998). (From Scientists' Working Group on Biosafety, *Manual for Assessing Ecological and Human Health Effects of Genetically Engineered Organisms*. Part Two: *Flowcharts and Worksheets*, The Edmonds Institute, Edmonds, WA, 1998. With permission.)

A second example of the Precautionary Principle at work in the Manual occurs when the user clearly lacks the information needed to evaluate a key question, such as a question regarding intended and unintended changes in traits of the GEO. Here the user is directed to risk management rather than being allowed to continue the assessment path leading to possible release. The rationale was that there is no merit in exposing the environment and its inhabitants to possible risks if the performance evaluation of the GEO is so incomplete that questions about it cannot be answered.

14.2.3.2 Examples of risk management recommendations that take a precautionary approach

The risk management recommendations of the Manual also take a precautionary approach in helping the user select and implement confinement measures to reduce or prevent an identified risk. The insistence on implementation of multiple types of barriers, not just one type of barrier or redundant barriers of the same type, aims to reduce the probability of total failure due to the inherent vulnerability of any given barrier. As described in the Manual, *physical barriers* induce 100% mortality through such physical alterations as imposing lethal water temperatures or pH to water flowing out of fish tanks or fishponds before the effluent is discharged to the environment. *Mechanical barriers* are devices, such as screens, that prevent any life stage of the GEO from leaving the project site. *Biological barriers*, such as induction of sterility in crop plants, trees, or fish, are those designed to prevent any possibility of the GEO reproducing or surviving in the natural environment.

Controversy is emerging over use of multiple barriers in risk management because it places additional technical development and financial burdens on the party proposing large-scale use of a GEO. In certain cases, it may also preclude commercial use of the GEO in common production systems. Consider, for example, current debates about minimum acceptable barriers for commercial aquaculture of transgenic salmon, if and when the U.S. Food and Drug Administration approves the first application under review (Hileman 2000, Reichhardt 2000). Physical barriers are not a viable option for net pen or cage farming of fish because such rearing units are suspended directly in a natural body of water and there is no "end of the pipe" effluent that can be so treated. Even mechanical barriers are highly problematic in cage and net pen farming. Materials such as extra predator barrier nets and rigid netting can help but cannot alone prevent large escapes of fish due to storm damage, predator damage, wear and tear, or human interference. The exclusive farming of all-female, triploid transgenic salmon that are functionally sterile is a feasible biological barrier for transgenic salmon although one needs to adequately address concerns about the success rate of triploid induction when treating a large batch of fish simultaneously and about the level of screening that should be required to identify failed cases.

The triploid condition must be induced in every generation of production fish. In experienced hands, one can expect rates of successful triploidy in the 90th percentile in large-scale production, but this will vary with fish strain, egg quality, age of spawners, and induction conditions. If regulatory agencies were ever to approve net pen farming of transgenic salmon, they should require screening of every production fish to confirm successful induction of the triploid condition (Kapuscinski 2000). This level of screening is essential because mechanical barriers are highly unreliable and physical barriers are impossible in net pen operations. Individual screening has long been required for large-scale stocking of grass carp in Florida (Griffin 1991, Wattendorf and Phillippy 1996). One effective screening method involves particle size analysis of fish blood samples with a Coulter Counter and Channelyzer that can be purchased for approximately U.S. $20,000 (Wattendorf 1986, Harrell and Van Heukelem 1998). Estimated labor and supply costs of small-scale production in 1986 were U.S. $0.08 to $0.20 per screened fish (Wattendorf 1986). It should be possible to maintain or lower this cost at year 2000 prices through economies of scale and the application of computer automation technology. In any event, the cost of individual screening is a small fraction of the market price of salmon smolts and adds a minor amount to the sale price of harvested and dressed fish.[9]

Even with individual confirmation of triploidy, the farming of transgenic salmon in net pens does not fit well with the principle of multiple barrier types. Land-based farming of transgenic salmon fits much better with this principle because it allows use of effective mechanical and physical barriers in addition to sterilization of production fish. The diversity and number of barriers may need to be higher in flow-through systems than in recirculating aquaculture systems. The risk of fish escaping is typically lowest in recirculating systems because no more than 10% of the rearing water is discharged daily and many upstream components of the system (such as solids removal) also act as mechanical barriers to fish escape.

14.2.3.3   Consistency between the Precautionary Principle and scientific concepts

Adherence to the Precautionary Principle is consistent with two tenets of scientific analysis. First, a precautionary approach fits with seeking to minimize type II statistical error. A type I error (false positive) occurs when the statistical analysis indicates that a particular treatment (in our case, the deployment of a GEO) has an effect (in our case, an adverse effect on the environment or human health) when, in fact, no such effect exists. A type II error (false negative) occurs when the statistical analysis indicates that the GEO has no adverse effect when in fact it does have such an adverse effect. Scientists in the fields of ecology, conservation biology, and natural resource management have become increasingly concerned about the tendency to downplay type II error in studies that aim to help inform environmental policy (Toft and Shea 1983, Peterman 1990, Lee 1993,Taylor and Gerrodette 1993, Hard 1995, Carpenter 1996, Shrader-Frechette 1996, Dayton 1998). The potential for harm is greater if conclusions commit a type II error compared with a type I error. Recovery from most harms to ecosystems or human health involves large time lags and some harms are irreversible. Type I errors, in contrast, are usually limited to short-term economic costs borne by developers and marketers of GEOs (Dayton 1998). Foremost, one should seek to minimize type II errors when designing risk assessment experiments for the laboratory or confined field trials, as well as when designing post-commercialization monitoring studies. One needs to determine the ability of the sampling design to allow detection of type II vs. type I statistical errors. It is important to estimate the costs of a sampling design that can reduce either error at a specified level, including the costs of remedial action if environmental harm occurs (Hard 1995). Hard reviews the factors that affect the trade-off between type I and type II error in field studies and long-term monitoring in the environment and recommends steps for reducing the latter.

Second, the Precautionary Principle realistically accounts for another understanding derived from ecological and systems research, notably, that some uncertainty is inherent in all biological systems (e.g., Holling 1982, 1995). Uncertainty arises in part from gaps in knowledge about the behavior of a GEO, the novelty of traits modified, variability of the environment, and limits to predicting the evolution of GEOs subsequent to release into the environment. Uncertainty is also inescapable because some responses of living systems will be surprises that simply are unknowable before the fact (Lee 1989). The advent of inserting multiple genes into organisms promises to increase uncertainty due to increases in potential interactions among inserted genes (Chen et al. 1998, Gelvin 1998). Rather than preventing all introductions of GEOs into the environment, a realistic way to cope with such inherent uncertainty is to implement an adaptive management approach to biosafety regimes for GEOs, as explained in Section 14.3.

Living systems involve complex, nonlinear dynamics, feedback interactions, and time lags in responses of different parts of the systems. The recent discovery of a potential Trojan gene effect in growth-enhanced transgenic fish is one example. In some transgenic fish lines, there is an unintended trade-off between two changes induced by insertion of a transgene construct into fish: an increase in mating advantage (due to larger size of

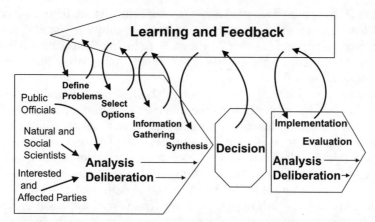

Figure 14.4 The risk decision process. (From Committee on Risk Characterization, Understanding Risk: Informing Decisions in a Democratic Society, National Academy Press, Washington, D.C., 1996. With permission.)

adults at sexual maturity) and a simultaneous decrease in viability of transgenic offspring (Muir and Howard 1999, Muir and Howard, Chapter 13 of this volume). If the risk assessment of such genetically engineered fish were limited to examining one trait at a time, as has been done in many risk assessment studies, one would likely miss the interaction uncovered by Muir and Howard (1999) and thereby miss a critically important piece of information. Emergent properties at larger scales of living systems cannot be fully predicted from knowledge about smaller scales of the system. The tendency to avoid consideration of and monitoring for such emergent properties lies at the root of many failed attempts to increase the productivity of one living natural resource (trees, fish, crop, etc.) embedded within a broader ecosystem (see reviews in Gunderson et al. 1995, Kapuscinski et al. 1999).

14.2.4   Who participates in risk analysis and decisions?

Finally, a major source of controversy involves the decision of who participates in making risk decisions. The approach of the Scientists' Working Group fits with the National Research Council's most recent framework of the risk decision process, as depicted in Figure 14.4 (Committee on Risk Characterization 1996). The wide arrow on the left side of Figure 14.4 represents risk characterization, a process that requires attention to analysis and deliberation. Analysis uses rigorous, replicable methods to arrive at answers to factual questions. This is where scientific studies and tools such as the Manual come into play. Deliberation uses processes such as discussion, reflection, and persuasion to communicate, raise and collectively consider issues, increase understanding, and arrive at substantive decisions. Public officials, scientists, and interested and affected parties participate through iterations of analysis and deliberation. Crucial to this framework is the mutual and recursive interplay between analysis and deliberation, for example, in reaching some degree of agreement among scientists, government, and affected and interested social groups over the set of possible ecological changes that should be treated as possible hazards.

Good science is *necessary and indispensable, but not sufficient* for good risk characterization. A broad cross section of the interested social actors is needed to participate in problem formulation, thereby assuring that the set of problems that will be addressed by scientists during the analytical phase is inclusive of all concerns that are relevant for the particular socioecological setting in which a GEO might be commercialized. Scientists and their methodologies and tools, such as the Manual, play primary roles in the analytical phase of risk characterization, presented earlier in terms of four systematic steps (see Table 14.1). The "full disclosure" forms of risk analysis, as illustrated by the Manual, should help to keep deliberations constructive, simply by greatly reducing the distrust and misinformation that is fueled when concerned parties lack access to information they consider critical to risk decision making.

Risk decisions are, ultimately, public policy choices. Scientists should contribute as much relevant knowledge as possible to the diverse set of participants, whose job it is to make value-laden decisions. Scientists also have a crucial responsibility to ensure risk characterization remains in the open. The process of peer review and publication of scientific results is the traditional means of keeping science open. But this tradition of open review of ecological information is being eroded in a number of ways in the biotechnology arena. For example, the U.S. Food and Drug Administration (FDA) is at present exercising authority over permit applications for commercial approval of transgenic fish and is poised to regulate other genetically engineered animals similarly (Kapuscinski and Hallerman 1994, Kaesuk-Yoon 2000). The agency has claimed authority to regulate the safety of the genetic engineering for animal health, human health, and the environment. The FDA justifies its authority on the basis that animal GEOs fit the definition of "new animal drug" in the Food, Drug and Cosmetic Act. It also points to its responsibility to review the ecological impacts of transgenic fish under the National Environmental Policy Act, which governs all federal actions that might affect the environment. The National Environmental Policy Act calls for public review of draft conclusions about environmental impact before an agency makes a final determination of safety/risk. However, under the Food, Drug and Cosmetic Act, FDA regulations forbid public review of information including ecological risk/safety evidence used to make permit decisions until after an approval is granted. So far, the FDA has chosen to follow the confidentiality provisions of the latter act. To date, the process of gathering and evaluating scientific information for ecological risk analysis of transgenic fish has happened in secret.

There are three compelling rationales — normative, substantive, and instrumental in nature — for broad participation in risk characterization. The *normative rationale* is that governments should obtain the consent of the governed. Citizens have rights to participate meaningfully in public decision making and to be informed about the bases for government decisions. The growing international trade (and likely transboundary escapes) of many GEOs heightens the importance and political sensitivity of this point. The *substantive rationale* is that scientists in the public and private sectors and public officials simply do not hold all the relevant wisdom. Participation by people with diverse experience will provide key information and insights to risk analysis. The *instrumental rationale* is that broad participation enhances the chances of reducing conflict and increasing general acceptance of or trust in risk decisions made by government agencies and international bodies.

Thus, when reaching a point of conclusion in the assessment pathway of the Manual, the reader encounters such phrases as "consider disallowing release" or "consider allowing release" instead of more prescriptive language. The Scientists' Working Group on Biosafety recognized that decisions about whether or not to allow a release might be influenced by socioeconomic considerations that vary across countries and are beyond the scope of the Manual. The group also recognized the Manual user's right to self-determination.

14.3   Adaptive biosafety assessment and management

The controversial issues discussed above are reminiscent of many past attempts to obtain human benefits from one type of living organism embedded in a complex environment. Biosafety programs for GEOs should learn from the documented errors (and lessons) of natural resource management regimes. Many of these regimes have treated the resource (e.g., timber) in an agricultural mode or focused on agro-ecosystems within the landscape examined. Gunderson et al. (1995) and authors therein reviewed common patterns of pathology in managed ecosystems, whereby resource utilization within too narrow a management framework leads to ecological, social, and institutional breakdown. They reviewed management histories of New Brunswick forests, the Everglades, Chesapeake Bay, the Columbia River, the Great Lakes, and the Baltic Sea. Kapuscinski et al. (1999) reviewed additional examples of this pattern of pathology in the annals of fisheries management. The first example involved the failure of an extensive Pacific salmon hatchery program in the U.S. Pacific Northwest to maintain high yields and satisfy multiple social needs in the face of negative impacts of dams, fishing, and other human activities on wild salmon populations. The second example summarized the cascade of dramatic, unintended, and interconnected ecological and social disruptions that resulted from introductions of Nile perch into Lake Victoria of East Africa. These two reviews are among an increasing number of analyses calling for reforms in managing human uses of living organisms in the environment, reforms along the lines of "Adaptive Environmental Assessment and Management" or "Adaptive Management."

According to Gunderson et al. (1995) and Kapuscinski et al. (1999), management programs that have failed to meet their objectives have:

1. Aimed to control a single species or a single component of ecosystems, largely ignoring critical ecological interactions;
2. Typically considered very narrow temporal and spatial scales of management;
3. Assumed that the surrounding environmental conditions would remain fairly constant and that scientists and managers have considerable abilities to predict future environmental conditions that might affect the single species of interest;
4. Thus, not surprisingly, did not make long-term monitoring an explicit part of the management program.

Biosafety programs for GEOs should not repeat these errors.

The inescapable implication of all the carefully examined cases of failures in living resources management is that the responsible institutions and the users were blindsided by surprising social and ecological feedbacks in the system or, to use the terminology of Senge (1994), "fixes that backfire." Major theoretical shifts, in both ecological and social science, have led to a major rethinking of the problems that environmental policies attempt to solve (Holling 1995). They also underscore the inescapable limits to our ability to predict co-evolutionary responses of social and ecological systems to human perturbations (Norgaard 1994). This new understanding has the following important implications for the design of biosafety policies:

1. Interdisciplinary and integrated modes of assessment and monitoring are needed for understanding effects of GEOs in the environment.
2. Biosafety assessment should include careful attention to temporal factors. It should focus on interactions between fast phenomena (e.g., escape of GEOs into the immediate environment due to failure of confinement measures or pollen-mediated gene flow) and slow phenomena (e.g., ecological consequences of transgene introgression into wild or weedy relatives or the adaptive evolution of GEOs in the environment, delayed transboundary movement of released individuals or propagules leading to unexpected establishment of a new GEO population far away from the locations of production, trade, or use). Furthermore, monitoring should focus on long-term, slow changes in driving variables (e.g., hydrological regimes affecting agriculture of GEOs or decadal changes in upwelling conditions in coastal waters that would affect the survival and dispersal of released marine GEOs).

3. Biosafety assessment should also carefully attend to spatial connections. The widespread connectivity of landscapes, international river basins, and particularly marine aquatic ecosystems combined with increased international trade of GEOs in a global economy imply a tremendous increase in the likelihood of unintentional transboundary movements of GEOs into distant ecosystems.

4. Biosafety policies that ignore the co-evolution, through feedback interactions, of human societies and the environment will fail (Norgaard 1994). Ignoring these interactions and system properties will lead to perpetual surprise, safety failures, and tremendous social distrust.

For example, consider the challenge of designing biosafety policies that can adequately address potential adverse environmental problems occurring decades after transgenic marine fish escape from an aquaculture operation and arrive in a distant coastal marine bay. The biosafety systems must prepare for situations in which invading transgenic fish breed in the new habitat, interact with the resident biota, possibly evolve increased fitness in response to natural selection, and displace native biota through mechanisms such as hybridization, competition, and predation. Such considerations underlie the idea of adaptive management or adaptive environmental assessment and management promulgated by Holling (1978) and developed by Lee (1993), Walters (1986), and others.

Inspired by the work of Holling (1978) and his colleagues, Kapuscinski et al. (1999) proposed a framework of adaptive biosafety assessment and management intended to be relevant to all levels of political jurisdiction, from the local community to the national and international. Emphasizing the improved social understanding of biosafety and the increased robustness of conclusions drawn from biosafety assessments possible when adaptive policies are implemented at all political levels, their framework requires effective mechanisms for information exchange and coordination of regulatory oversight across all levels of political jurisdiction.

Adaptive biosafety assessment and management are based on the recognition that knowledge of the societal-environmental systems into which GEOs will enter is always incomplete and therefore surprises regarding actual effects are inevitable (Lee 1989). Biosafety regimes, therefore, cannot be decomposed into independent phases of knowledge generation (or research), policy design, implementation, and long-term monitoring — a decomposition that addresses each phase outside of the context of the entire adaptive cycle. Instead, biosafety regimes are adaptive only when they consider interaction among these phases. Further, adaptive biosafety regimes cannot be reduced to a single passage through these different phases but rather involve a continual integrated consideration of all phases and iterative passes through all phases. Within such regimes, specific biosafety policies are designed and implemented to maximize learning from experience. Environmental, social, and economic dimensions of biosafety issues are integrated into all phases of adaptive biosafety assessment and management. Without such integration, biosafety regimes are doomed to failure (Holling 1978): lack of an adaptive learning approach increases the likelihood of either over- or underestimating the cautionary procedures needed for different GEOs and decreases the likelihood of broad social acceptance of policy decisions and actions regarding different GEOs.

Adaptive biosafety regimes must be transparent to the general public, and must involve deliberation among interested parties, including political and regulatory decision makers at critical points in decision making (see Figure 14.4). Implementation of the various stages of the adaptive cycle should involve political and regulatory decision makers, appropriate disciplinary specialists, methodologists, people with experiential knowledge about the social and environmental conditions, and all affected parties (e.g., Renn et al. 1993). This kind of civic engagement in the development of the knowledge

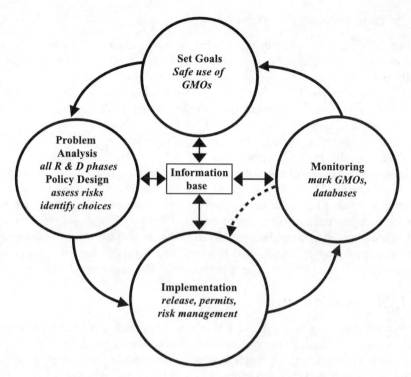

Figure 14.5 The interconnected phases of adaptive biosafety assessment and management for uses of GMOs in diverse societal-environmental systems. These phases should be applied iteratively and across multiple spatial scales from the local to global, with adequate provisions for information exchange among people implementing biosafety policies at different scales.

base and implementation of biosafety policies can assure that all phases of adaptive biosafety are broadly understood, built upon all relevant knowledge, and responsive to the concerns of the affected parties. Although even the best designed analytic-deliberative process cannot eliminate all controversy, this kind of transparency has been shown to increase the public's trust in diverse cases of risk decision making (Committee on Risk Characterization 1996a, 1996b). For example, new water quality regulations received broad social approval because the New Jersey Department of Environmental Protection involved diverse interest groups in a combination of deliberative, open methods that allowed citizen input on the process as well as the substance of rule making (Chess 1989, Committee on Risk Characterization 1996c).

Figure 14.5 represents the basic steps involved in adaptive biosafety assessment and management. These are summarized below and more fully elaborated by Kapuscinski et al. (1999). Although the limitations of human language constrain description of each phase of the cycle to one at a time, activities and decisions carried out in any phase must remain attendant to the integrative and contextual nature of the entire adaptive cycle.

14.3.1 Goals

Adaptive biosafety assessment and management begins with clear identification and statement of biosafety goals. At a minimum, goals should aim to anticipate/determine ecological risks posed by a given GEO and discern whether the GEO will yield the benefits it was designed to provide (but see the fuller description of biosafety goals in the introduction of this chapter).

14.3.2   Problem analysis and policy design

This stage of the adaptive biosafety framework involves clear definition of the ecological and social problems associated with attempts to meet stated goals. It involves bringing together disciplinary specialists, policy makers, methodologists, resource users, and other persons who have important experiential knowledge about the social and environmental contexts of the problems. This group considers a range of objectives, from unchecked release to a complete ban on uses of specific GEOs. Although these extremes may be unrealistic, they nonetheless provide the limits within which any realistic objective would fall; thus, they help the group to choose among policy options for testing.

This phase also involves addressing several important questions. What testing or assessment tools should be used to make predictions about potential environmental and food safety hazards of a GEO? What key indicators or driving variables of the ecosystem (e.g., suitable conditions for survival and reproduction of the GEO) and socioeconomic system (e.g., terms of global trade agreements) must be considered? How useful is current knowledge on risk assessment in formulating policies? What risk assessment experiments should be conducted in the laboratory, field, and in computer simulations? What additional data are required?

14.3.3   Implementation

Biosafety policies should be implemented at whatever spatial and political scales are likely to be impacted by GEOs. Local, national, and regional laws and regulations and international agreements, such as the Cartagena Protocol, may come into play to govern release and trade of GEOs. A comprehensive set of biosafety policies at the international scale would include measures for risk management (following risk assessment at the problem analysis phase), capacity building programs, national permitting of trade and uses of GEOs, advanced informed agreements on trans-national trade of GEOs, and an international system of liability and compensation for addressing future unfortunate cases in which harms do occur.

Whenever there are approvals, an adaptive approach would explicitly treat deliberate GEO releases as large experiments by requiring post-market monitoring for signs of problems in meeting biosafety goals as well as for unexpected problems, with institutional mechanisms in place to take prompt corrective action. (See the discussion of monitoring in Section 14.3.4 and of "follow-up" in Section 14.4 for further exposition of this point.) This is one of the most important distinctions between adaptive biosafety programs and some of the most prevalent, existing governance systems for GEOs. For example, in the United States, for the types of crop GEOs that fall under the jurisdiction of the U.S. Department of Agriculture, Animal and Plant Health Inspection Service, commercialization occurs when the product is *deregulated* from any further oversight for ecological effects. There is no systematic requirement or national program of long-term monitoring.

14.3.4   Monitoring

Monitoring is critically important to the continual learning process instituted in adaptive biosafety programs. The importance of monitoring follows from the recognition that, even with the best scientific process of pre-commercialization risk assessment, there may still be important blind spots that only long-term monitoring can uncover.

An adaptive approach requires that, once a GEO is approved for large-scale use, there is a systematic, technically, and financially feasible plan in place for post-release monitoring. The plan would stipulate indicators and variables to be monitored as well as agreed-upon threshold limits of ecological change (i.e., one category of "safety criteria," in the parlance of engineering industry safety programs, as outlined in Section 14.4). Those thresholds would derive from the biosafety objectives for case-by-case combinations of GEO and ecosystem. Reaching a threshold limit would trigger both a moratorium on use of the GEO until the problem is resolved and a thorough review of risk management or mitigation measures in place. Given that it will not be feasible to monitor each commercial application of a GEO, the monitoring plan would also include a statistically valid sampling design. Planning should include estimating the ability of the sampling design to detect type II and type I statistical errors and estimating the costs of a sampling design that can reduce either error at a specified level (e.g., Hard 1995). Monitoring the effects of GEOs released into the environment through large-scale production, trade, and use will also require marking or labeling of GEOs.[10] For example, there are many ways to identify DNA sequences that are diagnostic for specific groups of GEOs. The application of polymerase chain reaction (PCR) methods permits amplification of minute quantities of DNA from unpreserved, small, and non-lethally sampled tissues. Additionally, PCR methods allow long-term retrospective or prospective tracking of genetic changes, interbreeding, and geographical movements of GEOs through comparisons of genotypes resolved from unpreserved tissues collected at different points in time and in different places (e.g., Miller and Kapuscinski 1997).

The monitoring plan should also include controls or reference points to allow us to distinguish valid effects attributable to the GEO from background variability in the environment. True controls, such as those readily achieved in isolated laboratory environments, are rarely if ever possible in ecological field studies. However, strategies have been devised for establishing a reasonable reference point against which to compare the effects of a treatment (in our case, release of a GEO) in complex ecosystems (Eberhardt and Thomas 1991). One strategy fixes the location but varies the time of data collection, requiring pre-GEO-release measurements of critical ecological variables with which post-release changes can be compared. Another strategy fixes the time but varies the location of data collection, requiring parallel measurements at the GEO release site and in several reference sites carefully selected for similarities in key ecological variables (e.g., species composition).

As long as monitoring data indicate "normal" system behavior, the adaptive cycle runs between implementation and monitoring (depicted by the dotted line in Figure 14.5). When monitoring indicates "abnormal" system behavior, the resulting data are used to review and, if deemed necessary, re-define goals, then reanalyze problems, next review and revise implementation, and finally revise monitoring accordingly.

Monitoring will be nearly impossible without provision for tracking of GEOs within and across sovereign nation-states. This implies a need for national permitting systems and adequate sampling and information exchange among nations. Finally, the different participants involved in other phases of adaptive assessment and management need ready access to the results of monitoring. This is essential if we are to ensure that all the people who can affect the system gain an understanding of what was learned about biosafety from the adaptive process. Such openness is the best way to reduce social conflict over goals and policies. Tremendous disparity among nations in their capacity to adequately assess the risks of GEOs will pose an obstacle to adequate monitoring. That and other issues deserve a full discussion among scientists, public officials, and interested and affected parties.

14.3.5   Information base

The information base serves as a "source and sink" of the best available knowledge for each phase of the adaptive biosafety cycle. Continual updating and consultation of the information base also help to bypass or shorten the time spent at each phase by assuring that lessons learned from past experiences are incorporated into revised policies and what remains unknown is highlighted. Through continual interaction with the information base, adaptive biosafety regimes can continually unravel uncertainties and adaptively reformulate policies. This kind of actively adaptive biosafety assessment and management provides the most effective way for humans to address the uncertainty inherent in the interconnected natural and social systems (Lee 1989, Gunderson et al. 1995).

14.4   Industry-led safety programs

The core elements of adaptive biosafety, as outlined above, were drawn from the lessons of the checkered history of natural resource management and resemble some of the key elements of industry-led safety programs that also arose from hard lessons learned in engineering industries over the last century. Genetic engineering has emerged at the end of the 20th century, claiming precision and expertise matching that of other engineering industries. This industry has yet to adopt the lessons learned and scientific techniques applied by safety engineering professionals in other engineering industries, such as aerospace and automotive, that produce highly complex systems with the potential to harm people and the environment (Kapuscinski et al. 2000). The maturation of these industries has been marked by credible internal industry initiatives that use science to demonstrate product safety and incorporate government oversight and public involvement in ways that reinforce industry responsibility and responsiveness. Although safety programs for inanimate products are not directly transferable to the challenge of demonstrating safety of *self-replicating, living* systems, they provide a conceptual precedent and proven framework for shaping an industry-wide safety program for GEOs.

Consider the example set by the aerospace industry. Decades of debate over problems and catastrophes that could have been avoided stimulated this industry to take the lead in demonstrating the safety of its products to government and consumers, using rigorous and unambiguous government-approved safety criteria (Lloyd and Tye 1982, Anonymous 1996, Hann 2000a). Interdisciplinary safety thinking permeates the entire process of producing and operating components and entire airplanes, from blueprint designs through pre-market tests at key points in development and finally to ongoing, post-market monitoring of aircraft performance and maintenance. The rise of these comprehensive safety programs also involved the establishment of a profession of safety engineers who collaborate with engineers in other specialties, through an interactive and interdisciplinary process, to ensure company products comply with industry-wide safety standards. In the United States, for example, the Federal Aviation Administration certifies and de-certifies safety engineers, based on a program of training, on-the-job demonstration of competency, examinations, and in-service training at regular intervals. Government oversight then focuses on review of the safety verification data presented in applications to commercialize a new product; audits of company compliance with government-approved safety programs, certification of safety engineers; and, when problems and catastrophes occur, credible review of identified causes and corrective actions.

Major elements of a credible industry-wide safety program, assuming no erosion of existing national and trans-national regulatory regimes, include (Hann 2000a):

1. Criteria setting — Complete, rigorous, and unambiguous product safety design criteria established and vetted through interdisciplinary review at the outset of development of a new product;
2. Verification — Application of the best available science, from all relevant fields, in rigorous tests that fully challenge the product and unambiguously demonstrate that the product meets the pre-set and government-approved safety criteria;
3. Follow-up — In recognition that criteria-setting and verification cannot anticipate all problems, open-minded and scrupulous monitoring of the product in all its uses, with meaningful and timely corrective action upon discovery of problems;

4. Safety leadership — Rigorously trained and independently certified safety engineers and a style of company management that fosters broad thinking, application of the best science, self-imposed responsibility to make safe products, responsiveness to each real hazard and evidence of problems, and independent review of all aspects of the product safety program;

5. Government oversight — Thorough review of safety data submitted in commercialization applications for demonstration that the GEO product meets the pre-set safety criteria, government audits of compliance with government-approved industry safety programs, rigorous certification and de-certification of safety professionals, and oversight of industry efforts to learn from and promptly correct post-marketing problems;

6. Public participation — Mutually agreed-upon involvement of legitimate representatives of consumers and other potentially affected parties in industry-led formulation and periodic updating of safety criteria and in establishment and review of government oversight.

The genetic engineering industry, operating in different social and ecological contexts around the world, has yet to take a similar lead in demonstrating the safety of its products. Instead, the dominant approach to biosafety governance of GEOs involves a reactive approach of placing the burden on government to demonstrate risk. The United States, for instance, relies on a patchwork of risk-based government regulations (Office of Science and Technology Policy 1985, 1986, 1992). The focus on assessing risks occurs long after the multiple steps of design and development of a specific line of a GEO are completed. The appropriate federal government agency calls for and reviews risk assessments of the GEO at issue when a party seeks approval for field tests or seeks commercial approval; the assessments concentrate on demonstration of agronomic benefits and largely ignore rigorous testing for ecological safety.[11] At whatever point approval is sought, developers of the GEO have already invested considerable resources and are under pressure to gain market approval and generate new revenue.

In a further departure from credible, industry-wide safety demonstration, the United States, for example, lacks a system of post-market safety monitoring of GEOs and derived products, despite the existence of product monitoring in other engineering industries.

The world needs pro-active, industry-wide, and scientifically authoritative programs that place safety first in the development, pre-market testing, commercial approval, and post-market monitoring of GEOs. The private sector, with assistance from foundations, national governments, and multilateral institutions, and with legitimate participation by affected and interested parties, must direct its efforts toward establishing credible safety programs. Such safety programs will necessarily operate in a global economic context, raising the challenge that their structure and operation have to assure effectiveness in different social and ecological settings without exacerbating existing disparities among nations in their capacities to govern the production and uses of GEOs. Industry-wide safety programs, if properly implemented, could greatly reduce the incidence and severity of ecological and human health problems from genetic engineering. They cannot prevent all accidents that, as Perrow (1999) has explained, are "normal" for complex human–technological systems that bring together error-prone technologies with diverse and complicated human organizations.

14.5   Conclusion

The controversies discussed in this chapter are far from resolved, underscoring the continuing need to better address three overarching issues affecting the role of science in biosafety decision making as the Cartagena Protocol on Biosafety comes into force and new types of GEOs enter commercial use. First, there is a need to substantially increase the diversity of the types of biologists who play a major role in generating scientifically sound information that informs risk analysis, management, and monitoring. There should be significantly greater participation by those biologists who are most knowledgeable about complex biological phenomena and are experienced in integrative (whole system), rather than reductionist (component), modes of analysis. I am referring to the need to increase participation by such scientists as evolutionary and population biologists, ecologists, and public health biologists.

Second, industry must take the lead in demonstrating and monitoring the safety of its genetically engineered products through authoritative application of sufficiently broad scientific information. To achieve durable public credibility, industry-wide safety programs must also, at key points, incorporate government oversight and public participation to reinforce industry responsibility and responsiveness.

Third, there is a need to achieve broad participation in risk decision making. This is one of the greatest challenges that lies ahead. Many industry leaders and government regulators appear to fundamentally distrust the notion of involving citizens other than specific technical experts in decision making. Multiple legal, financial, political, and cultural obstacles to effective citizen participation exist in most countries, including such developed nations as the United States, situated at the "upstream" end of development and global promotion of GEOs (Blahna and Yonts-Shepard 1989, Slovic 1993, Lynn and Busenberg 1995, Committee on Risk Characterization 1996d, 1996e, Wondolleck and Yaffee 2000). These obstacles are often compounded in countries with little history of citizen participation in governance, strong cultural norms against challenging authority, and/or little capacity to carry out effective biosafety programs. Nevertheless, only by bringing together the right science, getting the science right, and integrating science into a participatory and open decision-making process will science play a legitimate and useful role in resolving biosafety controversies.

Acknowledgments

I thank Deborah Letourneau and Beth Burrows for inviting me to contribute to this volume, following an oral presentation I made at a symposium that they convened at the 1999 Annual Meeting of the Ecological Society of America. Deborah Brister provided crucial assistance with the preparation of figures, Brent Sewall with review of biosafety documents, and Laura Preus with references on public participation. Support came in part from a grant to the Institute for Social, Economic and Ecological Sustainability (ISEES) by the Graduate School of the University of Minnesota, the MacArthur Interdisciplinary Program on Global Change, Sustainability and Justice, the Minnesota Agricultural Experiment Station and Minnesota Sea Grant, Department of Commerce under Grant NOAA-NA86-RG0033, J.R. 474. The U.S. government is authorized to reproduce and distribute reprints for government purposes, not withstanding any copyright notation that may appear hereon.

Notes

1 Parties across these sectors who have found the Manual helpful in their work have also individually contacted me and other members of the Scientists' Working Group on Biosafety.
2 For example, Zhu et al. (2000) reported that, at no extra cost to farmers in Yunan Province, China, inter-cropping a second rice variety with the dominant variety over 3342 ha in ten townships nearly doubled yields, virtually eliminated the severe rice blast disease, and eliminated the need to apply fungicidal sprays.

3 Other terms have been used in different contexts to name novel organisms produced by modern biotechnologies. For example, the term *genetically modified organism* (GMO) was long promoted by biotechnologists in the United States and has been widely adopted in public discourse in English-speaking settings. The Cartegena Protocol on Biosafety, on the other hand, uses the term *living modified organism* (LMO). Each term bears a certain degree of ambiguity and is preferred by different groups of people for different reasons.

4 The U.S. Food and Drug Administration is at present consulting this earlier document to help guide its review of the first application to commercialize a transgenic fish (Eirkson 2000, Matheson 2000).

5 I was one of the instructors at this international workshop on biosafety and received this feedback from a number of participants.

6 The rationale for focusing on type II error appears below in the section on adherence to the Precautionary Principle.

7 The scientists had complete intellectual independence in determining the approach and content of the Manual. The staff of the Edmonds Institute provided editorial support to assure that the text would be logical and consistent and understandable to a diverse cross section of users, many of whom might not have expertise in one or all of the relevant biological sciences.

8 Most signatory parties to the Biosafety Protocol are also parties to the WTO. WTO rules, under its Agreement on Sanitary and Phytosanitary Measures, forbid import bans unless the *party of import can demonstrate the risk* of a product to health or the environment. In contrast, supporters of the Precautionary Principle see it as a means to require the *party of export to demonstrate safety* of the product. The final text of the protocol did not fully resolve whether or not rulings of an unfair trade barrier by the WTO could override a party's decision under the protocol to bar import on the basis of precaution (Anonymous 2000a, Mahoney 2000, Raloff 2000). Addressing this important international legal issue is beyond the scope of this chapter.

9 Consider a conservative assumption that individual screening on a large production scale will cost U.S. $0.10 to $0.20 per juvenile fish in relation to the prices of smolts and harvested adults (Forster 1995, Anonymous 2000b). Smolts sold by one company to growers in another company range from U.S. $1.50 to $4.00 per smolt whereas salmon farming companies that produce their own smolts spend U.S. $0.75 to $1.25 per smolt. The cost of individual screening would add U.S. $0.01 to $0.02 per pound to the cost of harvested and dressed adult fish for farmed Atlantic and chinook salmon that are typically harvested at 8 to 10 lb per fish in the round; 90% of this biomass is retained in fish dressed with head on and 80% in fish dressed without heads (Forster 1995).

10 The preferred mark for ecological monitoring will depend on: the ecological variables that are the foci of the monitoring plan; feasibility of non-lethal application of the mark; feasibility of maintenance of the integrity of the mark over time; and costs of application, recovery, and identification of the marks. Possible types of marks for ecological monitoring will range from heritable DNA sequences and protein polymorphisms, resolvable only through laboratory testing, to various types of non-heritable, often externally visible marks, for example, branding or addition of a benign physical tag to a living genetically engineered plant or animal. The best mark for ecological monitoring may have limited or no utility for human health monitoring. For example, printed labels on seed packages and processed foods would greatly facilitate the use of standard epidemiological procedures for monitoring human health effects of eating a genetically engineered food. An external label combined with segregation of a GEO throughout the food system will be more effective than sole reliance on internal DNA marks for epidemiological monitoring of human consumption of different retail foods derived from different GEOs.

11 For those plant GEOs that come under the jurisdiction of the U.S. Department of Agriculture, Animal and Plant Health Inspection Service, commercialization proceeds by default, once the agency approves a petition to *de-regulate* a GEO that it previously regulated during the phase of agronomic field testing. From this point, the government does not oversee the ecological effects of production or use of the GEO.

Literature cited

Agricultural Biotechnology Research Advisory Committee, Working Group on Aquatic Biotechnology and Environmental Safety, Performance Standards for Safely Conducting Research with Genetically Modified Fish and Shellfish, Parts I and II, Documents No. 95-04 and 95-05, USDA, Office of Agricultural Biotechnology, 1995.

Anonymous, *UNEP International Technical Guidelines for Safety in Biotechnology*, United Nations Environment Program, Nairobi, 1995.

Anonymous, Aerospace Recommended Practice, ARP4754, Society of Automotive Engineers, Warrendale, PA, 1996.

Anonymous, Swapping science for consensus in Montreal, *Nat. Biotechnol.*, 18, 239, 2000a.

Anonymous, "Victory for the environment": the Cartagena Protocol is based on the precautionary principle, *Bridges between Trade and Sustainable Development*, Year 4, No.1 (January–February), International Centre for Trade and Sustainable Development, Geneva, 17, 2000b, available at http://www.ictsd.org.

Anonymous, Market trends, *North. Aquacul.*, 6, 1, 2000c.

Balint, P. J., Colwell, R., Gutrich, J. J., Hite, D., Levin, M., Stenquist, S., Whiteman, H. H., and Zilinskas, R. A., Risks and benefits of marine biotechnology: conclusions and recommendations, in *Genetically Engineered Marine Organisms*, Zilinskas, R. A. and Balint, P. J., Eds., Kluwer Academic Press, Boston, 1998, chap. 7.

Barnes, T., Burgiel, S., and Yongo, T., International conference on biotechnology in the global economy: science and the precautionary principle, 22–23 September 2000, reported in *Sustainable Developments*, Winnipeg, 12, 2000, available at http://www.iisd.ca/linkage/sd.

Blahna, D. J. and Yonts-Shepard, S., Public involvement in resource planning: toward bridging the gap between policy and implementation, *Soc. Nat. Resourc.*, 2, 209, 1989.

Burgmann, M. A., Ferson, S., and Akcakaya, H. R., *Risk Assessment in Conservation Biology*, Chapman & Hall, New York, 1993.

Burrows, B. E., Director of the Edmonds Institute, personal communication, 1999.

Burrows, B. E., Director of the Edmonds Institute, personal communication, 2000.

Carpenter, R. A., Uncertainty in managing ecosystems sustainably, in *Scientific Uncertainty and Environmental Problem Solving*, J. Lemons, Ed., Blackwell Science, Cambridge, MA, 1996, chap. 4.

Chen, L., Marmey, P., Taylor, N. J., Brizard, J.-P., Espinoza, C., D'Cruz, P., Huet, H., Zhang, S., de Kochko, A., Beachy, R., and Fauquet, C. M., Expression and inheritance of multiple transgenes in rice plants, *Nat. Biotechnol.*, 16, 1060, 1998.

Chess, C., *Drafting Water Quality Regulations: A Case Study in Public Participation*, Center for Environmental Communication, Rutgers University, New Brunswick, NJ, 1989.

Committee on Risk Characterization, Understanding Risk: Informing Decisions in a Democratic Society, Stern, P. C. and Fineberg, H. V, Eds., National Academy Press, Washington, D.C., 1996a.

Committee on Risk Characterization, Deliberation, in Understanding Risk: Informing Decisions in a Democratic Society, Stern, P. C. and Fineberg, H. V., Eds., National Academy Press, Washington, D.C., 1996b, chap. 3.

Committee on Risk Characterization, Six cases in risk analysis and characterization, in Understanding Risk: Informing Decisions in a Democratic Society, Stern, P. C. and Fineberg, H. V., Eds., National Academy Press, Washington, D.C., 1996c, appendix A.

Committee on Risk Characterization, Common approaches to deliberation and public participation, in Understanding Risk: Informing Decisions in a Democratic Society, Stern, P. C. and Fineberg, H. V., Eds., National Academy Press, Washington, D.C., 1996d, 205.

Committee on Risk Characterization, The idea of risk characterization, in Understanding Risk: Informing Decisions in a Democratic Society, Stern, P. C. and Fineberg, H. V., Eds., National Academy Press, Washington, D.C., 1996e, 23.

Committee on Risk Characterization, Judgement in the risk decision process, in Understanding Risk: Informing Decisions in a Democratic Society, Stern, P. C. and Fineberg, H. V., Eds., National Academy Press, Washington, D.C., 1996f, 71.

Conference of the Parties to the Convention on Biological Diversity, Cartagena Protocol on Biosafety to the Convention on Biological Diversity, Montreal, 2000, available at www.biodiv.org/biosafe/Protocol/Protocol.html.

Convention on Biological Diversity, *Text and Annexes*, UNEP/CBD/94/1, Geneva, Switzerland, 1994.

Cosbey, A., The Cartagena Protocol on Biosafety: an analysis of results, An IISD Briefing Note, International Institute for Sustainable Development, Winnipeg, 2000, available at http://iisd.ca.

Dayton, P. K., Reversal of the burden of proof in fisheries management, *Science*, 279, 821, 1998.

Duvick, D. N., How much caution in the fields? *Science*, 286, 418, 1999.

Eberhardt, L. L. and Thomas, J. M., Designing environmental field studies, *Ecol. Monogr.*, 61, 53, 1991.

Eirkson, C., Food and Drug Administration involvement in aquaculture, presented at Encouraging Aquaculture Development in the U.S. — Critical Needs and National Initiatives, Aquaculture America 2000, February 2–5, New Orleans, 2000.

Forster, J., Cost trends in farmed salmon, A report prepared for the Alaska Department of Commerce and Economic Development, Juneau, 1995.

Gelvin, S. B., Multigene plant transformation: more is better! *Nat. Biotechnol.*, 16, 1009, 1998.

Griffin, B. R., The U.S. Fish and Wildlife Service's triploid grass carp inspection program, *Aquacult. Mag.*, January/February, 188, 1991.

Gunderson, L. H., Holling, C. S., and Light, S. S., Eds., *Barriers and Bridges to the Renewal of Ecosystems and Institutions*, Columbia University Press, New York, 1995.

Hann, S., Airline safety engineer, personal communication, 2000a.

Hann, S., Essential elements of a good product safety program, prepared for the Institute for Social, Economic and Ecological Sustainability, University of Minnesota, St. Paul, 2000b.

Hard, J., Genetic monitoring of life-history characters in salmon supplementation: problems and opportunities, in *Uses and Effects of Cultured Fishes in Aquatic Ecosystems, American Fisheries Society Symposium 15*, Schramm, H. L., Jr. and Piper, R. G., Eds., American Fisheries Society, Bethesda, MD, 1995, 212.

Harrell, R. M. and Van Helkelem, W., A comparison of triploid induction validation techniques, *Prog. Fish Cult.*, 60, 221, 1998.

Hegwood, D., The international regulatory environment for GMOs, presented at Agricultural Biotechnology: Food, Feed and Fiber, January 14–15, San Francisco, 1999.

Helmuth, L., Both sides claim victory in trade pact, *Science*, 287, 782, 2000.

Hileman, B, Biotech regulations under attack, *Chem. Eng. News*, May 22, 28, 2000.

Hodgson, J., Biosafety rules get thumbs up, *Nat. Biotechnol.*, 18, 253, 2000.

Holling, C. S., *Adaptive Environmental Assessment and Management*, John Wiley & Sons, Chichester, U.K., 1978.

Holling, C. S., Predicting the unpredictable: is it possible to identify the variables that trigger surprise and change? *UNESCO Courier*, 1982, 60, 1982.

Holling, C. S., What barriers? What bridges? in *Barriers and Bridges to the Renewal of Ecosystems and Institutions*, Gunderson, L. H., Holling, C. S., and Light, S. S., Eds., Columbia University Press, New York, 1995, 3.

Holmes, M. T., Ingham, E. R., Doyle, J. D., and Hendricks, C. W., Effects of *Klebsiella planticola* SDF20 on soil biota and wheat growth in sandy soil, *Appl. Soil Ecol.*, 326, 1, 1998.

Ingham, E. and Holmes, M., Biosafety Regulations: A Critique of Existing Documents, An Occasional Paper of the Edmonds Institute, Edmonds, WA, 1995.

Kaesuk-Yoon, C., Altered salmon leading way to dinner plates, but rules lag, *New York Times*, 1 May, A1, 2000.

Kapuscinski, A. R., Biosafety assessment of transgenic aquatic organisms: the case of transgenic salmon, in *Aquaculture and the Protection of Wild Salmon*, Gallaugher, P. and Orr, C., Eds., Continuing Studies in Science at Simon Fraser University, Speaking for the Salmon Workshop Proceedings, 56, 2000, available at www.sfu.ca/cstudies/science.

Kapuscinski, A. R. and Hallerman, E. M., AFS position statement: transgenic fishes, *Fisheries*, 15, 2, 1990.

Kapuscinski, A. R. and Hallerman, E. M., Benefits, Risks, and Policy Implications: Biotechnology in Aquaculture, prepared for the U.S. Congress, Office of Technology Assessment, Document PB96-107586, U.S. Department of Commerce, National Technical Information Service, Springfield, VA, 1994.

Kapuscinski, A. R., Nega, T., and Hallerman, E. M., Adaptive biosafety assessment and management regimes for aquatic genetically modified organisms in the environment, in *Towards Policies for Conservation and Sustainable Use of Aquatic Genetic Resources*, Pullin, R. S. V., Bartley, D. M., and Kooiman, J., Eds., ICLARM Conf. Proc. No. 59, International Center for Living Aquatic Resources Management, Makati City, Philippines, 1999, 225.

Kapuscinski, A. R., Jacobs, L., and Pullins, E., Safety first: active governance of genetic engineering for environment and human health worldwide, Call for March 2–3, 2001 Workshop, Institute for Social, Economic and Ecological Sustainability, University of Minnesota, St. Paul, 2000.

Lee, K. N., Columbia River Basin — experimenting with sustainability, *Environment*, 31, 7, 1989.

Lee, K. N., *Compass and Gyroscope: Integrating Science and Politics for the Environment*, Island Press, Washington, D.C., 1993.

Levin, M., Risk assessment for uncontained applications of genetically engineered organisms, in *Genetically Engineered Marine Organisms*, Zilinskas, R. A. and Balint, P. J., Eds., Kluwer Academic Press, Boston, 1998, chap. 1.

Lloyd, E. and Tye, W., *Systematic Safety*, Civil Aviation Authority, London, 1982.

Lynn, F. M. and Busenberg, G. J., Citizen advisory committees and environmental policy: what we know, what's left to discover, *Risk Anal.*, 15, 147, 1995.

Mahoney, R. J., Opportunity for agricultural biotechnology, *Science*, 288, 615, 2000.

Miller, D., Being an absolute skeptic, *Science*, 284, 1625, 1999.

Miller, L. and Kapuscinski, A. R., Historical analysis of genetic variation reveals low effective population size in a northern pike, *Esox lucius*, population, *Genetics*, 147, 1249, 1997.

Muir, W. and Howard, R. D., Possible ecological risks of transgenic organism release when transgenes affect mating success: sexual selection and the Trojan gene hypothesis, *Proc. Natl. Acad. Sci. U.S.A.*, 96, 13853, 1999.

Norgaard, R. B., *Development Betrayed: The End of Progress and a Coevolutionary Revisioning of the Future*, Routledge, New York, 1994.

Office of Science and Technology Policy, Executive Office of the President, Coordinated framework for the regulation of biotechnology; establishment of the Biotechnology Science Coordinating Committee, *Fed. Regis.*, 50, 47174, 1985.

Office of Science and Technology Policy, Executive Office of the President, Coordinated framework for the regulation of biotechnology, *Fed. Regis.*, 51, 23301, 1986.

Office of Science and Technology Policy, Executive Office of the President, Exercise of federal oversight within scope of statutory authority; planned introductions of biotechnology products into the environment, *Fed. Regis.*, 57, 6753, 1992.

Perrow, C., *Normal Accidents: Living with High Risk Technologies*, Princeton University Press, Princeton, NJ, 1999.

Peterman, R. M., Statistical power analysis can improve fisheries research and management, *Can. J. Fish. Aquat. Sci.*, 47, 2, 1990.

Pollack, A., Montreal talks agree on rules for biosafety, *New York Times*, 30 January, A1, 2000.

Raloff, J., Treaty nears on gene-altered exports, *Sci. News*, 157, 84, 2000.

Reichhardt, T., Will souped up salmon sink or swim? *Nature*, 406, 10, 2000.

Renn, O., Webler, T., Rakel, H., Dienel, P., and Johnson, B., Public participation in decision making: a three-step procedure, *Pol. Sci.*, 26, 189, 1993.

Rich, R., Airline system engineer, personal communication, 2000.

Scientists' Working Group on Biosafety, *Manual for Assessing Ecological and Human Health Effects of Genetically Engineered Organisms. Part One: Introductory Text and Supporting Text for Flowcharts. Part Two: Flowcharts and Worksheets*, The Edmonds Institute, Edmonds, WA, 1998, available at www.edmonds-institute.org/manual.html.

Senge, P. M., *The Fifth Discipline: The Art and Practice of the Learning Organization*, Currency Doubleday, New York, 1994.

Shrader-Frechette, K., Methodological rules for four classes of scientific uncertainty, in *Scientific Uncertainty and Environmental Problem Solving*, J. Lemons, Ed., Blackwell Science, Cambridge, MA, 1996, chap. 1.

Slovic, P., Perceived risk, trust and democracy, *Risk Anal.*, 13, 675, 1993.

Taylor, B. L. and Gerrodette, T., The uses of statistical power in conservation biology: the vaquita and northern spotted owl, *Conserv. Biol.*, 7, 489, 1993.

Tiedje, J. et al., The planned introduction of genetically engineered organisms: ecological considerations and recommendations, *Ecology*, 70, 298, 1989.

Toft, C. A. and Shea, P. J., Detecting community-wide patterns: estimating power strengthens statistical inference, *Am. Nat.*, 122, 618, 1983.

van Dommelen, A., *Hazard Identification of Agricultural Biotechnology: Finding Relevant Questions*, International Books, Utrecht, the Netherlands, 1999.

Walker, R., Global pact on GMOs approved, *The Christian Science Monitor*, 31 January, 6, 2000.

Walters, C. J., *Adaptive Management of Renewable Resources*, Macmillan, New York, 1986.

Wattendorf, R. J., Rapid identification of triploid grass carp with a Coulter Counter and Channelyzer, *Prog. Fish Cult.*, 48, 125, 1986.

Wattendorf, R. J. and Phillippy, C., Administration of a state permitting program, in *Managing Aquatic Vegetation with Grass Carp, a Guide for Water Resource Managers*, Cassani, J. R., Ed., American Fisheries Society, Bethesda, MD, 130, 1996.

Wondolleck, J. M. and Yaffee, S. L., *Making Collaboration Work: Lessons from Innovation in Natural Resource Management*, Island Press, Washington, D.C., 2000.

Zhu, Y., Chen, H., Fan, J., Wang, Y., Li, Y., Chen, J., Fan, J., Yang, S., Hu, L., Leung, H., Mew, T., Teng, P., Wang, Z., and Mundt, C. C., Genetic diversity and disease control in rice, *Nature*, 406, 718, 2000.

Zilinskas, R. A. and Balint, P. J., Eds., *Genetically Engineered Marine Organisms*, Kluwer Academic Press, Boston, 1998.

Index

A

α-viruses, 348

Abiotic factors
    effect on survival of GEMs, 226, 230–231, 239
    effect on survival of naked DNA, 235, 236, 237, 239
    mosquito distribution, 292, 293

ABRAC (Performance Standard for Safely Conducting Research with Genetically Modified Fish and Shellfish, 1995), 379, 394

"Absolute skepticism" (scientific inquiry), 391

Acetolactate synthase (ALS), 20

Achromobacter sp., 231

Acraea, W drive, 267

Acraea encedana, 267

Acraea encedon, 267

Actinomycetes, 212, 214

Adenoviruses, 340

Adsorption/binding
    Bt-toxins/clay minerals, 190–200, 203, 206–208, 215
    effects on DNA transfer, 235, 237, 238

Advanced Workshop on Biosafety, April 2000, 394

Aedes aegypti (yellow fever mosquito)
    distorter gene, 268–269, 270
    double translocation/sex distorter gene, 279–280
    refractoriness, 289–290, 291
    sex ratio/distorter gene, 268–269, 270
    transformation, 293, 316, 321

Aedes albopictus (Asian tiger mosquito)
    release, 319
    replacement experiment, 266
    Wolbachia infected, 284–285

Aedes mosquitoes, achiasmatic meiosis, 279

Aedes polynesiensis, 266

Aequorea victoria (jellyfish), 252

African river blindness, 274

African trypanosomiasis
    control of, 265, 278
    eradication in Zanzibar, 256

Agrawal, A. A., 76

Agreement on Sanitary and Phytosanitary Measures, 411n.8

Agricultural Resource Management Studies (ARMS), 38

Agrobacterium radiobacter, 199

Agrobacterium tumefaciens, 20

Agrotis ipsilon (black cutworm), 100

Alcaligenes eutrophus, 232

Algae, *Bt*-toxin effects, 199

Allelochemicals, 175–178

Allison, R. F., 334

Alphaviruses vaccine vector, 340

ALS (acetolactate synthase), 20

Alstad, D. N., 114

Alternative management options, 43, 45, *see also* Biological control; Natural enemies
    comparison to *Bt*-crops, 50–51
    integrated pest management, 170, 181–182, 256
    natural enemies as biological pest control, 170

Alvarez, A. J., 231

American Fisheries Society, 386

American Society of Tropical Medicine and Hygiene, 1986 symposium, 315–316

Andow, D. A., 119

Angle, J. S., 231

Animal and Plant Health Inspection Service, *see* APHIS

Animals, vaccines, 333

Anopheles albimanus
    compound chromosomes, 281
    malaria control, 318
    propoxur resistance, 280

Anopheles arabiensis
    climate conditions, 292
    control, 324
    subspecies hybrid fertility release, 277

Anopheles funestus, 324

Anopheles gambiae
    control, 323, 324
    genetic structure, 291–293
    genome-sequencing project, 293
    transformation, 293, 316, 319, 321
    Wolbachia microinjection, 285

Anopheles gambiae complex
    control, 323
    subspecies hybrid fertility release, 277

Anopheles melas, 277